AI and Digitalization in Energy Management

Other related titles:

You may also like

- PBPO247 | Alhelou, Hosseinzadeh, and Bahrani | Power System Strength: Evaluation methods, best practice, case studies, and applications | 2023
- PBPO228 | Motahhir | Digital Technologies for Solar Photovoltaic Systems: From general to rural and remote installations | 2022
- PBPO238 | Wang, Liu, Wang, Stroe, Fernandez, and Guerrero | AI for Status Monitoring of Utility Scale Batteries | 2022
- PBPO217 | Milano | Advances in Power System Modelling, Control and Stability Analysis, 2nd Edition | 2022

We also publish a wide range of books on the following topics:
Computing and Networks
Control, Robotics and Sensors
Electrical Regulations
Electromagnetics and Radar
Energy Engineering
Healthcare Technologies
History and Management of Technology
IET Codes and Guidance
Materials, Circuits and Devices
Model Forms
Nanomaterials and Nanotechnologies
Optics, Photonics and Lasers
Production, Design and Manufacturing
Security
Telecommunications
Transportation

All books are available in print via https://shop.theiet.org or as eBooks via our Digital Library https://digital-library.theiet.org.

IET ENERGY ENGINEERING 263

AI and Digitalization in Energy Management

Edited by
Antonio Sanfilippo, Sertac Bayhan and
Dragan Boscovic

Institution of Engineering and Technology

About the IET

This book is published by the Institution of Engineering and Technology (The IET).

We inspire, inform and influence the global engineering community to engineer a better world. As a diverse home across engineering and technology, we share knowledge that helps make better sense of the world, to accelerate innovation and solve the global challenges that matter.

The IET is a not-for-profit organisation. The surplus we make from our books is used to support activities and products for the engineering community and promote the positive role of science, engineering and technology in the world. This includes education resources and outreach, scholarships and awards, events and courses, publications, professional development and mentoring, and advocacy to governments.

To discover more about the IET please visit https://www.theiet.org/.

About IET books

The IET publishes books across many engineering and technology disciplines. Our authors and editors offer fresh perspectives from universities and industry. Within our subject areas, we have several book series steered by editorial boards made up of leading subject experts.

We peer review each book at the proposal stage to ensure the quality and relevance of our publications.

Get involved

If you are interested in becoming an author, editor, series advisor, or peer reviewer please visit https://www.theiet.org/publishing/publishing-with-iet-books/ or contact author_support@theiet.org.

Discovering our electronic content

All of our books are available online via the IET's Digital Library. Our Digital Library is the home of technical documents, eBooks, conference publications, real-life case studies and journal articles. To find out more, please visit https://digital-library.theiet.org.

In collaboration with the United Nations and the International Publishers Association, the IET is a Signatory member of the SDG Publishers Compact. The Compact aims to accelerate progress to achieve the Sustainable Development Goals (SDGs) by 2030. Signatories aspire to develop sustainable practices and act as champions of the SDGs during the Decade of Action (2020–2030), publishing books and journals that will help inform, develop, and inspire action in that direction.

In line with our sustainable goals, our UK printing partner has FSC accreditation, which is reducing our environmental impact to the planet. We use a print-on-demand model to further reduce our carbon footprint.

Published by The Institution of Engineering and Technology, London, United Kingdom

The Institution of Engineering and Technology (the "**Publisher**") is registered as a Charity in England & Wales (no. 211014) and Scotland (no. SC038698).

First published 2025

The Institution of Engineering and Technology
Futures Place,
Kings Way, Stevenage,
Herts, SG1 2UA,
United Kingdom

www.theiet.org

British Library Cataloguing in Publication Data

A catalogue record for this product is available from the British Library

ISBN 978-1-83953-979-4 (hardback)
ISBN 978-1-83953-980-0 (PDF)

Typeset in India by MPS Limited

Cover image credit: alvarez/E+ via Getty Images

Contents

About the editors

Antonio Sanfilippo is chief scientist at Qatar Environment and Energy Research Institute (QEERI), where he has led the energy management program and several projects funded by Qatar Research, Development and Innovation Council since 2014. While at QEERI, he has managed the establishment of several renewable energy and smart grid capabilities that have become national points of reference for local and international stakeholders, including a nationwide network of solar monitoring stations, a 100 kWp microgrid testbed and a network of Phasor Measurement Units in Education City, Doha, Qatar. Prior to QEERI, he was chief scientist at the Pacific Northwest National Laboratory in the USA, where he was awarded the Laboratory Director's Award for Exceptional Scientific Achievement. He has also held positions as research director in the private sector, senior consultant at the European Commission, and research supervisor and group manager at SHARP Laboratories of Europe. He holds MA and MPhil degrees from Columbia University in the USA and a PhD from the School of Informatics at the University of Edinburgh in the UK.

Sertac Bayhan is a principal scientist at Qatar Environment and Energy Research Institute (QEERI) and a Professor at Gazi University. His research encompasses power electronics and their applications in next-generation power and energy systems. He is the recipient of many prestigious international awards, such as the Teaching Excellence Award from Texas A&M University and a Research Fellow Excellence Award. He has been elected as a delegate of the Energy Cluster in IES and serves as an associate editor for several key journals.

Dragan Boscovic is a professor at Arizona State University's W. P. Carey Business School, USA. He also serves as the research director of the AZ Blockchain Applied Research Center and as CEO and founder of VizLore Group, a tech company specializing in pioneering IoT, data analytics, blockchain distributed computing and digital asset management. He has 24 patents issued to his name and a track record in conceiving strategies and overseeing development, investment, and innovation efforts.

Chapter 1

Introduction

Antonio Sanfilippo[1], Sertac Bayhan[1] and Dragan Boscovic[2]

As energy systems grow in complexity due to the increasing integration of renewable energy and the electrification of transportation, artificial intelligence (AI) and digitalization have arisen as the prime technologies to address emerging challenges. Machine learning, multi-agent simulations, optimization, and game theory methods are being successfully applied to a number of energy management tasks with the support of robotics, Internet of Things (IoT), and cloud/edge computing technologies. The goal of this book is to provide a venue for a deep dive into the AI and digitalization ecosystem for energy management through a review of the state-of-the-art and the identification of future directions. The book is structured in two parts. The first part focuses on the fundamentals of AI and digitalization in energy management, while the second part provides a review of representative applications.

1.1 The increasing complexity of energy management systems

Energy management emerged as a distinct discipline after the oil crisis of the 1970s [1,2]. The initial focus was on energy conservation in reaction to rising oil prices, but the endeavor gradually evolved as a more global and systematic approach to achieving economic sustainability in energy by ensuring economic competitiveness through the provision of efficient, resilient, and secure services that adhere to environmental standards to reduce air and water pollution [3].

With the introduction of renewable energy as a growing component of power systems, energy management has been used to address the variability and intermittency of solar and wind power generation [4]. Modern energy management covers a wide range of applications from wide area monitoring and control [5] to home energy management [6] and anything in between [7–9]. Accordingly, the use of the term energy management in this book reflects its broader application to qualify any system whose function is to monitor and control electricity networks that integrate renewable energy, energy storage, electric vehicles, and demand response to optimize and secure energy use. As shown in Figure 1.1, energy management systems can cover global power grids as well as local units of varying sizes, such as microgrids and nanogrids

[1]Qatar Environment and Energy Research Institute, Hamad Bin Khalifa University, Qatar
[2]W. P. Carey Department of Information Systems, Arizona State University, USA

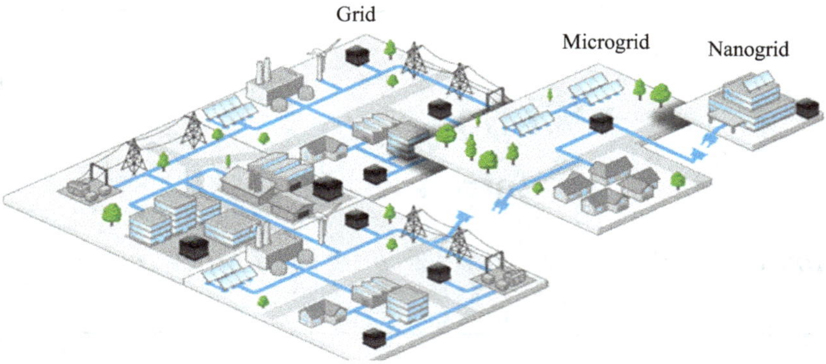

Figure 1.1 Application scope of energy management systems

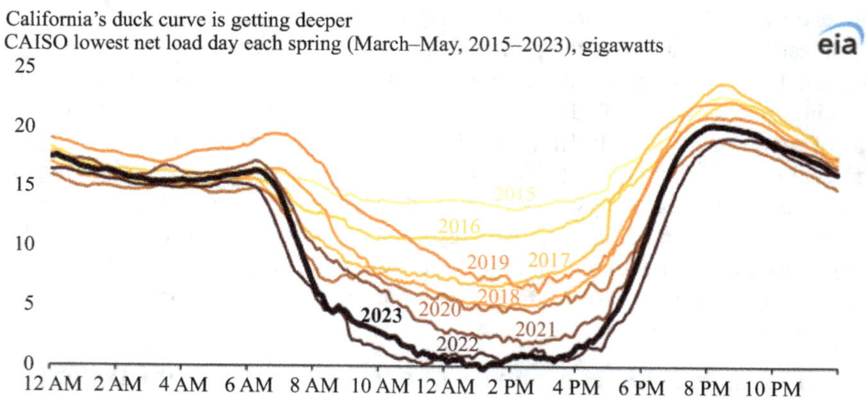

Figure 1.2 Daily duck curve in California's power system [13]. Data source: CAISO (https://www.caiso.com).

(e.g., single buildings/homes), which in turn can be operated as independent/islanded, grid-connected of hybrid power systems [10,11].

Rising shares of renewable energy introduce complexity in global and local power networks, which challenge network stability and resilience at both the technical and economic levels. For example, mismatches between renewable energy generation and energy demand can place significant stress on the grid and disrupt energy markets, as has been observed by the California Independent System Operator (CAISO) since 2012 [12] with reference to a power production curve that approximates the shape of a duck. The "duck curve" shown in Figure 1.2 indicates that daily net loads (electricity demand after subtracting available renewable energy) have been progressively decreasing in the middle of the day, when solar energy generation is peaking, as a function of increasing solar energy penetration year after year, with ensuing steep rises in the evenings when solar generation drops off. Due to its intermittent and variable nature, renewable energy is not dispatchable, and grid operators may have to resort to extreme and unsustainable measures such as curtailing (i.e., shutting down renewable energy generation) to mitigate oversupply when generation exceeds demand causing very low or even negative electricity prices. On the contrary, undersupply, as a result of sudden and unexpected drops in renewable energy generation, threatens the stability of the grid, whose system frequency must be managed within a tight band, i.e., around 50 or 60 Hz depending on the world region [14].

The increased adoption of electric vehicles (EV) is another factor that can have a significant impact on grid stability. More specifically, uncontrolled charging of EVs may give rise to grid disturbances, which in turn may lead to power outages. For example, if too many EVs are charging from the grid at the same time, the ensuing peak power demand may cause negative outcomes such as transformer heating and overloading, voltage instability, harmonic distortion, power quality problems, and power losses (see [15] and references therein).

1.2 AI and digitalization in energy management

The use of AI methods and IoT technologies through the digitalization of power system can help mitigate or solve energy transition challenges arising from factors including increased adoption of renewable energy and the electrification of the transportation sector. For example, the ability to forecast renewable energy productivity and energy demand can help grid operators and utilities maintain power system reliability [16]. AI-enabled inverters can integrate capabilities such as fault diagnostics and prognosis, device condition monitoring, adaptability to grid changes, fault tolerance, islanding detection, soft start, and ramp rate control to help prevent and mitigate grid disturbances leading power system instability [17–19]. Part II of this book examines more examples of how AI can address energy transition challenges, including solar forecasting (Chapter 12), real-time power system monitoring and control (Chapter 13), power conversion (Chapter 14), the operation of microgrids (Chapter 15), grid trust and resilience (Chapter 16), electric mobility integration (Chapter 17), electric vehicle adoption (Chapter 18), predictive photovoltaic (PV) maintenance strategies (Chapter 19), AI and robotic techniques for PV inspection (Chapter 20), and blockchain-enabled energy trading (Chapter 21).

The use of AI to enhance energy applications requires a set of concerted activities across data processes, algorithm development, and computing infrastructure engineering.

Data processes include data gathering, data generation, and data wrangling. Most data-gathering processes for energy applications result from the collection of measurements from sensors (Chapter 2) or the generation of synthetic data through simulations and digital twins (Chapters 3 and 4). Data wrangling focuses on the transformation of raw data into actionable evidence through the removal of questionable measurements (e.g., extreme outliers) and duplications, the provision of missing data, and other techniques, as discussed in Chapter 5 with reference to solar radiation measurements.

AI algorithms used to develop energy management applications include machine learning (Chapter 6), game theory (Chapter 7), optimization as discussed in Chapter

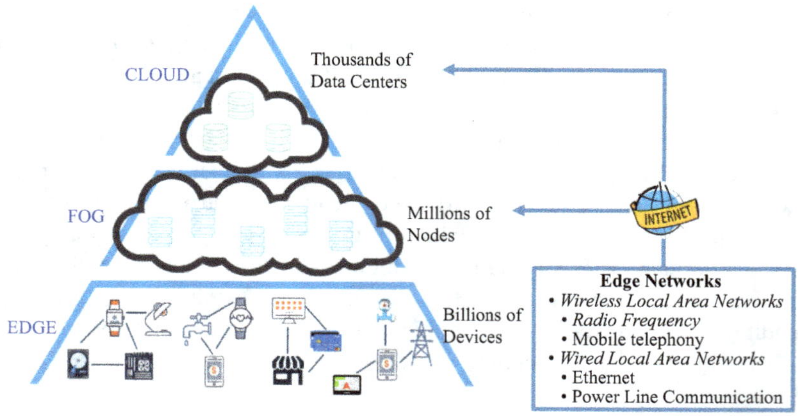

Figure 1.3 Edge to cloud computing infrastructure

18 with reference to electric vehicle adoption, and social simulation techniques such as agent-based modeling as described in Chapter 21 with reference to energy trade.

As shown in Figure 1.3 and discussed in Chapter 8, the computing infrastructure for energy management applications has evolved as an edge-to-cloud computing platform consisting of edge, fog, and cloud nodes [6]. Edge nodes record and communicate measurements from IoT sensors using wireless and wired local area networks to reach a final local gate from where the recorded measurements are sent to applications running on fog or cloud nodes using the Internet protocol. The Advanced Metering Infrastructure shown in Figure 1.4 provides an example of edge computing infrastructure where communication is articulated from a wired local area network reliant on power line communication to transmit smart meter data from households to data collecting units in electrical substations to the cloud using Wireless Internet where the application layer resides. Figure 1.5 exemplifies an edge computing infrastructure where radio-frequency communication is used to transmit sensor measurements taken inside a greenhouse to a WiFi-enabled gateway from where the data collected are sent to a fog/cloud node for further processing [20].

The global AI in energy management market was estimated at USD 24.4 bn in 2021, and current projections indicate that it is expected to grow at a CAGR of 21.2% from 2022 to 2030 to reach USD 113.61 bn [21]. A significant portion of the global AI energy management market focuses on renewable energy, with a 34% share in 2021, which is expected to grow twofold in 2030, as shown in Figure 1.6. The growing impact of AI and digitalization in the energy management market is discussed in Chapter 9.

Looking outward at the next frontier for AI in energy management, two likely avenues of development are federated learning in the short-to-medium term (Chapter 10) and hybrid quantum computing (Chapter 11) in the medium-to-long term. In federated learning, multiple edge resource-constrained computing devices jointly train a model under the orchestration of a central server using transfer learning without transferring data with ensuing benefits such as higher data privacy security and lower

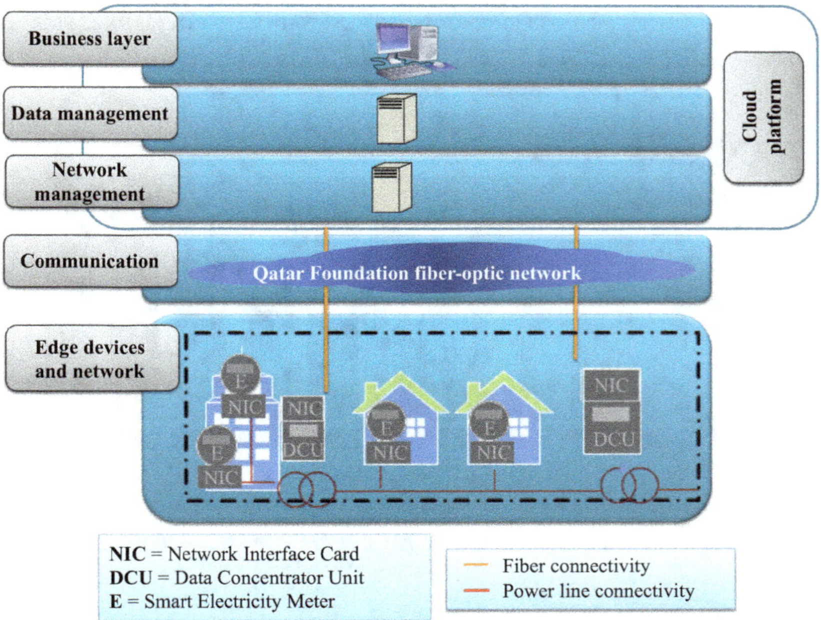

Figure 1.4 Advanced metering infrastructure: edge-to-cloud infrastructure based on power line communication and WiFi

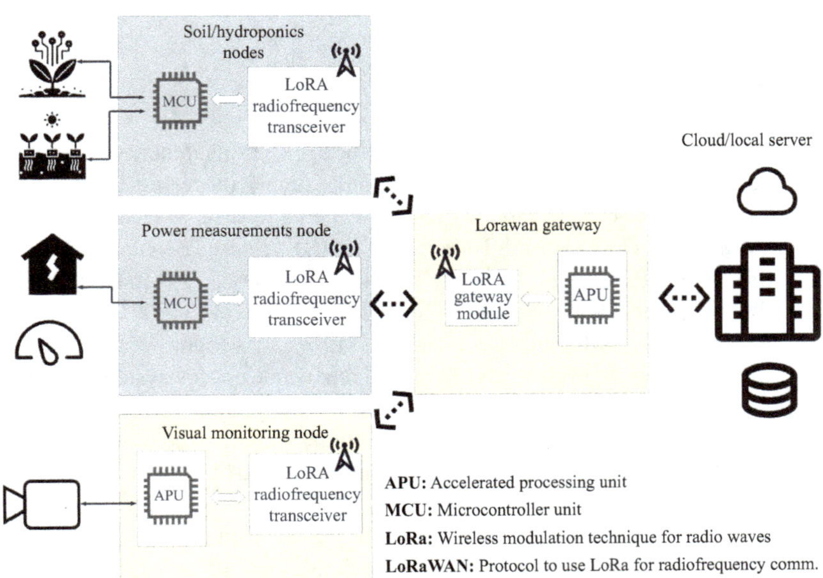

Figure 1.5 Resource management in indoor farming: edge-to-fog/cloud infrastructure based on radio-frequency and WiFi

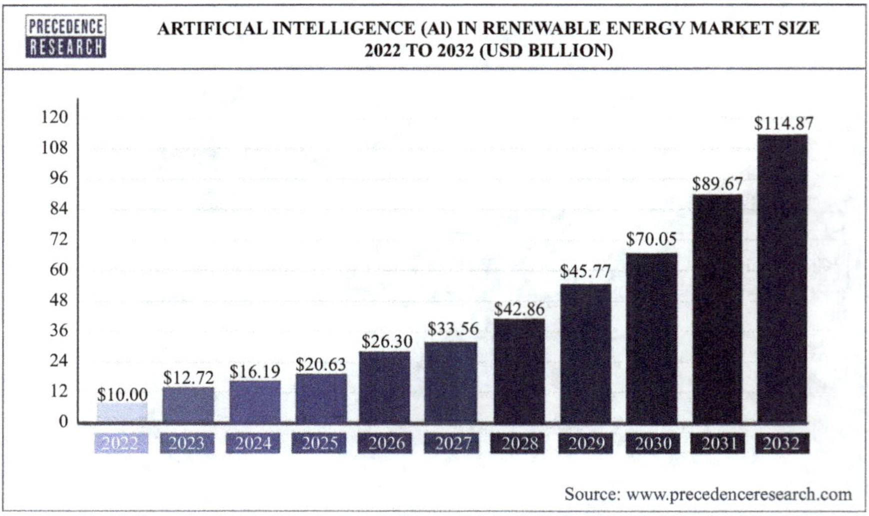

Figure 1.6 AI in renewable energy market size 2021–2030 [22]

cloud computing costs [23–25]. Hybrid quantum computing offers the promise of combining classical and quantum computing environments to solve large problems through decomposition efficiently [26,27].

References

[1] *Oil Crisis of the 1970s*. Energy Education, https://energyeducation.ca/encyclopedia/Oil_crisis_of_the_1970s#cite_note-2 (accessed: June 1, 2025).

[2] Corbett, M., 2013. *Oil Shock of 1973–74*. Federal Reserve History, 22.

[3] Roosa, S.A., Doty, S. and Turner, W.C., 2018. *Energy Management Handbook*. River Publishers. 9th edn. ISBN 9781138666979. First published in 1982 by Turner, W.C., Fairmont Press. ISBN 9780881734607.

[4] Olatomiwa, L., Mekhilef, S., Ismail, M.S. and Moghavvemi, M., 2016. Energy management strategies in hybrid renewable energy systems: A review. *Renewable and Sustainable Energy Reviews*, 62, pp. 821–35.

[5] Zima, M., Larsson, M., Korba, P., Rehtanz, C. and Andersson, G., 2005. Design aspects for wide-area monitoring and control systems. *Proceedings of the IEEE*, 93(5), pp. 980–96.

[6] Zafar, U., Bayhan, S. and Sanfilippo, A., 2020. Home energy management system concepts, configurations, and technologies for the smart grid. *IEEE Access*, 8, pp. 119271–86.

[7] Zia, M.F., Elbouchikhi, E. and Benbouzid, M., 2018. Microgrids energy management systems: A critical review on methods, solutions, and prospects. *Applied Energy*, 222, pp. 1033–55.

[8] Jamal, S., Tan, N.M. and Pasupuleti, J., 2021. A review of energy management and power management systems for microgrid and nanogrid applications. *Sustainability*, 13(18), p. 10331.

[9] Capehart, B.L., Turner, W.C. and Kennedy, W.J., 2020. *Guide to Energy Management*. River Publishers.

[10] Eltawil, M.A. and Zhao, Z., 2010. Grid-connected photovoltaic power systems: Technical and potential problems – A review. *Renewable and Sustainable Energy Reviews*, 14(1), pp. 112–29.

[11] Mehrizi-Sani, A. and Iravani, R., 2010. Potential-function based control of a microgrid in islanded and grid-connected modes. *IEEE Transactions on Power Systems*, 25(4), pp. 1883–91.

[12] What the duck curve tells us about managing a green grid, California ISO, https://www.caiso.com/documents/flexibleresourceshelprenewables_fastfacts.pdf (accessed: June 1, 2025).

[13] *As Solar Capacity Grows, Duck Curves Are Getting Deeper in California*. US Energy Information Administration, 2023. https://www.eia.gov/todayinenergy/detail.php?id=56880 (accessed: June 1, 2025).

[14] *Utility Frequency*, Wikipedia. https://en.wikipedia.org/wiki/Utility_frequency (accessed: June 1, 2025).

[15] Shariff, S.M., Alam, M.S., Hameed, S., *et al.*, 2022. A state-of-the-art review on the impact of fast EV charging on the utility sector. *Energy Storage*, 4(4), p. e300.

[16] Ahmad, T., Zhang, D., Huang, C., *et al.*, 2021. Artificial intelligence in sustainable energy industry: Status Quo, challenges and opportunities. *Journal of Cleaner Production*, 289, p. 125834.

[17] Xue, Y. and Guerrero, J.M., 2015, May. Smart inverters for utility and industry applications. In *Proceedings of PCIM Europe 2015; International Exhibition and Conference for Power Electronics, Intelligent Motion, Renewable Energy and Energy Management* (pp. 1–8). VDE.

[18] Zhao, X., Chang, L., Shao, R. and Spence, K., 2018. Power system support functions provided by smart inverters – A review. *CPSS Transactions on Power Electronics and Applications*, 3(1), pp. 25–35.

[19] Mirafzal, B. and Adib, A., 2020. On grid-interactive smart inverters: Features and advancements. *IEEE Access*, 8, pp. 160526–36.

[20] Sanfilippo, A., Kafi, A., Jovanovic, R., Ahmad, N. and Wanik, Z., 2024. Sustainable energy management for indoor farming in hot desert climates. *Sustainable Energy Technologies and Assessments*, 71, p. 103958.

[21] *AI in Energy Management Market Size, Market Share, Application Analysis, Regional Outlook, Growth Trends, Key Players, Competitive Strategies and Forecasts, 2022 to 2030*. Report. August 2022. Research and Markets. https://www.researchandmarkets.com/reports/5701958/ai-in-energy-management-market-size-market (accessed: June 1, 2025).

[22] *Artificial Intelligence (AI) in Renewable Energy Market Size, Share, and Trends 2024 to 2034*. Report. October 2023. Precedence Research.

https://www.precedenceresearch.com/artificial-intelligence-in-renewable-energy-market (accessed: June 1, 2025).

[23] Li, L., Fan, Y., Tse, M. and Lin, K.Y., 2020. A review of applications in federated learning. *Computers & Industrial Engineering*, 149, p. 106854.

[24] Khan, L.U., Saad, W., Han, Z., Hossain, E. and Hong, C.S., 2021. Federated learning for internet of things: Recent advances, taxonomy, and open challenges. *IEEE Communications Surveys & Tutorials*, 23(3), pp. 1759–99.

[25] Saputra, Y.M., Hoang, D.T., Nguyen, D.N., *et al.*, 2019. Energy demand prediction with federated learning for electric vehicle networks. In *2019 IEEE Global Communications Conference (GLOBECOM)* (pp. 1–6). IEEE.

[26] Varadharajan, V., Vidyasagar, S. and Shrivastava, A., 2022. *Quantum Computing – A Game Changer for the Energy and Utilities Sector*. INFOSYS. https://www.infosys.com/iki/perspectives/quantum-computing-game-changer.html (accessed: June 1, 2025).

[27] Perelshtein, M., Sagingalieva, A., Pinto, K., *et al.*, 2022. *Practical application-specific advantage through hybrid quantum computing*. arXiv preprint arXiv:2205.04858.

Part I

Fundamentals

Chapter 2

Sensor-based collection of solar and meteorological data

Dunia Bachour[1] *and Daniel Perez-Astudillo*[1]

Sensors are versatile devices used in a wide variety of applications, including scientific research, consumer electronics, industrial processes and automation, environmental monitoring, and medical practices. They are designed to detect changes in their environment and convert them into a measurable signal related to the specific type of physical phenomena, chemical quantity, or environmental condition. The measured signal can be a voltage change, resistance variations, or even light pulses depending on the sensor type and the measured quantity. In general, the measured signal requires some conditioning prior to further use, such as signal amplification, filtering, or calibration, and, if necessary, analog-to-digital conversion. After the pre-processing stage, the signal is eventually captured by a data acquisition (DAQ) system and converted to a suitable data format for storage and analysis. Overall, sensors act as the initial step in the data collection process, as they translate real-world phenomena into a digital representation that can be processed and analyzed to be used for various ever-growing applications.

In this chapter, we explore real-world examples of sensor-based data collection systems detailing the sensor body and specifications, the principle of operation, and the data collection process. Some figures in this chapter are adapted from the thesis work in [1]. The examples shown cover instruments and observations of atmospheric, solar, and meteorological parameters, including a lidar (LIght Detection And Ranging)-ceilometer measuring the backscatter profile of the atmosphere, solar radiation monitoring station with radiometers, a weather station for meteorological measurements, and an all-sky imager. We also describe the corresponding DAQ systems used and provide examples of data collected with these sensors.

2.1 Lidar-ceilometer

Lidar is a remote sensing technique used to measure and analyze a reflected laser signal to provide data about the distance and properties of distant objects. In the field of atmospheric research, lidars are used to detect atmospheric aerosol layers [2] through

[1]Environment Centre, Qatar Environment and Energy Research Institute, Hamad Bin Khalifa University, Qatar

the measurement of backscattered signals from specific heights in the atmosphere. The height is determined by the time interval between laser pulse emission and signal reception. The intensity of the backscattered light can also be measured.

A lidar-ceilometer is a compact, commercially available system designed for continuous operation with minimal maintenance. Ceilometers are commonly used in aviation and meteorology to detect cloud base heights, determine boundary layer structures [3], monitor dust and volcanic ash [4], and estimate particulate matter concentration [5]. This section will describe the CL51 unit, an active elastic ceilometer from Vaisala.

2.1.1 CL51 from Vaisala

The measurement unit of CL51, shown in Figure 2.1, primarily consists of a transmitter, a receiver, an optical unit, and electronics that digitize and process the reflected light beam to obtain the backscatter profile. The transmitter includes an indium gallium arsenide pulsed diode that emits short laser pulses (vertical or near-vertical) at a wavelength of 910 nm and energy of 3 μWs. The receiver uses a silicon avalanche photodiode to detect the backscatter resulting from the interaction of the transmitted pulsed light with obstructions in the atmosphere; detection is done at the same wavelength of the transmitted light, hence the name *elastic* ceilometer. The data measured by the ceilometer represents the signal strength of the backscatter at different height steps. The main specifications of the CL51 ceilometer are summarized in Table 2.1.

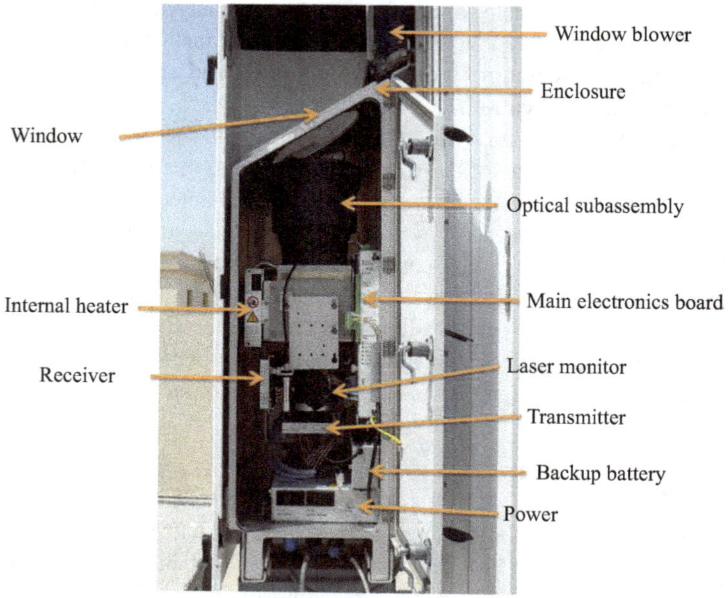

Figure 2.1 Main parts of the CL51 measurement unit installed at QEERI (Qatar Environment and Energy Research Institute)

Table 2.1 Specifications of the CL51 lidar-ceilometer

Property	Value
Backscatter range	15 km
Cloud range	13 km
Vertical resolution	10 m
Measuring interval	6 s
Reporting interval	36 s
Laser wavelength	910 nm
Pulse rate	6.5 kHz
Pulse duration	110 ns

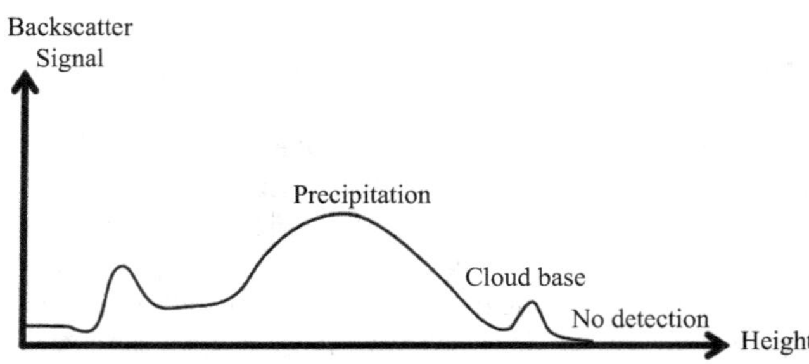

Figure 2.2 Ceilometer measurement signal (Adapted from the technical description document of the CL51 Vaisala Ceilometer available at https://docs.vaisala.com/viewer/book-attachment/JBhMEVJVu6DiLwZQTMMcxw/zl_DPKxB3KFjPrnctSXDQQ-JBhMEVJVu6DiLwZQTMMcxw)

CL51 operates by sending laser pulses into the atmosphere and measuring the time it takes for the light to be reflected back. The photodetector converts the returned optical signal into an electrical signal, which is then amplified for further processing. The signal is digitally sampled up to 100 μs and repeated every 67 ns, providing a spatial resolution of 10 m and an operational range of up to 15 km. To improve the signal-to-noise ratio, multiple laser pulses are transmitted, and the return signals are summed, allowing noise to partially cancel out. A typical measurement signal as a function of height is shown in Figure 2.2.

The basic equation relating the time delay t with the obstruction height h is given by

$$h = \frac{c * t}{2}, \tag{2.1}$$

where c is the speed of light, h is the distance traveled by the photons in the atmosphere, and the division by 2 accounts for the round trip of the photons (upward and downward). The lidar equation, which determines the instantaneous return signal power, is

$$P_r(Z) = E_0 \frac{c}{2} \frac{A}{Z^2} \beta(Z) \exp\left[-2 \int_0^Z \sigma(Z')dZ'\right] \tag{2.2}$$

where $P_r(Z)$ represents the instantaneous power received from distance Z, E_0 is the effective pulse energy considering all optics attenuation, c is the speed of light in, A is the receiver aperture, Z is the distance, $\beta(Z)$ is the volume backscatter coefficient at distance Z, and the exponential term accounts for the light attenuation over the round trip. By multiplying the return signal with the square of the height, the height dependence of the return signal power is eliminated. This equation helps derive the backscatter intensity by making certain assumptions.

2.1.2 Data collection

For data communication, the CL51 connects to a PC through a modem or serial cables (RS-232 and RS-485). Two manufacturer-provided software packages, CL-VIEW and BL-VIEW, enable data collection from the CL51. CL-VIEW is a graphical interface used "online" (receiving live data from the CL51) for collecting, storing, and displaying instantaneously the acquired data. Figure 2.3 shows an example of collected data, i.e. a cloud profile with the backscatter intensity collected at the experimental site in Doha, Qatar, showing time on the x-axis and altitude on the y-axis. The data points displayed in the graphical plot correspond to the location of the backscatter as a function of time, with the intensity of the backscatter visually identified by the darkness level in the color of the points. The saved backscatter data files can be exported in "offline" mode as comma-separated text files for further analysis of atmospheric backscatter coefficients. BL-VIEW is another software tool used to report the boundary layer height through an automated gradient method approach. It can also

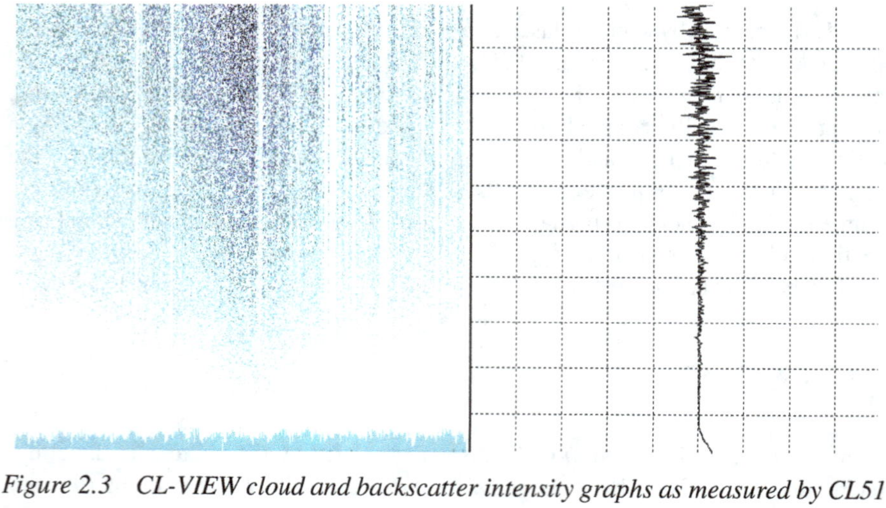

Figure 2.3 CL-VIEW cloud and backscatter intensity graphs as measured by CL51

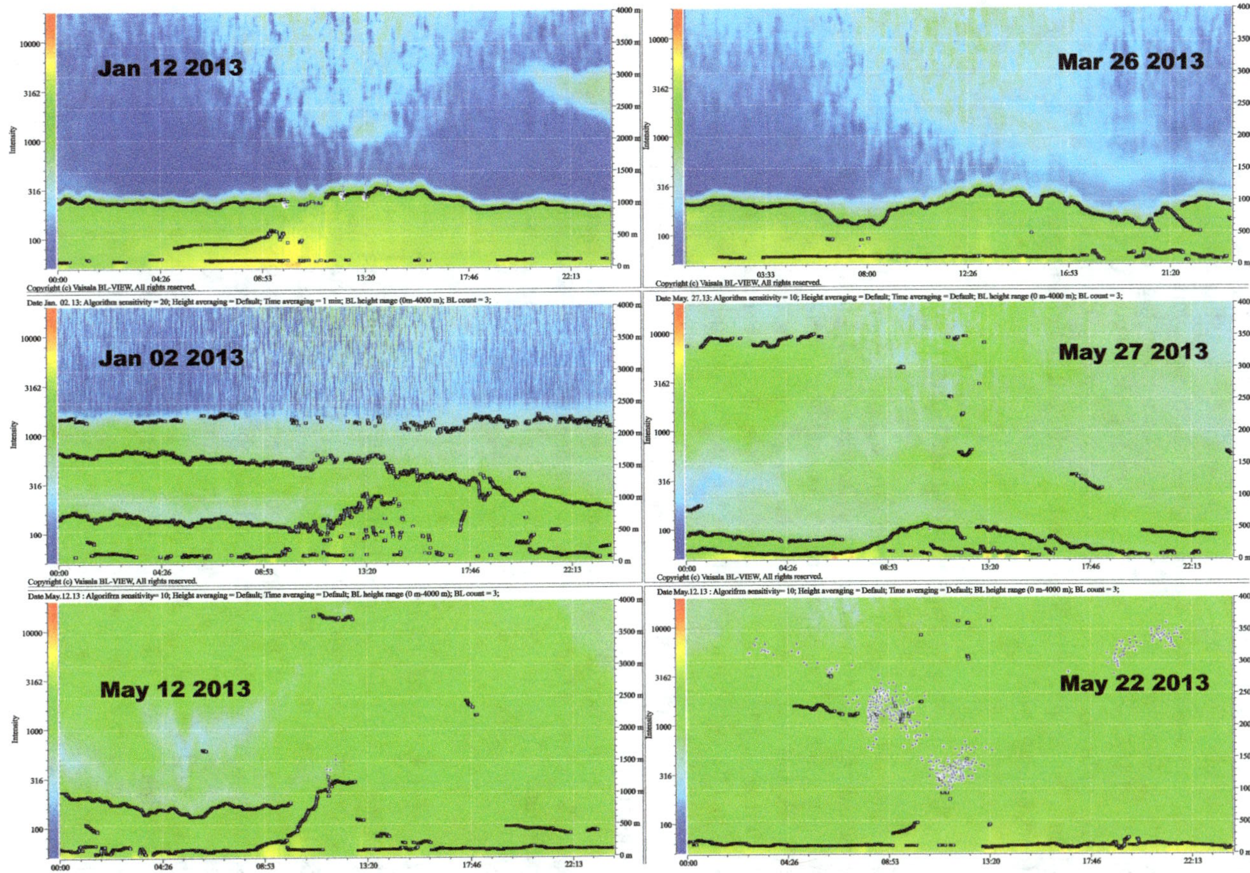

Figure 2.4 Samples of backscattered profile as displayed by BL-VIEW

be used for collecting, storing, and presenting the acquired data, but since it collects data only up to 4 km, it limits its use in the online mode when higher heights need to be analyzed. In BL-VIEW, the CL51 data are further processed using filters (cloud and precipitation), noise reduction, time- and height-averaging intervals, and signal detection thresholds in order to reduce noise and enhance the retrieval of the boundary layer evolution as well as the detection of additional layers of aerosols and clouds, especially in weather conditions with precipitation and fog [6]. Figure 2.4 presents CL51 BL-VIEW backscatter profiles for clear days in Doha, Qatar, showing varying dust levels without cloud presence. The backscatter signal evolution is depicted from midnight to midnight, with color scales on the left indicating signal intensity from low (dark blue) to high (red).

2.2 Solar radiation monitoring stations

A fully equipped solar radiation monitoring station is essential for accurate solar radiation measurements. A common solar station typically includes a sun tracker with the following sensors: a pyrheliometer for the measurement of direct normal irradiance (DNI), and two pyranometers—one shaded and one unshaded—for diffuse and global horizontal radiation, respectively. The sensors are mounted on the sun tracker in a way that the pyrheliometer follows the sun throughout the day and throughout the year, and the pyranometers are maintained horizontally to ensure stability and correct measurements. Additionally, a pyrgeometer may be included to measure broadband long-wave infrared irradiance.

This section describes a solar monitoring station from Kipp and Zonen, installed at QEERI's facility, which complies with Baseline Surface Radiation Network standards. Figure 2.5 illustrates the station with all sensors and accessories. Below is a description of each sensor and its operational principles.

2.2.1 SOLYS 2 sun tracker

The SOLYS 2 is a fully automatic, dual-axis sun tracker designed with platforms and accessories to accurately position solar sensors and ensure precise measurements. It performs horizontal (azimuth) and vertical (zenith) movements to follow the sun's arc, controlled by stepper motors and microprocessors. To ensure accurate measurements, the tracker is installed in an obstacle-free area, allowing unrestricted movement in both azimuth and zenith axes throughout the year. It is equipped with an integrated GPS receiver that sets the site's latitude, longitude, and time during the initial installation, with the time being regularly updated to prevent any internal clock drift during operation. The tracker uses the Michalsky algorithm to calculate the sun's position with a pointing accuracy of 0.1°. A sun sensor kit is also included with the SOLYS 2, providing additional tracking accuracy control. The sun sensor, featuring four-quadrant photodiode sensors, offers active tracking and corrections to the solar position algorithm when the tracker exhibits instability. The tracker is fixed on a firm and stable base through its tripod stand, leveled with an integrated bubble level situated on the top of the tracker. The pyrheliometer sensor is mounted on the side of the tracker on a plate with zenith adjustment for precise solar alignment, also holding the

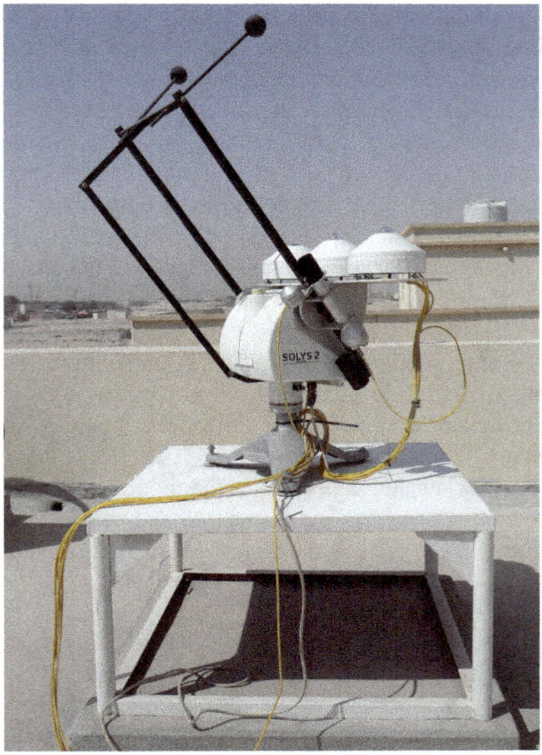

Figure 2.5 Solar radiation monitoring station as installed at the QEERI facility

Table 2.2 Main specifications of the SOLYS 2 sun tracker

Property	**Value**
Torque	20 Nm
Pointing accuracy	$<0.1^\circ$
Angular velocity	5°/s
Rotation	110° zenith and 540° azimuth
Operating temperature range	$-40\ ^\circ$C to $50\ ^\circ$C (AC power with heater)
Bubble level accuracy	$<0.1^\circ$

shading assembly to block direct radiation from reaching the diffuse radiation sensor. A horizontal platform is mounted on the top for hosting the other radiometers. The main specifications of the SOLYS 2 tracker are summarized in Table 2.2.

2.2.2 CHP1 pyrheliometer

The CHP1 pyrheliometer from Kipp and Zonen is designed to measure DNI and complies with the spectrally flat class A performance criteria for normal incidence pyrheliometers, as per the ISO 9060:2018 standard. The sensor is factory-calibrated and comes with a calibration certificate traceable to the World Radiometric Reference. The calibration process determines the sensor's sensitivity, which is the proportionality constant between the voltage output and an absolute level of irradiance. For example, the sensitivity of the sensor discussed here is 7.75 $\mu V/(W/m^2)$, valid at an ambient temperature of $22 \pm 2\ ^{\circ}C$ and a normal incident radiation of 500 W/m^2. This calibration is performed by comparing the sensor with a reference sensor of the same type using a horizontal parallel beam of light from a Xenon lamp. The CHP1 measures irradiance from a solid angle of 5°. Figure 2.6 illustrates the CHP1 installed on the SOLYS 2 sun tracker. The sensor features a flat quartz window, which ensures maximum transmission (up to 98%) of the solar radiation spectrum from 200 to 4,000 nm. This window is protected by a rain shield cover. Additionally, a screw-in drying cartridge filled with silica gel is mounted on the back of the sensor housing to prevent moisture from accumulating on the inner surfaces of the window. Accurate alignment of the sensor with the sun's direct beam is ensured using two sighting aids with an alignment accuracy of 0.5°. The CHP1 operates based on the thermoelectric effect. Its sensing element is a thermopile, which consists of multiple thermocouple junction pairs connected in series and coated with a black surface. When solar energy strikes the surface, the thermopile heats up, and the generated heat

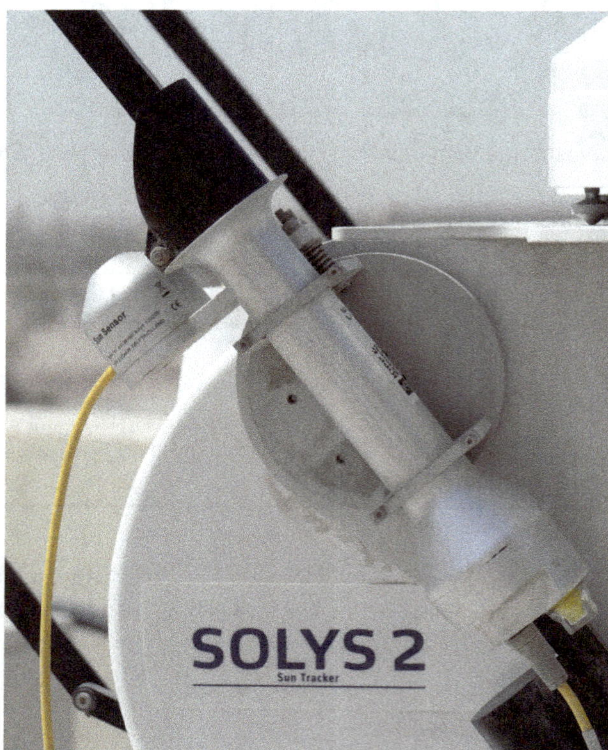

Figure 2.6 Kipp and Zonen CHP1 pyrheliometer installed at the QEERI facility

circulates through the thermal resistance inside the sensor, converting the temperature difference into a voltage. This voltage is linearly proportional to the absorbed solar irradiance, providing precise measurements of the DNI. Kipp and Zonen expect a maximum uncertainty of 2% for hourly totals and 1% for daily totals for the CHP1 pyrheliometer.

2.2.3 CMP11 pyranometer

CMP11 is a pyranometer from Kipp and Zonen used to measure global horizontal irradiance (GHI) and diffuse horizontal irradiance (DHI). It complies with the requirements of ISO 9060:2018 standard's Spectrally Flat Class A pyranometer, as shown in Figure 2.7. It is calibrated by the manufacturer according to ISO 9847 to obtain a calibration certificate. The CMP11 pyranometer is designed to measure the total solar irradiance on a plane surface in W/m^2 with the same principle of detection as the CHP1, which is the thermoelectric effect. It consists of a 32-junction thermopile sensing element, a non-spectrally selective black surface coating on the top, and a built-in passive temperature compensation system. Two glass domes protect the sensor and allow the transmittance of ~98% of the solar radiation spectrum, minimizing measurement errors at all sun incident angles. CMP11 is fixed horizontally on the mounting platform of the sun tracker with two of its three feet adjustable to properly align the sensor using the built-in bubble level. The housing of the sensor is covered with a sun shield to reduce direct solar heating. A removable drying cartridge filled with self-indicating silica gel desiccant is used to absorb the humidity within the sensor. The CMP11 does not require power to operate; the radiation absorbed by the sensing element produces a small analog output voltage. For diffuse measurements,

Figure 2.7 Kipp and Zonen CMP11 pyranometers used at the solar radiation monitoring station installed at the QEERI facility

a shading mechanism mounted on an articulated assembly and installed on the sun tracker casts a shadow (by means of a shading sphere) covering the pyranometer sensor completely. Both CMP11 pyranometers are fitted with ventilation units running continuously to help keep the domes clean and decrease the level of accumulated dirt and condensation on the dome, as well as to lower the thermal offset in the dome. All instruments are grounded for lightning and signal noise protection.

Based on many years of field measurements, Kipp and Zonen empirically determine uncertainty figures of their pyranometers, for typical operating conditions. For a high-quality pyranometer, the WMO expects maximum uncertainty totals of 3% in the hourly radiation. In the daily total, an uncertainty of 2% is expected, since some response variations cancel each other out if the integration period is long.

The CMP11 pyranometer at the QEERI station measures global radiation with a sensitivity of 8.78 $\mu V/(W/m^2)$ and diffuse radiation with a sensitivity of 8.47 $\mu V/(W/m^2)$. These values are obtained from the manufacturer by comparing the CMP11 to a reference pyranometer under stable artificial sunlight, with irradiance around 500 W/m^2. The reference pyranometers are calibrated outdoors at the WRC using the 'continuous sun-and-shade' method and are characterized for linearity, temperature dependence, and directional response. In the calibration procedure, both the test pyranometer and reference pyranometer are mounted side-by-side on a rotating table under a restricted light cone. The table rotates to interchange the positions of the two pyranometers, correcting for any light field inhomogeneity. Sensitivity is calculated by comparing readings, and a calibration certificate is issued accordingly.

2.2.4 Data collection

The solar measurement DAQ system utilizes a COMBILOG data logger provided by Kipp and Zonen and manufactured by Theodor Friedrichs & Co., as illustrated in Figure 2.8. The data logger is housed in a weatherproof enclosure suitable for outdoor applications and features eight analog signal inputs, six of which are used for the sensors presented here. It operates with a 12 V power supply, also enclosed, connected to the main AC power, and includes a solar panel as a backup supply with an internal battery. The recorded data are non-volatile, ensuring retention even if power is lost. An external memory card serves as a backup, as downloaded data are deleted from the logger's internal memory. Data are sampled every second and stored as 1-min averages. The COMBILOG connects to the PC for data download and monitoring either wirelessly via a GPRS/GSM modem or through serial cables. In this instance, an RS-232 serial cable is used due to the short distance (<10 m) between the PC and the logger. The data logger's clock is automatically synchronized with the computer's clock after each data retrieval. The PC clock is routinely verified for accuracy using another instrument's clock, which is updated by a GPS receiver.

For setting up and configuring the logger, specific MS Windows software called COMBILOG is used. This software allows live data viewing, although it does not store the data simultaneously. However, it provides the option for manual data download. Within COMBILOG, each measuring channel is defined as a variable in a variable table, with sensors selected from an integrated database. The software automatically linearizes the sensor signals. Another software package, COMGRAPH,

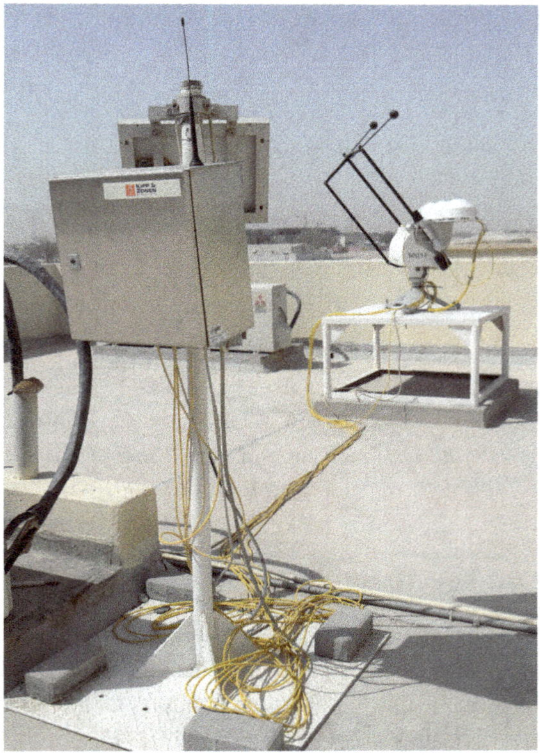

Figure 2.8 Data logger for the data collection and communication of the solar radiation monitoring station installed at the QEERI facility

is employed for data logging. COMGRAPH facilitates the scheduling of automatic data downloads and offers graphing capabilities and basic data analysis. The data are stored as text files. The data logger utilizes the calibration factors (sensitivity) embedded in the software library to convert the input voltages into radiation values W/m^2 according to

$$E = \frac{U}{S} \tag{2.3}$$

where U is the output signal of the radiometer in μV and S is the sensitivity in $\mu V/(W/m^2)$.

The COMGRAPH software provides quick visualization and daily checks of data, displaying each solar radiation component individually. For further analysis, all data are exported to text files for processing and cleaning to calculate hourly, daily, and monthly solar radiation averages after a thorough quality check [7]. Figure 2.9 shows an example of DNI, GHI, and DHI measurements over 3 days at QEERI's site.

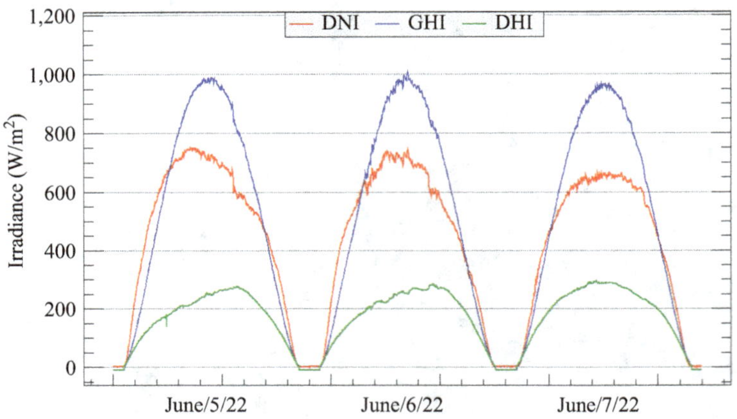

Figure 2.9 DNI, GHI, and DHI measured at the QEERI experimental site for three consecutive days (most night values are omitted)

2.3 Automatic weather station (AWS)

Similar to a solar monitoring station, an AWS autonomously collects, processes, and transmits meteorological data. AWSs typically include sensors for air temperature and relative humidity, wind speed, wind direction, precipitation, atmospheric pressure, and solar radiation. These sensors are connected to a data logger that records and stores measurements, often with a communication module for remote data transmission. AWSs are crucial for meteorology, climatology, hydrology, agriculture, and environmental monitoring, providing critical data for weather forecasting, climate research, and strategic decision-making.

This section describes an AWS installed at QEERI's facility. The AWS is from the LSI LASTEM Company. It is equipped with various sensors for measuring meteorological parameters: a thermohygrometer for temperature and relative humidity, a barometer (range: 800–110 hP) for atmospheric pressure measurements, and a direct signal output wind speed and direction sensor. The instruments are mounted on a 2-m-high meteorological pole, which is fixed to a concrete platform. The station is free from any obstacles that might obstruct or affect the wind measurements. The data logger used is the E-log data logger, featuring 12 input channels and housed in an IP65 enclosure, also fixed on the platform. The data logger is powered by a photovoltaic (PV) panel mounted on the AWS pole and includes a backup battery within its enclosure. An RS232 data cable connects the data logger to the control room, and the 3DOM (Datalogger Oriented Manager) host software is utilized for data collection. The station, complete with sensors and accessories, is illustrated in Figure 2.10. Below is a detailed description of each sensor and its operational principles.

2.3.1 CS215 thermohygrometer

The Campbell Scientific CS215 utilizes a digital humidity and temperature element based on Sensirion's CMOSens technology, employing both capacitive and resistive

Figure 2.10 AWS installed at QEERI's site

sensing principles. For temperature measurement, the sensor features a temperature-sensitive capacitor. Variations in temperature cause changes in the capacitance of this capacitor. The sensor then measures these changes in capacitance, which are directly proportional to the temperature. For relative humidity measurement, the sensor utilizes a polymer layer whose electrical resistance changes with the amount of water vapor in the air. This change in resistance is directly proportional to the relative humidity. The sensor is housed in a weatherproof enclosure and communicates via the SDI-12 protocol, ensuring compatibility with a wide range of data loggers and other SDI-12 compliant devices.

2.3.2 Barometric pressure Setra's Model 278

The Setra Model 278 is a precision barometric pressure transducer that operates on the principle of capacitive pressure sensing. It uses a ceramic sensor known for its stability and low hysteresis over a wide temperature range. The sensor consists of a thin, flexible diaphragm that acts as one plate of a capacitor. As atmospheric pressure changes, the diaphragm flexes, altering the distance between the plates and thereby changing the capacitance. The transducer's electronics measure these capacitance changes and convert them into a proportional voltage output representing the barometric pressure.

2.3.3 Anenometer DNA121C

The DNA121C anemometer is a dual-function sensor combining wind speed and direction sensing. It uses a three-cup anemometer to measure wind speed. Wind impinges on the cups, causing the anemometer to rotate at a rate proportional to the wind speed. This motion is converted into an electrical frequency signal, which is processed to determine the wind speed. Simultaneously, an integrated wind vane measures wind direction, translating its orientation into an analog voltage signal.

2.4 Sky imager

Ground-based all-sky imagers, also called all-sky cameras, are used to capture real-time images of the entire sky dome at their location, utilizing visible or infrared light. These devices are invaluable for visually assessing local sky conditions at any given time. The captured images are then processed to extract relevant sky features such as cloud cover, cloud type, and cloud motion. The data collected through an all-sky imager is useful for various applications, including weather forecasting, solar energy research and forecasting, and climate studies. Its ability to provide real-time observations of the sky helps scientists and researchers better understand and predict atmospheric conditions. In this section, we will describe the EKO SRF all-sky imager installed at QEERI's facility, illustrated in Figure 2.11.

2.4.1 EKO SRF imager

The EKO SRF all-sky imager is an instrument designed to capture high-resolution images of the entire sky dome. Its main function is to monitor cloud cover, classify cloud types, and observe other atmospheric phenomena. The imager uses a fish-eye lens with a 180-degree field of view, allowing it to obtain a comprehensive view of the sky. Equipped with a high-resolution digital camera, the SRF captures a sequence of visible-spectrum 2272 $\times$1702-pixel images at predefined intervals, typically every few minutes or less. These images are processed using specialized software that extracts information about cloud cover, cloud dynamics, and other atmospheric parameters. In [8], a custom software workflow was developed to automate the high-volume image processing and generate a "cloudiness index" for each image taken every minute from sunrise to sunset. This index is then compared with various solar resources and meteorological parameters to identify possible correlations. This serves as a foundation for developing solar nowcasting algorithms. Figure 2.12 shows a sample sky image taken by the EKO SRF all-sky imager installed at QEERI's site with the corresponding pixel brightness values.

Figure 2.11 All-sky imager installed at QEERI's site

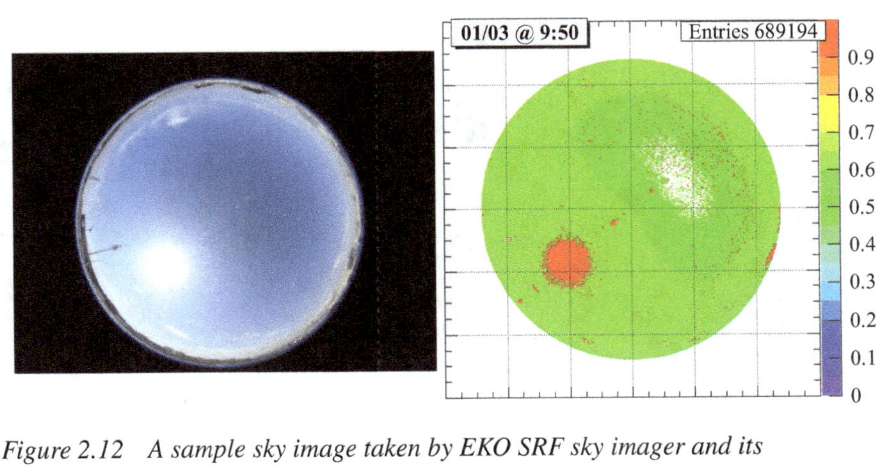

Figure 2.12 A sample sky image taken by EKO SRF sky imager and its corresponding brightness values

References

[1] D. Bachour, 2015. Measuring and estimating solar direct normal Irradiance using LIDAR, solar station and satellite data in Qatar. Doctoral Thesis, Universitat Autònoma de Barcelona. Departament de Física.

[2] C. Münkel, S. Emeis, and K. Schäfer, 2004. Aerosol concentration measurements with a lidar ceilometer: Results of a one year measuring campaign. *Proceedings of SPIE*. International Society for Optical Engineering.

[3] C. Münkel, 2006. Boundary layer and air quality monitoring with a commercial lidar ceilometer. *Proceedings of SPIE*, volume 6367, p. 20. International Society for Optical Engineering. DOI:10.1117/12.689775

[4] C. Münkel, K. Schäfer, and S. Emeis, 2013. Investigation of boundary layer dynamics, dust and volcanic ash clouds with laser ceilometer. *Remote Sensing of Clouds and the Atmosphere XVIII; and Optics in Atmospheric Propagation and Adaptive Systems XVI*, volume 8890, SPIE. International Society for Optics and Photonics.

[5] R.C. Muñoz and R.I. Alcafuz, 2012. Variability of urban aerosols over Santiago, Chile: Comparison of surface PM10 concentrations and remote sensing with ceilometer and Lidar. *Aerosol and Air Quality Research*, 12, 8–19.

[6] C. Münkel and R. Roininen, 2010. Automatic monitoring of boundary layer structures with ceilometer. *TECO-2010 – WMO Technical Conference on Meteorological and Environmental Instruments and Methods of Observation Helsinki*, Finland.

[7] D. Perez-Astudillo, D. Bachour, and L. Martin-Pomares, 2018. Improved quality control protocols on solar radiation measurements. *Solar Energy*, 169, 425–33.

[8] D. Perez-Astudillo and D. Bachour, 2023. All-sky Images for Solar Resource Assessment in Qatar. *Proceedings of EU PVSEC 2023*.

Chapter 3

Synthetic data generation through power hardware-in-the-loop (PHIL) simulations

Abdelbasset Krama[1], Anas Karaki[1], Ugur Fesli[2], Sertac Bayhan[1], Haitham Abu-Rub[3] and Mohd Zamri Che Wanik[1]

Over the past decade, artificial intelligence (AI) has played a pivotal role in driving significant advances in various technical fields, with energy management standing out as a key beneficiary of these advancements. The performance of AI and machine learning (ML) models is intricately tied to the quality and quantity of the training data; robust and stable model performance necessitates access to numerous amounts of high-fidelity data.

However, the acquisition of extensive real-world datasets that encompass the entire operational spectrum of an AI system can be loaded with difficulties. These challenges arise from the limited availability of data, the potentially expensive costs of data collection, and the risks or hazards that may be associated with certain data acquisition procedures.

Overcoming these obstacles, the synthesis of artificial data has emerged as a powerful solution. Synthetic data, artificially generated information that mirrors the statistical properties of real-world data, is a key tool. It is crafted using various techniques and models to ensure it closely resembles the structure, characteristics, and behavior of real data. Synthetic data is often used when real data is scarce, sensitive, or expensive to collect or when there is a need to augment existing datasets to improve the performance of ML models. The key advantage of synthetic data is its ability to be generated in large quantities while preserving privacy and confidentiality, as it does not contain any real or identifiable information. Synthetic data generation enables the augmentation of datasets with necessary additional data samples, thereby enhancing the performance and functionality of AI-based systems. This method effectively mitigates the constraints of real-world data collection, providing a controlled and adaptable approach to data generation that can be customized to meet the specific demands of the ML task at hand.

[1]Qatar Environment and Energy Research Institute, Hamad Bin Khalifa University, Qatar
[2]Department of Energy Systems Engineering, Faculty of Technology, Gazi University, Turkey
[3]College of Science and Engineering, Hamad Bin Khalifa University, Qatar

Hardware-in-the-loop (HIL) simulations are advanced testing methods that combine real physical hardware components with simulated electrical systems. This approach allows for the realistic evaluation of how new equipment or control strategies will perform in an electrical grid or energy system without needing full-scale, real-world testing.

The synergy between HIL simulations and AI in enhancing energy management is a powerful combination that leverages the strengths of both technologies to improve the efficiency, reliability, and sustainability of energy systems.

3.1 Synthetic data generation overview

The classification of synthetic data generation approaches is given in Figure 3.1, and it is classified into two main categories: model-based approaches and simulation-based approaches. Each of these categories is further subdivided into specific methods [1].

3.1.1 Model-based approach

3.1.1.1 Statistical-based generation

This method, centered on statistical modeling, is employed to produce synthetic data. The models used can be parametric, presupposing that the data adheres to a particular distribution, or non-parametric, which imposes fewer constraints on the data's distribution [2]. Statistical methods are often applied to generate synthetic load profiles and generation outputs, drawing on historical data for guidance.

Among the parametric approaches, Gaussian mixture models are utilized to model and generate data points that conform to a blend of multiple Gaussian distributions [3]. Conversely, Bayesian networks are employed to generate data that reflects the probabilistic relationships between different variables, offering a more flexible framework for capturing complex dependencies in the data.

3.1.1.2 Noise-based generation

Noise-based generation is a synthetic data creation method that adds noise to existing data points. This noise can be entirely random or follow a predetermined distribution,

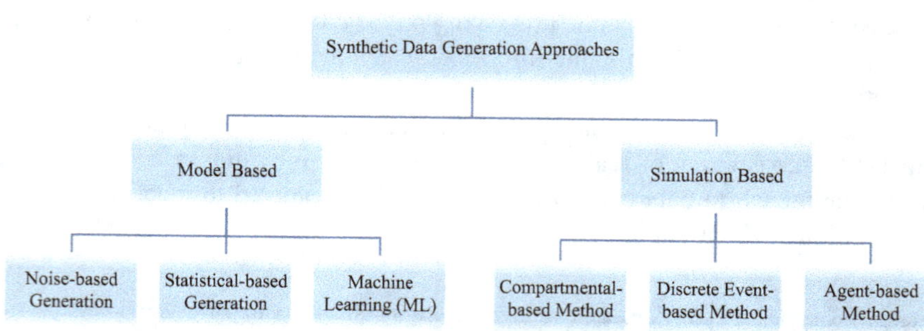

Figure 3.1 Synthetic data generation approaches

and the technique is frequently employed to enhance datasets or generate novel data variations. One approach within a noise-based generation is random noise addition, where random perturbations are introduced to real data to produce diverse variations. Alternatively, bootstrap sampling generates synthetic datasets by resampling the original dataset with replacement, effectively creating new data points from the existing ones.

3.1.1.3 ML-based generation

ML-based generation is a technique that utilizes ML models, specifically generative models, to produce synthetic data that closely resembles the distribution of actual data. Two popular methods in this domain are [4]

- **Generative adversarial networks (GANs)**: This approach involves a pair of neural networks composed of a generator and a discriminator. The generator is tasked with creating synthetic data, and the discriminator judges the authenticity of this data by comparing it to realistic data samples [5]. Through iterative training, the generator becomes an expert at producing synthetic data that is increasingly difficult for the discriminator to distinguish from real data.
- **Variational autoencoders (VAEs)**: VAEs are neural networks that compress input data into a compact latent space representation and then reconstruct it to generate new synthetic instances [6]. This encoding–decoding process allows VAEs to learn the underlying structure of the data and produce variations that mimic the original data distribution.

In the context of power systems, ML-based generation finds applications in

- **Synthetic time series data**: This technique is used to generate synthetic time series data for load, generation, and prices. Such data is invaluable for training and testing ML models, as it can provide additional data points representative of real-world patterns and trends.
- **Anomaly detection**: By creating synthetic anomalies, ML-based generation can enhance the detection capabilities of monitoring systems. These synthetic anomalies serve as training examples that help algorithms learn to identify unusual patterns or behaviors in power system operations.

The model-based approach to synthetic data generation presents several key advantages that make it a valuable tool in various domains. First, the flexibility inherent in statistical and ML models allows their adaptation to a wide range of data types and specific application needs. This versatility is particularly useful when dealing with complex or unique datasets.

Moreover, the accuracy of model-based synthetic data is a significant advantage. These models are capable of discerning and replicating intricate patterns and dependencies present in the original data, resulting in synthetic datasets that closely mirror real-world information. This realism is crucial for applications where the synthetic data must substitute or augment real data effectively.

The use cases for model-based synthetic data are extensive, with particular emphasis on sectors where data privacy is of utmost importance. In healthcare, finance, and customer data analytics, for example, synthetic data enables the protection of sensitive information that allows for the development and testing of models and systems.

Despite its benefits, the model-based approach is not without its challenges. The quality of the synthetic data generated is directly dependent upon the quality and quantity of the original data used to train the models. Insufficient or low-quality training data can lead to synthetic data that does not adequately represent the desired characteristics or may even introduce biases.

Another consideration is the computational complexity of advanced models such as GANs and VAEs. These sophisticated models often require significant computational resources and a high level of expertise to implement and fine-tune. This complexity can pose a barrier to entry for some organizations, particularly those without the necessary technical infrastructure or skilled personnel.

3.1.2 Simulation-based approach

3.1.2.1 Agent-based methods

Agent-based methods are simulation approaches that focus on the interactions of autonomous agents, such as consumers, generators, or market participants, within a specified environment. These agents are programmed with predefined rules and behaviors, and the synthetic data generated through their interactions reflects the emergent behavior of the overall system [7]. This technique applies to various domains, including social simulation for studying societal phenomena like disease spread or market dynamics and traffic simulation for analyzing traffic flow and congestion. Within power systems, agent-based methods are particularly useful for modeling the interactions between distributed energy resources (DERs) and the grid, simulating the behavior of participants in electricity markets to understand market dynamics and price formation, and examining consumer demand response to price signals or incentives that affect electricity usage patterns.

3.1.2.2 Compartmental model-based methods

Compartmental model-based methods are a simulation approach that segments a system into distinct compartments or segments, with predefined rules dictating the transfer of entities between these compartments. While these methods are prevalent in epidemiology, they can also be adapted in other fields, including power systems [8]. In the context of energy management, compartmental models might be employed in specialized scenarios where the system naturally lends itself to compartmentalization, such as in outage modeling, where different states of a component (e.g., operational, failed, under maintenance) and the transitions between these states are represented. Another potential application is in modeling energy storage systems, particularly in capturing the charging and discharging states of batteries.

3.1.2.3 Discrete event-based methods

This approach, known as discrete event simulation, models systems as a series of distinct events that occur at precise moments in time. It is particularly well suited for systems where changes are punctuated rather than ongoing, such as queueing systems in customer service or manufacturing processes [9,10]. In queueing systems, events might include customer arrivals, service times, and departures in manufacturing, and they could represent the initiation and completion of production stages.

Within power systems, discrete event simulation applies to various aspects of operation and planning, such as simulating equipment failures, maintenance activities, and switching operations of power converters. It is also used in grid simulation to model the power grid's operation, including load fluctuations, generation dispatch, and fault occurrences.

Examples of applications of discrete event simulation in power systems include contingency analysis, where the impact of various disruptions like line outages or generator failures on the power system is simulated. Additionally, event-driven market simulations model how specific market events, such as bids and trades, influence electricity prices and overall market dynamics [11].

Simulation-based approaches to synthetic data generation offer a suite of advantages that make them particularly well suited for certain applications. These methods are ideal for modeling complex systems where understanding the interactions and dynamics is crucial, such as in energy management and power systems. They excel in scenario analysis, allowing researchers and practitioners to explore different scenarios and gauge the impact of various variables on the system's behavior. Such approaches are frequently employed in fields like science and engineering fields, where they enable the simulation and analysis of potential outcomes, aiding in informed decision-making.

However, simulation-based approaches also come with their own set of challenges and considerations. Producing accurate and reliable simulation models can be a complex and time-consuming process, demanding deep field knowledge. Moreover, validating these models to ensure that the synthetic data faithfully represents real-world phenomena is a significant challenge.

Both model-based and simulation-based approaches provide powerful mechanisms for generating synthetic data, each with unique strengths and appropriate use cases. The decision to use one over the other depends on the specific demands of the task, including the data characteristics, the imperative for privacy, and the complexity of the system being modeled. A comprehensive understanding of these methods and their applications can greatly augment the efficacy of data-driven decision-making processes.

3.2 Simulation methodologies

Digital simulation is engineered to optimize the trade-off between accuracy and adaptability, meticulously facilitating a comprehensive replication of the dynamics of real-world physical systems under diverse operational conditions. These simulations are categorized into several types, see Figure 3.2, each with varying levels of integration between software models and physical hardware. Software-based digital simulation (SDS) relies entirely on software models, offering a low-cost, low-risk solution with moderate validity, ideal for early-stage analysis [12].

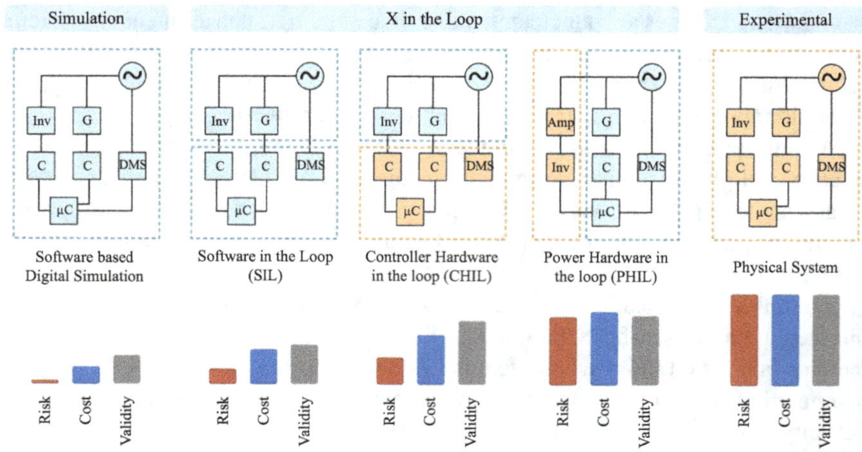

Figure 3.2 Characteristics of simulation methodologies

Software-in-the-loop (SIL) enhances SDS by integrating software algorithms with simulated environments, allowing for near real-time performance assessment [13]. Controller hardware-in-the-loop (CHIL) introduces actual control hardware, providing more accurate system dynamics at the cost of increased complexity. PHIL further extends this by incorporating real power hardware, significantly enhancing realism but also raising risks and costs [14,15]. These approaches differ from fully experimental setups, which involve testing complete physical systems and offer the highest validity but at maximum risk and expense.

3.2.1 *Software-based digital simulation*

Software-based digital simulation is a method that models a system's components and their interactions entirely within a software environment without the need for physical hardware. This approach is particularly beneficial during the initial design and testing phases due to its flexibility and cost-effectiveness. SDS is conducted using offline simulation software on a dedicated computer, which allows for the analysis of the process's dynamic behavior, the display of control variables, and the adjustment of model parameters [16]. Developers can design, manage, and execute multiple simulation scenarios within this software environment to evaluate the model's various functionalities and attributes. Furthermore, a trial-and-error approach can be utilized to refine the model parameters iteratively, striving to achieve optimal performance outcomes.

3.2.2 *SIL*

Known as model-in-the-loop, processor-in-the-loop, or real-time simulation (RTS), when the entire system (including protection, controller, and other accessories) is modeled and executed in real-time, synchronizing the simulation's time step with real clock time. This methodology allows for near-real-time performance assessment and facilitates testing of complex systems through high-performance computing platforms. The SIL can be achieved through a central processing unit or field programmable gate array (FPGA) that provides much more computational capabilities to assess and refine the functional operation of the model with a cost-effective platform [16]. The main advantage of SIL over offline simulation [modeling the software using advanced modeling environments like (MATLAB®/Simulink®)] is carrying out a simulation of a complex system in a short time, which requires a longer time when using offline simulation. In addition to the ability to conduct parametric studies on the model before going to experimental testing involving hardware components [17]. Moreover, SIL does not require any hardware other than the real-time simulator, which makes the implementation easier. However, ignoring some components' specifications and details in the simulated model may cause model inaccuracy, which has a significant effect on the dynamics of the entire system. However, ignoring some component's specifications and details, such as non-idealities introduced by physical power devices, controllers, and input/output (I/O) interfaces in the simulated model, may cause model inaccuracy, which has a significant effect on the entire system dynamics.

SIL simulation is designed to simulate the model by solving mathematical equations at a fixed time step while the time step is synchronous with real clock time. Unlike offline simulation, the simulation time can be faster or slower than real-time. In SIL, all the computational processes need to be done within the same time duration, including input reading, solving mathematical equations, and generating the outputs. An overrun can happen if the computation process is not successfully achieved within the given time step, which could affect the result's accuracy. To ensure exact and proper representation of a system transient response and frequency response, an accurate time step length must be computed. Unlike the accelerated mode in offline simulation, the idle time between two time steps, shown in Figure 3.3, remains idle in RTS.

SIL is used in control systems for industrial automation, automotive applications, and power systems, where testing needs to be conducted in a virtual environment to assess system performance and optimize control strategies before any real-world deployment.

3.2.3 HIL

HIL simulation diverges from SIL by incorporating actual hardware into the testing process, transcending the limitations of software-based simulations [18]. HIL is an advanced testing methodology that integrates real physical hardware into a simulated environment. It enables engineers and researchers to test and validate control systems in real-time by coupling physical components with simulation models. HIL simulations allow for accurate system performance testing under near realistic operating conditions without requiring the entire system to be physically constructed.

HIL differs from SIL simulation methodology, which involves purely software-based simulations without physical hardware. SIL is often used in the early stages

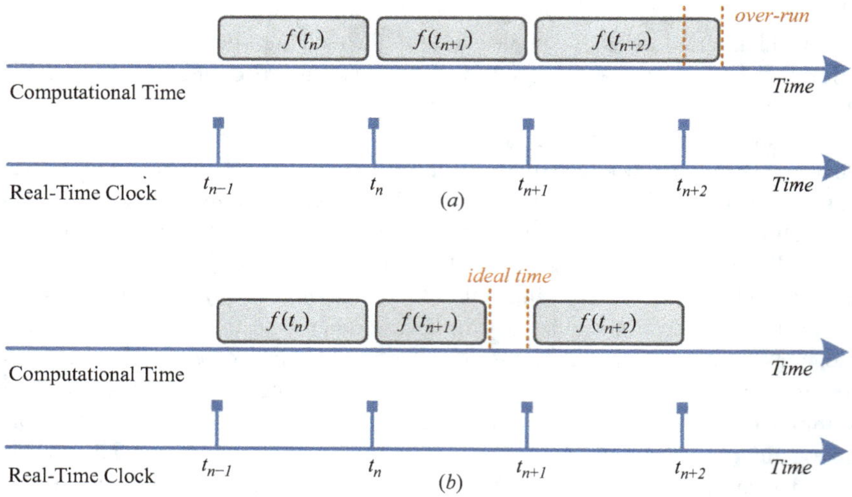

Figure 3.3 Time synchronization in a real-time simulation illustrating (a) overrun and (b) idle time

of system design and development, where models of controllers or system dynamics are tested within a purely virtual environment. In SIL, the focus is on verifying the logic and functionality of the system by synchronizing the simulation with real-time computational constraints. While SIL offers a cost-effective and risk-free way to evaluate control strategies, it lacks the fidelity that comes from integrating real-world hardware into the simulation process.

HIL builds upon the foundation provided by SIL by introducing real hardware into the testing loop, as illustrated in Figure 3.2. This allows for more comprehensive validation of system performance, as it accounts for hardware specific non-idealities such as signal delays, noise, and power flow issues that are absent in purely virtual simulations. The transition from SIL to HIL typically occurs after the control algorithms have been validated in software and require real-world testing to ensure robustness and reliability before deployment. This approach can be categorized into two main types [14]:

3.2.3.1 CHIL

CHIL involves a physical controller interacting with a simulated power plant within a hardware simulator (Figure 3.4). This method combines the efficiency of software-based simulation with the high fidelity of hardware prototyping. In a CHIL setup, only low-voltage signals are exchanged between the simulated power plant and the controller. The power plant is modeled virtually within the simulator, and the hardware controller communicates with the power plant via I/O interfaces, with no high power exchange occurring (typically within a ± 15 V range). The absence of high-power exchanges makes CHIL less complex and safer. The physical controller processes feedback signals from the simulator to generate outputs that actuate the simulated

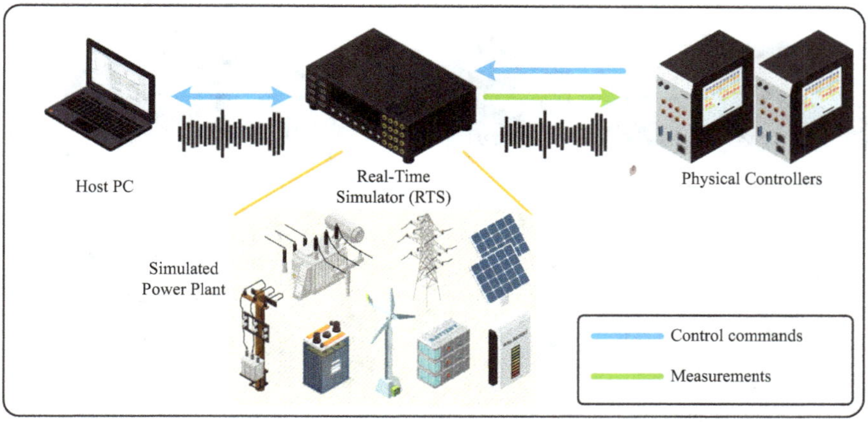

Figure 3.4 CHIL fundamental concept

model. CHIL allows for including real hardware components and signal processing delays, which are not accounted for in SIL simulations.

CHIL provides a cost-effective and efficient method for testing real-world controllers in a simulated environment. One of its main advantages is the ability to conduct realistic control testing by capturing non-idealities, such as signal delays and inaccuracies, which are often not accounted for in purely software-based simulations. This ensures that the controller performs reliably under real-world conditions. Moreover, CHIL reduces complexity by excluding high-power devices, making the simulation setup simpler and safer than full PHIL testing, which involves high-power exchanges. This methodology also supports controller optimization by allowing engineers to adjust and fine-tune control parameters based on reliable real-time feedback from the simulation. This iterative process enables precise tuning of the controller's performance, ultimately improving its overall functionality. By offering these benefits in a more controlled and less expensive environment, CHIL provides a practical solution for controller development while minimizing both complexity and cost. This HIL methodology is used in power systems, renewable energy integration, and automotive and aerospace industries to test control systems that must interface with complex physical environments.

3.2.3.2 PHIL

PHIL signifies the presence of high-power signals in the setup, achieved by integrating real hardware devices to simulate part of the power plant. A power amplifier (PA) mediates the interaction between high-power devices and the low-power-rated I/O of the simulator, enabling high-power flow between the actual hardware and the simulated model (Figure 3.5). This amplifier acts as either a sink or a source, depending on the reference signal from the simulated model, generating current or voltage for the hardware under test (HUT) [19]. Measurements from the HUT are then fed back to the simulator/controller to close the loop.

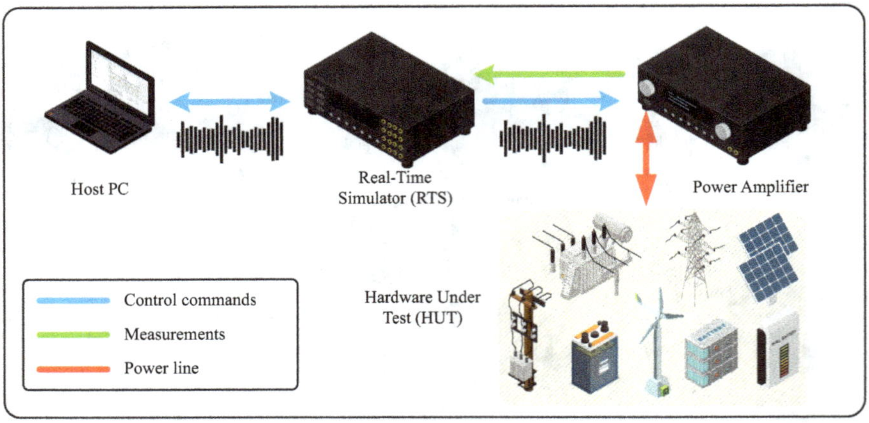

Figure 3.5 PHIL fundamental concept

The appearance of multi-core processor-based technology and FPGAs has facilitated the simulation and testing of complex power systems using HIL methodologies, with time steps as fine as tens of nanoseconds [16]. The effectiveness of HIL testing has been demonstrated across various industries, including power systems, renewable energy, mechanical engineering, manufacturing, automotive, and aerospace [19–21]. HIL enables a thorough examination of power control system functionality, performance, and design, as well as failure testing. Manufacturers are increasingly adopting HIL testing due to the growing complexity and interactivity of power systems, along with other benefits such as reduced simulation time, the ability to test under extreme conditions, and cost savings.

Ideally, PHIL is a viable alternative to a full-scale hardware prototype, allowing engineers and researchers to focus on developing new control techniques or designs rather than on the time consuming and safety critical construction of experimental prototypes. HIL methodologies have gained widespread use in recent years, replacing conventional electromechanical test benches and providing a user-friendly, near-realistic testing environment. This simulation approach is widely used in power, electric vehicles (EVs), and battery management systems. It is valuable in industries where safety and performance are critical.

3.3 Characteristics of PHIL components

3.3.1 Power amplification

Power amplification in PHIL simulation is a critical component that enables the interaction between high-power hardware devices and the simulated environment running on a real-time simulator. The PA acts as an interface that converts low-power signals from the simulator into high-power signals that can drive the HUT or vice versa [22]. This allows for realistic testing of power systems, components, or control strategies without the need for a full-scale physical prototype.

The PA must be capable of handling the required power levels and dynamics of the system being tested. It should have a high bandwidth to reproduce the waveforms generated by the simulator accurately and must be able to operate in real-time to maintain synchronization with the simulation [23,24]. The performance characteristics of the amplifier, such as its gain, slew rate, and total harmonic distortion, are crucial for ensuring the fidelity of the test results [25].

In a PHIL setup, the real-time simulator models the behavior of the power system or components can generate control signals. These signals are typically within a safe, low-power range ($\pm$15 V). The PA then scales these signals to the appropriate power level to interact with the high-power hardware. For instance, if the HUT is a power converter, the PA will provide the necessary current and voltage to drive the power converter according to the simulator's commands.

The PA can operate in different modes depending on the test requirements. It can act as a source, providing power to the HUT, or as a sink, absorbing power from the HUT [26]. This bidirectional capability is essential for simulating various operating conditions and transients that the hardware might encounter in real-world applications.

To ensure stable and safe operation, the PA is often equipped with protection features such as over-current, over-voltage, and over-temperature shutdowns [24] These safeguards prevent damage to the amplifier, the HUT, and the surrounding equipment in case of unexpected events or system faults. The following sections describe three different types of power amplification [27].

3.3.1.1 Switched mode power amplification

Switched mode devices have emerged as the preferred choice for power amplification in PHIL simulation, even at the megawatt scale. These devices, which operate by rapidly switching between on and off states, offer high efficiency and are well suited for the demanding requirements of PHIL testing.

A typical configuration for such a PA in PHIL simulation is an AC/DC/AC converter. This setup includes a front-end rectifier that converts AC grid power to DC and a back-end inverter that inverts the DC power back to AC for application to the HUT. This bidirectional converter allows for seamless coupling with the utility grid, enabling power to be sourced from the grid to the HUT or absorbed from the HUT and fed back into the grid.

The RTS plays a pivotal role in this setup by sending a low-level reference signal to the inverter. This signal dictates the voltage or current the inverter will apply to the HUT, utilizing an appropriate control algorithm. Furthermore, the system can also operate as a DC amplifier when required.

Various control strategies, such as proportional integral (PI), proportional integral derivative (PID), and model predictive control (MPC), have been developed and fine-tuned for PA control. These control approaches are complemented by various output filter configurations, including LC and LCL filters, to meet the specific needs of the PA.

Different semiconductor topologies are employed for three-phase PAs. The standard three leg DC/AC inverter is commonly used, but for scenarios requiring unbalanced operation, four leg inverters are also utilized. Additionally, to ensure that the PA can accurately follow the rapid dynamics of the RTS, a high switching frequency for the semiconductors is recommended. This high-frequency operation allows for precise control and minimizes the impact of switching transients, ensuring that the PHIL simulation accurately reflects real-world conditions.

3.3.1.2 Generator-type power amplification

This method involves the utilization of a three-phase synchronous generator. This generator is powered by a DC motor and accompanied by a separate rotating exciter system. This setup is integrated with a real-time simulator that conducts a simulation of a power system at the PCC that interfaces with the generator's output.

The synchronous generator in this arrangement is regulated in torque mode, allowing for precise manipulation of the generated power to match the requirements of the simulated power system. This level of control is vital for emulating the complex behaviors of real-world power systems and for subjecting power hardware to rigorous and realistic testing scenarios.

A detailed examination of this setup, including its operational characteristics and control mechanisms, can be found in [28]. It delves into the derivation of the closed-loop transfer function for the motor torque input, which is a fundamental aspect of the system's control dynamics.

The control system for this setup utilizes proportional integral differential acceleration (PIDA) controllers, a variant of the conventional PID controllers that includes an additional acceleration term. This control strategy, though unconventional, was found to be necessary for maintaining stability in the controller's output. The absence of the acceleration element led to instability, pointing out the need for specialized control approaches in complex power systems. The control parameters for the PIDA controllers, derived through meticulous analysis, are presented in the form of equations. These equations encapsulate the intricate tuning required to achieve stable and effective control of the synchronous generator within the PHIL environment.

3.3.1.3 Linear power amplification

The architecture of the linear amplification phase is designed with a number of linear metal–oxide–semiconductor field-effect transistor (MOSFETs) operating in parallel, ensuring that their combined characteristics mimic those of a single discrete MOSFET in its linear region. This approach allows for the key attributes of the linear amplifier to be precisely identified.

The amplifier's control system can be operated using either of its two internal oscillators or an external input signal. This external analog voltage signal, with a maximum range of 5 V_{peak} (3.535 V_{rms}), is especially beneficial for PHIL testing systems. The amplifier supports two output voltage ranges from 135 to 270 V, making it suitable for various low voltage experiments. In PHIL applications, the 270 V range is often employed, where an input voltage signal of 3.011 V_{rms} corresponds to an output voltage of 230 V_{rms} [27]. Additionally, an anti-aliasing filter may be added to smooth the RTS signal's stepped appearance, which results from the simulator's discrete nature.

3.3.2 Interface algorithm (IA)

An IA is essential in a PHIL system, determining the types of signals such as voltage, current, power, or torque that are transmitted and how they are processed [26]. This section will examine five existing IAs. Although PHIL simulations can involve multiple types of variables, voltage and current are the most used signals in power system applications. Consequently, the following algorithms will concentrate exclusively on voltage and current as the primary interface signals [14].

3.3.2.1 Ideal transformer model (ITM)

The ITM algorithm is a conventional and simple method for implementing PHIL simulations. This algorithm can be adapted to different types of signals, making it either a voltage-type ITM or a current-type ITM, depending on the signals being amplified [14]. The voltage-type ITM, depicted in Figure 3.6, is used when the focus is on voltage signals, while the current-type ITM, illustrated in Figure 3.7, is employed for current signals [23].

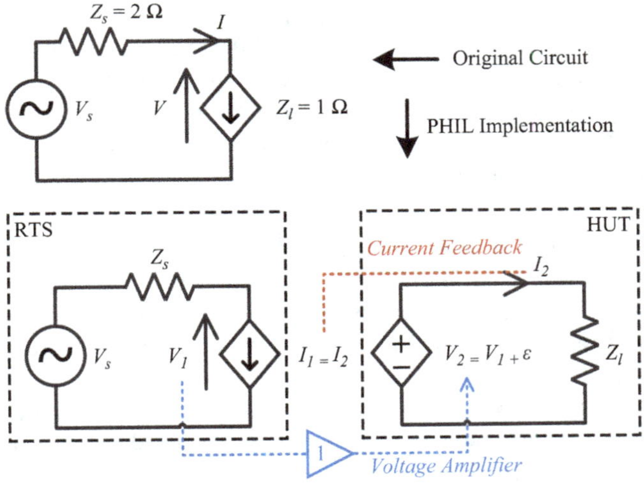

Figure 3.6 A PHIL implementation of a voltage divider circuit

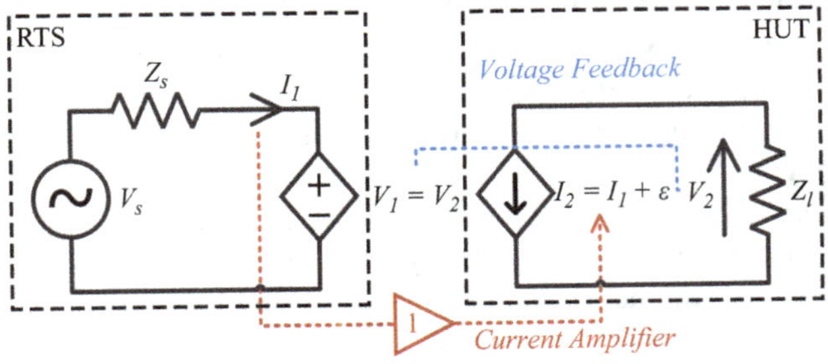

Figure 3.7 An alternative PHIL implementation of the voltage divider circuit

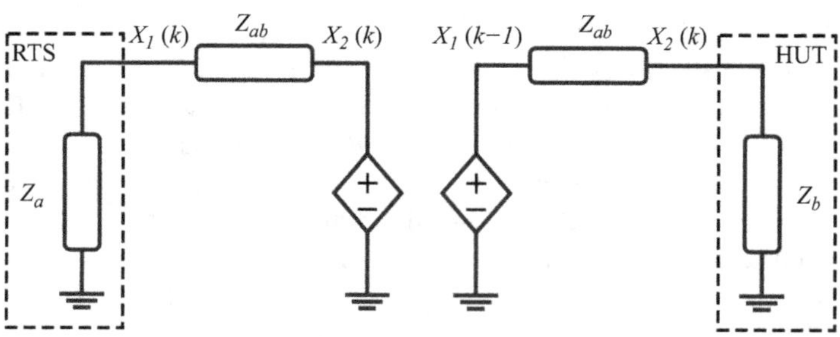

Figure 3.8 Depiction of the PCD interface in a PHIL experiment

Consider a voltage divider circuit where the load impedance Z_l is a real hardware resistor, while the rest of the circuit, including the voltage source V_s and source impedance Z_s, is simulated. In this setup, a voltage amplifier replicates the simulated voltage V_1 as voltage V_2 and applies it to the load resistor. The real current I_2 drawn by the resistor is then measured and fed back into the simulated circuit via a modeled current source I_1.

However, if the circuit is implemented differently, as shown in the PHIL system in Figure 3.7, the outcomes can be unexpectedly different. In this alternative setup, the circuit current I_1 (rather than the voltage V_1) is amplified, and the voltage V_2 (instead of the current I_2 is measured and fed back to the simulator. These two topologies differ significantly in terms of stability, with the current-based approach generally providing a more stable simulation.

3.3.2.2 Partial circuit duplication (PCD)

The PCD method is derived from the relaxation method used in large circuit simulation software like SPICE. When applied to sizable circuit simulations, the PCD method divides the original circuit into smaller sub-circuits (Figure 3.8). It then employs iterative methods, such as the Gauss–Seidel method, to find the solution. As long as the iterations converge, the solution can be considered accurate [14]. This approach prevents the need to solve the enormous set of ordinary differential equations for the entire circuit all together, thereby significantly reducing computation time.

3.3.2.3 Damping impedance method (DIM)

The DIM algorithm is a hybrid approach that combines elements of both the ITM and the PCD algorithms. The key innovation of the DIM is the insertion of a damping impedance, denoted Z^*, into the PHIL implementation. This unique feature is illustrated in Figure 3.9.

In the DIM algorithm, the damping impedance Z^* serves to enhance the stability and accuracy of the simulation. By strategically placing Z^* within the circuit, the algorithm can effectively dampen oscillations and improve the convergence of the iterative solver used in the PCD technique. This results in a more stable and reliable simulation, especially for large and complex circuits [29].

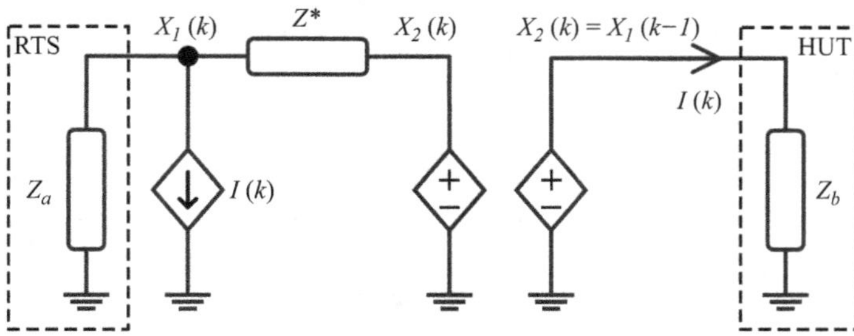

Figure 3.9 Simplified topology of the DIM system

The PHIL implementation of the DIM algorithm leverages the strengths of both the ITM and PCD methods. It retains the simplicity and effectiveness of the ITM in handling signal amplification and transformation while also benefiting from the computational efficiency and scalability of the PCD technique. The addition of the damping impedance (Z^*) further refines this hybrid approach, making the DIM algorithm a powerful tool for accurate and stable circuit simulations [30].

3.3.2.4 Transmission line model

The transmission line model (TLM) algorithm is particularly effective for decoupling large systems, thanks to its ability to represent a linking inductor or capacitor as a TLM. This approach is further enhanced by using the Bergeron line model, which separates the resistive Norton or Thevenin circuits on either side, allowing each side to interact only with the previous value from the other. This separation facilitates parallel computation, making the TLM algorithm highly efficient [14].

In PHIL simulations, as shown in Figure 3.10, the TLM algorithm is implemented by replacing the linking inductor L or capacitor C with a Bergeron TLM. The numerical stability of this algorithm is ensured by its reliance on the trapezoidal approximation, making it a robust choice for such simulations.

Power consumption: In the algorithm, the linking inductor and capacitor are substituted with a resistor R_{link}. However, in high-power applications, this resistor can result in considerable power consumption, raising efficiency concerns that may make this approach impractical.

Flexibility and maintenance: The value of R_{link} is dependent on the specific characteristics of the linking components. Consequently, R_{link} must be recalibrated whenever there is a change in the simulated system. This need for adjustment reduces system flexibility and can increase maintenance costs.

Despite these challenges, the TLM algorithm's stability and suitability for parallel computation make it a valuable method for certain power HIL simulation applications.

3.3.2.5 Complexity concerns of PHIL

In the realm of PHIL simulations, the lack of a standard approach for nonlinear HUT devices introduces a unique set of challenges and opportunities [31]. The

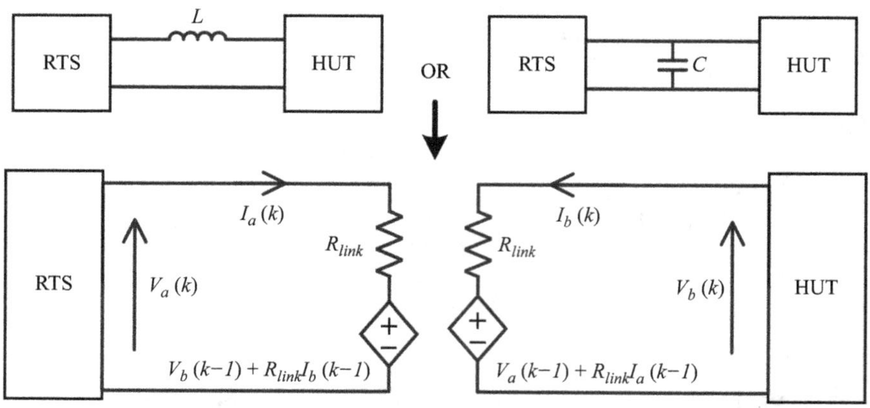

Figure 3.10 PHIL setup utilizing the TLM method for the IA

complexity and resource demands of PHIL simulations from an equipment perspective are significant, often making software simulations a more cost-effective alternative. However, the ability of PHIL simulations to closely mimic actual system behaviors in a controlled environment is unparalleled. This capability is attributed to integrating cutting-edge RTS technologies, high-performance PAs, and sophisticated software/hardware IAs, which collectively ensure stability and accuracy across a broad frequency spectrum [32].

The necessity for real hardware to assess system-wide impacts, particularly when HUT models are unavailable or need to be developed through PHIL simulations, underscores the value of PHIL as a systematic testing tool. For a wide array of electric power systems, PHIL emerges as a promising technology, especially when hardware implementation for testing is either prohibitively expensive or impractical. Moreover, the urgency of conducting tests and impact analyses before system installation is a critical consideration from both safety and practicality standpoints.

The software centric nature of PHIL simulations, with a significant portion executed on the RTS, offers substantial benefits. Users can easily modify the parameters of the simulated system model to create new configurations, constraints, and behaviors, a feat that is far more challenging to achieve in hardware experiments. The flexibility to transition subsystems between hardware and software realms is not only feasible but often essential for enhancing simulation stability or achieving higher stability margins.

Incorporating PAs and IAs into PHIL simulations introduces a layer of complexity. Yet, the diversity of interface topologies available allows for informed selections based on prior knowledge and experimental requirements. The trade-offs between system accuracy, bandwidth, and stability margins are linked to the initial configuration of hardware components. This flexibility in choosing the most appropriate interface for a given PHIL setup is crucial for preventing systematic errors.

3.4 Power HIL accuracy and stability evaluation

3.4.1 Understanding delays in PHIL simulation

In PHIL simulations, the real-time simulator plays a pivotal role. It consists of processor cards for real-time processing and an interface, either analog or digital, for communication with external components. The processor cards execute the simulation at a discrete time step, while the interface handles the exchange of setpoints and feedback signals between the RTS, PA, and HUT [33,34].

3.4.1.1 Simulation platform delay

The time step of the RTS, denoted as T_s, is a critical parameter influenced by several factors:

Complexity and number of nodes: The complexity and size of the simulated power system affect the time step. Specific Application Requirements: Certain applications may demand a specific time step.

Platform limitations: The RTS hardware may impose limitations on the achievable time step.

Typical time steps range from 1 to 200 μ s for systems using powerful processors [33]. Advanced platforms utilizing FPGA technology can achieve even lower time steps, enabling sub-microsecond simulations [16]. This time step introduces a processing delay in the RTS, which is constant and constitutes the time delay introduced by the processing stage of the RTS. Additionally, the simulated component used for electrical coupling with the power system, usually a controlled current or voltage source, can introduce further delays, denoted as T_{SPD} .

3.4.1.2 Communication delay

The communication delay, TComD, arises from the interfacing of the RTS with external components like the PA and HUT. This delay includes the feed-forward path from the RTS to the PA and the feedback path from the HUT to the RTS [28].

- **Analog communication delay**: In analog interconnections, digital-to-analog converters (DAC) and analog-to-digital converters (ADC) are necessary. These components introduce inherent delays, T_{ADCD} and T_{DACD}, respectively. The time required for information to pass from the processor card to the ADC or DAC is part of the total delay of each conversion component. An anti-aliasing filter at the ADC can further delay the process. The measurement of the HUT response can also introduce additional delays in the feedback path. Digital filters, which are used for noise reduction, also introduce a certain delay.
- **Digital communication**: High-speed serial protocols like peripheral component interconnect express (PCI Express) and Aurora enable faster communication without the need for DACs and ADCs, reducing overall delay. The delay, in this case, depends on the latency of the optical fiber and the processing delay of the communication algorithm.

3.4.1.3 PA delay

As mentioned previously, there are several methods for power amplification, with switched mode power amplification being a prominent choice in power system studies. These devices typically include an active rectifier for grid connection and a power converter that amplifies signals from the RTS unit. The AC/DC/AC converter in these systems can operate in all four quadrants of the power plane, enabling it to both supply and absorb real and reactive power [35]. The delay introduced by the switched mode PA, denoted as T_{PAD}, is an inherent characteristic of the digital control loop architecture of power converters. This delay consists of two primary components:

Sample period: This is the time interval during which the controller can only update the duty ratio at the start of each switching cycle.

PWM signal comparison: An additional delay equivalent to half a time step is required to compare the PWM signal with the carrier waveform.

Hence, the time delay in PHIL setups can be calculated as follows [33]:

$$PHIL_{Timedelay} = T_{SPD} + T_{ComD} + T_{PAD} \tag{3.1}$$

Typically, the RTS and PA units are major contributors to the overall delay in PHIL simulations, with additional delays arising from communication interfaces and filtering components. Managing these delays is essential for ensuring the accuracy and stability of PHIL simulation results.

3.4.2 Stability analysis of PHIL system

When conducting a PHIL simulation, ensuring closed-loop stability is crucial. Instability in the system can lead to inaccurate results and potentially cause costly damage to both the testing equipment and the hardware being evaluated. Power HIL simulations are particularly susceptible to instability due to inherent challenges such as time delays, limited bandwidth, and harmonic injections from the PA. This risk is amplified in large-scale power applications [32,34].

Consider, for example, a basic power HIL setup shown in Figure 3.6. The system includes a voltage divider circuit where the load impedance (Z_l) is represented by a physical resistor, while the voltage source and source impedance are simulated components. To interface between the simulated and physical elements, a voltage amplifier converts the simulated voltage V_1 into a physical voltage V_2, which is then applied to the load resistor. The actual current I_2 flowing through the resistor is measured and fed back into the simulation as a current source I_1. Even if the original circuit is stable in theory, implementing it in a PHIL simulation introduces potential instability. This can be understood by recognizing that RTSs are typically performed in a discrete manner, which introduces additional complexity and challenges to maintaining stability in the system [36].

Assume that at time t_k, an error ε arises during the amplification process of the voltage. This error leads to a subsequent deviation in the current I_2, given that $\Delta V_2(t_k) = \varepsilon$ and $I_2 = V_2/Z_l$, then

$$\Delta I_2(t_k) = \varepsilon/Z_l \tag{3.2}$$

When the current I_2 is fed back into the simulator, it interacts with the system's existing dynamics, leading to an additional discrepancy in the voltage. This feedback loop introduces a new layer of complexity, as the error in V_1 is not static but evolves based on the current feedback as follows: $V_2(t_k) = V_s(t_k) - Z_s * I_1(t_k)$ becomes

$$\Delta V_1(t_{k+1}) = -(Z_s/Z_l)\varepsilon \tag{3.3}$$

As the system continues to operate, this initial error can propagate and amplify, potentially destabilizing the simulation if not properly managed. The ongoing interaction between the fed-back current and the simulated environment underlines the importance of precise control and calibration within the simulation loop to ensure that errors are minimized, and the system behaves as intended.

As the updated value of V_1 is transmitted in the subsequent time step, the initial error is amplified by the ratio Z_s/Z_l. When the condition $Z_s/Z_l > 1$ holds, this error persists and grows steadily, leading to an escalation in its magnitude with each iteration. This increasing oscillation in error will persist until it hits the constraints imposed by the hardware.

If the circuit is configured differently, as illustrated in the PHIL system in Figure 3.7, the outcome changes significantly. In this alternative setup, the circuit amplifies the current I_1 rather than the voltage V_1 and measures the voltage V_2 instead of the current I_2 to feed back into the simulator. By applying the same analytical approach used previously, we observe that the error amplification factor now becomes $(-Z_l/Z_s)$. Since the magnitude of this factor is less than 1, the error diminishes over time, leading to a stable simulation. This contrast underscores the significant impact that the configuration of the PHIL system has on both the stability and accuracy of the simulation. Consequently, carefully selecting IAs ensures a stable and reliable PHIL setup.

3.5 Conclusion

PHIL simulations have become indispensable in the field of energy management. By integrating real power hardware with advanced simulation environments, PHIL enables the accurate testing and validation of energy systems under realistic operating conditions. This capability is crucial for generating synthetic data that mimic real-world scenarios, including the dynamic behaviors of DERs, grid stability under various loads and conditions, and the impact of renewable energy integration. The synthetic data generated through PHIL simulations is vital for optimizing energy management strategies, improving grid resilience, and ensuring the reliability and efficiency of modern power systems.

PHIL simulations have significantly contributed to the ongoing digitalization and AI revolution within the energy sector. By providing a reliable platform for the testing and deployment of digital twins, advanced control algorithms, and AI-driven energy management systems, PHIL bridges the gap between theoretical models and practical implementation. The ability to simulate various grid conditions and scenarios in a controlled environment allows for the refinement of AI algorithms, leading to smarter, more adaptive energy systems. PHIL also facilitates the integration of IoT devices and smart grid technologies, supporting the transition to a more digital, data-driven energy landscape.

As the energy sector continues to evolve, the role of PHIL simulations is expected to expand further. The growing complexity of energy systems, driven by the increased adoption of renewable energy sources, EVs, and decentralized generation, will necessitate even more advanced PHIL platforms. Future advancements in PHIL technology will likely focus on enhancing simulation fidelity, expanding the range of real-time testing capabilities, and integrating more advanced AI and ML techniques. These developments will improve the accuracy of synthetic data generation and enable more proactive and predictive energy management strategies, paving the way for a resilient, sustainable, and intelligent energy future.

Acknowledgment

This publication was made possible by ARG01-0428-230023 from the Qatar Research, Development, and Innovation (QRDI). The statements made herein are solely the responsibility of the authors.

References

[1] Wang Z, Draghi B, Rotalinti Y, *et al. High-fidelity synthetic data applications for data augmentation*. 2024. [online]. Available from: www.intechopen.com (accessed: June 1, 2025).

[2] Pezoulas VC, Zaridis DI, Mylona E, *et al.* Synthetic data generation methods in healthcare: A review on open-source tools and methods. Comput. Struct. Biotechnol. J. 2024;23:2892–2910.

[3] Pezoulas VC, Tachos NS, Gkois G, *et al.* Bayesian inference-based gaussian mixture models with optimal components estimation towards large-scale synthetic data generation for in silico clinical trials. IEEE Open. J. Eng. Med. Biol. 2022;3:108–114.

[4] Lu Y, Chen L, Zhang Y, *et al. Machine learning for synthetic data generation: A review*. [online]. 2023. Available from: http://arxiv.org/abs/2302.04062 (accessed June 1, 2025).

[5] Cusworth S, Gkoutos GV, Acharjee A. A novel generative adversarial networks modelling for the class imbalance problem in high dimensional omics data. BMC Med. Inform. Decis. Mak. 2024;24(1):1–17.

[6] Desai A, Freeman C, Wang Z, *et al. TimeVAE: A variational auto-encoder for multivariate time series generation* [online]. Available from: http://arxiv.org/abs/2111.08095.

[7] Monti C, Pangallo M, De Francisci Morales G, *et al.* On learning agent-based models from data. Sci. Rep. 2023;13(1):1–13.

[8] Gatto A, Accarino G, Aloisi V, *et al.* Limits of compartmental models and new opportunities for machine learning: A case study to forecast the second wave of covid-19 hospitalizations in Lombardy, Italy. Informatics. 2021;8(3):1–21.

[9] Capocchi L, Santucci JF, Pawletta T, *et al.* Discrete-event simulation model generation based on activity metrics. Simul. Model Pract. Theory. 2020;103:102–122.

[10] Chan KC, Rabaev M, and Pratama H. Generation of synthetic manufacturing datasets for machine learning using discrete-event simulation. Prod. Manuf. Res. 2022;10(1):337–353.

[11] Paulista CR, Peixoto TA, and de Assis Rangel JJ. Modeling and discrete event simulation in industrial systems considering consumption and electrical energy generation. J. Clean Prod. 2019;224:864–880.

[12] ND Vuong, TM Lim, and G Yang. *Simulation and Offline Programming for Contact Operations*. London: Springer; 2015.

[13] Faruque MO, Strasser T, Lauss G, *et al.* Real-time simulation technologies for power systems design, testing, and analysis. IEEE Power Energy Technol. Syst. J. 2015;2(2):63–73.

[14] Ren W, Steurer M, and Baldwin TL. Improve the stability and the accuracy of power hardware-in-the-loop simulation by selecting appropriate interface algorithms. IEEE Trans. Ind. Appl. 2008;44(4):1286–1294.

[15] Steurer M, Edrington CS, Sloderbeck M, *et al.* A megawatt-scale power hardware-in-the-loop simulation setup for motor drives. IEEE Trans. Ind. Electron. 2010;57(4):1254–1260.

[16] Cao S, Lin N, and Dinavahi V. Faster-than-real-time hardware emulation of transients and dynamics of a grid of microgrids. IEEE Open Access J. Power Energy. 2023;10:36–47.

[17] Kotsampopoulos PC, Lehfuss F, Lauss GF, *et al.* The limitations of digital simulation and the advantages of PHIL testing in studying distributed generation provision of ancillary services. IEEE Trans. Ind. Electron. 2015;62(9):5502–5515.

[18] Liu Y, Steurer M, and Ribeiro P. A novel approach to power quality assessment: Real time hardware-in-the-loop test bed. IEEE Trans. Power Deliv. 2005;20(2):1200–1201.

[19] Kim M, Kwak SK, Kim KA, *et al.* Enhanced computation performance of photovoltaic models for power hardware-in-the-loop simulation. IEEE Trans. Ind. Electron. 2021;68(8):6952–6961.

[20] Banda JK, Mota DDS, and Tedeschi E. A physics-informed scaling method for power electronic converters in power hardware-in-the-loop test beds. IEEE Open J. Ind. Appl. 2024;5:1–14.

[21] Kotsampopoulos PC, Kleftakis VA, and Hatziargyriou ND. Laboratory education of modern power systems using PHIL simulation. IEEE Trans. Power Systems. 2017;32(5):3992–4001.

[22] Upamanyu K and Narayanan G. Improved accuracy, modeling, and stability analysis of power-hardware-in-loop simulation with open-loop inverter as power amplifier. IEEE Trans. Ind. Electron. 2020;67(1):369–378.

[23] Constantine J, Lian KL, Fan YF, *et al.* New power interface based on multi-dimensional golden section search algorithm for power-hardware-in-the-loop applications. IEEE Access. 2024;12:14487–14498.

[24] Marks ND, Kong WY, and Birt DS. Stability of a switched mode power amplifier interface for power hardware-in-the-loop. IEEE Trans. Ind. Electron. 2018;65(11):8445–8454.

[25] Mota DDS, Banda JK, Adeyemo AA, *et al.* Harmonic-invariant scaling method for power electronic converters in power hardware-in-the-loop test beds. IEEE Open J. Ind. Appl. 2023;4:139–148.

[26] Lauss GF, Faruque MO, Schoder K, *et al.* Characteristics and design of power hardware-in-the-loop simulations for electrical power systems. IEEE Trans. Ind. Electron. 2016;63(1):406–417.

[27] Lehfuss F, Lauss G, Kotsampopoulos P, *et al.* Comparison of multiple power amplification types for power hardware-in-the-loop applications. In: *2012 Complexity in Engineering (COMPENG)*; 2012.

[28] Roscoe AJ, MacKay A, Burt GM, *et al.* Architecture of a network-in-the-loop environment for characterizing AC power-system behavior. IEEE Trans. Ind. Electron. 2010;57(4):1245–1253.

[29] Viehweider A, Lauss G, and Lehfuss F. Interface and Stability Issues for SISO and MIMO Power Hardware in the Loop Simulation of Distribution Networks with Photovoltaic Generation. In: *Conference Proceedings*; 2012.

[30] Riccobono A, Liegmann E, Pau M, *et al.* Online parametric identification of power impedances to improve stability and accuracy of power hardware-in-the-loop simulations. IEEE Trans. Instrum. Meas. 2017;66(9):2247–2257.

[31] Mihalič F, Truntč M, and Hren A. Hardware-in-the-loop simulations: A historical overview of engineering challenges. Electronics 2022;11(15):2462.

[32] Lauss G, Feng Z, Syed MH, *et al.* A framework for sensitivity analysis of real-time power hardware-in-the-loop (PHIL) systems. IEEE Access. 2022;10:101305–101318.

[33] Guillo-Sansano E, Syed MH, Roscoe AJ, *et al.* Characterization of time delay in power hardware in the loop setups. IEEE Trans. Ind. Electron. 2021;68(3):2703–2713.

[34] Lauss G and Strunz K. Accurate and stable hardware-in-the-loop (HIL) real-time simulation of integrated power electronics and power systems. IEEE Trans. Power Electron. 2021;36(9):10920–10932.

[35] Markou A, Kleftakis V, Kotsampopoulos P, *et al.* Improving existing methods for stable and more accurate power hardware-in-the-loop experiments. *2017 IEEE 26th International Symposium on Industrial Electronics*; 2024.

[36] Ren W, Steurer M, and Baldwin TL. An effective method for evaluating the accuracy of power hardware-in-the-loop simulations. IEEE Trans. Ind. Appl. 2009;45(4):1484–1490.

Chapter 4

Data generation through digital twins

Dragan Boscovic[1], Antonio Sanfilippo[2] and Sertac Bayhan[2]

4.1 Introduction

Digital twins represent a revolutionary technological advancement, offering unprecedented capabilities in monitoring, analyzing, and optimizing physical assets and systems. These virtual replicas provide a real-time, detailed view of their physical counterparts, enabling comprehensive understanding and control that drive efficiency and reduce downtime [1,2]. Learning about digital twins is essential for staying competitive in various industries that are increasingly embracing digitalization. By understanding this technology, professionals can enhance decision-making through data-driven insights, optimize maintenance schedules to reduce costs, and improve overall operational performance. The interdisciplinary nature of digital twins means that skills in this area are highly versatile and applicable across sectors such as manufacturing, healthcare, transportation, and more.

Specifically, digital twins offer transformative benefits in the energy industry that can significantly enhance energy production, reliability, and sustainability. They enable the optimization of energy-producing assets like wind turbines, solar panels, and power plants, leading to increased efficiency and output, which is critical in sustainably meeting the growing energy demands [3]. Digital twins also play a crucial role in enhancing reliability by predicting equipment failures and scheduling preventive maintenance, thereby reducing unexpected downtimes and ensuring a consistent energy supply. Moreover, they contribute to sustainability by monitoring and optimizing energy usage, reducing waste, and ensuring compliance with environmental regulations [4]. The ability to integrate with renewable energy sources, balance supply and demand, and support the development of smart grids and other innovative solutions positions digital twins as a strategic tool for long-term planning and innovation in the energy sector. Understanding and leveraging digital twin technology can drive significant advancements in energy management, contributing to more efficient, reliable, and sustainable energy practices.

[1]W. P. Carey Business School, Arizona State University, USA
[2]Qatar Environment and Energy Research Institute, Hamad Bin Khalifa University, Qatar

Digital threads refer to the seamless, integrated flow of data throughout an asset's entire lifecycle, from design through production, operation, and decommissioning. They ensure data continuity and traceability, linking all phases of an asset's lifecycle to create a comprehensive dataset. This interconnected data stream supports various asset lifecycle stages, enhancing operational efficiency and strategic decision-making.

The power of digital twin technology lies in its ability to generate and integrate vast amounts of data. This capability is crucial for several reasons:

Digital twins facilitate the continuous generation of data through real-time monitoring. Sensors and Internet of Things (IoT) devices collect vast amounts of data on various parameters such as temperature, pressure, vibration, and more. This constant data stream provides an up-to-date view of the asset's condition and performance, enabling timely interventions and optimizations [3].

Integrating data from various sources and stages into a unified digital twin model ensures a comprehensive view of the asset or system. This integration is essential for:

Informed Decision-Making: A holistic view of operations, performance, and potential issues supports better decision-making. For example, integrating maintenance logs, operational data, and performance metrics helps predict and prevent failures [4].

Improved Efficiency: Seamless data integration reduces redundancy, streamlines processes, and enhances operational efficiency. For instance, automating data collection and analysis through digital twins minimizes manual intervention and speeds up response times [2].

Real-Time Insights: Continuous data integration enables real-time monitoring and analytics, facilitating timely interventions and optimizations. This capability is critical in dynamic environments where quick responses are necessary to maintain operational continuity [4].

Scalability: Effective data integration supports the scalability of digital twin implementations, allowing for the management of complex and large-scale systems. As the volume of data grows, integrated digital twins can efficiently handle the increased data load and complexity [3].

The integration of digital twin and digital thread technologies represents a significant advancement in data generation, integration, and utilization. These technologies drive efficiency, reliability, and innovation improvements across various sectors by providing continuous, real-time insights and enabling predictive analytics. This chapter explores the detailed mechanisms of data generation through digital twins, highlighting their stages, advantages, and practical applications, underscoring the transformative potential of these technologies in modern industry.

4.2 Digital twin

4.2.1 Definition and stages

A digital twin is a virtual representation of a physical entity or system designed to mirror its real-world counterpart accurately. This representation involves a bidirectional data exchange between the physical and virtual entities, enabling the virtual model to operate simultaneously and interactively with the physical object. Digital twins are pivotal in smart manufacturing and Industry 4.0, facilitating seamless integration between cyber and physical spaces and enhancing design, analysis, and operational efficiency [1,2].

Digital twins evolve through several stages, each contributing uniquely to data generation and utilization.

The concept of digital twins encompasses various stages, including

1. *Digital Twin Prototype (DTP):* This stage involves creating a virtual prototype before the physical asset is constructed. It enables designers and engineers to test and refine designs, identify potential issues, and optimize performance [5].
2. *Digital Twin Instance (DTI):* Once the physical asset is built, a DTI is created to mirror the actual object in real time. Continuous data feeds from sensors enable real-time monitoring, performance analysis, and predictive maintenance [6].
3. *Digital Twin Aggregate (DTA):* At this stage, multiple DTI are aggregated to provide a holistic view of a complex system or network, facilitating comprehensive analysis, optimization, and decision-making [7].

The DTP focuses on virtual prototyping and design optimization. The DTI enables real-time monitoring and predictive maintenance of the physical asset. The DTA provides a comprehensive view of a complex system by aggregating multiple instances, facilitating system-wide analysis and optimization. This progression enhances the ability to design, monitor, and optimize physical assets and systems, leading to improved efficiency and performance in various industries.

4.2.2 Advantages of data generation

Benefits of using data generated through digital twins to enhance business operations and decision-making in the energy industry

- **Speed Prototyping:** Digital twins enable rapid prototyping and testing, significantly reducing the time and cost associated with physical prototyping. Energy companies use digital twins to virtually test new designs of solar panels and wind turbines, allowing engineers to iterate quickly and address potential issues before physical models are built [8].
- **Cost-Effectiveness:** Digital twins help minimize operational costs and resource wastage by identifying issues early and optimizing performance. Wind farm operators use digital twins to monitor turbine performance, predict maintenance needs, and reduce unexpected breakdowns, thus lowering maintenance costs [6].
- **Forecasting Problems:** Predictive analytics powered by digital twins can foresee potential failures and maintenance needs, preventing costly downtime and enhancing reliability. Power plants use digital twins of their equipment to predict and address maintenance needs before they become critical, ensuring continuous operation and preventing costly shutdowns [9].
- **Optimizing Performance:** Continuous data analysis optimizes real-time performance, improving efficiency and productivity. Digital twins of energy grids

help optimize the distribution of electricity, balancing supply and demand in real time and improving overall grid efficiency [10,11].

- **Remote Accessibility:** Digital twins provide remote access to asset data and control, facilitating decentralized management and decision-making. Utility companies use digital twins to monitor and manage power grids remotely, ensuring efficient energy distribution and quick response to issues, even from distant locations [12].
- **Safety:** Digital twins enhance safety protocols and risk management by simulating hazardous conditions and stress tests. Oil and gas companies use digital twins to simulate drilling operations and offshore platform conditions, identifying potential hazards and improving safety measures [13].
- **Waste Reduction:** Optimized processes and resource management reduce waste, contributing to sustainable practices. Refineries use digital twins to optimize their production processes, minimizing energy waste and reducing their operations' environmental impact [7].
- **Documentation:** Digital twins automatically document changes and updates, ensuring accurate and up-to-date records. Energy infrastructure projects, such as pipeline construction, use digital twins to keep track of design changes and construction progress, providing a reliable record for future reference and maintenance [14].
- **Communication:** Enhanced data visualization and sharing improve communication and collaboration among stakeholders. Renewable energy projects use digital twins to visualize wind farm layouts and solar panel placements, facilitating better collaboration among engineers, project managers, and regulatory bodies [15].
- **Training:** Virtual simulations provide immersive training experiences, preparing employees for real-world scenarios. Power plant operators use digital twins to simulate emergency scenarios, training staff on how to handle potential crises and improving their readiness for actual events [16].

In summary, by leveraging the data generated through digital twins, energy companies can make more informed decisions, optimize their operations, and achieve significant cost savings, ultimately enhancing their competitiveness and sustainability in the market.

4.3 Digital threading

Digital threading refers to the seamless, integrated flow of data across the entire lifecycle of an asset, from initial design through production, operation, and eventually decommissioning. In the context of digital twin applications, digital threading ensures continuous data continuity and traceability, enabling a holistic view of the asset's history and status [17,18]. This integration allows for real-time decision-making, improved efficiency, and enhanced operational management by providing a comprehensive, interconnected data stream that supports various stages of the asset's lifecycle [19].

The digital twin concept involves creating a virtual representation of a physical asset, enabling real-time monitoring, analysis, and optimization. When combined with digital threading, it provides a powerful tool for industrial operational management. Digital Threading ensures an integrated view of asset data across its lifecycle, facilitating data continuity and traceability from design through production, operation, and decommissioning [20].

In an industrial context, digital threading enhances operational management by linking data from various asset lifecycle stages. This integration enables seamless information flow, improving visibility, efficiency, and responsiveness. For instance, in a manufacturing plant, the digital thread connects design data with production processes, quality inspections, and maintenance records. This continuous data generation and analysis ensure that every component is tracked and managed effectively throughout its lifecycle [4].

By integrating digital twin and digital threading, industrial operations can achieve significant improvements in operational efficiency. The digital twin's real-time monitoring and predictive maintenance, combined with the comprehensive data traceability provided by digital threading, allow for proactive decision-making and optimized asset performance. This approach enhances the reliability and efficiency of industrial processes and ensures that every aspect of the asset's lifecycle is meticulously managed and optimized [8].

4.4 Data integration necessity

4.4.1 Importance of digital twin data integration

The data generated by digital twins is vast and complex, requiring robust integration mechanisms to be effectively utilized. Integration ensures that data from various sources and stages is harmonized, providing a comprehensive view of the asset or system. This is particularly critical in the energy industry, where the efficient and reliable operation of infrastructure is paramount [21].

4.4.2 Enhanced decision-making

In the energy industry, informed decision-making is crucial for optimizing operations, ensuring safety, and reducing costs. Integrated data supports informed decision-making by providing a complete picture of operations, performance, and potential issues. For instance, energy companies can integrate data from various sensors, maintenance logs, and performance metrics to predict equipment failures, optimize maintenance schedules, and improve energy distribution strategies [16,22]. This leads to more accurate forecasting and better resource allocation, ultimately enhancing the reliability and efficiency of energy supply.

4.4.3 Improved efficiency

Streamlining processes and reducing redundancy are key drivers for cost reduction and productivity improvement in the energy sector. Seamless data integration streamlines processes, reduces redundancy, and enhances operational efficiency. For example, integrating digital twin data with enterprise information systems (EIS) can automate routine inspections and maintenance tasks, reduce manual data entry errors, and ensure that all departments have access to up-to-date information. This improves coordination between different teams, speeds up decision-making, and reduces operational bottlenecks [15,23].

4.4.4 Real-time insights

Timely interventions and optimizations are vital to maintaining an uninterrupted energy supply and managing dynamic energy demands. Continuous data integration enables real-time monitoring and analytics, facilitating timely interventions and optimizations. For instance, energy companies can monitor the performance of power plants and renewable energy sources in real-time, quickly identifying and addressing any deviations from optimal performance. This prevents costly downtimes and allows for dynamic adjustments to energy production and distribution based on real-time demand, enhancing overall system efficiency [4,24].

4.4.5 Scalability

The energy industry often deals with large-scale and complex systems that require scalable solutions to manage growing data volumes and increasing operational demands. Effective data integration supports the scalability of digital twin implementations, allowing for the management of complex and large-scale systems. For example, as an energy company expands its infrastructure, integrated digital twin data can scale seamlessly to incorporate new assets and systems without disruption. This ensures that the company can maintain high levels of performance and reliability even as it grows, supporting long-term operational goals and strategic expansion [3,19].

Example scenario: Consider a wind farm operator who utilizes digital twin technology to monitor and manage their turbines. By integrating data from each turbine's digital twin into their enterprise information system, the operator can

- **Enhance Decision-Making:** Analyze performance data across all turbines to identify patterns and optimize maintenance schedules.
- **Improve Efficiency:** Automate reporting and reduce manual data processing, allowing technicians to focus on high-priority tasks.
- **Gain Real-Time Insights:** Monitor turbine performance in real-time, enabling immediate response to issues and adjustments to optimize energy production.
- **Ensure Scalability:** Easily add new turbines and integrate their data into the existing system, ensuring consistent management and monitoring as the wind farm grows.

4.5 Shortcomings of traditional ERP systems

Traditional enterprise resource planning (ERP) systems often struggle to handle the volume, velocity, and variety of data generated by digital twins. They may lack the flexibility and real-time capabilities required for dynamic data environments. Common shortcomings include

4.5.1 Data silos

Traditional ERP systems often suffer from data silos, which significantly impede seamless data integration and organizational visibility. Different departments and processes operate using separate databases and systems in a traditional ERP setup. This fragmentation results in isolated data pools where information from one department, such as maintenance, is not readily accessible to others, like operations. For example, maintenance data detailing equipment status and repair logs may be housed in a specialized system, while operational data regarding production schedules and output metrics is stored in another. This lack of integration can lead to delays, miscommunication, and inefficiencies, as critical information is not easily shared across departments. The result is a fragmented view of operations, making it challenging to coordinate activities and optimize overall performance [5,15].

In contrast, the digital twin approach integrates data from various sources into a unified virtual model, providing a comprehensive and cohesive view of the entire asset or system. Digital twins effectively break down data silos by consolidating information from production, maintenance, and environmental monitoring into a single, holistic representation. This integration enables real-time visibility and cross-functional collaboration, fostering a more efficient and informed decision-making process. For instance, a digital twin can aggregate data from the energy industry's turbines, solar panels, maintenance schedules, and environmental sensors. This consolidated view allows stakeholders to see the complete picture of operations, identify potential issues promptly, and optimize maintenance schedules and production processes based on real-time data. The ability to view and analyze all relevant information in one place eliminates the inefficiencies and communication barriers inherent in traditional ERP systems, driving enhanced operational performance and strategic planning.

4.5.2 Limited real-time processing

A significant limitation of traditional ERP systems is their inability to effectively support real-time data processing and analytics. These systems typically rely on periodic data updates and batch-processing methods. Data collected from various sources is processed at scheduled intervals rather than continuously, leading to potential data availability and analysis delays. This periodic approach can result in outdated information being presented to decision-makers, which is particularly problematic in dynamic and fast-paced environments. For instance, in manufacturing or supply chain management, where conditions can change rapidly, the inability to access real-time data hampers timely interventions and decision-making. Consequently, organizations may experience slow response times, reduced operational efficiency, and an increased risk of missing critical opportunities or failing to address urgent issues promptly [23].

On the other hand, the digital twin approach revolutionizes real-time data processing and analytics by creating a virtual replica of physical assets, processes, or systems. Digital twins are designed to continuously collect and analyze data from their physical counterparts through sensors and IoT devices, providing immediate insights and enabling prompt action. This real-time capability is a game-changer for industries that require constant monitoring and swift responses to changing conditions. For example, in the energy sector, digital twins can continuously monitor the performance of power plants and grid systems. They gather real-time data on various parameters such as temperature, pressure, and energy output. By doing so, digital twins can detect anomalies, predict potential failures, and optimize operations instantaneously. This capability enhances operational efficiency, reduces the risk of unexpected failures, and ensures a more reliable energy supply. The immediate feedback loop provided by digital twins allows for proactive maintenance, reducing downtime and extending the lifespan of critical assets.

4.5.3 Complex integration

Integrating traditional ERP systems with digital twin data sources can be a highly complex and resource-intensive process. ERP systems are often built with rigid architectures that require significant customization to accommodate new data inputs from digital twins. This customization involves modifying existing ERP modules or developing new interfaces to handle the continuous data streams generated by digital twins. Such modifications demand substantial time and financial investment and necessitate ongoing maintenance to ensure compatibility and functionality over time. Furthermore, the integration efforts are complicated by the lack of standardization and interoperability between different systems. ERP systems from various vendors might not easily communicate with the digital twin platforms, necessitating bespoke solutions for data exchange. These bespoke solutions often lead to higher costs and prolonged implementation times, as each integration point must be meticulously engineered and tested. The resultant system might also face challenges in scalability and flexibility, limiting the organization's ability to adapt to future technological advancements or changes in operational requirements [20,24].

In stark contrast, digital twin technology is inherently designed to integrate seamlessly with a wide array of data sources and systems. Digital twins leverage standardized protocols and interfaces, such as IoT standards and APIs, which facilitate smooth and efficient data exchange between disparate systems. This standardized approach significantly reduces the complexity associated with system integration, enabling faster and more cost-effective implementations. For example, digital twins can effortlessly integrate data from diverse sources such as wind turbines, solar panels, and energy storage systems in renewable energy projects. This capability allows for the creation of a unified platform that provides comprehensive monitoring and management of the entire energy infrastructure. By using standardized data formats and communication protocols, digital twins ensure that data from various sources can be aggregated and analyzed without extensive customization.

4.5.4 Scalability issues

Traditional ERP systems often face significant challenges when it comes to scalability, particularly in environments characterized by rapidly growing data volumes and increasing complexity. These systems are typically designed with predefined structures and capacities, making it difficult for them to adapt seamlessly to the expanding demands of modern data-driven operations. As the number of IoT devices and sensors proliferates, the sheer volume of data generated can quickly overwhelm traditional ERP systems. This influx of data leads to performance bottlenecks, where the system's ability to process and manage data efficiently is compromised. Consequently, users may experience reduced system responsiveness, slower transaction processing times, and a general decline in operational efficiency. These scalability limitations can hinder an organization's ability to adapt to evolving business needs and technological advancements, ultimately impacting overall productivity and competitiveness [25].

In contrast, digital twin technology is inherently designed to be highly scalable and easily manage vast amounts of data and complex interactions. Digital twins create dynamic, virtual replicas of physical assets, systems, and processes, continuously updated with real-time data from IoT devices and sensors. This technology's architecture allows it to expand fluidly, incorporating new assets, systems, and data sources without compromising performance or efficiency. For instance, in the energy industry, digital twins enable companies to effectively manage extensive networks of power plants, distribution systems, and renewable energy installations. As these networks grow, digital twins can seamlessly integrate additional data points, ensuring that operations remain efficient and strategic planning is well-informed by comprehensive, real-time insights. This scalability ensures that businesses can maintain high levels of operational efficiency, even as their data and system complexities increase, positioning digital twins as a superior solution for modern, data-intensive environments.

In summary, while effective for standard business operations, traditional ERP systems often fall short in dynamic, data-intensive environments like the energy industry. The digital twin approach addresses these shortcomings by providing integrated, real-time, and scalable solutions that enhance operational efficiency, strategic decision-making, and overall business performance. Integrating digital twin data into enterprise information systems transforms the way companies manage their assets, paving the way for a more responsive, efficient, and innovative operational strategy.

4.6 Using blockchain technology for digital twin data integration

Blockchain technology offers a robust and innovative solution for securing and managing the vast and complex data generated by digital twins. By providing a decentralized, immutable ledger, blockchain enhances data integrity, transparency, and trust, which are critical for the effective operation of digital twins [4,14].

4.6.1 Data security

Blockchain ensures data integrity and security through cryptographic hashing and decentralized consensus mechanisms. Each transaction is recorded in a block and linked to the previous block, creating a secure chain of data. In the energy industry, digital twins of power plants can leverage blockchain to securely record operational data, ensuring that performance metrics and maintenance logs are tamper-proof and reliable. This capability is crucial for maintaining accurate records and preventing unauthorized data modifications [4].

4.6.2 Transparency

Blockchain provides a transparent and tamper-proof record of all transactions, enhancing traceability and accountability. Every data entry is visible to authorized stakeholders and cannot be altered without detection. Renewable energy projects can use blockchain to track the production and distribution of energy. This transparency ensures that stakeholders can verify the source of energy, promoting trust and compliance with regulatory standards. For instance, blockchain can document the entire lifecycle of energy production, from generation to consumption, providing clear visibility into the process [14].

4.6.3 Decentralization

Blockchain eliminates the need for a central authority, reducing the risk of single points of failure and enhancing resilience. Decentralized networks distribute data across multiple nodes, ensuring continuous operation even if some nodes fail [25]. Decentralized energy grids can use blockchain to manage data from multiple sources, such as solar panels and wind turbines, ensuring consistent and reliable energy distribution without relying on a single control center. This decentralized approach can improve the grid's robustness and reduce vulnerabilities associated with centralized systems.

However, the technology also presents certain challenges that must be considered.

4.6.4 Scalability

Blockchain networks can face scalability challenges, particularly with high transaction volumes and data sizes. The process of adding transactions to the blockchain can become slow and inefficient as the network grows. To address scalability, energy companies can explore advanced blockchain protocols such as proof-of-stake (PoS) or Layer 2 solutions that offer higher throughput and faster transaction processing. These solutions can significantly enhance the scalability of blockchain networks, making them more suitable for large-scale digital twin implementations [4].

4.6.5 Energy consumption

Some blockchain protocols, especially proof-of-work (PoW), consume significant amounts of energy, raising sustainability concerns. The high computational power required for mining can be a major drawback. Adopting energy-efficient blockchain protocols like PoS or delegated proof-of-stake (dPoS) can significantly reduce energy consumption, aligning with the energy industry's sustainability goals. Therefore, these protocols require less computational power and are more environmentally friendly [4].

4.6.6 Complexity

Implementing and maintaining blockchain solutions can be complex and require specialized expertise. The integration process may involve significant technical challenges and require ongoing maintenance. Energy companies can collaborate with blockchain experts and invest in training to build internal capabilities. Additionally, leveraging blockchain-as-a-service (BaaS) platforms can simplify deployment and reduce complexity. BaaS providers offer ready-to-use blockchain infrastructure, making it easier for companies to integrate and manage blockchain technology [4].

Blockchain technology presents a promising solution for integrating and securing data generated by digital twins, offering benefits such as enhanced data security, transparency, and decentralization. These advantages make blockchain a suitable technology fabric for digital twin data integration in the energy industry, where data integrity, traceability, and operational resilience are paramount. However, challenges related to scalability, energy consumption, and complexity must be carefully managed to fully realize the potential of blockchain in this context. By adopting advanced blockchain protocols and leveraging specialized expertise, energy companies can overcome these challenges and enhance their digital twin implementations, driving improved efficiency, reliability, and sustainability.

4.7 Data integration technologies and models

Digital twin environments leverage an array of advanced technologies to facilitate data collection, mapping, processing, and transmission. These technologies are integral in creating a dynamic, real-time representation of physical assets, processes, or systems, enabling predictive maintenance, optimization, and improved performance [3,26].

4.7.1 Technology enablers

4.7.1.1 Sensors and IoT devices

Sensors and IoT devices are the primary tools for data collection in digital twin environments. They capture real-time data from physical assets by measuring various parameters such as temperature, pressure, vibration, and rotational speed. In nuclear power plants, sensors are installed on critical components to monitor operational conditions continuously. This data is essential for identifying patterns that may indicate potential failures, thus enabling timely maintenance [26].

4.7.1.2 Geographic information system (GIS)

GIS technology helps map the collected data to the physical locations of assets. It provides spatial context, which is crucial for understanding the physical layout and environmental conditions affecting the assets. GIS can map data collected from sensors on wind turbines, providing insights into how location-specific factors like wind patterns and terrain affect turbine performance [17].

4.7.1.3 Edge computing

Edge computing involves processing data locally, near the source of data collection, to reduce latency and bandwidth requirements. This allows for real-time analytics and immediate response to detected anomalies. In wind farms, edge devices process data from sensors on-site, enabling real-time adjustments to turbine blade angles to maximize efficiency based on current wind conditions [12].

4.7.1.4 Cloud computing

Cloud platforms provide scalable storage and computational resources necessary for managing and analyzing large volumes of data. They offer the flexibility to perform extensive data analytics and long-term data storage. Data from multiple wind farms can be aggregated and analyzed in the cloud to identify broader performance trends and optimize overall energy production strategies [17].

4.7.1.5 Big data analytics

Big data analytics tools process and analyze vast amounts of data to extract valuable insights and identify patterns. These tools are essential for handling the high volume, variety, and velocity of data generated in digital twin environments. Big data analytics can analyze historical and real-time data from nuclear power plants to predict potential failures and optimize maintenance schedules, reducing the risk of costly downtimes [9,26].

4.7.1.6 Communication network

Robust communication networks are necessary for transmitting data from sensors and edge devices to central processing units or cloud platforms. These networks ensure reliable and secure data transfer. High-speed, low-latency networks such as 5G or dedicated fiber-optic connections facilitate the seamless transmission of real-time data from remote wind turbines to central control systems [4].

4.7.1.7 Security protocols

Data security during transmission and storage is critical. Encryption and secure access controls protect sensitive information from unauthorized access and cyber threats. Data encryption and secure access protocols in nuclear power plants safeguard operational data and ensure that only authorized personnel can access and modify the digital twin data [24,27].

4.7.1.8 Continuous learning models

Machine learning (ML) models continuously learn from incoming data, improving their accuracy and performance over time. They are trained to recognize patterns and make predictions based on historical and real-time data. ML models can optimize wind turbine performance by analyzing data on wind conditions and turbine responses. Over time, these models learn to adjust turbine blades more precisely to maximize energy output under varying wind conditions [9].

Integrating these technologies in digital twin environments transforms how data is generated, mapped, processed, and transmitted. This comprehensive approach enhances real-time monitoring and predictive maintenance and drives significant improvements in efficiency, reliability, and overall operational performance. As these technologies continue to evolve, the potential for digital twins to revolutionize various industries, particularly those reliant on complex and critical infrastructure, becomes increasingly apparent.

4.8 Digital twin data integration models

The integration of technologies such as IoT devices, edge computing, cloud computing, big data analytics, artificial intelligence (AI), and ML, combined with robust security measures, is essential for harnessing the full potential of digital twins. These technologies enable the continuous generation, processing, and analysis of vast amounts of data, ensuring that digital twins remain accurate and valuable tools for decision-making and optimization. By employing these technologies, organizations can tailor their digital twin implementations to meet specific needs, from basic spatial visualization to comprehensive, data-driven management and optimization of assets and systems. This technological synergy enhances decision-making, improves efficiency, and supports scalable and secure operations, particularly in complex environments like the energy industry.

Digital twins are virtual representations of physical assets, systems, or processes created to mirror their real-world counterparts in real-time. They are modeled at varying levels of complexity, each offering distinct capacities for data generation and integration. Understanding these models is crucial for leveraging their full potential in various industrial applications, including the energy sector.

4.8.1 3D models

3D digital twins are basic models that provide spatial and geometric data of physical assets. They focus primarily on visual and structural representation, capturing the physical dimensions and configurations. 3D models generate data related to an asset's geometric and spatial configuration. This data includes dimensions, shapes, and the relative positions of different components. The data is typically gathered through sensors, computer-aided design (CAD) drawings, and 3D scanning technologies. The data from 3D models is integrated into digital platforms that allow for visualization and basic structural analysis. This integration helps identify spatial relationships and potential design issues [15].

A 3D wind turbine model includes the geometric and spatial configuration of its components, such as the blades, nacelle, and tower. This model enables visualization of the turbine's physical structure and assists in basic analyses, such as identifying potential physical interferences and assessing overall design integrity.

4.8.2 5D models

5D digital twins build upon 3D models by incorporating additional time and cost dimensions. This enhancement allows for detailed analysis and planning, considering both the temporal aspects and financial implications of asset management. In addition to spatial and geometric data, 5D models generate data related to the timeline of construction, installation, maintenance schedules, and cost estimates. This data is typically obtained from project management systems, financial databases, and historical records. The data from 5D models is integrated into comprehensive project management platforms that facilitate detailed planning and budgeting. This integration supports project progress tracking, cost control, and scheduling [10].

A 5D solar power plant model includes the spatial data of the solar panels and supporting structures, the installation timeline, and cost estimates. This model helps plan the construction phases, accurately budget the project, and optimize resource allocation throughout the plant's lifecycle.

4.8.3 7D models

7D digital twins are the most comprehensive models, incorporating additional sustainability and operational data dimensions. These models support holistic management and optimization, focusing on performance, environmental impact, and sustainability assessments. 7D models generate data related to environmental factors, operational performance metrics, and sustainability indicators. This includes emissions data, energy efficiency metrics, resource consumption rates, and lifecycle assessments. The data is collected through advanced IoT sensors, environmental monitoring systems, and performance tracking tools. The integration of 7D model data involves combining operational, environmental, and sustainability data into a unified platform. This allows for in-depth analysis, continuous optimization, and informed decision-making to achieve both operational excellence and sustainability goals [7].

A 7D model of a hydroelectric dam includes spatial data, construction timelines, cost information, additional layers of environmental impact data, operational performance metrics, and sustainability assessments. This model provides a comprehensive view that supports optimizing the dam's performance and environmental compliance, ensuring that the operations are sustainable and efficient.

4.9 Digital twins for energy: practical applications and examples

Following the increasing digitalization of the energy sector through the use of IoT technologies, cloud computing, and ML methods [36], there has been an exponential growth in the development of energy digital twins in the last 5 years. According to do Amaral *et al.* [28], the number of publications on the use of digital twins in the energy industry has grown more than tenfold from 2019 through 2023 in major energy science journals with studies addressing issues in control and operations, design and sizing, scheduling and coordination, forecasting, fault detection, and situation awareness in a variety of energy systems and subsystems including microgrids, heating, ventilation, and air conditioning, buildings, photovoltaic (PV) and other power plants, smart cities, generation, distribution, transmission, consumption, and storage. This level of growth is corroborated by Sleiti *et al.* [29].

The application of digital twins has made an impact on all phases of energy assets lifecycles including design, operation, and services [30]. Efforts on the design phase have focused on optimization data collection and virtual evaluation, with reference to operational processes the emphasis has been on monitoring, control prediction, optimization, and production planning, while services have addressed predictive maintenance, fault detection and diagnosis, and virtual testing [30]. Several energy digital twin software architectures have been proposed based on AI and ML technologies supporting IoT sensing, data mining, supporting predictive modeling [31–35].

Power plant operation and energy saving are the two key application areas where the use of digital twins has made the strongest impact [29,36]. The major challenge across power plants, including fossil fuel, nuclear and renewable energy plants, is the improvement of operational flexibility. Digital twins address this challenge through a multi-tier approach where IoT sensing, modeling, and simulation enable situation awareness and prediction for condition monitoring, fault diagnosis, and performance improvement [29,36]. For example, Xu *et al.* [37] show how a digital twin that integrates the boiler island, steam-turbine island, and emission control models embedded in the Thermoflow software (https://www.thermoflow.com) can optimize operations in a 320-MWe coal-fired thermal power plant to increase cost-effectiveness by reducing coal consumption to achieve fuel-cost savings of approximately 4 million RMB. In terms of commercial applications, Mitsubishi [38] Hitachi Power Systems has developed a boiler digital twin that uses measurement data of pressure, temperature and flow rate and ML techniques to provide optimal settings to control a real boiler. This digital twin enabled an annual cost reduction of 100 million yen for the 800MW Linkou Power Plant Unit 2 (https://www.gem.wiki/Linkou_power_station). With reference to energy savings, some of the key applications include process engineering, manufacturing, the monitoring equipment, buildings, and pumping and ventilation systems [29,36]. In the energy saving area, a key digital twin application has been energy management in smart buildings. For example, Francisco *et al.* [39] present a digital-twin enable energy management system that uses smart meter electricity data to develop strategic daily demand profiles for university buildings segmented in terms of school and seasonal activities to provide a measure of building efficiency that enables a more specific and precise identification of building retrofit strategies and near-real-time energy efficiency.

Transportation and energy storage are two additional areas where digital twin technologies has been successfully applied [36,40]. For transportation, digital twins offer an effective environment to manage the integration of mechanical, electrical, and software components to address issues such as fuel efficiency across the automotive, railway, and aviation industries, see [36] and references therein. With reference to energy storage, batteries are by far the technology most frequently integrated with digital twins, alongside thermal storage, pumped hydro, supercapacitors, and fuel cells [36,40]. As applied to batteries, digital twins help determine efficiency and lifetime through the combined analysis of factors such as operating parameters, system design, and battery materials to provide intelligent monitoring and control battery systems. For example, Qu *et al.* [41] describe a digital twin that evaluates the performance degradation of a lithium-ion battery through estimation of the battery actual capacity in dynamic operating conditions. The digital twin extracts a health indicator from the measurable battery parameters that reflects battery performance degradation and model the relationship between the cell voltage and the cell state-of-charge using long short-term memory algorithm where the health indicator functions as a temporal measurement.

Finally, an area of recent interest is the use of blockchain in energy digital twins [42–44]. For example, Boumaiza and Sanfilippo [45] present a testing framework for energy trade in microgrid communities that integrate a transactive energy blockchain with a GIS environment and an agent-based modeling simulation platform to provide a realistic analysis of peer-to-peer energy transactions in local housing community markets. The emerging digital twin framework capture the impact of factors such as population density, energy market prices, and household proximity on residential PV energy trading providing a decision-support platform for designing and implementing decentralized trading systems based on blockchain technology.

4.10 Conclusion

Data generation through digital twins is a transformative process that fundamentally enhances the understanding, management, and optimization of physical assets and systems. This technology represents a significant shift from traditional methods, offering unprecedented visibility and control over operational processes. By creating a dynamic and interactive digital replica of physical entities, digital twins provide real-time insights and enable predictive analytics, fostering a proactive approach to maintenance and operations [1,5].

Integrating digital twin technology with advanced data analytics and secure data management practices unlocks new levels of efficiency and reliability. Advanced analytics tools process vast amounts of data, extracting valuable insights that drive informed decision-making. Secure data management ensures the integrity and confidentiality of the data, safeguarding it against unauthorized access and cyber threats. This combination not only optimizes performance but also extends the lifecycle of assets, reduces operational costs, and enhances overall productivity [4,21].

Industries adopting digital twin technology witness significant improvements in various areas:

- **Efficiency:** Real-time monitoring and optimization streamline processes, reduce waste, and enhance resource utilization [8].

- **Reliability:** Predictive maintenance and anomaly detection minimize unplanned downtimes and extend asset lifespans [7].
- **Innovation:** Continuous data integration and analysis foster innovation, driving the development of new solutions and improvements [19].

The future of digital twins lies in their continued evolution and integration with emerging technologies such as AI, ML, blockchain, and the IoT. These technologies will enhance the capabilities of digital twins, enabling even more sophisticated data-driven decision-making. For instance, AI and ML algorithms can enhance predictive analytics, while blockchain can provide secure and transparent data management. The integration with IoT will expand the scope of data collection, providing more comprehensive insights into physical systems [4,21].

As digital twins evolve, they will increasingly shape the future of various sectors, including manufacturing, energy, healthcare, and transportation. In manufacturing, digital twins will drive smart factory initiatives, optimizing production lines and supply chains. In the energy sector, they will enhance the management of power plants and renewable energy sources, improving efficiency and sustainability. Digital twins of medical devices and even human organs in healthcare will revolutionize patient care and treatment. In transportation, digital twins of vehicles and infrastructure will enable smart transportation systems, enhancing safety and efficiency [3,7].

In conclusion, digital twins are a technological innovation and a strategic asset that can transform industries. By harnessing the power of data generation and advanced analytics, digital twins provide a comprehensive and real-time understanding of physical assets, driving efficiency, reliability, and innovation. The ongoing integration with emerging technologies will further enhance their capabilities, making digital twins an indispensable tool for the future. As industries continue to adopt and innovate with digital twin technology, they will pave the way for a data-driven and optimized future [1,4,5,21].

References

[1] Grieves, M. (2014). "Digital Twin: Manufacturing Excellence through Virtual Factory Replication." White Paper.

[2] Boschert, S., and Rosen, R. (2016). "Digital twin – The simulation aspect." In *Mechatronic Futures* (pp. 59–74). Springer, Cham.

[3] Tao, F., Cheng, J., Qi, Q., Zhang, M., Zhang, H., and Sui, F. (2018). "Digital Twin-driven product design, manufacturing and service with big data." *International Journal of Advanced Manufacturing Technology*, 94(9–12), 3563–3576.

[4] Fuller, A., Fan, Z., Day, C., and Barlow, C. (2020). "Digital Twin: Enabling Technologies, Challenges and Open Research." *IEEE Access*, 8, 108952–108971.

[5] Kritzinger, W., Karner, M., Traar, G., Henjes, J., and Sihn, W. (2018). "Digital Twin in manufacturing: A categorical literature review and classification." IFAC PapersOnLine.

[6] Uhlemann, T.H.J., Lehmann, C., and Steinhilper, R. (2017). "The digital twin: Realizing the cyber-physical production system for Industry 4.0." *Procedia CIRP*.

[7] Negri, E., Fumagalli, L., and Macchi, M. (2017). "A Review of the Roles of Digital Twin in CPS-based Production Systems." *Procedia Manufacturing*.

[8] Tao, F., Zhang, H., Liu, A., and Nee, A.Y.C. (2019). "Digital twin in industry: State-of-the-art." *IEEE Transactions on Industrial Informatics*, 15(4), 2405–2415.

[9] Qi, Q., and Tao, F. (2018). "Digital twin and big data towards smart manufacturing and Industry 4.0: 360 degree comparison." *IEEE Access*, 6, 3585–3593.

[10] Wright, L., and Davidson, S. (2020). "How to tell the difference between a model and a digital twin." *Advanced Modeling and Simulation in Engineering Sciences*, 7(1), 1–13.

[11] Kang, H.S., Lee, J.Y., Choi, S., *et al.* (2016). "Smart manufacturing: Past research, present findings, and future directions." *International Journal of Precision Engineering and Manufacturing-Green Technology*, 3(1), 111–128.

[12] He, B., and Bai, K.J. (2021). "Digital twin-based sustainable intelligent manufacturing: A review." *Advances in Manufacturing*, 9 (1): 1–21.

[13] Shafto, M., Conroy, M., Doyle, R., *et al.*, (2010). "NASA digital twin: Modeling and simulation for aircraft safety and life cycle management."

[14] Batty, M. (2018). "Digital twins." *Environment and Planning B: Urban Analytics and City Science*.

[15] Schleich, B., Anwer, N., Mathieu, L., and Wartzack, S. (2017). "Shaping the digital twin for design and production engineering." *CIRP Annals*, 66(1), 141–144.

[16] Parrott, A., & Warshaw, L. (2017). "Industry 4.0 and the digital twin." *Deloitte Insights*.

[17] IEEE Standards Association. (2020). "IEEE P2806 – Standard for Digital Representation of Engineering Information for Digital Thread and Digital Twin Framework."

[18] International Organization for Standardization (ISO). (2020). "ISO/DIS 23247-3: Automation systems and integration – Digital Twin framework for manufacturing—Part 3: Digital representation of manufacturing elements."

[19] Jones, D., Snider, C., Nassehi, A., Yon, J., and Hicks, B. (2020). "Characterising the digital twin: A systematic literature review." *CIRP Journal of Manufacturing Science and Technology*, 29, 36–52.

[20] Rosen, R., Von Wichert, G., Lo, G., and Bettenhausen, K.D. (2015). "About The Importance of Autonomy and Digital Twins for the Future of Manufacturing." IFAC-PapersOnLine.

[21] Rasheed, A., San, O., and Kvamsdal, T. (2020). "Digital twin: Values, challenges and enablers from a modeling perspective." *IEEE Access*, 8, 21980–22012.

[22] Leng, J., Zhang, H., Yan, D., Liu, Q., Chen, X., and Zhang, D. (2019). "Digital twin-driven manufacturing cyber-physical system for parallel control and management of smart workshop." *Journal of Ambient Intelligence and Humanized Computing*, 10(1), 1–14.

[23] Hopkinson, K.M., Wang, X., Giovanini, R., Thorp, J.S., and Birman, K.P. (2006). "EPOCHS: A platform for agent-based electric power and communication simulation built from commercial off-the-shelf components." *IEEE Transactions on Power Systems*, 21(2), 548–558.

[24] Dutton, W.H. (2014). "Putting things to work: Social and policy challenges for the Internet of Things." Info.

[25] Bolton, R., Butler, D., and Dagli, C. (2018). "A new approach to systemic risk management. In *Complex Systems Design & Management* (pp. 3–15). Springer, Cham.

[26] Gabor, T., Belzner, L., Kiermeier, M., Beck, M.T., and Neitz, A. (2016). "A simulation-based architecture for smart cyber-physical systems." *IEEE International Conference on Autonomic Computing (ICAC).*

[27] Berkel, F., Hitz, C., and Ehm, H. (2017). "Data integrity and analytics of digital twin-based manufacturing in the semiconductor industry." In *24th IEEE International Conference on Industrial Engineering and Engineering Management (IEEM).*

[28] do Amaral, J.V.S., Dos Santos, C.H., Montevechi, J.A.B., and De Queiroz, A.R. (2023). Energy digital twin applications: A review. *Renewable and Sustainable Energy Reviews*, 188, 113891.

[29] Sleiti, A.K., Kapat, J.S., and Vesely, L. (2022). Digital twin in energy industry: Proposed robust digital twin for power plant and other complex capital-intensive large engineering systems. *Energy Reports*, 8, 3704–3726.

[30] Yu, W., Patros, P., Young, B., Klinac, E., and Walmsley, T.G. (2022). Energy digital twin technology for industrial energy management: Classification, challenges and future. *Renewable and Sustainable Energy Reviews*, 161, 112407.

[31] Blume C., Blume S., Thiede S., and Herrmann C. (2020). Data-driven digital twins for technical building services operation in factories: A cooling tower case study. *Journal of Manufacturing and Materials Processing*, 4(4), 97. https://doi.org/10.3390/JMMP4040097

[32] Huang, J., Zhao, L., Wei, F., and Cao, B. (2021). The application of digital twin on power industry. *IOP Conference Series: Earth and Environmental Science* 647(1), 012015.

[33] Min, Q., Lu, Y., Liu, Z., Su, C., and Wang, B. (2019). Machine learning based digital twin framework for production optimization in petrochemical industry. *International Journal of Information Management*, 2019;49, 502–519. https://doi.org/10.1016/j.ijinfomgt.2019.05.020.

[34] Ors E., Schmidt R., Mighani M., and Shalaby M. (2020). A conceptual framework for AI-based operational digital twin in chemical process engineering. In *Proceedings of the 2020 IEEE International Conference on Engineering, Technology and Innovation (ICE/ITMC 2020)*; pp. 1–8. Cardiff, UK

[35] Steindl G., Stagl M., Kasper L., Kastner W., and Hofmann R. (2020). Generic digital twin architecture for industrial energy systems. *Applied Science*, 10, 8903. https://doi.org/ 10.3390/app10248903

[36] Ghenai, C., Husein, L.A., Al Nahlawi, M., Hamid, A.K., and Bettayeb, M. (2022). Recent trends of digital twin technologies in the energy sector: A comprehensive review. *Sustainable Energy Technologies and Assessments*, 54, 102837.

[37] Xu, B., Wang, J., Wang, X., *et al.*, (2019). A case study of digital-twin-modelling analysis on power-plant-performance optimizations. *Clean Energy*, 3(3), 227–234.

[38] Mitsubishi Heavy Industries, (2018). Boiler digital twin applying machine learning. *Mitsubishi Heavy Industries Technical Review* 55(4), 1–7.

[39] Francisco, A., Mohammadi, N., and Taylor, J.E., (2020). Smart city digital twin-enabled energy management: Toward real-time urban building energy benchmarking. *Journal of Management in Engineering*, 36(2), 04019045.

[40] Semeraro, C., Olabi, A.G., Aljaghoub, H., *et al.*, (2023). Digital twin application in energy storage: Trends and challenges. *Journal of Energy Storage*, 58, 106347.

[41] Qu, X., Song, Y., Liu, D., Cui, X. and Peng, Y., (2020). Lithium-ion battery performance degradation evaluation in dynamic operating conditions based on a digital twin model. *Microelectronics Reliability*, 114, 113857.

[42] Yaqoob, I., Salah, K., Uddin, M., Jayaraman, R., Omar, M., and Imran, M., (2020). Blockchain for digital twins: Recent advances and future research challenges. *IEEE Network*, 34(5), pp. 290–298.

[43] Borowski, P.F. (2021). Digitization, digital twins, blockchain, and industry 4.0 as elements of management process in enterprises in the energy sector. *Energies*, 14(7), 1885.

[44] Teng, S.Y., Touš, M., Leong, W.D., How, B.S., Lam, H.L., and Máša, V. (2021). Recent advances on industrial data-driven energy savings: Digital twins and infrastructures. *Renewable and Sustainable Energy Reviews*, 135, 110208.

[45] Boumaiza A. and Sanfilippo A. (2024). A testing framework for blockchain-based energy trade microgrids applications. *IEEE Access*, 12, 27465–27483.

Chapter 5
Data wrangling

Daniel Perez-Astudillo[1] *and Dunia Bachour*[1]

Data wrangling is a practice by which data is processed from an original, 'raw' source format or values, to a 'usable' form, more suitable for a given application or further applications. The process can include from relatively simple formatting or conversion from one format to another, to quality assessment, clean-up, gap filling, aggregation and even visualisation.

For solar energy applications, in particular, solar radiation data is the primary source of information to be further processed and input to all kinds of models. Thus, in this chapter, we describe a basic but complete workflow for solar radiation data, and more specifically, solar radiation measurements at a surface level; satellite-derived solar radiation data is not explicitly included here, as users normally obtain these data from online services (free or paid) and the provided data have already been processed, with no further checks needed from the end user (with the possible exception of verification or site validation of the values, but this is a topic beyond the scope of this chapter); however, formatting of the files provided by these satellite data services may be needed, so parts of this chapter may be relevant for these cases too.

A word of caution: the procedures described in this chapter will alter the contents of the data files, although (important to stress) not the measured values, but the format and handling/representation of missing or 'bad' data. For numerous reasons, from practical to traceability and even legal in some cases, it is not a good idea to modify the original ('raw') data files, so any and all changes that may be needed should be done in separate copies, never overwriting the original files, which should be kept safe.

5.1 Direct measurements of solar radiation

Surface-level solar radiation is measured through radiometers. Ideally, the three so-called components of broadband solar radiation, namely direct normal, global horizontal and diffuse horizontal radiations (DNI, GHI and DHI, respectively), should be measured by three separate sensors (for direct irradiance, the sensor is called pyrheliometer, and for global and diffuse irradiances, it is called pyranometer), although it is possible to measure two of them and derive the third one from these two (with a

[1]Environment Centre, Qatar Environment and Energy Research Institute, Hamad Bin Khalifa University, Qatar

rotating shadow band radiometer, for example, the global and diffuse components are measured with one sensor, and the direct component is then calculated from these), thanks to the formula relating the three

$$GHI = DNI * \cos(\theta_z) + DHI \tag{5.1}$$

where θ_z is the sun's zenith angle at the time of the measurement and is measured along the observer's vertical direction, with zero overhead and 90° at the horizontal.

As the sun is constantly moving across the local sky, the pyrheliometer and the shading device (to allow measurement of diffuse irradiance) must be mounted on a sun tracker to ensure that the sun trajectory is correctly followed at all times. This, however, introduces more complexity into the solar monitoring system and thus, more possible points of failure, and since monitoring stations are commonly operated at remote locations, not always supervised except for maintenance visits (which could be of high or low frequency, sometimes with weeks between visits), the direct measurements of solar radiation must not be automatically believed to be completely reliable. A well-maintained station will provide the most accurate measurements, but this requires a maintenance regime that is not always possible in practice, and sometimes the maintenance information is unknown to the data users, so a quality assessment should always be performed before using this data.

Radiometric measurements are collected from the sensors with the help of a data logger, which digitises the sensor outputs (measurements) if needed and converts the raw values (usually voltages or currents) from the sensors into irradiances in W/m^2, by using the calibration constants of each individual sensor. The data logger also adds a timestamp to each record and then stores and possibly sends the measurements to another destination.

5.2 Formatting

5.2.1 File formatting

Whether the user has some control over the collection of solar radiation data (if the user operates a monitoring station) or not (solar data obtained from a third party), one of the first steps is the determination of the format in which the data will be stored and used.

As described in the previous section, data loggers provide the first processing of measurements, adding timestamps and storing the data in some way, which can vary greatly depending on the brand and sometimes even the model of data logger, but in general the data can be saved in ASCII files containing one record (row) for each summarisation interval (e.g. every minute, or every hour, depending on the configuration), plus some additional information, usually at least in the form of headers indicating the parameter represented by each column and its units. Generally, the user only has limited control on the output format from commercial data loggers, so the files are formatted as per pre-defined options (if any) given by the manufacturer.

Some data loggers, or logging software, generate separate files for every year, or for every month or day, or for even smaller intervals. Keeping the data in a large number of files is not a practical solution, so in these cases, one of the first data management steps should be to compile the data into larger periods, for example, into monthly or annual files. In the case of ASCII source files, this in most cases requires the removal of headers and any other non-data lines from the source files (except perhaps for the first one of the period, or the header can be manually added to the new file/s) and the concatenation of the lines, which ideally should be done carefully to ensure that the files/lines are saved in the correct chronological order.

Remark 1. *It is important to point out that one should not automatically assume that a received dataset, especially when it comes from a third party, is in correct chronological order; the records should be checked and properly sorted if needed. Similarly, duplicated records – with the same timestamp – should be checked for and dealt with as well.*

Another common data storage option is to input the measurements (either automatically or manually read from the data logger files, using some custom workflow) into a database. Databases have several advantages over simple files; for example, data can be queried to obtain data within certain time intervals, and the output can be chronologically sorted through the query even if the records were not inserted in that order into the database. In addition, databases are stored in binary form and compressed, saving storage space.

Other alternatives are also possible, in which data is stored into other binary (generally compressed) files used by different software, such as in the case of netCDF[2] files or ROOT[3] files, in which data can be organised in structures within the files (in the form of arrays in the case of netCDF, or so-called Trees and other forms in ROOT). Compressed, binary files can be a good solution, especially when sharing data, thanks to their reduced size – database files, though, are not easily shared; instead, what is shared is the *access* to the database. For offline use, however, data are most commonly stored (or exported from a database) in text files that can be easily read by analysis software or scripts.

5.2.2 Data formatting

Just like file format can vary greatly between data loggers, the format of the data and even of the timestamps inside those files can also vary widely. When combining data files from different sources, this can become an additional source of errors if data are not read correctly from the corresponding files. It is therefore necessary to determine a format that simplifies the next processing steps and will possibly be used for the 'final' files that will ultimately be used. This requires a pre-processing of the input files where, most importantly (at the ASCII file stage), the following points are considered and applied uniformly to all sets of data:

- All timestamps are (converted to, if necessary) in one same format. Some considerations: determine the order for year, month and day, the separator between them, number of digits for each part of the date (year, month, day) and time (hour, minute, second, etc.) and whether zero padding will be used for values

[2]https://doi.org/10.5065/D6H70CW6

[3]The format used by CERN's ROOT data analysis framework (https://root.cern/).

less than 10; and determine which of these parts of the timestamp, as well as timezone, will be included.

- For solar data, an additional point to treat carefully is whether the timestamps are in Local Standard Time (LST) or in Universal Time (UT or UTC). Independently of the choice, the timestamps of all data from different sources should be converted, if necessary, so that all are in the same time reference in order to avoid unexpected time shifts.
- Similarly, the timestamps representing temporal summarisation (averaging) intervals should also be carefully understood and set to a common convention. There are no general rules, and some data sources assign the beginning time of a summarisation interval as timestamp, while other sources assign the ending time of the interval, and yet others assign the middle of the interval. Thus, for example, an hourly averaged value marked as 10:00 a.m. may correspond to the interval 9 to 10, or 10 to 11, or 9:30 (or 29) to 10:30 (or 29); the user should first obtain the information of which convention is used for each data source, and then adjust as necessary so that all timestamps follow the same convention. The study in [1] highlights the significance of time averaging conventions. It examines errors that arise when users assign incorrect time intervals to timestamps at different averaging scales, ranging from 5 to 60 minutes, using real measurements.
- Set a field separator to use (e.g. comma, space, tab, semicolon), and whether text fields will be in quotes.
- Determine whether comment lines should be present or not, and if so, decide on a comment line indicator (a # at the beginning of the line is a commonly used indicator).
- Decide what headers, if any, will be included and their format (will the header be a name? units? both? etc.).
- Determine how to handle missing data from fields (columns) in a record (line); see section 5.3.

An important, additional consideration at the time of reading the data files is the knowledge of the type of variable for each field; that is, whether a column should be read as integer, boolean, float or double, etc. While some software and programming/scripting languages can automatically try to identify variable types, this is not always infallible; sometimes, for instance, decimals are not included for floating point fields when a value's decimal part is exactly zero (e.g. if 9.0 is recorded simply as 9), and the column might then be incorrectly thought to be integer, resulting in errors in the best case (since it helps identify the issue), and in the worst case, the precision might be silently lost (the floats converted to integers, discarding the decimals of other records) without the user noticing.

5.3 Missing data

Datasets are not always complete, especially those consisting of measurements; from equipment malfunctions or downtime to data communication issues, there are multiple reasons why some of the data may be missing from the raw stage. While some methods exist to attempt filling gaps in data, at initial stages it is best not to introduce such data into the datasets. Missing raw data may be divided into two types:

Missing fields Or partially missing records. This is the case where several variables are measured at the same timestamp, and one or more of them, but not all, are not present for a given timestamp.

Missing records Those in which no data is recorded at all for a timestamp, and the timestamp itself is not present in the dataset. This is typically the case, for example, when a data logger completely loses power for some time, so nothing is recorded during that time, and chronologically, the timestamps of existing records show a jump, sometimes even with missing fields in the last/first records before/after the power loss.

It is a good practice to avoid leaving missing measurements without a value. In ASCII CSV (comma-separated values) files, for example, these gaps are usually found as two consecutive commas when they are not at the end of a line, or as a comma with no value to its right, if found at the end of a line. A recommended approach to avoid these empty fields is to introduce a known, pre-defined value in those fields. While in some cases, the missing values are filled with a 'not a number' indicator such as NaN, this is not universally and automatically read as a numerical value – for instance, if reading data from an ASCII file with C++ and a number is expected, this results in an error that crashes the code if unprotected. Therefore, it is best to use a numerical value to substitute all present NaNs. This value should be chosen such that, while being of the same *type* as expected from that variable, is a value that would not be produced by a real measurement of that variable; that is, it should be a value outside and *well beyond* the possible values of that variable. Typical values used to this end are formed by the number 9 repeated as many times as needed to be outside the possible range, and positive or negative, such as −999, −99, −9, 9999, etc. (naturally, only one of these should be chosen and used throughout). Very roughly, measurements of solar irradiance, for instance, could be expected, allowing for some instrument uncertainties, not to be outside the range from -20 to $1500\ \mathrm{W/m^2}$; thus, missing entries could be filled with a value of −99 or −999, for example. With such filled-in values, it is very easy to identify and exclude or select, for counting or any statistical purposes, the missing entries by simply looking for that known value.

For the case when complete records are missing, one must decide whether the missing records will be introduced (not by gap-filling, but with all fields containing the 'missing value' values) so that all timestamps exist in the dataset, or if the time gaps will be left as they are. In a database, for example, leaving the time gaps is not generally an issue and might even be desired in some cases, but as mentioned above, it is a good practice to have all records present at least in the 'final' datasets, that is, those that will be used for an application or delivered to other users.

5.3.1 Gap filling

Working with a dataset as complete as possible is crucial for improving forecasting accuracy, so it is important to try to reduce data gaps while, at the same time, minimising the errors introduced by this practice. Indeed, the 'artificial' introduction of missing data (either not collected or discarded by quality checks – see next section) also introduces additional uncertainties in the dataset, and the errors usually increase along with the size of the data gap, so the benefits of introducing gap-filling techniques should be evaluated before deciding when to use them (e.g. they could be limited to filling gaps of shorter lengths, avoiding longer gaps that introduce larger errors, etc.).

Many gap-filling methods for solar radiation data have been studied; a 2021 report by NREL [2] compared eight different techniques based on time series with varying gap lengths and found all RMSE percentages in the double digits, which provides an idea of the kind of errors introduced. Recent advances in the use of machine learning techniques may improve accuracies, but this is a field currently in fast development. In any case, the simplest and probably safest gap-filling method is when two of the irradiance components are present, and only one is missing; in this case, equation (5.1) can be used to calculate the missing component. As long as the two measured irradiances have passed their own quality checks, the calculated component should also be correct, so it is recommended to do the gap-filling after the existing data have been quality-checked, but it can also be done before, applying afterwards the quality checks on the full dataset. With most other gap-filling techniques, however, being based on other methods, the quality of the introduced values may not be automatically ensured, and it is best to always perform a data quality assessment on the resulting dataset, even if the measured irradiances (used or not in some way or another to obtain the gap-filling values) were already quality-checked before the procedure.

5.4 Solar radiation data quality

As previously described, measuring solar radiation is a complex task, with different instruments and parts that can present issues resulting in measured data of varying quality, and sometimes even erroneous measurements. Especially when running the monitoring system for long periods, even with high-quality equipment and adherence to operation and maintenance best practices, the occurrence of bad measurements can be reduced but not completely eliminated. Given the importance of having accurate solar radiation measurements (as input for models, analyses, PV performance monitoring, bankability studies, forecasting, among other uses), it is crucial that the correctness of the measured data is assessed in some way, allowing to understand and determine their reliability, as well as enabling a meaningful comparison among different datasets.

The output of quality checks on solar radiation data consists of data quality flags that identify suspicious or erroneous measurements. It is important to note that the measured values are not modified here; the purpose is to assign to each measured value one or more quality flags so that, following some given criteria, it can be decided whether the measurement will be used or not in further applications or processing (like temporal averaging into hourly, daily, etc., averages).

Although there are no established rules, some of the basic and most commonly used quality checks are the ones recommended by the Baseline Surface Radiation

Network (BSRN) [3]. The BSRN checks consist of equations (limits or equalities) which the irradiances measured at a given time should comply with in order to be flagged as 'good'; otherwise, they are either not flagged (if the test could not be applied) or flagged as suspicious or 'bad' quality; these checks are divided into three categories (see [3] for the equations):

- Physically possible limits. For each irradiance, these are upper and lower limits on the values that can be possible at any time given the available (incoming at the top of the atmosphere) solar irradiance.
- Extremely rare limits. These are more stringent upper and lower limits, representing values that, although physically possible, should not be frequently observed.
- Comparisons. While the physical and extreme tests above are applied separately to each irradiance component, the comparison tests require two or three of the irradiances, as they use them to calculate quantities that can then be compared against each other (e.g. if the three irradiance components are measured, the BSRN checks include equation (5.1) within some limits).

One important feature of these tests is that they rely only on the irradiance measurements being checked and on the solar position at the time of each measurement; as the solar position is calculated, this means that no additional data (from measurements or models, for instance) is needed in order to apply the BSRN tests.

Various other sets of tests have been proposed, and most of them build upon the BSRN tests, incorporating additional checks to cover issues not identified by these. Some sets, however, depend on additional measurements (e.g. meteorological) and/or on modelled values (like the clear-sky irradiances), which conditions their applicability to having access to these additional parameters; in the case of modelled irradiances, for instance, additional uncertainties are introduced in these model calculations, so the users should evaluate whether it is acceptable to assess the correctness of measurements against these.

Another point to consider when using a set of data quality checks is that in most cases the result, i.e. the quality flags, indicate whether the measurement is suspicious, but ultimately the user has to decide whether the measurement can be used (if it is actually a 'bad' measurement) or not; and when two or more irradiances are used in one check and the check fails, it is not usually clear whether all the irradiances involved should be marked as bad, when not all of them may be causing the test to fail. So the user should study the quality results carefully before filtering the data.

A comprehensive set of tests was proposed in [4], in which none of the tests require external data, and the only modelled values are clear-sky global and beam irradiances, which are calculated with a fixed Linke turbidity parameter. In addition, the study provides clear descriptions on which irradiances are evaluated by each test. In the same paper, all irradiances involved in failed tests were filtered out, and the results of this combination of tests and filtering were studied, showing the final effect on three full-year datasets from two different locations.

A review of the latest advances in solar data quality checks is also included in Chapter 4 of the IEA PVPS report of NREL's 4th Edition (2024) of its *Best*

Practices Handbook for the Collection and Use of Solar Resource Data for Solar Energy Applications [5].

Ultimately, the result of the quality checks should be a quality flag for each measurement; that is, each measured value of each irradiance must have its own quality flag (at least one 'final' flag, or a set of flags, one for each test). The data can then be delivered in different ways:

1. Deliver the measurements and their quality flags, so that the end user decides what to do with the information. In this case, a clear description of what each quality flag represents should be included as well.
2. If the data will be provided 'ready-to-use', the data failing the quality tests should be removed from the provided dataset, but replaced by the 'missing' value previously defined. This must also be indicated so the users know that the data have been already quality-assessed and filtered.

References

[1] Perez-Astudillo, D., and Bachour, D. 2020. Importance of time averaging conventions. In J. M. Cardemil, K. Guthrie, and R. Ruther (Eds.), *Proceedings of the ISES Solar World Congress 2019 and IEA SHC International Conference on Solar Heating and Cooling for Buildings and Industry 2019* (pp. 1996–2000). International Solar Energy Society. https://doi.org/10.18086/swc.2019.42.07

[2] Denhard, A., Bandyopadhyay, S., Habte, A., and Sengupta, M. 2021. *Evaluation of Time-Series Gap-Filling Methods for Solar Irradiance Applications.* Technical Report NREL/TP-5D00-79987. Available from https://www.nrel.gov/docs/fy22osti/79987.pdf (accessed: June 1, 2025).

[3] Long, C.N., and Dutton, E.G. 2002. *BSRN Global Network Recommended QC Tests*, V2.0. Available from http://epic.awi.de/30083/1/BSRN_recommended_QC_tests_V2.pdf (accessed: June 1, 2025).

[4] Perez-Astudillo, D., Bachour, D., and Martin-Pomares, L. 2018. Improved quality control protocols on solar radiation measurements. *Solar Energy* 169, 425–33.

[5] Jensen, A.R., Saint-Drenan, Y.-M., Gueymard, C.A., *et al.*. 2024. Data quality assessment and control (Chapter 4). In M. Sengupta, A. Habte, S. Wilbert, *et al.*, (Eds.), *Best Practices Handbook for the Collection and Use of Solar Resource Data for Solar Energy Applications.* 4th edn. Report IEA-PVPS T16-6:2024. Available from https://doi.org/10.69766/ENEH5295 (accessed: June 1, 2025).

Chapter 6
Machine learning

Joseph Cazier[1] *and Dragan Boscovic*[1]

Artificial intelligence (AI), data analytics, and machine learning (ML) are transforming energy management, and its impact will continue to grow as this age of AI and digitalization advances. However, to take advantage of this transformative potential, organizations need to understand the different paths to value, including those on both the time and action dimensions.

For the first dimension of time, we are looking for insight into either the past or future. Traditional algorithms often focus on either describing and visualizing, known as descriptive analytics, or testing and understanding historical data, which includes many statistical tests and/or experimental approaches to determine cause and effect. These are forms of diagnostic analytics that go beyond descriptions to build knowledge, theories, and understanding. Both descriptive and diagnostics are important forms of hindsight, which is critical to running an organization effectively.

Foresight is also essential to guiding the organization to future value and activities and includes both predictive and prescriptive analytics. Predictive analytics helps you understand what is likely to happen, whereas prescriptive analytics helps you optimize future situations. Foresight is incredibly valuable, but without good hindsight, the path to foresight is murky and both hindsight and foresight are needed to be efficient today and remain competitive tomorrow.

Of course, our insights from both hindsight and foresight are only as valuable as the actions you can take from them, which leads us to the next dimension which is action. For this dimension, our insights from both hindsight and foresight need to be converted into actions that add value to our organization and hopefully society.

This can be done by a person guessing or hypothesizing what actions should be taken based on insights. Alternatively, it can be done by taking actions that have been validated and tested by the data, which are much more likely to result in the desired outcomes. This includes the use of diagnostic analytics to uncover cause and effect, better understanding what is driving what, or with prescriptive analytics that can recommend optimization strategies leading to desirable outcomes.

Many of these processes can also be automated once validated, reducing the guess work with data supported actions and allowing for faster reaction times and use at scale. However, both hypothesizing ideas and approaches as well as getting to

[1]Center for AI and Data Analytics, Arizona State University, USA

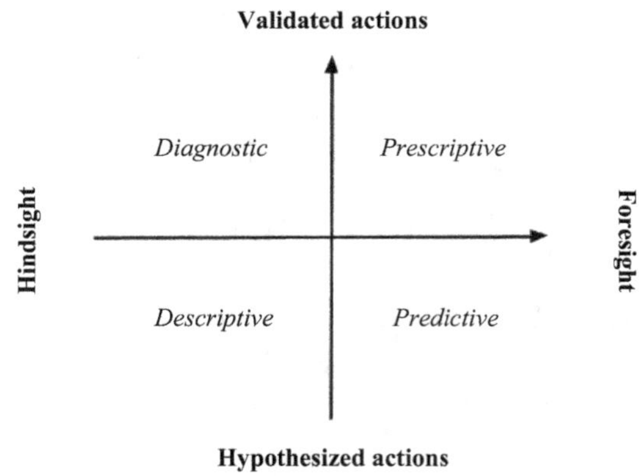

Figure 6.1 Categories of analytics

validated actions are important, and it is difficult to get to validated actions without first hypothesizing as part of building a mature analytics process.

For organizations to achieve the full value of analytics, they need to use all types of analytics, descriptive, diagnostic, predictive, and prescriptive. This means that they need both hindsight and foresight while leveraging their ability to grow from hypothesized actions into validated actions. If we place the time dimension on a horizontal axis (hindsight and foresight), and the hypothesized actions vs validated actions on the vertical axis, we can nicely organize the four common categories of analytics, as shown in Figure 6.1.

All of these dimensions are on a continuum, typically starting low on the one-way arrows from hindsight to foresight and from hypothesized actions to validated action. However, this does not mean you stop considering the past, but rather that you see the past with more clarity and understanding as you move toward the future. You also continue to hypothesize forever so you can learn and discover more validated actions that work. It is not one or the other, but a natural growth and progression from one to the other as you mature in your approach to analytics.

6.1 Mini-case study

As an example of how these types of analytics can work in tandem, consider the work of one of our authors, who has led several energy modeling projects. Here is a simplified summary of some of the work he did for a mid-size non-profit public utility in collaboration with the American Public Power Association (APPA) as it relates to the framework above.

6.1.1 Case introduction

Our midsize utility, like many across the United States, purchased power from a large energy producer and resold it to the customers in their district. Most of the time the purchase price was fairly nominal, around $0.04 per kWh. They then delivered and resold most of this power to their customers for an average price of around $0.08/kWh.

However, they also had peak-demand charges. This meant that during certain time periods, unknown in advance, they were charged significantly more per kWh, as much as 200 times more, with a very constrained ability to pass that cost on to customers, who were still paying the same very low price per kWh. These periods were also not known until the end of the peak season, often months later, making them hard to predict. This peak-demand charge was so significant that it resulted in a surcharge roughly equal to 30% of their annual revenue, making reducing that demand charge an important goal of the utility managers.

Additionally, they had significant costs in running their operations and delivering reliable power to their customers, but their "profits" as a non-profit public utility were controlled and set by regulators after accounting for their costs. This meant that their primary goal had historically been reliability and customer service, as the "profits" were set or limited by regulators and ultimately given back to their customers.

However, they were also facing growing political pressure to be more sustainable and efficient, with an additional new emphasis on using less energy and having a greater proportion of their energy be renewable. Additionally, there was a growing focus on energy equity, that is keeping costs down for customers, particularly those of lower and fixed incomes who often faced a personal energy crisis during times of extreme weather. As a public utility, they embraced this mission to better serve their customers and the environment, and they asked the author's analytics team to help, which we did with additional funding and support from APPA.

6.1.2 Descriptive

After taking time to understand the key goals and objectives of the utility, our team started collecting potentially relevant data. This included data such as records of previous peak charge times, a variety of weather and astronomical data, industrial and residential usage patterns, including seasonality and hourly/weekday/weekend/holiday usage patterns and major event data as well as core operations data.

We spent some time analyzing and visualizing this data as it is related to the utilities' core goals, providing important insights to management about the most important areas to focus on leading to our next steps. It was determined that the value that would have the most imminent impact on their goals was predicting and mitigating demand charges. This would level electricity usage, free up resources for more sustainable power options, and lower operations cost which would, and did, lower prices for their customers.

6.1.3 Diagnostic

We conducted a deeper analysis of the descriptive data we collected and supplemented it with additional data that we thought could help explain what was driving peak load and usage patterns, including customer meter data, demographic data, and home size and value information from publicly available tax records. To ensure appropriate ethical implications were addressed, we also sought and got approval from an Institutional Review Board (IRB) for the analysis.

Here, we used a variety of statistical tools to better understand and explain relationships in the data, as well as potential causal variables that we might be able to predict or influence.

6.1.4 Predictive

Next, we were able to use the insights from our diagnostic analysis to develop better predictive models around demand use and mitigation strategies. This led to considerable savings for the firm, reducing peak-demand charges by approximately 5% using internal load management strategies. Interestingly, with the understanding of drivers from our diagnostic analysis our models became accurate enough that we predicted, correctly, a demand charge on a major holiday.

Unfortunately, management did not believe us, as they had never had a peak-demand charge on a major holiday since much of the industrial usage base shut down, reducing load. Therefore, they did not exercise their mitigation strategies on that day. However, when the peak load was determined a few months later, and they saw that we were right, trust soared, and it took our relationship to a new level. They all wanted to use data now.

6.1.5 Prescriptive

Once we had exhausted what could be done internally to mitigate demand charges, be more sustainable, and reduce inequity in power, we then turned to focusing on prescriptive ways we could invite our customers to work with us to further mitigate peak charges, invest in sustainable solar energy, which was well suited in our area, and time shift or reduce usage at critical times.

To do this effectively, we reviewed best practices and theories for influencing behaviors and partnered with large landlords to develop property initiatives around conservation as tied to rent, especially near peak periods. The already validated theories guided us on the design, followed by a large controlled experiment to determine impact.

Additionally, we used data to identify those homes that were the least energy efficient for their size, age, and building characteristics and reached out to the owners to help find ways to improve. We also explored those that were struggling with bill payment and proactively worked with them and various government and non-profit groups to help with their energy and financial security.

All told the multi-year project saved energy, saved money, developed more sustainable energy practices, reduced energy prices, and helped energy accessibility and equity.

In the next section, we introduce various classes of algorithms that have shown their worth in various aspects of the energy sector.

6.2 Energy management algorithm classes by category

Each of these categories of analytics is relevant and useful for energy management. Some have been extensively used for many years, while others are new, exciting, and uncertain in their impact and applicability. In this chapter, we briefly summarize the widely known and extensively used categories and algorithms while delving deeper into the new and less well-known approaches as they apply to energy management.

For those involved in energy management, it is more important to understand the tasks these algorithms can perform and how these tasks can benefit their organization, rather than understanding the deep inner workings of various algorithms.

This categorization was inspired by prior work by Provost and Fawcett [1] that organizes algorithms by their tasks and further categorizes a few of the more useful, newer ones by the types of analytics they support (descriptive, diagnostic, predictive, and prescriptive). In each section, we will describe the tasks these algorithms can perform for you and list common tools or algorithms that can help you accomplish them. The analytics team can then find the best algorithm for the specific application of the task that needs to be done.

6.2.1 Descriptive analytics tools

- **Numeric reporting** – Data presented and shared in numerical reports. Common examples include numerical data such as energy consumption metrics, financial reporting, or emission reports.
- **Descriptive statistics** – Descriptive statistics uses statistical tools to summarize key variables with common metrics for quick understanding.
- **Multi-dimensional cubes** – Multi-dimensional cubes allow you to look at data in different ways, making relationships in areas of interest easier to spot and manage. This is a hallmark of the data warehouse era and can be done in pivot tables or with more complex software. For example, one might analyze energy usage across different dimensions such as time, geography, and customer segment.
- **Visual reporting** – Descriptive data presented in an intuitive easy to understand visual format. Common examples might include energy dashboards, visualizations of financial metrics, or trend analysis.

Descriptive analytics focuses on understanding current data to inform actions, though it is increasingly becoming more automated. How timely, intuitive, accurate, granular, and relevant this information is presented can make a big difference to success. This field is fundamental to all the other types of analytics. Without capturing quality descriptive information, further analytical work cannot be effectively built. Common tools in this category include numeric reporting, visual reporting, multi-dimensional cubes, descriptive statistics, and visual statistics, as summarized in Table 6.1.

6.2.2 Classical diagnostic tools

Diagnostic analytics involves processing and analyzing data to provide insight and understanding. The better you become at diagnosis, the more effective and successful

Table 6.1 Descriptive tool class summary

Tool class	Function/job	Common tools
Numeric reporting	Numeric reporting provides information about observable events in a way that someone, or an algorithm, can act on them	Databases Spreadsheets Monitoring systems
Descriptive statistics	Descriptive statistics focuses on describing the shape of the data with measures of mean, median, mode, interquartile range, distributions, and similar metrics	SAS R Excel
Multi-dimensional cube	The multi-dimensional cube presents data and reports on useful dimensions of the data that are pre-identified highly relevant to the business. A dimension is a way of looking or segmenting the data such as by gender, location, or time. These dimensions are carefully planned in advance and set up with a data structure that facilitates easy, self-service access to information a person would need to manage better	Excel (Pivot tables) OLAP Business intelligence Power BI Tableau SAS R
Visual reporting	Visual reporting, such as with charts and graphs, shows similar information to reporting, but generally summarizes it in a way that allows people to more easily process it and then act	Excel Power Point Power BI SAS Tableau

you will be at guiding your actions in a way that has a positive impact on the organization. Not only do diagnostic analytics tools help you better understand your data or a given situation, but they are also helpful with more advanced modeling. Advanced diagnostic techniques generate inputs for predictive and prescriptive models, often used in ML.

This step is essential for mature and reliable analytics and provides valuable insight and understanding. While some predictive algorithms can generate predictions based on raw data or with little diagnostics being done first, these models are less stable over time, less generalizable to new situations, and more likely to have hidden ethical issues. It is almost always worth the time to explore and understand your data with descriptive and diagnostics techniques before jumping to prediction or prescription.

Listed here are some common tools used for diagnostic analytics.

6.2.2.1 Inferential statistics

Inferential statistics can determine whether there is a significant difference between two variables, and to what degree, guiding your action with greater confidence and precision. Numerous tools exist in this domain, some of which have been around for decades or even centuries. The key strength of these tools is that they provide more certainty by separating out the true signal from the random noise that is often seen in descriptive statistics, allowing more insight and confidence in actions to take.

For example, a utility company might use inferential statistics to analyze the energy consumption data from a group of households before and after implementation of energy-efficient appliances. By using methods like hypothesis testing or confidence intervals, the company can infer the reduction in energy usage across the broader customer base and decide whether to expand the initiative. This approach helps in making data-driven decisions about energy conservation strategies and the scalability across a larger population.

Includes all the elements of inferential statistics, but represents them visually and often more interactively to make them more intuitive and easier to digest and explore.

6.2.2.2 Causal analysis

By actively collecting data under various designed scenarios, experimentation goes beyond just applying inferential statistics to proving causation with purposefully designed experiments. Sometimes, causal analysis can also be done by identifying natural experiments – situations not intentionally designed but sufficiently similar, with a key difference that allows determinations of cause and effect.

While experimentation is the most well-known type of causal analysis, there are also other techniques that can be used to determine cause and the extent to which one variable influences another. One option is to build a model, such as with structural equation modeling (SEM), another is through the use of counterfactual analysis common in economics. Many of these approaches start out as diagnostic exercises, but if something useful is discovered, it can lead to models that are predictive and sometimes even prescriptive.

In the energy sector, SEM has been used to assess the impact of triple transformations on environmental, social, and governance (ESG) performance. This form of causal analysis enables researchers to understand and predict the factors influencing the adoption and diffusion of new technologies and overall sustainability of organizations within the sector.

For example, SEM helped analyze data from various energy companies to evaluate how changes related to decarbonization, decentralization, and digitalization impact ESG and economic performance over time [2].

Table 6.2 presents a variety of diagnostic tools by class and function.

Unsupervised ML

ML in general, and in particular, unsupervised ML, offers valuable descriptive and diagnostic insights. The unsupervised part of these algorithms means that they do not have a particular target, such as predicting who will do something. Rather, there is more freedom for the algorithm to explore the data looking for certain types of patterns that give insight and understanding. These algorithms are termed "unsupervised" as they are mostly exploratory aiming at discovering hidden patterns. In contrast, many of the predictive ML algorithms are labeled as "supervised" because we assign them a particular target to predict.

Table 6.2 Statistical diagnostic tool class summary

Tool class	Function/job	Common algorithms
Inferential statistics	Inferential statistics includes many algorithms and techniques, many of which focus on proving if something is statistically different from something else	T-Test ANOVA Chi-Square
Visual statistics	Visual statistics shows important information about the shape of the data, including key changes, visually. This visualization allows for quicker understanding and increases the likelihood of action	Tableau SAS R
Causal analysis	While diagnostic experimentation can prove causation, causal analysis often goes a step farther by measuring the strength and breadth of impact of that causation, improving precision and eventually allowing for robust prediction after adequate experimentation and analysis	Real and natural experiments Regression and structural equation modeling Counterfactual analysis

Unsupervised ML tools can deliver critical insights and understanding that enrich other models and inform business strategy, and outreach by aiding human understanding of cause and effect. They analyze descriptive data to uncover deeper meanings, similar to inferential statistics and experiments, but at a scale and scope most other algorithms cannot match.

Here are some key applications of unsupervised ML, each encompassing various algorithms suitable for different tasks.

Clustering

Many are familiar with the concept of clustering, which segments data into groups based on observed similarities in each record. Organizations use clustering to better understand their customers and other data. Besides providing executives with valuable insights, natural clusters can serve as inputs into predictive models by passing on these labels to another algorithm or by suggesting additional data mining tasks and approaches.

In the energy sector, clustering analysis has been applied to assess energy consumption across the countries. By analyzing energy consumption data, similarities and differences among these countries were identified, helping to understand sector-specific energy usage and guide energy policy and economic transformation strategies toward greater efficiency and sustainability [3].

6.2.2.3 Similarity matching

While clustering examines the entire dataset to identify general patterns, similarity matching focuses on finding records that most closely match a specific record of interest. Rather than identifying a small number of big clusters for the whole group, this algorithm looks very closely at thousands of micro-clusters to identify records that most closely match behaviors to the record of interest.

Similarity matching can have a similar effect and impact as prediction, and a credible case could be made to include it in that category. However, its primary function is to find instances in the data that are similar to known instances of interest and then handle these instances similarly. It can be seen as an ML application of clustering at a micro level, identifying similar patterns in existing data rather than predicting future outcomes. This technique is particularly useful because it provides actionable insights at a micro level more immediately than many other methods.

In the energy sector, similarity matching has been effectively used for schema matching within the building energy domain. This approach maps the raw data from building energy systems to a generic data model, which is crucial for integrating heterogeneous data sets. It enhances the understanding and management of building energy consumption by aligning disparate data sources, making it easier to analyze and predict energy usage patterns. This method not only identifies similar instances in data but also treats these instances in a similar manner to enhance data management and analysis strategies [3].

6.2.2.4 Co-occurrence grouping

Co-occurrence grouping is conceptually similar in approach to some of the other techniques discussed here, but rather than focus on customers, the focus is on the products and services businesses sell. For example, by identifying which products are often sold together, such as peanut butter and jelly, a store might be able to take advantage of that association in various ways.

In the energy sector, co-occurrence grouping has been applied to analyze patterns and associations in energy consumption data. For instance, a study conducted on energy consumption used clustering and co-occurrence methods to explore the relationships between different sectors and their energy use patterns. The analysis helped to identify how similar or different these countries are in terms of sector-specific energy consumption, which is vital for shaping energy policies and strategies for efficiency improvements and sustainability efforts [3].

6.2.2.5 Data reduction

It is easy, even for machines, to get lost in a sea of highly detailed data, missing important patterns and observations, especially when data and computing power are limited. Techniques such as factor analysis and concept stemming can help reveal latent or larger hidden concepts driving individual variable's actions.

For example, data reduction has been used to focus on identifying the underlying factors influencing energy poverty. The analysis revealed significant associations between energy poverty and factors like housing quality and household size, which then informed targeted policy interventions aimed at reducing energy poverty by improving housing conditions and energy efficiency [4].

Although data reduction involves some loss of information, it often reveals the bigger picture patterns, providing deeper understanding and more generalizable actions. By focusing on the forest instead of just the trees. Data reduction minimizes

Table 6.3 ML diagnostic tool class summary

Tool class	Function/job	Common algorithms
Clustering	Clustering breaks data into groups that are similar to each other, allowing for deeper understanding of natural breaks and affinities in the data and providing insights and strategies for working with those groups	K-means Hierarchical clustering Self-organizing maps
Similarity matching	Similarity matching finds data points that are similar to a chosen data point of interest. It is different from clustering in that it is not putting them into natural groups, but identifying an individual data point, then looking for others that have similar characteristics to the one described	Euclidean distance Cosine similarity Manhattan distance Pearson correlation
Co-occurrence grouping	Co-occurrence grouping looks for patterns of events that often occur together, either at the same time or with a time dependency. By identifying items or events that occur together, businesses can seek to capitalize on this value	Market basket analysis Latent semantic analysis Association rule mining
Data reduction	Data reduction reduces data to core elements by identifying the most important features to model. While there is some loss of information, it allows for focus on what is most important and enables new insights to often emerge	Principal components analysis Factor analysis

the risk of multicollinearity – where similar items skew results – making analysis easier to manage and more effective at identifying broader trends.

Table 6.3 shares a few unsupervised ML tools, functions, and algorithms for diagnostic analytics.

6.3 Predictive analytics tools

While some predictive tasks were achievable without computers, such as simple regression analysis and the application of proven formulas or laws (i.e. gravity), it was not until recent advances in computational power and increased data availability that these tools could be widely adopted. The shift is primarily due to how supervised ML, the primary tool of prediction, operates.

Both traditional analytics and ML require data, but they differ in their algorithm and data usage. Traditional analytics takes a known algorithm, applies it to data, and produces a one-time result. The process usually stops there.

ML, however, operates differently. It learns by example and often improves continuously. In this process, you start with the known results for a problem of interest, where you have labeled data to highlight the attribute of interest, such as identifying profitable energy customers. By labeling results of interest in your data, you can then train and build a custom algorithm, unique to your data, that can be applied to similar new data for prediction.

This is essentially learning by example. You provide numerous examples of the desired and undesired results, use these results to create a new algorithm, and then apply this algorithm to generate more of the desired results. You then use the new results for the desired purpose, assess their performance, and then improve the algorithm based on the new data. A few prominent categories of these algorithms are described below and summarized in Table 6.4.

6.3.1 Classification

Classification algorithms sort items in your data into two or more classes based on their attributes. For example, suppose you were rolling out a new energy services product nationwide and have data from a region where you already tested the product. In the test group, some customers had turned out to be very profitable, but others were not. As you expand the new service, you want to selectively offer the service to those customers who are likely to be profitable, and avoid offering it to those who are likely to be unprofitable.

In this scenario, the algorithm examines the key attributes of the customers who were profitable versus those who were not. It then analyzes the differences between

Table 6.4 Predictive tool class summary

Tool class	Function/job	Common algorithms
Classification	ML techniques predict in which class a new instance of data is likely to belong. New multi-objective classification models are maturing, allowing for greater granularity and efficiency in classification	Logistic regression Decision trees Support vector machines Naive Bayes
Regression	More than classic regression, the name has expanded in general use to mean assigning a number, as opposed to a class, to an item of interest	Regression Decision trees Neural networks
Link prediction	Link prediction focuses on predicting connections or links to other people or objects that should be there, but are not currently recorded in the data. It is useful for recommending new connections	Random walk Common neighbors Preferential attachment
Behavioral profiling	Behavioral profiling allows you to get rich granular data on a person or object to establish a baseline of normal behavior and then flag any unexpected deviations from expected behaviors. It is useful in security and fraud detection	Markov chain analysis Collaborative filtering

these two groups and develops an algorithm to classify new clients into the identified groups (profitable or unprofitable). This allows you to selectively market to customers who are most likely to be profitable and avoid those who are not.

Generally, with sufficient data and labeled groups (e.g., historically profitable versus less profitable), an algorithm can be built to distinguish between these groups in new data. The algorithm's quality and usefulness depend on the data's relevance and quality, but ML models excel at this type of classification.

The advantage of ML is that it learns by examples, allowing you to develop a custom algorithm tailored to your data, making it highly effective for tasks like the one described here. However, building a useful algorithm requires accurate and often extensive labeling, which takes time and costs money. The result is a predictive label on new data that mirrors the attribute of interest. This can be achieved by assigning new labels to a classification group, or by giving a probability score for belonging to that group, allowing you to see the strength of the prediction for each row in the data.

6.3.2 Regression

Most of you will be familiar with classic linear regression. It was one of the first prediction techniques that were easy to calculate, interpret, and understand. Linear regression effectively predicts many linear relationships and provides a numerical prediction value rather than just a label on whether it belongs to a given class. This feature distinguishes it from classification and makes it valuable for various applications.

There are many different types of regression, but not all of them are "regression" in the classic sense. In data science, additional algorithms have emerged that use different approaches to estimate values similar to regression. However, we use the term regression more broadly to refer to any algorithm that provides a numerical prediction estimate, not just a classification category.

6.3.3 Link prediction

Link prediction uses a variety of tools to forecast connections, making it particularly useful for social media recommendations. For instance, if you know customer x is adopting a new energy platform and x knows y, and y knows z, then x is more likely to know z than someone without a connection to y. The more mutual connections, the stronger the recommendation.

This capability is equally beneficial for product recommendations based on social connections. If many of your "friends" or connections like a particular product or service, you are more likely to like it as well. Based on the old adage "birds of a feather flock together," recommending a product or service that you do not currently use makes sense if your connections are already using it.

6.3.4 Behavioral profiling

Behavioral profiling focuses on actions of a device, individual or small group, rather than their personal attributes or characteristics, but how they behave over time. It gathers data on how a person, computer devices, or these groups behave over time, eventually developing a unique behavioral profile for that entity. When behavior deviates from this established norm, it can flag it as an anomaly, prompting further investigations or immediate action.

Unlike demographic profiling, which is based on attributes like age, gender, or race, behavioral profiling is rooted in individual observed behaviors. These algorithms build a baseline from this behavioral data and flag any behavior that deviates too far from that individual baseline.

Many of us have experienced this first hand, such as when a credit card is flagged for fraud and temporarily suspended due to an unusual activity. This technology is also effective in detecting cyberattacks and any other unusual behavioral change, or detecting energy leaks or energy siphoning.

Behavioral profiling can categorize and describe the complex behaviors of individuals, small groups, segments, or clusters. This method provides deeper insights than demographic descriptions, revealing not just what something is, but how it behaves, which can be highly valuable depending on the context.

By monitoring behaviors and detecting changes, these systems generate alerts whenever there is a shift in the status quo. For example, changes in cell phone usage might indicate a need for additional services or signal potential customer churn, allowing companies to respond proactively.

Another advantage of these systems is reliance on rich data from what is being profiled, instead of the often shallow data from many different sources. This makes them easier to deploy for custom and unusual applications, allowing for greater tailoring and customization. It is also a little more robust than many other ML systems and is able to add value with fewer data records.

6.4 Prescriptive analytics tools

Prescriptive analytics tools help shape the future you want by providing insights and guided actions based on data, algorithms, and a clear vision for the desired outcome. Positioned in the upper right quadrant from Figure 6.1, these tools do more than just predict the future; they actively shape it through action driven by analytics (Table 6.5).

Next, we briefly explore a few classes of prescriptive tools.

6.4.1 Known algorithms

The next class of prescriptive tools consists of algorithms designed to guide and optimize business decisions. Many of these algorithms originate from operations research and are focused on enhancing business efficiency and effectiveness.

These prescriptive tools typically address generic business problems with known solutions and apply these algorithms to specific situations. Examples include routing (traveling salesman problem), product/portfolio optimization, pricing, scheduling, and network modeling for supply chain management. They utilize optimization techniques such as linear, integer, mixed, and constraint programming, among others, to provide prescriptive solutions that maximize value for these types of problems.

In the energy sector, prescriptive analytics is increasingly used to optimize the energy supply chain. Energy companies, for instance, use prescriptive analytics to integrate the planning of all supply chain components. This approach allows them to visualize the outcomes of various operational options all the way through to financial statements. By optimizing business functions within the supply chain, energy companies can ensure that these optimizations positively affect financial reporting. This application of prescriptive analytics supports more informed decision-making, enhancing efficiency and profitability across the entire supply chain [5].

6.4.2 Experimentation

In addition to the established frameworks and known algorithms described here, there are a few techniques that are useful for generating prescriptive guidance. The first technique is based on mass experimentation, which generally focuses on gathering diagnostic information or brute-force optimization, as discussed earlier. However, when combined with important economic impact variables, it can be used as a tool for prescriptive analytics.

By examining probabilities derived from prediction models and/or experimentation and linking them to economic impact variables based on probable outcomes, objective-maximizing solutions like profit optimization can be developed. In the energy sector, real-time pricing is used in smart microgrids with energy consumption schedules. These schedulers use heuristic optimization algorithms like binary particle swarm optimization, wind-driven optimization, and genetic algorithms to optimize energy consumption based on real-time pricing. This strategy not only helps reduce energy expenses and the peak-to-average ratio but also ensures user comfort by effectively managing energy demands and availability within the grid [3].

6.4.3 Simulation

Another technique in this area is simulation. By using known information about a problem and probable outcomes, a solution can be simulated thousands of times to find the most likely range of results based on available inputs. One advantage of simulation is that it allows for the exploration of various ranges of inputs and what–if scenarios, which can better inform decision making. As computing power and data availability continue to increase, the application and usefulness of these tools will expand.

Simulation techniques are extensively used to optimize energy systems. For example, Modelon Impact is a cloud platform that enables the design, simulation, and analysis of energy systems, including renewable energy integration and power plant performance optimization. This platform allows engineers to simulate thousands of scenarios to determine the most efficient configurations for energy storage systems, helping to optimize the size and operation schedules for components like batteries in integrated energy systems [6].

6.4.4 Reinforcement learning

Most ML tools are static because they build a model based on a snapshot of data and then applied to future applications. While there are some adjustments made over time to maintain and improve these models, they are fundamentally built on mostly static data.

In contrast, reinforcement learning is designed and built for continual learning and adaptation to new and dynamic circumstances. Reinforcement learning operates like a game of hot and cold, providing feedback for every action taken, with rewards (hot) and punishment (cold) based on the outcome. It incorporates an important time dimension in the way it acts, gets feedback from the outcome, then it reevaluates. This process allows reinforcement learning to be potentially quite prescriptive in its effect [6].

For these reasons, it is considered a new, third type of ML, in addition to supervised and unsupervised learning. It is best used where continual learning and automation are important. For example, this is one of the core technologies behind both self-driving cars and winning ML video game applications, such as Alpha Go. It is the continual learning and adaptations-based feedback received in the terms of rewards and punishment that can make it useful in prescriptive analytics. Because it learns from the environment, it can quickly adapt to new situations and has potential in automation and optimization.

The integration with dynamic time, as opposed to the other one-shot categories of algorithms, has many advantages, but also presents challenges. One significant challenge is that it is often hard to know which action caused the desired impact, after thousands or millions of actions. Another challenge is that in real life and business, you do not get many do-overs, so a mistake early on in reinforcement learning is difficult to overcome. Machines can do things very quickly, but generally lack judgment. For a video game, where there can be infinite attempts at resetting the game, this works well, but for business applications we usually just get one try, unlike video games.

In the energy sector, simulation techniques are increasingly used for optimizing systems and processes. One such application involves using dynamic simulation to understand and optimize energy storage systems in poly-generation setups, where renewable sources like wind power play a significant role. This modeling helps balance and manage energy flows within microgrids, enhancing energy efficiency and reliability. Such simulations are essential for planning and operating energy systems that include diverse sources and complex interactions, like those involving hydrogen buses and district heating systems [7].

6.4.5 Generative AI

Generative AI has possibly become one of the most widely used prescriptive analytics tools in history. It did so quicker than any other to date, with ChatGPT having reached the milestone of 100 million users in a short 2 months. This is compared to the 4.5 years it took Facebook to reach that number of users [8].

What makes generative AI prescriptive is the way it works. It starts with taking most of the world's accessible data, including text, video, music, location, and more depending on the application. It then analyzes, transforms, and tokenizes the data, building models on how the data relates to itself and other data. This is typically done through multiple rounds of building and tuning models in deep neural networks, inspired by how humans think. This allows the models to respond to queries in a very human-like manner, but with the knowledge, and some of the liabilities, of the world's data.

What makes it fit in the prescriptive analytic category is its ability to respond intelligently to very specific requests, tailored to their needs. While the response may not always be perfect, it rivals and can exceed what many humans can do. This includes generating text responses in a chat used for customer service, creating marketing or communications materials, including text, graphics and videos. It can also hold an interactive voice or video conversation.

In addition to the business operations, it can also dramatically increase innovation. For example, a recent article [9] demonstrated how generative AI could be used to better predict thermal properties of new materials, allowing engineers to design more efficient energy systems by reducing the 70% of electricity that turns into heat waste.

Many more relevant applications are likely as the technology advances. However, as the capabilities increase there are also emerging risks to prepare for, including deep fakes, hallucinations or incorrect information, use of illegal or inappropriate data, others using AI to probe vulnerabilities in energy systems and more. Additionally, AI is incredibly energy intensive, likely increasing energy demand and use beyond what earlier models predicted. Care should be taken to plan and prepare for these possible negative consequences.

Table 6.5 presents classes of prescriptive tools, their functions, and common algorithms.

6.5 Limits of analytics tools

There are many analytics tools with real-world applications that help solve business problems, but they have, and likely always will have, their limits. For example, the US Geological Survey states no one can predict earthquakes [10], which can dramatically impact energy production and service. While we can identify high risk areas, we lack the necessary data, understanding. and systems in place to predict them, and likely never will. Earthquakes are rare "black swan" style events that are thankfully rare.

Prediction is based on past data. When past data is sparse, it is challenging to properly diagnose and understand it to make precise (useful) predictions. To predict earthquakes, one might imagine the need for sensors deep within the earth collecting highly granular and relevant data, which is difficult to achieve. Additionally, due to the data sparseness, even if available, developing accurate algorithms would still be challenging. At best, we could hope for earlier warning of imminent earthquakes.

Similar to earthquakes, there are many low-frequency, low-probability events that, if they occur, could have a dramatic impact on the world, such as the COVID pandemic. Scientists anticipated an event like this, but predicting its exact timing, location, and intensity is beyond our capabilities, and perhaps always will be.

There are too many random elements that are nearly impossible to predict. The best we might achieve is risk assessment for these types of events, along with impact modeling and simulations that can help us prepare for these events appropriately. This risk assessment is still very valuable, but not the same as knowing the future. It

Table 6.5 Prescriptive tool class summary

Tool class	Function/job	Common algorithms
Known algorithms	There are many known and proven algorithms that can apply to many business problems. Many are from the category of operations research and can be incredibly valuable	Linear programming Simulated annealing Genetic algorithms
Experimentation	We distinguish two types of experimentation in this chapter. The first is aimed at understanding and the second is aimed at optimizing. Understanding often leads to optimizing, but it can be done with experimentation through trial and error alone. I do not recommend it, as these optimizations are often inefficient and shaky, but if you only care about short-term optimization, it might be worth it	A/B testing at scale for optimization of outcomes
Simulation	Simulations look at probabilities, interrelationships, and likely interactions, then simulates an event happening thousands or millions of times. This repetition gives a better sense of the shape of likely outcomes for different scenarios	Monte Carlo Discrete Event System Dynamics
Reinforcement learning	Most ML algorithms are considered one-shot learning, built and run. By contrast, in reinforcement learning, learning occurs continuously as new data comes in and constantly adapts to new environments and situations, often finding novel and useful solutions that were previously missed	Q-learning Multi-armed bandits Deep Q networks
Generative AI	Harvest data and uses it to create text, images, video and more in a near human like fashion. Has the ability to accelerate knowledge work, discoveries, customer interactions and business operations	OpenAI's ChatGPT Google's Bard Salesforce's Einstein Deepminds Alphacode

is important to realize that the data is only as useful as the extent to which the future resembles the past. In a changing world, there will always be limits to what analytics can do. We should use these tools but understand their strengths and limitations to manage them appropriately.

Additionally, these tools are evolving at a rapid pace. For example, as we go to press, generative AI has captured the public's imagination and holds immense potential for the energy sector. This branch of AI not only generates new insights based on vast datasets from the world's libraries and the internet but can also drive automation at an unprecedented scale. In particular, generative AI can streamline the creation of digital twins—virtual models of energy infrastructure—enabling more accurate simulations and predictive maintenance. Moreover, it can translate complex analytical data into actionable insights, providing operators with real-time recommendations and offering customers personalized energy solutions. While much of its impact is still speculative, its transformative potential for improving operational efficiency and decision-making is undeniable, provided the risks are managed effectively.

References

[1] Provost, F., and Fawcett, T. (2013). *Data Science for Business: What You Need to Know About Data Mining and Data-Analytic Thinking*. O'Reilly Media.

[2] Gyimah J, Fiati MK, Nwigwe UA, Vanessa AE, and Yao X. (2023). Exploring the impact of renewable energy on economic growth and carbon emissions: Evidence from partial least squares structural equation modeling. *PLoS ONE* 18(12): e0295563.

[3] Gostkowski, M., Rokicki, T., Ochnio, L. *et al.* (2021). "Clustering analysis of energy consumption in the countries of the Visegrad Group" *Energies* 14(18):5612.

[4] Taltavull de La Paz P., Juárez Tárrega F., Su Z., and Monllor P. (2022). Sources of Energy Poverty: A Factor Analysis Approach for Spain. Front. *Energy Research* 10:847845.

[5] CGI Global. (2024). Using prescriptive analytics to optimize your energy supply chain: How to decide "what's best" with integrated business planning and analytics. Available from https://www.cgi.com/sites/default/files/2018-07/cgi-prescriptive-analytics-to-optimize-energy-supply-chain-whitepaper.pdf [CGI accessed 19 July 2024]

[6] Christian, B. (2020). *The Alignment Problem: Machine Learning and Human Values*. W. W. Norton.

[7] Saxena, V., Kumar, N., and Nangia, U. (2024). Computation and optimization of BESS in the modeling of renewable energy based framework. *Archives of Computational Methods in Engineering*, 31(3): 2385–2416.

[8] Carter, N. (2023). ChatGPT hits 100m users after only two months. Available from https://www.graphicnews.com/en/pages/43884/tech-chatgpt-is-fastest-growing-internet-app [Retrieved 22 July 2024]

[9] MIT News. (2024, July 16). AI method radically speeds predictions of materials' thermal properties. *MIT News*. Available from https://news.mit.edu/2024/ai-method-radically-speeds-predictions-materials-thermal-properties-0716 [Retrieved 22 July 2024]

[10] US Geological Survey. (n.d.). *Can you predict earthquakes?* US Department of the Interior. Available from https://www.usgs.gov/faqs/can-you-predict-earthquakes [Retrieved 21 July 2024]

Chapter 7

Game theory and AI for strategic energy management

Anas Karaki[1], Khaled Abedrabboh[1] and Luluwah Al-Fagih[1]

In an increasingly complex and interconnected energy system, decisions made by one party can significantly impact the outcomes for others. Game theory (GT) helps to model these interactions, enabling stakeholders to anticipate the actions of others, identify optimal strategies, and achieve desired outcomes. GT provides a mathematical framework for analyzing and optimizing the interactions between these stakeholders, such as energy producers, consumers, grid operators, and regulators. The integration of GT into energy networks has been driven by the advancement of smart grid technology, which includes a communication layer that facilitates the active management of grid assets and distributed energy resources (DERs) such as renewable energy sources (RESs), energy storage systems (ESSs), and electric vehicles (EVs). In this chapter, we present the basic principles of GT and investigate its applications to energy management.

This chapter examines the potential of integrating GT and AI in the context of energy management. The focus is on understanding how the strategic decision-making frameworks of GT, combined with the capabilities of AI, can enhance the efficiency and effectiveness of energy systems. An analysis of relevant applications from the literature is presented to provide insights into the current state of research and practice at this intersection.

Section 7.1 presents the principles of GT and explains the motivation behind its adoption for energy management. In Section 7.2, we analyze the areas of the smart grid where GT has been applied, introduce the concept of smart grid communities (SGCs), and investigate their characteristics. The game-theoretic approaches to energy management are investigated in Section 7.3. Section 7.4 explores the potential of combining AI and GT for energy management and presents a review of recent literature. Finally, conclusions and future research opportunities are discussed in Section 7.5.

[1]College of Science and Engineering, Hamad Bin Khalifa University, Qatar

7.1 Game theory

Before we delve into the various applications of GT to energy systems, let us first explore the principles of GT. Game theory is a scientific approach used to study human conflict and interactions. It applies mathematical tools to analyze situations that involve both competition and collaboration. In a game, *players*—whether individuals or organizations—make choices from a range of possible options, known as *strategies*. The strategies selected by the players lead to different *outcomes*, which represent the consequences of their decisions. It is assumed that the players have *preferences* for certain outcomes over others. GT is concerned with studying the rational choice of strategies to achieve the most desirable outcomes.

The significance of GT lies in its ability to mathematically analyze the interactive decision-making where involved players consider their actions and the actions of others. This mathematical framework explicitly helps model strategic actions considering other players' interactions and analyzing the expected reward accordingly.

In game theory, a *normal-form game* is represented as a tuple (N, A, u), where

- $N = \{1, \ldots, n\}$ is a finite set of players (or agents).
- $A = A_1 \times \cdots \times A_n$ is the set of action profiles, where A_i is the set of actions available to player i.
- $u = (u_1, \ldots, u_n)$ is the profile of utility functions, where $u_i : A \to \mathbb{R}$ represents the utility that player i receives.

This formalism allows for the analysis of various strategic interactions where each player's optimal strategy may depend on the actions chosen by others. Assuming that players are rational and self-interested, each player would choose the strategy that maximizes their own utility. This optimal strategy is referred to as the *best response*, where

Best response

For a player i, let a_{-i} represent the action profile of all players other than i. The action $a_i^* \in A_i$ is a best response for player i to the action profile a_{-i} if

$$u_i(a_i^*, a_{-i}) \geq u_i(a_i', a_{-i}) \quad \forall a_i' \in A_i \tag{7.1}$$

This equation indicates that player i prefers a_i^* over any other possible action a_i' given the actions of the other players a_{-i}.

Nash equilibrium

A *Nash equilibrium* is a situation where every player selects their best response to the strategies of the other players. In other words, no player has an incentive to unilaterally deviate from their chosen strategy, as doing so would not lead to a higher payoff.

An action profile $a = (a_1, a_2, \ldots, a_n)$ is a *pure Nash equilibrium* if, for every player $i \in N$; a_i is a best response to a_{-i}. In mathematical terms, a is a Nash equilibrium if

$$u_i(a_i, a_{-i}) \geq u_i(a_i', a_{-i}) \quad \forall a_i' \in A_i \quad \forall i \in N \tag{7.2}$$

7.1.1 Classification of game theory based on player interactions

GT can be broadly classified into three categories based on the nature of interactions between players: *non-cooperative*, *cooperative*, and *evolutionary* GT. Each of these approaches offers different perspectives on how players make decisions and interact with one another. The basic concepts of these approaches are discussed below.

Non-cooperative GT

In non-cooperative GT, players make decisions independently, with the primary focus on individual payoffs rather than collective outcomes. The core idea is that each player seeks to maximize their own utility without any binding agreements or collaboration with other players. A non-cooperative game is typically represented as a tuple (N, A, u), where

- N is the set of players.
- $A = A_1 \times A_2 \times \cdots \times A_n$ is the set of action profiles.
- $u = (u_1, u_2, \ldots, u_n)$ is the set of utility functions, where each $u_i : A \to \mathbb{R}$ represents the payoff for player i given the action profile $a \in A$.

Each player i aims to maximize their utility $u_i(a_i, a_{-i})$ by choosing the best possible action a_i given the strategies of the other players a_{-i}.

Cooperative GT

In cooperative GT, players can form coalitions and make binding agreements to improve their collective outcomes. The focus shifts from individual payoffs to the distribution of payoffs among members of a coalition. A cooperative game can be represented by a pair (N, v), where

- $N = \{1, 2, \ldots, n\}$ is the set of players.
- $v : 2^N \to \mathbb{R}$ is the *characteristic function* that assigns a value $v(S)$ to every coalition $S \subseteq N$, representing the total payoff that the coalition S can achieve by cooperating.

The challenge is to determine how the total value $v(S)$ of the coalition should be fairly distributed among its members. This allocation of payoffs can be determined using the *Shapley value*

$$\phi_i(v) = \sum_{S \subseteq N \setminus \{i\}} \frac{|S|!(n - |S| - 1)!}{n!} (v(S \cup \{i\}) - v(S))$$

where $\phi_i(v)$ represents the Shapley value for player i, indicating their fair share of the total payoff.

Evolutionary GT

Evolutionary GT studies strategic interactions in populations of players who adapt their strategies over time. This approach is particularly useful in understanding the dynamics of behavior in multi-agent systems. The key concept is the *replicator equation*, which models the evolution of strategy frequencies in a population over time

$$\dot{x}_i = x_i \left(u_i(x) - \bar{u}(x)\right)$$

where

- x_i represents the proportion of the population adopting strategy i.
- $u_i(x)$ is the payoff for strategy i given the population distribution x.
- $\bar{u}(x) = \sum_j x_j u_j(x)$ is the average payoff in the population.
- $\dot{x}_i$ represents the rate of change in the proportion of strategy i over time.

Evolutionary GT can be applied to understand how strategies evolve over time based on their success, where successful strategies become more prevalent in the population.

7.2 The smart grid

The smart grid represents an advanced and complex infrastructure. At its core, the smart grid integrates a network of diverse technologies that include hardware, software, communications, and control systems, that collectively empower both producers and consumers with enhanced visibility and control over the generation, distribution, and consumption of electricity. Designed to be both adaptable and resilient, the smart grid is not only built upon current technologies but also structured to accommodate future innovations. This flexibility allows it to meet evolving energy demands, address unforeseen challenges, and incorporate new technological solutions as they emerge.

A significant feature of the smart grid is its incorporation of DERs, including photovoltaic (PV) arrays, ESS, wind turbines, and EVs. Such components are essential for tapping into renewable energy, providing environmentally friendly power while alleviating dependency on conventional generation sources, such as gas turbines. However, the variable output from renewables presents challenges in maintaining power quality and stability. To address this, the smart grid incorporates robust control systems designed to optimize energy flow, regulate frequency, and stabilize voltage levels, all of which are critical to harnessing renewable resources effectively. The smart grid's capability to integrate renewable energy also hinges on advanced control and communication strategies that ensure these resources are effectively utilized, while safeguarding the grid from cyber threats and operational disruptions.

In addition to physical and control infrastructures, the smart grid utilizes sophisticated energy trading and demand-side management (DSM) mechanisms. These tools are designed to optimize power distribution and ensure an efficient and balanced power flow across heterogeneous systems that include multiple user types and generation profiles. These mechanisms support sustainable energy transition by enhancing DER utilization and reducing reliance on fossil fuels. Ultimately, the smart grid is designed to meet complex and dynamic requirements by combining resilient technology, sophisticated control systems, and advanced communication networks.

Through these capabilities, the smart grid can provide stable, efficient, and green energy solutions that align with modern socioeconomic and environmental priorities.

7.2.1 SGCs

With the penetration of RES, the grid is no longer a unidirectional network where customers (electric loads) purchase electricity directly from the upstream grid. Hence, the power network consists of producers, consumers, and prosumers interacting and trading energy between each other. Electricity providers are considered producers since they use conventional energy sources or renewable resources to generate electricity. Due to regulations and economic considerations, the number of producers in an energy market is sufficiently small, and it is a single entity in a few markets. Most players in an energy market are consumers, classified into residential, commercial, and industrial, based on their characteristics and size [1]. In general, one can conclude that a large, and sometimes very large, number of players are participating in the energy market. This enormous interaction imposes a trading challenge for producers since they need to balance their generation with the total energy demand at any given moment in time. In the last two decades, a new type of player emerged, i.e., prosumers, who invest in renewables to cover their local demand and sell excess energy in local energy markets. This mix of players represent an SGC, illustrated in Figure 7.1, where the three types of players: consumers, producers, and prosumers interact and trade in a local energy market.

The concept of SGC includes bidirectional power lines and communication infrastructure to facilitate energy trading between players. The transition from the conventional centralized power network scheme toward a modern distributed structure requires grouping neighboring players. This process, intrinsically, enhances the stability at the community level (locally) and at the grid level (globally). Moreover, the SGC structure reduces the requirement for long transmission lines, thus, reducing power losses. Shifting toward smaller networks enables each SGC to trade energy

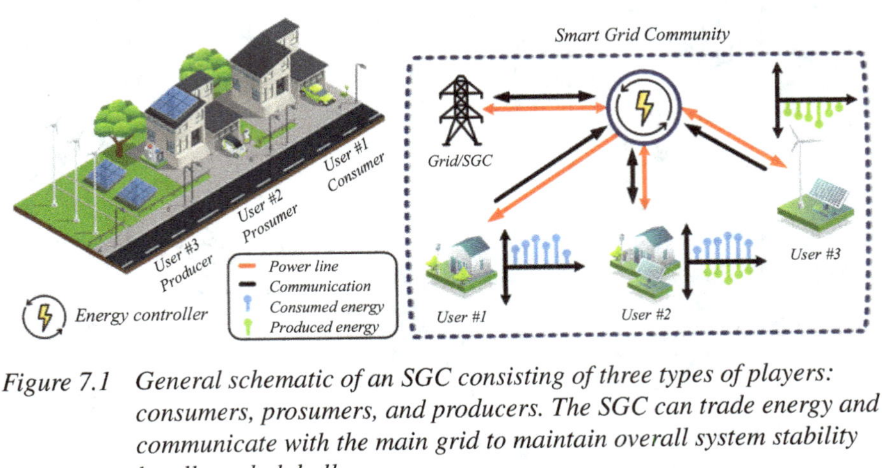

Figure 7.1 General schematic of an SGC consisting of three types of players: consumers, prosumers, and producers. The SGC can trade energy and communicate with the main grid to maintain overall system stability locally and globally

locally when the SGC is operating off-grid. Also, it allows energy trading with the upstream grid or other SGCs. The latter type of trading is beneficial for the seller and the buyer. Sellers can benefit economically and profit from harvesting renewable energy. At the same time, buyers can cover their demand and can have additional ancillary services to improve power quality.

There exist different types of SGC where the players have unique characteristics considering energy generation and consumption. One ought to keep in mind that no two SGCs are the same. This difference, minor or significant, is due to players' energy mix and demand profiles within each SGC. A modern grid structure composed of different SGCs is presented in Figure 7.2, where SGCs incorporate centralized producers and/or decentralized prosumers.

Number of players in an SGC

The size of a community is a challenge that appears when considering the different characteristics of each SGC. Since each SGC has a unique structure and features distinguishing it from other SGCs, there is a gap in defining the population size of SGC. There is currently no clear definition of an SGC population, more specifically, the size (i.e., the number of players). Therefore, in the context of an SGC, a population can be defined as a group of participants interacting with each other in a particular geographic area via communication means and capable of performing power exchange which affects the generation and distribution of electricity between them. The population size is defined by the number of participants and the total number of communities considered within a power grid. Figure 7.3 depicts the different population categories for SGC based on the number of players. A small population is considered to have less than 100 participants, a medium-size comprises 100 and less than 10,000 players, and a large population contains more than 10,000 participants. In many articles, the consumers are grouped into aggregated players since they tend to behave similarly,

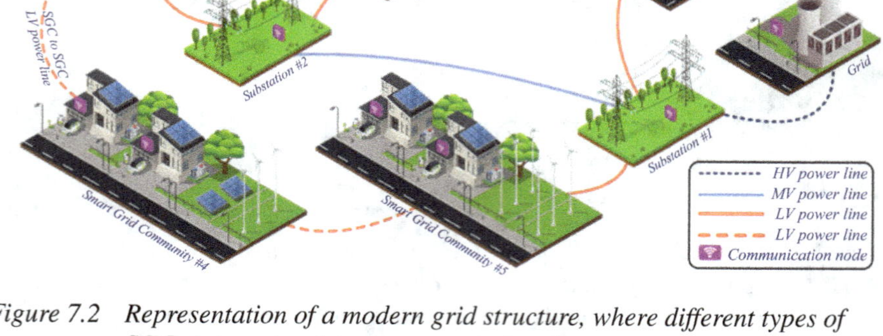

Figure 7.2 Representation of a modern grid structure, where different types of SGCs interact and trade energy in local energy markets

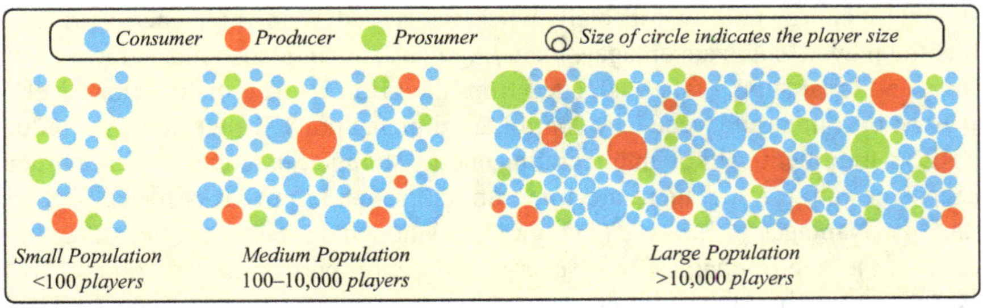

Figure 7.3 Population categories based on the number of players

which is not a realistic scenario considering selfish behavior as a characteristic of many players, mainly prosumers, and the stochastic generation and demand of each player. Nevertheless, consumers prefer to purchase energy at a lower cost or shed some non-critical loads to save energy. In many GT papers, each player represents an entire community or the utility in some games. Most presented case studies consider only SGC with a small population, which could deliver a feasible solution for the game during each trading window. To elaborate, solving the game based on a relatively small number of players significantly reduces the complexity of solving the game in real-time. However, these methods could not reach find the optimal trading strategy for when having a large number of players in real-time EM.

7.2.2 Energy management of SGCs

Green energy is an essential element considering the environment and the continuously growing energy demand. There exist multiple challenges when considering energy trading with the penetration of renewable energy; the main challenge is the stochastic energy output of the RES which relies on fluctuating inputs, i.e., irradiance, temperature, and wind speed. There are many advanced methods for forecasting these inputs to help estimate RES output. The effect of intermittent RES power on trading in an energy market could be critical on the local community or global grid level. Numerous research has been dedicated to DSM approaches to overcome the challenge of RES fluctuating energy output [2]. A crucial element in DSM is storage systems; implemented to store any excess of energy when the demand is lower than the generation and discharge when the demand is higher. However, a realistic model for energy trading to predict the seasonal effects of renewable energy and consumers' behavior becomes necessary. To build such models, short-term and long-term forecasts with quantitative propositions should be considered. With this, many players can benefit from the penetration of renewable energy and interaction with others while ensuring power reliability. Accurate forecasting also enables them to participate in the energy market with certain bids without jeopardizing the market's reliability. However, the forecasting models can be computationally expensive and time-consuming, thus increasing overall cost and negatively affecting trading within each window. ESSs play a significant role in compensating for any surplus and shortage of RES due to their stochastic outputs, acting as an energy buffer. Also, ESS is essential for energy trading, especially when discharged during high energy prices. It offers the players some advantages, including peak shaving, cost management, time of use, and local supply when islanded from the grid [3]. At a distribution level, it can assist in voltage and frequency control, minimizing power congestion, and reducing RES curtailment. Although integrating energy storage can reduce the intermittent RES power output issue, there are some limitations and drawbacks for ESS. Some of the disadvantages of ESS are (1) high installation and maintenance cost, (2) limitation on the rate of charge and discharge, (3) dire effects on the environment, (4) power-electronic interface with the network, and (5) safety considerations [3]. However, there are environmental considerations when deploying ESS concerning the hazardous, toxic, and corrosive materials like lithium, cadmium, cobalt, and mercury which contained in batteries. Hence, they cause a severe impact on the environment when they are discarded. The transition toward more sustainable and greener energy must account for the disposal of all equipment used. There exist some drawbacks when considering the trading aspects, including the limitation on SoC of batteries, mainly when experiencing a power outage.

7.3 Game theory in energy management

Game theory has found diverse applications within smart grids, effectively addressing various complexities by modeling interactions between stakeholders, such as producers, consumers, and prosumers, under changing energy and load conditions. Illustrated in Figure 7.4, these applications span four primary areas: energy trading, power balancing, system planning, and grid reliability. Each area leverages game-theoretic tools to optimize decision-making and resource allocation while considering the individual behaviors of grid participants and the broader requirements of a reliable and efficient energy system.

(1) Energy trading:

Energy trading within smart grids involves dynamic interactions between players seeking optimal strategies for buying, selling, and distributing electricity. Game theory provides a framework for developing bidding strategies, setting energy prices, managing profits, and controlling loads. For instance, bidding in energy markets, where electricity prices often vary stochastically based on locational

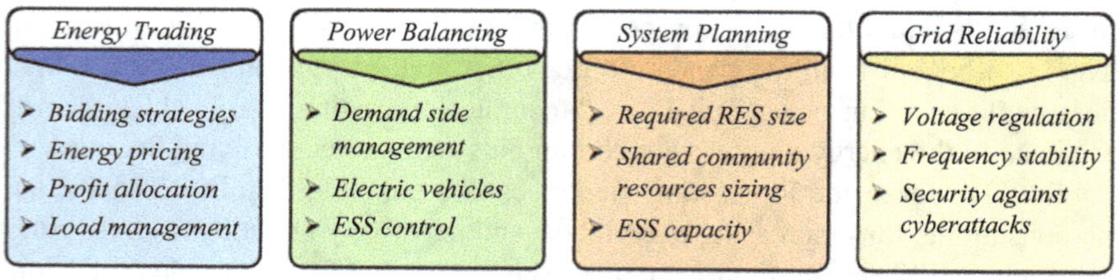

Figure 7.4 Categorization of game-theoretic applications in power systems

marginal prices, can be modeled as a Markov decision process. Here, game-theoretic strategies enable producers to maximize profits, while consumers minimize costs in a competitive setting. This application also considers the role of prosumers, who may act as either buyers or sellers, depending on their energy surplus or deficit. Cooperative games can be applied when players form coalitions to stabilize prices, while non-cooperative games capture the competitive strategies among individual participants aiming for personal gain. This dual nature of cooperation and competition offers a flexible mechanism for managing energy exchanges in real-time, where players adjust bids to respond to fluctuating demands and prices.

(2) Power balancing:
In power balancing, game-theoretic methods are used to enhance DSM, optimize the operation of ESSs, and coordinate the integration of EVs. These applications focus on balancing the load in the grid by incentivizing consumers to adjust their energy usage according to system needs. For example, DSM encourages consumers to shift their usage patterns to off-peak times, thus reducing peak loads and improving system efficiency. ESSs and EVs further support grid stability by storing surplus energy during low-demand periods and releasing it during high-demand times. Game theory models these interactions by representing consumers as players whose strategies influence the overall demand curve. Non-cooperative games, in particular, help simulate how individual participants, motivated by cost savings, adjust their consumption patterns to achieve balanced power demand and supply.

(3) System planning:
Game theory also plays a pivotal role in system planning, specifically in DER sizing, community resource allocation, and energy storage capacity optimization. System planning games address how resources are allocated across multiple players to ensure the grid's long-term sustainability and efficiency. For instance, DER sizing and the allocation of shared resources among community members are treated as cooperative games where participants collaborate to determine optimal capacity allocations based on anticipated demand and available resources. The shared resource sizing within communities also takes into account diverse user preferences and consumption habits. Furthermore, energy storage capacity optimization helps reduce the dependency on grid power and maximizes renewable energy utilization. In this context, game theory facilitates decisions that consider both the private benefits for individual players and the collective benefits for the community, promoting both economic and environmental sustainability.

(4) Grid reliability:
Game-theoretic approaches enhance grid reliability by addressing critical issues such as voltage regulation, frequency stability, and protection against cyberattacks. Voltage regulation and frequency stability are essential for maintaining power quality and preventing disturbances across the grid. In these applications, game theory provides mechanisms for players, such as distributed generators and storage systems, to

adjust their outputs to help stabilize voltage and frequency levels dynamically. Additionally, the smart grid's reliance on information and communication technologies makes it vulnerable to cyber threats, which can disrupt operations. Game-theoretic models here act as a defense strategy, modeling interactions between attackers and defenders as a security game. In such games, the defender (grid operator) employs strategies to detect and mitigate potential cyberattacks, while the attacker seeks to exploit vulnerabilities within the system. By simulating these adversarial dynamics, game theory supports robust security planning and operational resilience.

These game-theoretic applications in the smart grid facilitate a transition to a decentralized, resilient, and efficient power system. By modeling diverse interactions among stakeholders with varying objectives, game theory contributes to optimized energy management and strategic planning, which are essential for the sustainability of modern energy systems.

7.3.1 Non-cooperative game theory in energy management

Non-cooperative game theory has become a critical tool for tackling the complexities of modern energy systems, especially in decentralized SGCs. As the energy landscape shifts toward renewable integration and smart grid systems, players such as prosumers and consumers actively participate in energy markets with unique consumption and production profiles. Traditional centralized energy management approaches struggle to address the needs of these self-interested agents effectively. Non-cooperative game theory provides a framework for modeling the decision-making processes of these agents by allowing them to act independently while reaching optimal strategies through equilibrium-based approaches, such as Nash and Stackelberg equilibria. By capturing the autonomous and often conflicting goals of agents, non-cooperative game theory supports more adaptive, efficient, and economically viable solutions in energy management, ultimately contributing to cost reduction and increased sustainability. Table 7.1 lists the recent applications of non-cooperative game theory in energy management.

In the context of demand response, non-cooperative game theory enables the coordination of flexible energy consumption by modeling the behavior of prosumers under peak and off-peak pricing schemes [4–6]. Prosumers are encouraged to modify their demand based on real-time energy prices and DR incentives. Using distributed optimization games, non-cooperative frameworks allow prosumers to adjust their energy use independently, which enhances grid efficiency and reduces peak load [7]. These models are designed to facilitate real-time adjustments in energy consumption, making demand response programs more responsive to dynamic pricing and demand fluctuations [8]. By applying Nash equilibrium strategies, prosumers balance between minimizing personal costs and contributing to overall supply-demand balance, which is particularly beneficial in renewable-integrated grids where generation variability poses challenges to grid operators [4,9].

Microgrids, operating as localized networks with both energy generation and consumption capabilities, benefit significantly from non-cooperative game theory in energy trading and management. Non-cooperative games facilitate optimal energy exchange among microgrids, allowing each entity to maximize its self-interest while

Table 7.1 Non-cooperative game theory applications in energy management

Ref	Application	Method	Contribution
[4]	Microgrid coordination	Non-cooperative game	Efficient resource allocation among renewable and non-renewable generators
[10]	Residential demand response	Non-cooperative game	Incentivizes user participation in DR, optimizing load distribution
[5]	EV charging pricing	Stackelberg game	Price differentiation to manage demand and reduce grid strain
[11]	Renewable integration	Nash equilibrium	Balances power output variability between wind and solar
[7]	Home energy management	Nash bargaining	Reduces household costs through strategic load shifting
[8]	Power distribution network	Dynamic game	Manages grid congestion via user competition for resources
[9]	P2P trading in smart grids	Stackelberg game	Enhances energy autonomy via competitive trading
[6]	Adaptive demand response	Nash bargaining	Real-time DR pricing to balance grid supply and demand
[12]	Community energy sharing	Leader–follower	Cooperative incentives for renewable resource sharing
[13]	Multi-agent ESS management	Non-cooperative game	Optimizes ESS use, balancing supply and demand
[14]	Building-to-building (B2B) grid	Non-cooperative game	Optimizes welfare in B2B grid, accounting for ESS degradation
[15]	P2P trading market	Stackelberg game	Reduces carbon emissions, incentivizes prosumers

collectively improving energy balance within the network [10,11]. Cournot competition and Nash equilibrium are commonly used to coordinate the distribution of generated power from renewables, optimizing the allocation of resources across multiple microgrids and stabilizing local supply-demand dynamics [6]. Prosumers within a microgrid can participate in peer-to-peer (P2P) energy trading and share their surplus energy efficiently, reducing costs associated with grid imports [14]. The non-cooperative approach also allows microgrids to autonomously trade energy while maintaining local control, which is essential for managing renewable energy variability and enhancing resilience in isolated or rural areas [12]. In P2P trading specifically, Stackelberg and Nash-based models are employed to determine dynamic pricing structures and trading volumes, ensuring cost-effective energy transactions between prosumers and microgrids [8,13]. Dynamic energy pricing models can reduce dependence on the main grid and incentivize the use of local renewable resources, thus supporting a more sustainable and community-focused energy network [15].

Non-cooperative game theory also plays a pivotal role in managing EV charging schedules, particularly as EV penetration increases the demand burden on conventional grids. By applying Stackelberg game models, energy providers set time-of-use (TOU) pricing structures that guide EV users to schedule charging during off-peak hours, thereby flattening demand peaks [6,11]. The non-cooperative approach enables EV owners to autonomously determine their charging schedules based on cost and convenience, balancing personal benefits with grid stability [7]. In multi-agent environments where numerous EVs are connected, Stackelberg games help establish hierarchical structures in which grid operators or aggregators can manage large-scale EV loads without compromising individual decision-making freedom [13]. Similarly, non-cooperative game theory offers a practical framework for optimizing ESS dispatch in response to TOU pricing [5,9].

7.3.2 Cooperative game theory in energy management

Cooperative game theory has gained traction in energy management for optimizing shared resources and costs within smart grids and integrated energy systems (IESs). These models can facilitate efficient resource-sharing agreements and thus enhance energy management across diverse energy systems, enabling optimal cost allocation, system reliability, and emission reduction. This section explores applications of cooperative game theory across networked energy hubs (NEHs), integrated energy communities, smart grid interactions, and more, highlighting contributions from multiple studies. Table 7.2 lists the recent applications of non-cooperative game theory in energy management.

NEHs form a significant application area, enabling connected facilities to optimize their resources through cooperative models. These hubs benefit from cooperative scheduling to reduce overall costs and carbon emissions [16]. For instance, a study uses Shapley value to allocate coalition gains based on hub contributions, coordinating their functions to reduce costs and CO_2 emissions by approximately 3% and 1.8%, respectively [16]. Similar frameworks in other NEHs apply stochastic scenario generation and Monte Carlo simulation to manage renewable uncertainties and facilitate efficient energy trading among distributed hubs [16,17]. These cooperative frameworks underscore how coordinated management across NEHs supports both economic and environmental objectives, serving as a strategic pathway for interconnected hub networks.

In energy communities and multi-microgrid systems, cooperative game theory is applied to optimize energy distribution, incentivize DSM, and ensure fair cost-sharing. For instance, community members can share renewable resources costlessly using Shapley value allocation, promoting equitable access and community-wide

Table 7.2 Cooperative game theory applications in energy management

Ref	Application	Method	Contribution
[16]	NEHs management	Shapley value, stochastic scenario generation	Optimized cost and emission reduction for connected energy hubs
[17]	Multi-park energy system cooperation	Nash bargaining theory	Joint dispatching of hydrogen systems, maximizing social benefits
[18]	RES sharing in communities	Shapley value-based allocation	Costless renewable distribution
[19]	P2P energy trading with blockchain	Coalition formation game with blockchain	Secure trading among prosumers, enhanced by Proof of Energy Generation
[20]	Multi-building IES coordination	Nash bargaining theory	Reduced grid interaction and operational fluctuations
[21]	Regional IESs	Nash bargaining with multi-phase structure	Optimal bidding and revenue sharing under uncertain market conditions
[22]	Smart building energy management	Shapley value with coalition model	Optimized surplus energy sharing
[23]	Community-level energy scheduling	Bi-level game theory, load shifting	Coordinated scheduling, peak load minimization
[24]	Demand response in microgrids	Cooperative game theory with blockchain	Optimized demand response, secure data management for microgrid users
[25]	Multi-stage smart grid trading	Nash bargaining, multi-stage pricing	Fair P2P pricing, improved prosumer engagement
[26]	HEV cooperative speed-energy management	Nash equilibrium-based optimization	Improved fuel efficiency and safety in HEV speed and energy management
[27]	Coalition formation in DERs	Shapley value, nucleolus	Fair profit allocation among DERs, coalition stability
[28]	Incentive mechanisms in P2P markets	Game-theory driven incentive model	Peak shaving and enhanced prosumer satisfaction in P2P energy trading
[29]	Multi-carrier energy systems' TEC	Nash bargaining, stochastic programming	Enhanced flexibility in managing multi-carrier energy under uncertainty
[30]	Hierarchical energy management	Stackelberg game, Nash–Harsanyi bargaining	Optimal DSO-pricing, efficient hydrogen–electricity trading coordination

engagement in demand response programs [18]. These setups lead to cost reductions for communities and enhance system sustainability. In addition, demand response incentives within microgrids managed through blockchain create a robust, decentralized energy exchange, while coalition formation and blockchain enhance P2P trading, ensuring fair rewards for active participants [19,24]. These applications demonstrate how cooperative frameworks drive effective local energy trading and ensure optimal benefit-sharing across microgrids and community setups.

IESs within public and commercial sectors use cooperative models for operational efficiencies in building-to-grid interactions and energy storage management. Studies propose multi-building IESs, utilizing Nash bargaining theory to optimize energy exchanges, minimize grid interactions, and reduce operational fluctuations [20]. Additionally, the cooperation of public and industrial buildings under variable renewable sources and carbon pricing regimes underscores the system's stability, leading to enhanced cost savings and demand-side flexibility [23]. Cooperative frameworks here allow facilities to share resources adaptively, balancing operational costs while reducing carbon footprints.

P2P trading and demand response within smart grids also benefit from cooperative models. Nash bargaining approaches enable fair price setting in P2P markets, motivating prosumers to engage actively in energy exchange and benefit from optimized load shifting [28]. For instance, cooperative frameworks that integrate incentives in local energy communities allow for peak shaving and cost savings, balancing individual utility with grid reliability [28]. The collaboration across community participants further supports flexible load management, allowing users to leverage their energy assets optimally based on real-time pricing and demand response incentives.

Smart building clusters employ cooperative game theory for real-time energy management, where coalition models optimize surplus energy allocation between buildings based on demand. Using Shapley value for allocation, the energy management systems in smart buildings efficiently balance surplus and deficit demands, minimizing energy waste [22]. Similarly, regional integrated energy systems (RIES), comprising multi-energy microgrids, use collaborative models to optimize day-ahead bidding and real-time dispatch, with Nash bargaining to ensure fair revenue sharing among microgrid participants [21]. These applications reveal how cooperative strategies in smart buildings and RIES drive energy efficiency and cost-sharing while supporting grid stability.

7.3.3 Evolutionary game theory (EGT)

EGT has certain advantages over classical game theory models. These include (i) its robustness to players' irrational decision making, (ii) its ability to handle incomplete information games efficiently, and (iii) its inherent capability of preserving the players' data privacy. EGT was first proposed by Maynard Smith and Price [31]. Compared with classic games, EGT takes the population as the environment where players can take actions and gain rewards, where it assumes that the players are bounded rational and have limited information. EGT finds the evolutionary stable strategy (ESS) that can be an equivalent best response to describe the studied behavior in the electricity market. For example, it is challenging to describe the bidding behavior in an electricity market. Via learning an ESS, players can achieve an overall greater reward when executing actions at each time window. The market behavior constantly changes due to the stochastic nature of SGCs. Thus, EGT is more in line with the realistic markets since it provides the best strategy for players considering the dynamic environment they interact with and avoids strict rational assumptions. Replicator dynamics (RD) is a fundamental concept in EGT, primarily used to describe the evolution of population behavior and predict players' decision-making patterns. RD provides an approximation of the expected values of different strategies within a population, measuring each strategy's fitness (reproduction rate) against competing strategies. This evolutionary process is driven by two main components: selection and mutation [32]. Selection favors certain strategies over others, while mutation introduces diversity into the population. RD highlights the role of selection in shaping population evolution by modeling how a system changes over time as players adopt various strategies. Importantly, the application of EGT extends beyond single-population interactions; it can also model interactions among two or more players from different populations, known as asymmetric games. An asymmetric game occurs when the game structure is heterogeneous, as described in Complex Network Theory, or involves a centralized scheme, as in control theory. In these cases, reward parameters are distributed asymmetrically. When different populations have varying numbers of neighbors, the network is termed heterogeneous [33]. Conversely, if each player in the population has the same number of neighbors, the network is homogeneous, and reward parameters are symmetrically distributed among all players and populations. In evolutionary games, each node can represent a player or a population comprising players with different strategies. Figure 7.5 illustrates the network structures for symmetric and asymmetric situations in evolutionary games. The advantage of learning an ESS among players in different populations is explored in [34], where a two-population, multi-strategy evolutionary game is applied to study bidding behavior in energy markets.

EGT applications in energy management

EGT has been widely applied in energy management for analyzing strategic interactions among entities in smart grids, microgrids, and P2P trading frameworks. EGT is particularly valuable in environments where participants are bounded by rationality, adapting strategies based on payoff feedback. Table 7.3 presents the primary applications of EGT in energy management.

In DSM systems, EGT facilitates optimal decision-making among users, utilities, and DERs. For example, a hierarchical game approach combines EGT with non-cooperative games to optimize plug-in hybrid electric vehicle (PHEV) charging, balancing user demands with grid stability by incentivizing users to align consumption with reliability-differentiated pricing [35]. Similarly, in a multiple utility scenario, a two-level game approach is employed where utilities compete non-cooperatively, while residential users adopt evolutionary strategies to minimize peak loads through dynamic adaptation to pricing signals [36]. These applications emphasize EGT's role

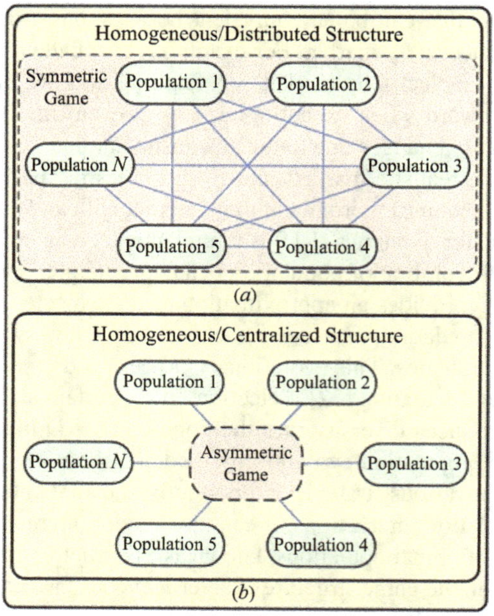

Figure 7.5 Illustration of an evolutionary game between N populations in (a) symmetric evolutionary game with a homogeneous structure and (b) an asymmetric evolutionary game with a heterogeneous structure

in managing decentralized resources by allowing users to evolve their consumption patterns based on real-time incentives and constraints.

EGT also plays a critical role in P2P energy trading, where communities trade energy generated by distributed sources, such as solar panels and battery storage systems. A two-stage P2P trading model integrates EGT to optimize shared energy storage (SES) leasing fees, considering bounded rationality and market dynamics to ensure fair distribution of benefits across participants [37]. Additionally, a community energy management model applies EGT to balance interactions between energy consumers and vendors, reaching Stackelberg and Nash equilibria to maximize community welfare and reduce grid dependency [38]. These applications demonstrate how EGT helps achieve stable and fair outcomes in community-level energy trading by allowing participants to adapt trading strategies and prices over time.

In hierarchical microgrid control, EGT is used for dynamic population games to optimize load sharing across distributed generators, improving reliability and efficiency in microgrid operations [39]. Another study applies EGT to optimize dispatch strategies in microgrids with task-sharing frameworks, where each generator autonomously adapts to load demands, enhancing system stability and operational flexibility [40]. These applications illustrate EGT's potential to manage decentralized microgrids by enabling adaptive strategies that respond to real-time operational needs, ensuring both reliability and economic efficiency.

Table 7.3 Summary of EGT applications in energy management

Ref	Application	Method	Contribution
[35]	Reliability management in DSM	Hierarchical game with EGT	Optimized PHEV charging for grid reliability
[36]	Demand response with utilities	Two-level game with EGT	Minimized peak load through adaptive pricing
[37]	P2P energy storage leasing	Two-stage EGT model	Fair SES leasing fees in P2P trading
[38]	Community P2P energy trading	EGT with Stackelberg equilibrium	Maximized welfare and reduced grid dependency
[39]	Hierarchical microgrid control	Dynamic population games	Improved reliability and load sharing in microgrids
[40]	Microgrid load dispatch optimization	Task-sharing EGT model	Enhanced microgrid stability and flexibility
[41]	Sustainable PV poverty alleviation	Tripartite EGT model	Promoted stakeholder collaboration for sustainability
[34]	Long-term bidding in deregulated markets	Two-population, multi-strategy EGT	Stable bidding equilibria in competitive markets

7.4 GT and AI in smart grids

The intersection of AI and GT holds significant promise in advancing energy management within SGCs. AI's ability to handle vast amounts of data and learn from complex, dynamic environments complements GT's frameworks for modeling decision-making among multiple agents with potentially conflicting objectives. This integration is particularly beneficial in decentralized energy systems, where independent actors like prosumers need to coordinate their decisions to optimize local energy use, balance demand, and ensure grid stability. Recent research papers have demonstrated various applications of AI-enhanced game-theoretic methods, especially in managing DERs and P2P energy trading, as well as in addressing challenges such as renewable energy intermittency, grid flexibility, and dynamic pricing strategies.

One promising application area of AI–GT in energy management is local energy trading and demand-side flexibility in decentralized networks. AI-driven reinforcement learning (RL) is frequently combined with game-theoretic approaches to develop autonomous, adaptive trading systems that learn optimal strategies through repeated interactions within the market. For instance, multi-agent reinforcement learning (MARL) enables prosumers to engage in decentralized energy markets, dynamically adjusting their energy generation, storage, and consumption strategies based on real-time pricing and demand signals [42]. This has been used in frameworks where P2P trading is structured as a continuous double auction, creating a market mechanism that allows agents to optimize their individual utility while benefiting the grid as a whole [43]. Furthermore, advanced AI models incorporating Q-learning have been used to improve P2P pricing accuracy, enabling fair exchanges and maximizing individual and collective welfare by dynamically adjusting to real-time supply and demand data [44]. Beyond economic gains, AI and GT support environmental benefits by promoting local energy use over grid reliance. In community energy systems, where multiple prosumers interact, AI-enabled GT frameworks have been shown to incentivize prosumers to prioritize local renewable energy use [45]. Here, Q-learning algorithms learn optimal trading behaviors that align with sustainability goals, encouraging prosumers to share resources within the community and helping the grid manage renewable variability without resorting to fossil-fuel backups [46].

In DSM schemes, AI and GT work in tandem to optimize energy consumption during peak and off-peak hours. This is essential for DER systems, where energy demand and supply can vary unpredictably. For example, Stackelberg game models allow grid operators (leaders) to set TOU prices, while prosumers (followers) respond by adjusting their consumption patterns based on real-time energy rates [47]. AI-driven RL algorithms further enable prosumers to learn optimal consumption strategies based on these TOU prices. Over time, prosumers adapt to fluctuating prices, helping to reduce peak demand and optimize energy use without sacrificing comfort [48].

Grid flexibility services are also enhanced by AI–GT, as distributed devices can coordinate their responses to grid signals. By using MARL and hierarchical games, distributed devices provide demand-response services autonomously, adjusting load and offering flexibility to the grid. This approach allows devices to independently assess demand signals, calculate their flexibility, and provide services based on real-time incentives and feedback from the grid [49]. The integration of RL with GT frameworks supports responsive DSM systems, enhancing grid reliability and reducing the need for costly grid upgrades. In microgrids, AI and GT help to optimize the allocation and dispatch of renewable resources. AI-based learning algorithms, such as deep Q-learning, can dynamically adjust generation and storage parameters to meet demand while maintaining system stability [50]. When combined with Nash equilibrium models, these systems enable coordinated energy flows, balancing supply and demand in real-time. For instance, MARL in a competitive game setting allows DERs to reach stable operating points, maximizing the economic benefits of energy storage and generation within the microgrid [51].

In hybrid microgrids where multiple energy sources coexist, EGT combined with RL has been shown to optimize energy dispatch by encouraging cooperation among DERs. For instance, ESS and solar generation units can engage in cooperative games that account for weather variability, ensuring stability in generation and consumption even under fluctuating conditions [52]. AI models predict energy availability, while GT provides optimal strategies for storing, dispatching, or sharing energy across interconnected microgrids, improving both grid resilience and resource efficiency. AI–GT approaches help address the uncertainty of RES by using machine learning to forecast renewable output and game-theoretic models to predict players' decisions. In one application, AI-based weather forecasting is integrated with Stackelberg games to optimize grid operations, where grid operators predict renewables' output and adjust grid control strategies accordingly [53]. This process, often referred to as "predictive dispatching," allows the grid to use renewables effectively, minimizing reliance on non-renewable sources and reducing overall carbon emissions. Additionally, dynamic load forecasting—powered by machine learning—enables GT models to plan demand response in anticipation of renewables' variability. This is particularly useful in microgrids, where load and generation patterns are highly localized [54].

The integration of AI and GT is also crucial in scheduling ESS. ESS must be managed carefully to balance grid supply and demand, as overuse can lead to battery degradation and reduce storage capacity. AI-enabled RL models learn optimal ESS dispatch strategies, considering both short-term gains (e.g., peak shaving) and long-term goals (e.g., battery health) [55]. For instance, Nash bargaining games allow ESS owners to negotiate dispatch schedules that align with grid needs while preserving battery life. When integrated with predictive models, ESS management systems optimize energy release during high-demand periods, helping to stabilize prices and provide emergency reserves for the grid [56].

In addition to practical applications, the intersection of AI and GT in energy management opens opportunities for addressing critical challenges and creating new research directions. Some of these research directions are outlined in Table 7.4.

7.5 Conclusion

In this chapter, we have explored the foundational concepts of GT and its significant applications in the realm of energy management. As energy systems become increasingly decentralized and interconnected, GT provides a powerful framework for modeling interactions among various stakeholders, including energy producers, consumers, and grid operators. The applications of non-cooperative, cooperative, and evolutionary GT in energy management demonstrate the capability of GT to enhance DSM, microgrid coordination, P2P energy trading, and DER sharing in SGCs.

Furthermore, the intersection of AI and GT reveals new possibilities for the intelligent automation of decision-making in energy systems. AI-driven game-theoretic models enabled adaptive responses to real-time data, improved resource allocation, and enhanced system stability in decentralized markets. This integration offers promising solutions to address emerging challenges, such as renewable variability, real-time load balancing, and optimal energy trading.

For future research, the synergy of AI and GT opens pathways for innovative research and development, particularly in areas like cybersecurity, policy alignment, and sustainable energy market design. Through the development of these tools, energy systems can evolve to become more sustainable, economically efficient, and responsive to the needs of all stakeholders. This chapter provides a foundation for understanding the transformative role of GT in energy management and serves as a guide for future research and applications in this dynamic field.

Table 7.4 Future research opportunities for AI–GT in energy management

Area	Challenges	Opportunities
Decentralized energy markets	Privacy concerns in trading data; balancing stability with autonomy in P2P markets; preventing unfair advantages among prosumers	Development of secure, privacy-preserving trading protocols; enhanced scalability and responsiveness in autonomous energy markets
Demand response and flexibility	Real-time decision-making under high uncertainty; difficulty coordinating diverse flexibility resources in large networks	Enhanced grid reliability; optimized, autonomous response to demand peaks; more effective use of DR resources in decentralized settings
Microgrid optimization	Managing complex multi-agent interactions; scalability issues in multi-layered coordination; balancing computational efficiency with accuracy	Greater resource autonomy in distributed grids; improved local renewable usage; scalable models adaptable to varying microgrid setups
Renewable integration	Addressing intermittency without relying on fossil fuel reserves; limited predictive control for real-time operations	Predictive dispatch for effective renewables use; real-time decision-making algorithms for grid stability; reduced environmental impact
Energy storage management	Balancing battery life with dispatch needs; developing long-term optimization strategies for community-based storage	Prolonged battery lifespan; increased ESS cost-effectiveness; optimal storage utilization across diverse load and market conditions
Cybersecurity in energy grids	Vulnerability to cyber threats in decentralized systems; protecting data privacy in AI-driven markets	More secure, decentralized grids; real-time detection of security threats; enhanced data protection for trusted AI-based energy systems
Policy and regulation	Rapid technological changes challenge regulatory frameworks; balancing regulation with market freedom in decentralized settings	Policy alignment for sustainable growth; supportive frameworks for AI and GT innovations in energy markets; enhanced policy flexibility
Explainability in AI models	Lack of transparency in AI decisions; interpreting model outputs in multi-agent, game-theoretic contexts	Increased transparency; improved regulatory acceptance; higher user trust in AI-driven decisions, supporting sustainable market growth

References

[1] Yang H, Zhang J, Qiu J, *et al.* A practical pricing approach to smart grid demand response based on load classification. *IEEE Transactions on Smart Grid*. 2018;9(1):179–190.

[2] Iqbal S, Sarfraz M, Ayyub M, *et al.* A comprehensive review on residential demand side management strategies in smart grid environment. *Sustainability*. 2021;13(13):7170.

[3] Faisal M, Hannan MA, Ker PJ, *et al.* Review of energy storage system technologies in microgrid applications: Issues and challenges. *IEEE Access*. 2018;6:35143–35164.

[4] Wu C, Gu W, Yi Z, *et al.* Non-cooperative differential game and feedback Nash equilibrium analysis for real-time electricity markets. *International Journal of Electrical Power & Energy Systems*. 2023;144:108561.

[5] Zhao B, Duan P, Fen M, *et al.* Optimal operation of distribution networks and multiple community energy prosumers based on mixed game theory. *Energy*. 2023;278:128025.

[6] Gao X and Zhang X. Robust collaborative scheduling strategy for multi-microgrids of renewable energy based on a non-cooperative game and profit allocation mechanism. *Energies*. 2024;17(2):519.

[7] Liu X, Ji Z, Sun W, *et al.* Robust game-theoretic optimization for energy management in community-based energy system. *Electric Power Systems Research*. 2023;214:108939.

[8] Yang J, Xu W, Ma K, *et al.* Integrated demand-side management for multi-energy system based on non-cooperative game and multi-energy pricing. *Sustainable Energy, Grids and Networks*. 2023;34:101047.

[9] Chen Y and Yi P. Multi-cluster aggregative games: A linearly convergent Nash equilibrium seeking algorithm and its applications in energy management. *IEEE Transactions on Network Science and Engineering*. 2024;11(3):2797–2809.

[10] Zhang D, Han R, Wan Y, *et al.* Robust optimal energy management with dynamic price response: A non-cooperative multi-community aggregative game perspective. *International Journal of Electrical Power & Energy Systems*. 2023;154:109395.

[11] Li K, Ye N, Li S, *et al.* Distributed collaborative operation strategies in multi-agent integrated energy system considering integrated demand response based on game theory. *Energy*. 2023;273:127137.

[12] Yang J, Zhao B, Ma K, *et al.* A carbon integrated energy pricing strategy based on non-cooperative game for energy hub in seaport energy system. *Energy*. 2024;309:133009.

[13] Párraga M, Fajardo A, Rodríguez-Urrego L, *et al.* A non-cooperative game approach on isolated water-energy microgrids. *Sustainable Energy, Grids and Networks*. 2024;38:101392.

[14] Gebremariam MB, García P, Ángel Navarro-Rodríguez, *et al.* Energy exchange optimization for a building-to-building grid: A non-cooperative

game-theory method. *Sustainable Energy, Grids and Networks*. 2024; 39:101420.

[15] Yu X, Pan D, and Zhou Y. A Stackelberg game-based peer-to-peer energy trading market with energy management and pricing mechanism: A case study in Guangzhou. *Solar Energy*. 2024;270:112388.

[16] Akbari S, Hashemi-Dezaki H, and Martins J. Cooperative stochastic energy management of networked energy hubs considering environmental perspectives. *Energy Reports*. 2024;12:1638–1654.

[17] Wang LL, Xian RC, Jiao PH, *et al.* Cooperative operation of industrial/commercial/residential integrated energy system with hydrogen energy based on Nash bargaining theory. *Energy*. 2024;288:129868.

[18] Gomes L and Vale Z. Costless renewable energy distribution model based on cooperative game theory for energy communities considering its members' active contributions. *Sustainable Cities and Society*. 2024;101:105060.

[19] Moniruzzaman M, Yassine A, and Benlamri R. Blockchain and cooperative game theory for peer-to-peer energy trading in smart grids. *International Journal of Electrical Power & Energy Systems*. 2023;151:109111.

[20] Duan P, Zhao B, Zhang X, *et al.* A day-ahead optimal operation strategy for integrated energy systems in multi-public buildings based on cooperative game. *Energy*. 2023;275:127395.

[21] Wang Y, Zheng Y, and Yang Q. Nash bargaining based collaborative energy management for regional integrated energy systems in uncertain electricity markets. *Energy*. 2023;269:126725.

[22] Saeed MA, Eladl AA, Alhasnawi BN, *et al.* Energy management system in smart buildings based coalition game theory with fog platform and smart meter infrastructure. *Scientific Reports*. 2023;13:2023.

[23] Naji El Idrissi R, Ouassaid M, and Maaroufi M. Game theory approach for energy consumption scheduling of a community of smart grids. *Journal of Electrical Engineering & Technology*. 2023;18(4):2695–2708.

[24] Bai F, Zhang C, and Zhang X. Intelligent optimal demand response implemented by blockchain and cooperative game in microgrids. *International Transactions in Operational Research*. 2024;31(6):3704–3731.

[25] Devi NN, Thokchom S, Singh TD, *et al.* Multi-stage bargaining of smart grid energy trading based on cooperative game theory. *Energies*. 2023;16(11):4278.

[26] Zeng X, Gao H, Chen Z, *et al.* Cooperative optimization of speed planning and energy management for hybrid electric vehicles based on Nash equilibrium. *Journal of Power Sources*. 2023;571:233070.

[27] Moafi M, Ardeshiri RR, Mudiyanselage MW, *et al.* Optimal coalition formation and maximum profit allocation for distributed energy resources in smart grids based on cooperative game theory. *International Journal of Electrical Power & Energy Systems*. 2023;144:108492.

[28] Lilliu F and Reforgiato Recupero D. A cooperative game-theory approach for incentive systems in local energy communities. *Sustainable Energy, Grids and Networks*. 2024;38:101391.

[29] Alizadeh A, Esfahani M, Dinar F, *et al.* A cooperative transactive multi-carrier energy control mechanism with P2P energy + reserve trading using Nash bargaining game theory under renewables uncertainty. *Applied Energy*. 2024;353:122162.

[30] Cai P, Mi Y, Xing H, *et al.* Hierarchical coordinated energy management strategy for electricity-hydrogen integrated charging stations based on IGDT and hybrid game. *Electric Power Systems Research*. 2023;223:109527.

[31] Taylor PD and Jonker LB. Evolutionary stable strategies and game dynamics. *Mathematical Biosciences*. 1978;40(1):145–156.

[32] Bloembergen D, Tuyls K, Hennes D, *et al.* Evolutionary dynamics of multi-agent learning: A survey. *Journal of Artificial Intelligence Research*. 2015;53:659–697.

[33] Zaballos A, Vallejo A, and Selga JM. Heterogeneous communication architecture for the smart grid. *IEEE Network*. 2011;25(5):30–37.

[34] Cheng L, Chen Y, and Liu G. 2PnS-EG: A general two-population n-strategy evolutionary game for strategic long-term bidding in a deregulated market under different market clearing mechanisms. *International Journal of Electrical Power & Energy Systems*. 2022;142:108182.

[35] Tan J and Wang L. Enabling reliability-differentiated service in residential distribution networks with PHEVs: A hierarchical game approach. *IEEE Transactions on Smart Grid*. 2016;7(2):684–694.

[36] Chai B, Chen J, Yang Z, *et al.* Demand response management with multiple utility companies: A two-level game approach. *IEEE Transactions on Smart Grid*. 2014;5(2):722–731.

[37] He Y, Wu H, Wu AY, *et al.* Optimized shared energy storage in a peer-to-peer energy trading market: Two-stage strategic model regards bargaining and evolutionary game theory. *Renewable Energy*. 2024;224:120190.

[38] Fernandez E, Hossain MJ, Nawazish Ali SM, *et al.* An efficient P2P energy trading platform based on evolutionary games for prosumers in a community. *Sustainable Energy, Grids and Networks*. 2023;34:101074.

[39] Mojica-Nava E, Macana CA, and Quijano N. Dynamic population games for optimal dispatch on hierarchical microgrid control. *IEEE Transactions on Systems, Man, and Cybernetics: Systems*. 2014;44(3):306–317.

[40] Zhu B, Xia X, and Wu Z. Evolutionary game theoretic demand-side management and control for a class of networked smart grid. *Automatica*. 2016;70:94–100.

[41] Shan H and Yang J. Sustainability of photovoltaic poverty alleviation in China: An evolutionary game between stakeholders. *Energy*. 2019;181:264–280.

[42] Zhang S, May D, Gül M, *et al.* Reinforcement learning-driven local transactive energy market for distributed energy resources. *Energy and AI*. 2022;8:100150.

[43] Ye Y, Papadaskalopoulos D, Yuan Q, *et al.* Multi-agent deep reinforcement learning for coordinated energy trading and flexibility services provision in local electricity markets. *IEEE Transactions on Smart Grid*. 2023;14(2):1541–1554.

[44] Qiu D, Wang J, Dong Z, *et al.* Mean-field multi-agent reinforcement learning for peer-to-peer multi-energy trading. *IEEE Transactions on Power Systems*. 2023;38(5):4853–4866.

[45] May R and Huang P. A multi-agent reinforcement learning approach for investigating and optimising peer-to-peer prosumer energy markets. *Applied Energy*. 2023;334:120705.

[46] Zhang B, Hu W, Ghias MYM, Xu X, and Chen Z. Multi-agent deep reinforcement learning based distributed control architecture for interconnected multi-energy microgrid energy management and optimization. *Energy Conversion and Management*. 2023;277:116647.

[47] Monfaredi F, Shayeghi H, and Siano P. Multi-agent deep reinforcement learning-based optimal energy management for grid-connected multiple energy carrier microgrids. *International Journal of Electrical Power & Energy Systems*. 2023;153:109292.

[48] Mahmoud M and Slama SB. Peer-to-peer energy trading case study using an AI-powered community energy management system. *Applied Sciences*. 2023;13(13):7838.

[49] Arumugham V, Ghanimi HMA, Pustokhin DA, *et al.* An artificial-intelligence-based renewable energy prediction program for demand-side management in smart grids. *Sustainability*. 2023;15(6):5453.

[50] Ali ANF, Sulaima MF, Razak IAWA, *et al.* Artificial intelligence application in demand response: Advantages, issues, status, and challenges. *IEEE Access*. 2023;11:16907–16922.

[51] Qiu S, Li Z, Pang Z, *et al.* Multi-agent optimal control for central chiller plants using reinforcement learning and game theory. *Systems*. 2023;11(3):136.

[52] Maghami MR and Mutambara AGO. Challenges associated with Hybrid Energy Systems: An artificial intelligence solution. *Energy Reports*. 2023;9:924–940.

[53] Yang N, Han L, Liu R, *et al.* Multiobjective intelligent energy management for hybrid electric vehicles based on multiagent reinforcement learning. *IEEE Transactions on Transportation Electrification*. 2023;9(3):4294–4305.

[54] Joshi A, Capezza S, Alhaji A, *et al.* Survey on AI and machine learning techniques for microgrid energy management systems. *IEEE/CAA Journal of Automatica Sinica*. 2023;10(7):1513–1529.

[55] Khabbouchi I, Said D, Oukaira A, *et al.* Machine learning and game-theoretic model for advanced wind energy management protocol (AWEMP). *Energies*. 2023;16(5):2179.

[56] Nie Y, Liu J, Liu X, *et al.* Asynchronous multi-agent reinforcement learning-based framework for bi-level noncooperative game-theoretic demand response. *IEEE Transactions on Smart Grid*. 2024;15(6):5622–5637.

Chapter 8
Edge to cloud

Milenko Tosic[1], *Fábio Coelho*[2,3], *Luis Ferreira*[4], *David Rua*[4] *and Dragan Boscovic*[5]

8.1 Introduction

While the rest of this book deals with the application of artificial intelligence (AI) in the energy sector, this chapter will cover the infrastructure enablers for the proper realization of AI-based solutions. Training complex AI models (e.g. deep learning models) requires immense and often specialized computational resources [graphical processing unit (GPUs), tensor processing units (TPUs), etc.]. In addition, powerful AI models require massive datasets for training and validation, often spanning terabytes of data. This calls for high-performance computing infrastructure with virtually unlimited scalable storage. The cloud computing paradigm is capable of addressing this need for "infinitely" scalable storage. Conversely, AI applications often require real-time decision making (e.g. autonomous driving, Internet of Things (IoT)-based automation, healthcare monitoring, etc.) where even a minimal latency can have serious consequences. Additionally, some data streams impose strict privacy protection requirements, which limits the possibility of transferring them to a remote location over a public network for further analysis. Limitations in network bandwidth may also limit the amount of data that can be forwarded from edge to cloud nodes. Performing AI inference closer to the sources of data can result in avoiding latencies introduced by communication networks, lowering bandwidth use, and addressing privacy concerns. The edge processing approach provides a solution for these cases.

Proper implementation of complex AI-based processes in domains such as industry, agriculture, transportation, healthcare, and energy requires exploitation of the full potential offered by both cloud and edge computation systems [1]. The cloud continuum entails the integration of computing and data storage resources spanning across centralized data centers, cloud systems, and edge/fog computing nodes closer to the actual source of data [2]. The continuum is tasked with dynamically shifting loads between centralized and distributed resources depending on the overall workload,

[1]VizLore Labs Foundation, Serbia
[2]High Assurance Software Laboratory, Institute for Systems and Computer Engineering, Technology and Science, Portugal
[3]University of Minho, Portugal
[4]Centre for Power and Energy Systems, Institute for Systems and Computer Engineering, Technology and Science, Portugal
[5]W. P. Carey Business School, Arizona State University, USA

complexity, and required latency limitations. When it comes to AI, large-scale data analytics and model training are to be performed in the cloud, leveraging its computational power, while real-time decision making and AI model inference occur at the edge, thus minimizing the latency. This approach optimizes resource utilization and enhances responsiveness and privacy by reducing reliance on centralized systems. In this chapter, we delve into details of the cloud continuum, technological enablers, possibilities, and limitations and reflect on its role in enabling AI-based energy services.

8.2 Understanding the cloud–edge continuum

In this section, we cover the history of the cloud and edge computing models, provide detailed explanations of their current forms and features, and list the main technological enablers for their practical realization.

8.2.1 Brief history behind the cloud continuum concept

The cloud computing concept, that we know today, took several decades to come to fruition. The developments were driven by advances in processing technologies, communication networks, and the growing demand for scalable, on-demand resources [3,4]. The first foundations were set in the 1960s and 1970s when time sharing and virtualization were introduced as technologies for managing access to mainframes [5]. Time-sharing systems allowed multiple users to access a mainframe concurrently, thus laying the basis for a key concept of cloud computing – resource sharing. Next, technologies behind modern virtual machines (VMs) were introduced by IBM. The goal was to enable multiple software systems to run on the same underlying computational infrastructure.

The 1980s and 1990s marked the era of networking and the Internet. First came the emergence of client–server models where centralized servers provided resources and services to multiple clients. The development of ARPANET and its transition to the modern Internet in the 1980s and 1990s made it possible to access remote computing resources over wide-area networks. At that time, the concept of grid computing emerged which enabled pooling distributed computing resources to establish more capable processing environments. The first decade of the new millennium can be considered the era when the concept of cloud computing was introduced to the masses in its, more or less, current form and purpose. While the exact origin of the term is debated (some date back to 1996 when Compaq Computer Corporation introduced the term in their internal planning), Google CEO Eric Schmidt used it in a public context in 2006, referring to data and services accessed over the Internet, which had the first big impact of the cloud computing term on wider tech public. At that time, Amazon launched AWS Elastic Compute Cloud, with its ability to provide on demand scalable resources to the integrators. This marked the formal start of modern cloud computing as a commercial service and paved the way for other major tech companies to introduce their versions of the cloud computing model. By the mid-2000s, cloud computing was defined by characteristics like on-demand self-service, resource pooling, rapid elasticity, and measured service provision (pay-as-you-go).

At around the same time, two different ideas for placing resources closer to the end user to provide advantages in latency, bandwidth, and privacy, among others, were presented. The first was introduced by Mark Weiser in 1991, as ubiquitous computing [6], and then further explored by Noble, Satyanarayanan, and colleagues, renaming it pervasive computing [7]. The idea was to place computing-capable devices very close to the end users (e.g. in the same room) so that the objects we interact with themselves would work transparently to improve our lives. The second was introduced by Akamai in 1999, having launched the first Content Delivery Network, which in contrast to a centralized web server deployed web servers in various geographical locations to achieve lower response latency [8].

In the 2010s, the cloud ecosystem experienced exponential growth, with providers introducing concepts like Software as a Service, Platform as a Service, and Infrastructure as a Service that led to general adoption of the paradigm. Hybrid cloud models and multi-cloud strategies, combining private and public cloud infrastructures, came into focus of major organizations. With the proliferation of IoT-based automation services, the need for real-time processing and decision making raised a new set of requirements, which led to an increase in the popularity of edge and fog computing models. Combining edge and cloud computing paved the way for the cloud continuum which is redefining how distributed systems operate, particularly in mission-critical sectors like IoT, autonomous vehicles, and energy.

Regarding present and future trends, cloud computing is subject to constant innovation targeting improved performance, efficiency, and approachability for developers and integrators. Cloud-native applications and services are becoming a new paradigm in designing and implementing digital systems with scaling capabilities required by complex AI solutions. This is heavily enabled by technologies like containers and serverless computing. Finally, cloud providers are now focusing on energy efficiency and sustainability of their massive infrastructures aimed at achieving carbon neutrality and net zero emissions in the near future, while some studies indicate that cloud computing and data centers will continue to use more and more energy, leading to 8% of global carbon emissions by 2030 [9].

8.2.2 Cloud computing

Cloud computing became a mainstream technology in recent years, enabling access to ubiquitous, highly available, and elastic services to businesses of all sizes. The paradigm made access to all kinds of media (e.g. documents, presentations, photographs, videos, or other raw formats) widely accessible due to the immediate, continuous, and configurable service level agreement. The storage, structuring, and indexing of Petabyte scale data volumes for efficient access and processing was a major business opportunity, but also a great challenge for cloud service providers. At the same time, the high-availability standards and competitive offerings for cloud storage services and processing capabilities allure an increasing number of businesses to migrate their systems into the cloud, reducing maintenance and ownership costs. This adds up not only to the sheer volume of data to be managed by cloud

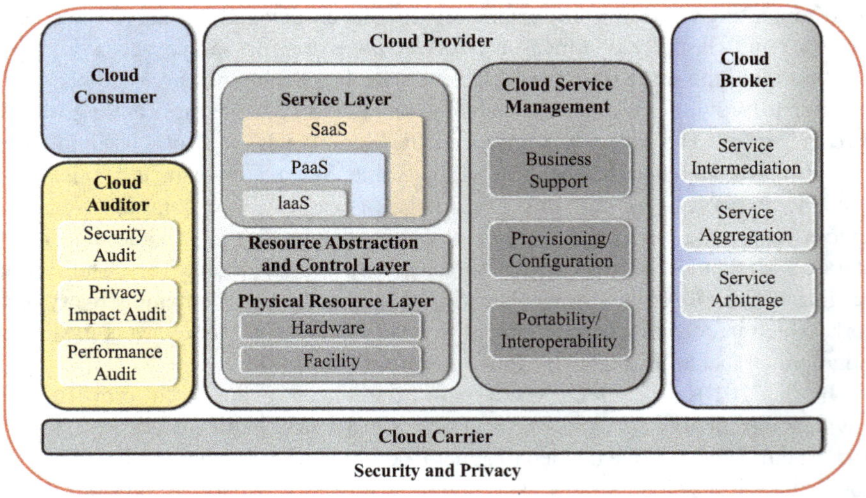

Figure 8.1 NIST cloud computing reference model

providers, but also to the diversity of solutions that are currently offered. Figure 8.1 presents a reference model for cloud computing introduced by the National Institute of Standards and Technology (NIST) in Special Publication 500-292 "NIST Cloud Computing Reference Architecture." It showcases that the cloud computing ecosystem is very complex and goes beyond data centers and virtualization services. Cloud providers are central entities in this model with their service offerings executed over resource abstraction and control enablers for efficient and scalable utilization of interconnected physical resources. Opening cloud services to the wider public requires a complex set of control and management services. Public access also entails strict regulation and auditing. Finally, a wide range of standards and regulations for the security and privacy of cloud systems are introduced (ISO/IEC 27017, ISO 27018, NIST 800-53), ensuring that the complex distributed systems are properly addressing cybersecurity risks and handling sensitive data in line with rules and regulations.

Expectations are that by 2025, 49% of all data in the world may be stored in public cloud environments [10], together with data structuring and processing engines such as databases (regardless of their primary data paradigm). The reason why engines as databases are often placed in cloud data centers is that they allow delegating hardware management and a variable degree of software responsibilities to the cloud provider. Furthermore, the cloud provides a seemingly infinite source of resources, which are elastically assigned to an application based on its workload at each point in time. There are, however, some downsides to cloud offloading. Since cloud data centers are often at a significant distance from the end devices that produce and generate data, applications can be affected by the consequent network latency [11]. In addition, failure or interruption in network connectivity between end-devices and the cloud can either lead to interruptions of service [12] or to an even greater latency penalty, for the case of geo-replicated scenarios, where requests are diverted to a different data center [13]. Finally, delegating data to a public cloud environment or third party can be considered a liability in terms of privacy [14,15]. Private cloud data centers are a possible alternative, but not all companies possess the scale to justify that approach [16].

8.2.3 Edge and fog computing

Edge and fog computing are a rising alternative to systems that purely offload data and tasks exclusively to a cloud provider. The closer placement of computation and storage nodes to data-generating devices, provides several advantages: shorter distance to the origin of requests, lower latency responses, and possibly increased privacy. The terms edge and fog are often mentioned in parallel with the cloud; the expression edge actually predates the mainstream cloud concept, pertaining to a location closer to end devices where tasks are executed. The distinction between edge and fog computing varies in three main definitions.

Figure 8.2 depicts the scenario and assumes the edge and fog computing to be disjoint concepts [17–19]. The edge computing paradigm places systems to move all tasks to edge nodes, effectively replacing the centralized resources with the edge. The fog computing paradigm combines the cloud with edge nodes to extract advantages from both planes. Other definitions assume no difference and thus both can be used interchangeably [20,21] representing the computation that occurs beyond a centralized location. Another approach considers edge, fog, and cloud as hierarchical layers that co-exist and differentiate by their placement in the network [22,23]. Distance to the end devices increases as we go up to the cloud, as do the amount of resources. This definition considers the fog as an intermediate layer in terms of distance and resources.

Both edge and fog computing place resources closer to end devices, overcoming the limitations of centralization. Developers must decide how to best distribute storage and processing tasks among planes. Depending on use case characteristics, users must choose, for each layer, whether to keep data temporarily or permanently, with what granularity (e.g. raw, filtered, or aggregated), the level of consistency to ensure between layers, and where to perform processing tasks such as filtering and query optimization.

Edge and fog computing have renewed their popularity in the past few years [24,25], emerging as an alternative or complement to the cloud that can help tackle evolving application demands. More capable edge hardware enables more complex tasks to be undertaken by small factors and low-powered devices that can easily be deployed at the edge plays an important role in the growing interest. The focus for this

Figure 8.2 Edge definition as an option to the cloud

environment is often centered on connectivity and its impacts on the energy footprint of the devices, but little attention is devoted to data processing and storage. Moreover, the storage plane is often neglected in favor of the processing plane. Most often the edge is used to produce, filter, aggregate, and compress data [18,19,26] and to produce automated responses according to most recently read values. Storage and more complex data analysis are delegated to the cloud due to its vast amount of resources and ready-to-use software solutions. As identified in the literature [17], placing data at the edge should take into consideration four main axis:

- **Privacy**: Data protection acts such as General Data Protection Regulation (GDPR) [27], Lei Geral de Proteção de Dados [Brazil's General Personal Data Protection Law] (LGPD) [28], South Africa's Protection of Personal Information Act (POPIA) [29], Canada's Personal Information Protection Act (PIPA) [30], and Swiss Data Protection Act (DSG) [31] are a growing requirement for data privacy. Often, the term privacy refers to authentication or encryption, i.e., security concepts. From a security perspective, edge nodes are more susceptible to physical attacks, and constrained resources equate to limitations in supported security algorithms [32]. However, the edge also provides an important advantage in data placement. The ability to deploy databases in fog/edge nodes massively increases the options for geographical location and infrastructure ownership. This provides greater flexibility in scenarios where data is under strict regulations or where user privacy preferences are of concern [14].
- **Latency**: Lower latency may lead to a better user experience or provide an important advantage over competitors, thus justifying placing databases at the edge [33–35]. However, quantifying the minimum latency gain that constitutes an advantage is non-trivial. The difference between edge-enabled and cloud-only latencies may be very small, making the decision to place the database at one location or the other very difficult [21].
- **Scalability**: Data is being generated at an unprecedented scale, and IoT applications are expected to generate up to 80% of all data by 2025 [36]. Uploading it all to the cloud would cause major network bottlenecks [37]. Storing data at the edge; applying deduplication, filtration, and aggregation; and only uploading data on an as-needed basis mitigates this bottleneck and contributes to reducing the amount of computation to be done at cloud servers, allowing systems to scale further.
- **Energy efficiency**: In scenarios where edge nodes are battery powered, and communication is one of the most energy-consuming stages [38], diminishing the number of messages exchanged equates to a longer lifetime of the edge nodes' battery. Keeping data stored in the nodes that generate it, and only retrieving it on a per-query basis, can lead to a significantly longer lifetime for the sensors.

There are other computational paradigms that fall under the umbrella of edge computing [39]:

- **Cloudlet model** recognizes three system layers – mobile device, cloudlet, and cloud. In this model, the cloudlet is acting as a computation and storage off-loader for mobile devices and logically exists between the cloud services and mobile devices.

Table 8.1 Comparison of computational models and key characteristics

Characteristic	Cloud computing	Fog computing	Edge computing
Deployment	Centralized	Distributed	Distributed
Location awareness	Not aware	Aware	Aware
Resources (physical)	Unlimited	Limited	Limited
Distance to data	Far	Close	Very close
Response time	Slow	Fast	Very Fast

- **Dew computing model** includes end-user computers and mobile devices into the edge computing frameworks.
- **Mist computing model** pushes data processing to sensors and actuators.
- **Osmosis computing model** uses the opportunistic placement of data processing micro-services at the edge or at the cloud depending on the load/intensity.

8.2.4 Cloud continuum

Let us start by comparing the three main computational models and their characteristics [40] (Table 8.1).

It is not uncommon to see these computational models as competitors, but the truth is that they complement each other perfectly. Only with their combined capabilities the modern AI-based decision making systems can be enabled. Hence, the cloud continuum, as the model that combines the edge and cloud computing, is the new norm in computational systems. This continuum is powered by key technological enablers such as virtualization, containers, 5G (and beyond) connectivity, edge AI hardware, and intelligent orchestration platforms [39,40]. It provides the foundation for innovative applications in industries like energy, healthcare, manufacturing, and smart cities by optimizing resource allocation and enhancing performance. The continuum has the potential to break down the inherited technological siloes of the cloud and edge computing systems, ensuring the complementary features can be properly utilized, especially for training complex AI models and real-time decision making based on them.

The cloud continuum has a unique feature reflected in the dynamic allocation of computational tasks to the cloud or edge resources based on the expected workload and latency requirements. The cloud computing infrastructure would take on computationally intensive and data-hungry tasks like AI model training and validation. Conversely, fog and edge computing would be responsible for performing AI inference closer to the data sources and real-time decision making required by latency-sensitive tasks. The resource utilization is also optimized in the continuum. Data replication is minimized, and transferring unnecessary data to remote locations can also be avoided while maintaining the same level of contextual insight into the problem space at hand. This saves both storage and communication bandwidth. Finally, the cloud continuum improves security and privacy by allowing sensitive data to remain at the edge devices, while still enabling cloud-based insights. The continuum systems are better aligned with strict privacy regulations like the GDPR since

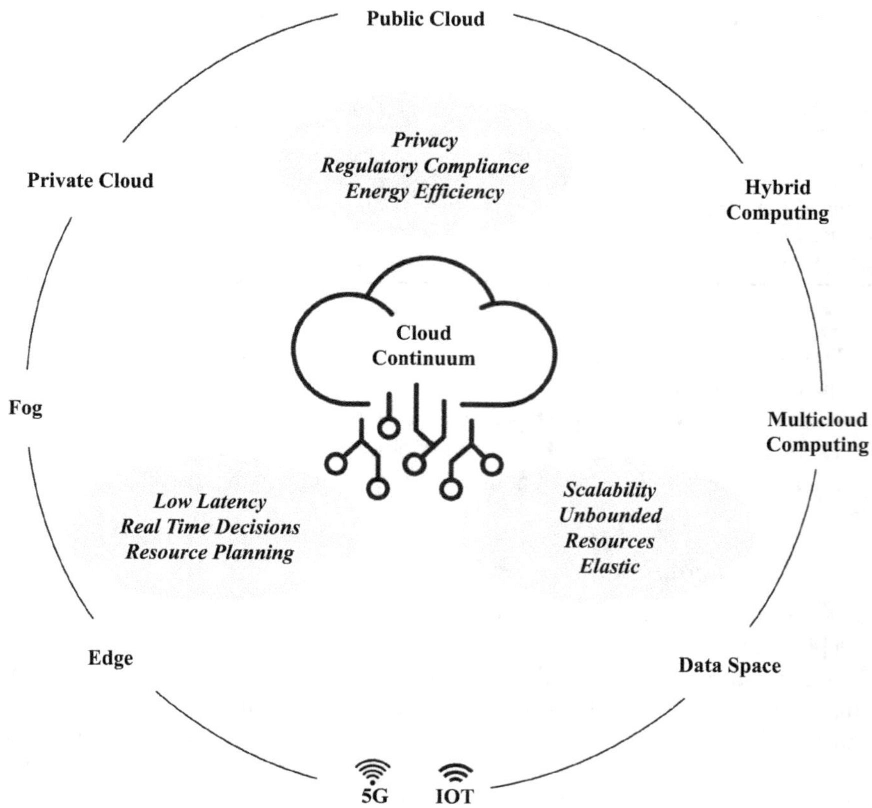

Figure 8.3 High level illustration fo the cloud continuum

they can ensure that data stays at the source and maximally avoid sending sensitive data to remote locations.

Figure 8.3 depicts the landscape of the cloud continuum paradigm.

8.2.5 Technological enablers for the cloud continuum

The cloud continuum is made possible by a set of enabling technologies for high-performance communication, resource and system virtualization, predictive and reactive workload distribution, and ever-increasing computational and data storage capabilities of both servers and edge devices. Key enablers include [41–43]:

- **Virtualization and containers** VMs provide hardware abstraction, enabling multiple software services/systems to run on shared infrastructure. Containers offer lightweight, logically portable environments for deploying services across cloud and edge.

- **Distributed computing frameworks** technologies like Apache Kafka, Apache Spark, and KubeEdge support the processing and management of distributed workloads across the continuum.
- **Hybrid cloud architectures** Tools like VMware, OpenShift, and Anthos support seamless operation across private, public, and edge clouds. Micro-service technologies enable flexibility in deploying and scaling applications across the continuum.
- **Networking enablers:**
  - **5G** high-speed, low-latency connectivity enables real-time communication between and across edge and cloud infrastructures.
  - **SDN (software-defined networking)** Ensures dynamic routing and efficient resource allocation in communication networks.
  - **Edge Gateways** Bridge local devices and cloud systems and represent optimal points for handling data pre-processing and filtering.
- **AI for workload orchestration** enables predictive and reactive workload placement and dynamic resource optimization across cloud and edge.
- **Federated learning** allows AI models to train collaboratively across distributed edge and cloud environments with global and local model syncing logic. An important feature of this AI model training paradigm is data privacy preservation.
- **Hardware accelerators placed at the edge** solutions like NVIDIA Jetson or Google Edge TPU are capable of performing AI inference on end devices.
- **IoT devices and controllers** act as endpoints in the continuum, generating and consuming data for localized AI processing.
- **Data lakes and streams** enable the collection, transformation, and integration of data from multiple sources.
- **Data spaces** ensure data sovereignty, provenance, and trustworthiness across the continuum and between stakeholders offering and consuming data.
- **Zero trust architectures** Ensure secure interactions across distributed systems.
- **Homomorphic encryption and secure multiparty computation** Allow sensitive data processing across the continuum without compromising privacy.
- **Interoperability enablers** syntactic and semantic interoperability are seen as key enablers for executing cloud continuum models across cloud and edge systems of different vendors and stakeholders. Cross-domain interoperability is also an important enabler for the continuum deployment in complex systems like the energy grid. The tech domain is introducing standards for interoperability: EdgeX Foundry, Cloud Native Computing Foundation, and OpenFog. Projects like H2020 InterConnect [44] successfully address cross-domain interoperability challenges between cloud and edge/IoT systems of the energy and smart building domains.

8.3 Applications in the energy sector

The energy sector is subject to massive research and innovation efforts across the globe. Resolving energy system stability and addressing the ever-increasing demand is one of the main tasks set for the whole society. The complexity of the electric grid cannot be overstated. Renewable sources, electric vehicles, microgrids, distributed energy storages, energy communities, smart appliances, and other concepts introduced as possible solutions for the energy crisis are posing challenges for proper service delivery and efficient resource utilization. The regulations are ever more stringent, putting pressure on the grid operators and large energy consumers (like data center operators) to lower energy consumption, increase usage of renewable energy and drastically reduce carbon emissions. Ensuring grid stability, performing real-time monitoring, and properly managing distributed energy resources (DERs) are actively being tackled with the state-of-the-art technologies like AI, blockchain, and cloud continuum. The complexity of the electric grid systems is well illustrated in the CEN/CENELEC Smart Grid Architecture Model (Figure 8.4) [45]. It illustrates the multi-dimensionality of the problem space.

The energy sector innovators are introducing new solutions to help address the ever-increasing complexity of the energy system challenges. The answer lies in the data, which means that the main challenge of the energy system, as is the case for all complex systems nowadays, is efficient management of heterogeneous data streams and derivation of actionable insights through big data analytics. Innovative applications and services in the energy sector are the subject of other chapters of this book. Here, we briefly list the main trends/challenges from the energy domain which can benefit from proper integration of the cloud and edge computing models:

- Management of DERs and their effective inclusion into a wider grid.

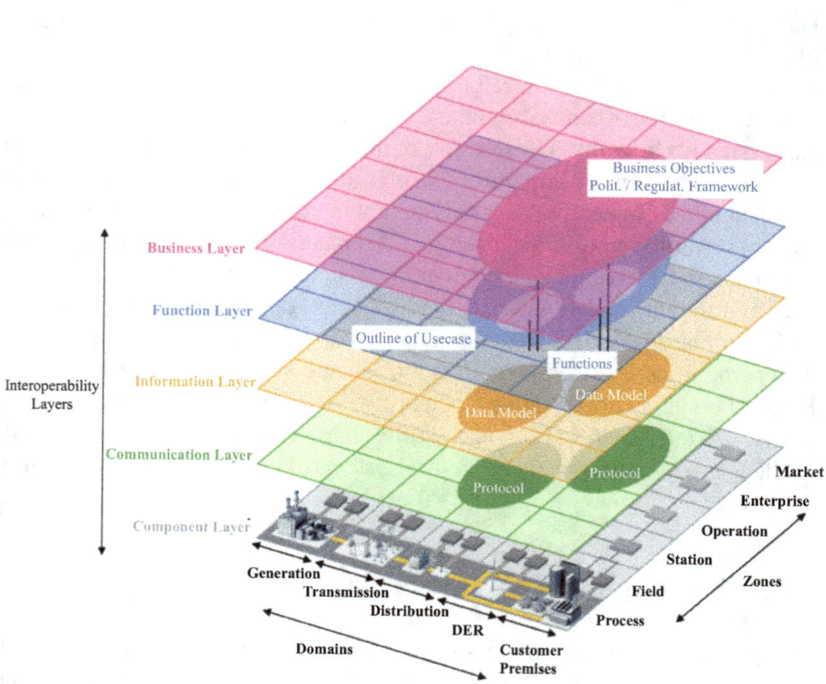

Figure 8.4 Smart grid reference architectural model

- Electric vehicle integration through vehicle-to-grid technologies and as mobile energy storage systems.
- Management of distributed energy storage with batteries and heating systems.
- Real-time monitoring of consumption patterns and consumption profiling through smart meters and IoT systems.
- Predictive maintenance of transmission and distribution networks based on historic data and real-time monitoring.
- Demand-side flexibility aggregation and management relying on smart appliances and home and building energy management systems.

AI is capable of streamlining the integration of DERs into energy grid operations by providing predictive and reactive methods for planning, real-time monitoring, and resource/service management. A comprehensive review in [46] highlights that AI has a transformative potential for grid operation. Where there is a need for AI-based decision making, there is a need for powerful and scalable computing infrastructure capable of handling heterogeneous data streams and analyzing massive amounts of data.

8.3.1 Why the cloud and edge computing matter in the energy sector?

We argue that the diverse systems that represent smart grids are a solid basis for implementing cloud continuum paradigms. First, we will present the main characteristics of the smart grids problem space which can be tackled with the cloud and edge processing – in line with previously described roles of cloud computing (storing and crunching massive amounts of data with unbounded resources) and edge computing (analyzing data at the point of origin, minimizing latency, and preserving privacy).

- Smart grids generate massive amounts of data thanks to the smart metering infrastructure and other real-time monitoring techniques. A single smart meter recording energy usage every 15 min produces 96 data points per day. In an energy grid with 1 million smart meters, this equates to 96 million readings daily [47]. Large amounts of smart metering data allow utilities to perform detailed analyses of consumption patterns, understand consumer and prosumer behavior, and manage the grid accordingly. Handling such large quantities of data presents a challenge for storage, analytics, and general scalability of the decision making in the grid systems [48]. And this covers only the smart metering systems. The modern smart grid operators are seeking to collect and manage data streams from various DERs and wider ecosystem elements (like smart buildings, smart cities, and transportation systems) to get an even deeper understanding of the consumption trends and to properly integrate DERs into the grid balancing and stability operations.
- Latency in data transfer, data processing, and decision-making can significantly impact the performance and reliability of smart grids by delaying critical actions, thus impacting stability and efficiency. Inability to act in real-time can cause power outages and cascading grid failures. Delayed decisions in demand response systems can lead to poor DERs integration, unbalanced grid loads, and

potentially even outages. According to a study [49], minimizing latency in smart grid communication and control systems is vital for enabling rapid and real-time decision making required by highly dynamic distributed systems.

- Data privacy is a significant concern in smart grid systems. Smart metering data can infer energy consumption patterns, which in turn can be used to recognize consumer behavior patterns. Personal information about consumers, such as daily routines, appliance usage, and occupancy status, can potentially be derived from smart metering data [50]. The integration of advanced metering infrastructure increases the vulnerability of consumer data to cyberattacks and unauthorized surveillance. The edge processing capabilities of smart buildings and smart home systems should be leveraged to mitigate this risk.

Let us also list some characteristics of the smart grids and their sub-systems, which provide a solid basis for the realization of the cloud continuum enablers (illustrated in Figure 8.5).

- DERs (renewable energy sources, energy storage systems, microgrids, home and building energy management, electric vehicles) usually include digital subsystems capable of conducting computational and data generation/management tasks. There are various sensory, actuation, and on-premise control modules integrated into DER systems and these are interconnected with centralized control and monitoring systems. Their edge computation potential is substantial.

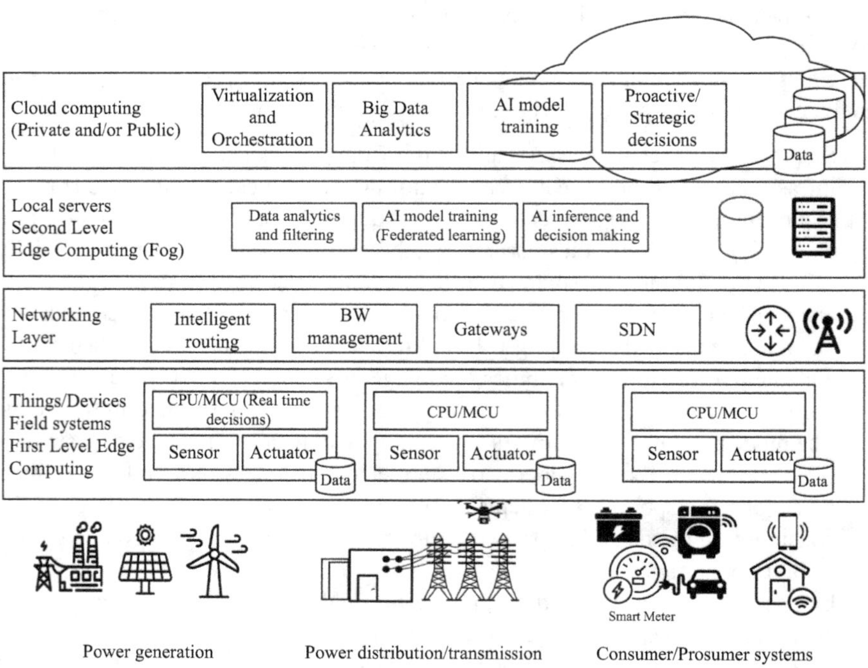

Figure 8.5 High-level depiction of the computational and system layers enabling smart grid services

- Smart grid operators migrate many of their digital services toward private and public cloud infrastructures. The big data nature of complex smart grids is widely recognized, and the main stakeholders in the domain are readily using scalable data storage and computation capabilities of cloud systems [48].

With that in mind, the need and potential for both cloud edge processing is evident. The cloud continuum has potential to address energy grid requirements, which are the result of a highly distributed infrastructure and the dynamic nature of consumption and production loads. Once more we argue that proper integration of DERs requires complex data collection and analysis mechanisms and infrastructure, which can be supported only with cloud continuum computational models. Cloud computing supports collection and storage of vast amounts of grid data (e.g. smart metering data, renewable generation forecasts, market trends, and system performance metrics) and using them for AI model training and validation [51]. Being able to store historical data and crunch them in search of patterns allows grid operators to execute advanced management mechanisms like predictive maintenance, load balancing relying on demand-side flexibility and integrated DERs, and intelligent investment planning. Managing risks in a predictive manner is a key for a stable energy grid without outages [52].

Edge computation allows grid operators to process collected data locally resulting in real-time decision making and grid operation adjustments. This immediate response to unforeseen grid conditions is the biggest challenge of the smart grid systems. It requires complex orchestration of deployed edge and fog computing and data collection resources and also relies on other adjacent systems (e.g. IoT systems, smart buildings, and EV fleet management) for quick inference of field conditions and execution of corrective actions. Sudden power fluctuations (more and more prevalent with ever-increasing number of DERs) and grid faults require near zero latency in decision making so that the grid remains stable [53].

Being able to analyze data closer to the source (e.g. smart building energy management system access and analysis of smart metering data) improves privacy protection in smart grids making them less prone to regulatory penalties. The ability to scale resources dynamically across the continuum ensures that grids can adapt to changing energy demands and expand infrastructure efficiently as energy ecosystems grow [54].

8.3.2 *Case for demand-side flexibility management*

The integration of cloud and edge computing in smart grids significantly enhances demand-side flexibility management (challenge is illustrated in Figure 8.6) by combining analytical insight from centralized data processing with localized, real-time decision-making. Cloud computing systems are capable of analyzing extensive data collected from various grid points, including smart appliances and other flexibility-offering systems. This facilitates the development of predictive models and optimization strategies for demand-side flexibility management. Edge computing processes data near its source, allowing for immediate responses to local events and reducing latency, which can be critical for proper aggregation and execution upon flexible

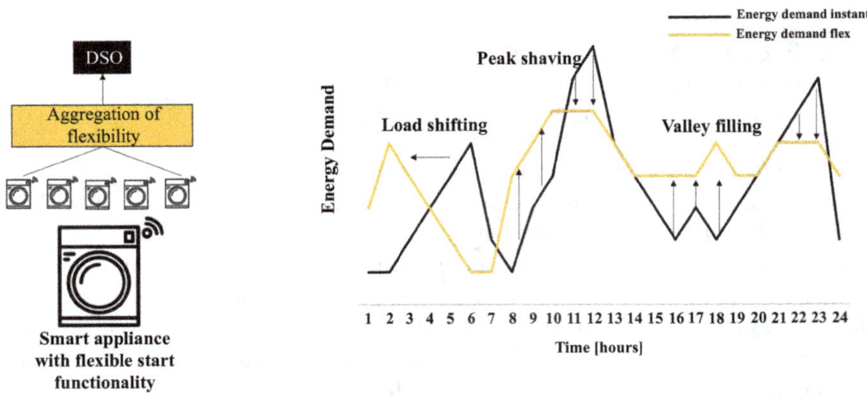

Figure 8.6 Concept of demand-side flexibility aggregation for load balancing tasks

resources [55]. This synergy ensures proper alignment between distribution and real-time grid demands. The demand-side flexibility is largely an untapped potential in the management of energy systems. With the proliferation of smart appliances and other connected DERs, there is a large pool of flexible resources ready for exploitation. The simplest example of demand-side flexibility can be a delayed start of a washing machine cycle. The end user can indicate that the washing machine is ready to start and that the cycle must finish in the next 8 h. This information can be used by the energy management system as a resource to be offered to the DSO, or a wider energy market. Aggregating such flexibility on a building, community, or even city level provides great opportunities in the energy market and for the grid operators to balance and manage resources more efficiently. However, the challenge lies in collecting a large amount of data coming from heterogeneous systems, analyzing it for aggregation strategies, and adapting to the real conditions in real time.

The study conducted in [56] highlights the benefits of integrating smart energy management systems with IoT devices (e.g. smart appliances and home and building management systems), indicating that this integration enables efficient management of demand-side flexibility by collecting energy consumption data from smart devices (from homes and industry) and relying on both edge and cloud computation to infer insights from the data. Finally, research conducted within [57] emphasizes potential for data privacy protection through cloud continuum while ensuring that all crucial insights represented within privacy-sensitive data are inferred and used in demand-side flexibility aggregation and execution.

8.3.3 Case for predictive maintenance

The integration of cloud and edge computing facilitates AI-based predictive maintenance in smart grids by combining centralized data analysis with real-time, localized processing. Storing massive amounts of collected grid data in scalable cloud storage supports the development of sophisticated AI models that can predict equipment failures and maintenance needs across the entire grid infrastructure. At the same time, edge-based decentralized processing reduces latency and enables real-time interventions that prevent equipment failures and enhance grid reliability even in the face of unforeseen conditions.

An anomaly detection-based power-saving scheme is introduced in [58] for predictive maintenance using edge computing (ADEPOS). The research postulates that by employing approximate computing throughout a machine's lifecycle, it is possible to achieve significant energy savings while maintaining predictive maintenance capabilities. This approach highlights the effectiveness of edge computing in reducing data transmission and energy consumption, which are critical factors in the efficient operation of smart grids.

8.4 Challenges and limitations

The cloud and edge computing systems are key enablers for any data-intense domain space, including the energy grid. However, integrating such technologies into legacy systems while maintaining overall stability, integrity, and security is a costly challenge. In order to perform a good risk assessment and SWOT analysis, there are a number of limitations and open issues behind the cloud continuum that need to be understood.

8.4.1 Note on standardization

Overall, a clear standard in what refers to system evaluation is lacking in the context of edge computing and, consequently, the cloud continuum. The assessment of data usage considers workloads that are not realistic, and considers few nodes at each processing layer. There are no suitable benchmarking solutions for the presented application domains [17], which leads evaluations to differ considerably, even among publications targeting the same application domain. In dense scenarios such as sensor networks, node simulations, and emulation tools are used due to the high number of nodes involved, but those simulations often only accurately simulate certain aspects of a node, e.g. its network communication component.

- Research should develop relevant benchmarks focusing on database performance evaluation for the application domains presented, workload traces in some publications provide valuable insights. The availability of more workload traces would also benefit system evaluation.
- Research should target the development of specialized simulation tools that enable node simulations that are both accurate and able to scale to significant sizes.
- Research should compare newly developed systems against already existing systems targeting the same application domain.

Other technologies could be analyzed as to their applicability to edge-enabled data management systems, e.g. delegating databases tasks, such as transaction execution, to network devices, a technique currently being explored on cloud infrastructure [59], and the use of persistent memories on edge devices, as faster mediums for write operations requiring persistence [60,61].

8.4.2 Note on security and privacy

Cloud computing systems with their centralized data management have inherent data privacy challenges. Also, security of such systems is subject to many standards and regulations. Here, we will focus more on the case for edge computing. Privacy at the edge holds two perspectives. First, edge devices present a security disadvantage, as they are deployed in locations with greater physical exposure, and their often limited resources could compromise the quality of algorithms used. Second, they present an advantage from the data placement perspective. Multiple layers holding data at increasingly shorter distances from data sources allow data to be stored closer to its origin, and with different degrees of data aggregation between layers. Thus, the edge can help answer data regulations restricting the location of data, as well as empower data owners with greater control over how their data is handled. End-devices in many edge applications carry detailed and sensitive information. For example, smart meter information may allow attackers to infer various living habits, such as current TV channels, and specific audiovisual content being watched. Nonetheless, only six surveyed solutions incorporate or mention the use of authentication and encryption mechanisms [12,21,32,38]. Furthermore, hardware limitations and the increased physical exposure of devices to possible attackers present increased difficulty in ensuring the security of data being received, processed, and transmitted on these devices. Future research should consider authentication and encryption mechanisms as a core element of their designs. Furthermore, the trade-off between computational efficiency and the resilience of authentication and encryption solutions, in edge database environments, should be better understood.

8.4.3 Note on latency

With latency often being an argument for placing computation and storage at the edge, the difference between cloud and edge deployments is sometimes only a few milliseconds. The latency gain between edge-enabled and cloud-only deployments is not clear. Users expect increasingly lower latency values. Statistically, a difference of 250 ms or more between competing web services and search-engines can provide an advantage in user adoption, but this may have a different impact depending on the use-case. For interactive systems, as map panning applications, the expected latency value can be as low as 100 ms. For autonomous vehicles, 100 ms is the time a vehicle has to react.

- Research should compare cloud, edge, and fog deployment latency to measure the latency gain incurred for data at the edge, in systems using latency as their main driver.
- Research should establish a common methodology for latency evaluation, enabling direct comparison of competing solutions' latency.

- Research should focus on quantifying latency requirements in critical systems, and for competitive advantage purposes, to understand required latency improvement needed to provide an actual impact, in reaction time for critical systems, and response time.

8.4.4 Note on energy efficiency of the cloud and edge computing

Several studies currently explore the possibility of measuring, collecting, and analyzing the power profile of applications, particularly those that are central processing unit (CPU) or GPU intensive, or both, which are running on servers in data centers. [62] compares different software-based power meters for CPU and GPU processors to evaluate their performance, concluding that it is not trivial to choose the right tool for a given application, as each tool has its own strengths and weaknesses. However, they also found that some tools perform better than others in certain scenarios. Two main interfaces to collect energy profiling dominate the available proposals as the underlying source of information, namely running average power limit (RAPL) and performance application programming interface.

While previous works cover the collection of energy profiling information on local hardware, the plethora of cloud providers and services running on top of virtualized computing resources led Khan *et al.* [63] to evaluate the effectiveness of RAPL interface for measuring power consumption in cloud computing servers, concluding that is a powerful tool for measuring power consumption in distributed computing systems, with promising accuracy and granularity, but still sensitive to system noise. Moreover, it shows a lack of support for certain hardware components, requiring further improvements. Murwantara and Yugopuspito [64] discuss the impact from running workloads in cloud computing services by measuring the energy usage of three different virtualization technologies, namely virtual machines, Docker, and Kubernetes in OpenStack – an open-source cloud computing environment. Docker, virtual machines, and Kubernetes respond differently to a similar workload that influences the consumption of energy. Particularly regarding Docker, Murwantara and Yugopuspito [64] discusses on power usage of Docker, concluding that there is a significant increase in energy consumption. It is important to keep in mind that despite the increased power consumption of using virtualization systems, a testing tool using these profilers should be able to isolate what is the actual consumption of the applications running within Docker. As long as a common reference is applied, consecutive tests will render comparable results. In Murwantara and Bordbar [65], the authors propose a method of measurement of energy consumption in the virtualized environment such as the cloud or laptops with virtualization support and show that it is possible to measure energy consumed by individual processes, components, servers, and applications using PowerTOP [66] that collects data from Advanced Configuration and Power Interface.

Edge nodes are said to be energy limited. Energy limitations are not well understood. While a battery-powered node needs to prolong lifetime, many nodes such as gateways, points of presence, roadside units, sink nodes, and others are usually powered by main power in surveyed publications, without compelling proof that the case will be the same in real-life deployments, as they may be limited by energy harvesting scenarios. Not all sensor network-oriented use cases take into account the energy-limited nature of sensors, namely the SN-SN and SN-SE use cases.

- Research should expand energy consumption concerns to other use cases, such as in sensor networks. Energy limitations of other types of devices, e.g. points of presence, gateways, and sink nodes should be evaluated, and specific energy optimization mechanisms be developed.

8.4.5 Note on interoperability

Interoperability is a significant challenge in the energy domain due to the complexity and diversity of systems, devices, and stakeholders involved in modern energy networks. Energy ecosystems comprise a wide range of components, including generation assets (e.g. solar panels, wind turbines, and traditional power plants), distribution and transmission infrastructure, smart meters, IoT systems, and energy storage systems. These components are often developed by different manufacturers using proprietary technologies, protocols, and different standards, creating compatibility issues. Since the cloud continuum model requires utilization of the edge resources specific to the domain, the lack of interoperability and vendor lock-ins are inherited from the energy systems. This makes it difficult to effectively and efficiently utilize cloud and edge processing resources. Additionally, regional variations in regulations and standards further complicate interoperability. For example, grid communication protocols may vary across countries or even states, making it difficult to scale solutions globally. The lack of a unified framework for data formats, APIs, and communication standards leads to inefficiencies, higher integration costs, and delayed adoption of innovative technologies like AI, IoT, and the cloud continuum. Most digital systems and platforms from the energy domain are expected to be interoperable at a syntactic level. In reality, this is not the case. They might use the same communication protocol and common data formats, but data models are usually proprietary. Figure 8.7 depicts interoperability levels recognized by the GridWise Architecture Council.

Once syntactic interoperability is achieved, the next level is semantic interoperability. Semantic interoperability is the ability of digital systems to exchange data with unambiguous, shared, and agreed meaning. The meaning of data is set in an ontology that is accepted by all parties participating in data exchange. Semantic interoperability is a requirement to enable semantic reasoning and knowledge inference, knowledge discovery, and data federation between digital systems.

Addressing interoperability requires industry-wide collaboration to establish common standards and frameworks, enabling seamless communication, data sharing, and system integration. Without this, achieving the full potential of smart grids through proper utilization of the cloud continuum remains a formidable challenge.

8.4.6 Note on costs

While initial investments in infrastructure and integration may be significant, the cloud continuum enables utilities and energy companies to optimize their resources,

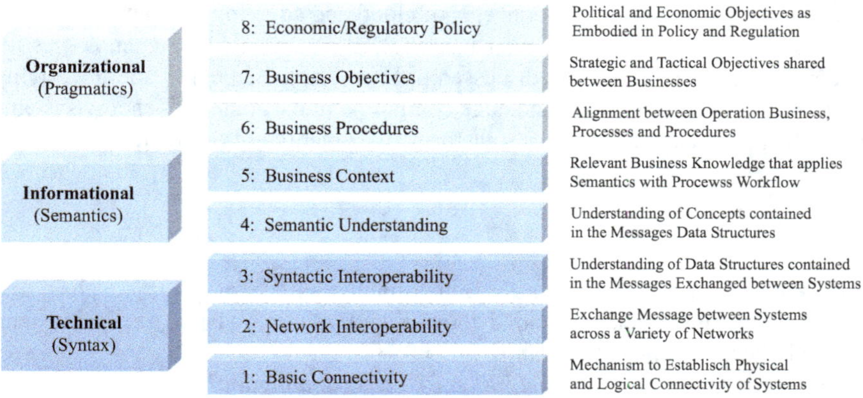

Figure 8.7 GridWise Architecture Council – levels of interoperability

reduce waste, and unlock new revenue streams, making it a compelling economic choice. By leveraging the scalability of cloud computing, energy companies can process vast amounts of data from IoT sensors, smart meters, and renewable energy assets without the need for substantial on-premises infrastructure investments into local servers and data centers. Edge computing minimizes bandwidth costs by performing localized data processing and real-time decision-making. Additionally, predictive maintenance powered by AI in the cloud continuum prevents costly equipment failures, while optimized grid operations and renewable energy integration improve overall efficiency and reliability. Challenges like upfront costs and managing distributed systems are significant and require detailed research for understanding the long-term savings, risks of process automation, overall impact on energy efficiency, and ability to scale with demand [67].

8.5 Conclusion

AI requires cloud and edge computing to address the diverse computational, storage, and latency demands of modern smart grids. Cloud computing provides the immense computational power and scalable storage necessary for training complex AI models on massive datasets, making it ideal for resource-intensive tasks such as deep learning and data analytics required for predictive maintenance of the grid. An important economic driver is cloud's pay-per-use model which allows fine tuning of costs and time/resource investments from the side of grid operators. Edge computing brings AI capabilities closer to data sources, enabling real-time inference with minimal latency, which is crucial for the energy domain. By preprocessing data at the edge, smart grid operators can address ever-increasing privacy protection demands from both end consumers and ruling regulatory bodies. Together, cloud and edge computing form a synergistic framework that supports the AI lifecycle, from model training in the cloud to low-latency deployment and inference at the edge, ensuring efficiency, scalability, and responsiveness needed by the smart grid systems. In this chapter, we have presented the basics behind the concepts of cloud and edge/fog computing and how they form the cloud continuum. We have reflected on the enabling technologies for achieving cloud continuum and discussed open research challenges and underlying issues. This chapter puts cloud and edge computing in the context of energy systems and applications and discusses the transformative potential of proper integrations.

References

[1] Gill SS, Tuli S, Xu M, *et al.* Transformative effects of IoT, blockchain and artificial intelligence on cloud computing: Evolution, vision, trends and open challenges. *Internet of Things*. 2019;8:100118.
[2] Moreschini S, Pecorelli F, Li X, *et al.* Cloud continuum: The definition. *IEEE Access*. 2022;10:131876–131886.
[3] Surbiryala J, Rong C. Cloud computing: History and overview. In: *2019 IEEE Cloud Summit*; 2019. p. 1–7.
[4] Arutyunov VV. Cloud computing: Its history of development, modern state, and future considerations. *Scientific and Technical Information Processing*. 2012;39(3):173–178.
[5] Denning PJ. Origin of virtual machines and other virtualities. *IEEE Annals of the History of Computing*. 2001;23:73.
[6] Weiser M. The computer for the 21st century. *Scientific American*. 1991;265(3):94–105.
[7] Satyanarayanan M. Pervasive computing: Vision and challenges. *IEEE Personal Communications*. 2001;8(4):10–17.
[8] Akamai. Company history; 2024. Available from: https://www.akamai.com/company/company-history (accessed: January 30, 2025).
[9] Cao Z, Zhou X, Hu H, *et al.* Toward a systematic survey for Carbon Neutral Data Centers. *IEEE Communications Surveys & Tutorials*. 2022;24(2): 895–936.
[10] Reinsel D, Gantz J, and Rydning J. Data Age 2025: The digitalization of the world from edge to core. International Data Corporation (IDC); 2018.
[11] Mizutani M, Tsukada M, Iida Y, *et al.* 3D maps distribution of self-driving vehicles using roadside edges. In: *Eighth International Symposium on Computing and Networking Workshops (CANDARW)*. IEEE; 2020. p. 40–45.
[12] Wu F, Qiu C, Wu T, *et al.* Edge-based hybrid system implementation for long-range safety and healthcare IoT applications. *IEEE Internet of Things Journal*. 2021;8(12):9970–9980.
[13] Lin Y, Kemme B, Patino-Martinez M, *et al.* Enhancing Edge Computing with Database Replication. In: *26th IEEE International Symposium on Reliable Distributed Systems (SRDS 2007)*. IEEE; 2007. p. 45–54.
[14] Sonbol K, Özkasap Ö, Al-Oqily I, *et al.* EdgeKV: Decentralized, scalable, and consistent storage for the edge. *Journal of Parallel and Distributed Computing*. 2020;144:28–40.
[15] Beedkar K, Brekardin D, Quiané-Ruiz JA, *et al.* Compliant geo-distributed data processing in action. *Proceedings of the VLDB Endowment* 2021;14(12):2843–2846.

[16] Zhang M. How Much Does it Cost to Build a Data Center?; 2022. Available from: https://dgtlinfra.com/how-much-does-it-cost-to-build-a-data-center/ (accessed: January 30, 2025).

[17] Meruje Ferreira LM, Coelho F, Pereira J. Databases in edge and fog environments: A survey. *ACM Computing Surveys* 2024;56(11):1–40.

[18] Yousefpour A, Fung C, Nguyen T, *et al.* All one needs to know about fog computing and related edge computing paradigms: A complete survey. *Journal of Systems Architecture*. 2019;98:289–330.

[19] Hong CH, Varghese B. Resource management in fog/edge computing: A survey on architectures, infrastructure, and algorithms. *ACM Computing Surveys*. 2019;52(5):1–37.

[20] Shi W, Cao J, Zhang Q, *et al.* Edge computing: Vision and challenges. *IEEE Internet of Things Journal*. 2016;3(5):637–646.

[21] Rahmani A, Gia T, Negash B, *et al.* Exploiting smart E-Health gateways at the edge of healthcare internet-of-things: A fog computing approach. *Future Generation Computer Systems*. 2018;78:641–658.

[22] Dustdar S, Avasalcai C, Murturi I. Edge and fog computing: Vision and research challenges. In: *IEEE International Conference on Service-Oriented System Engineering (SOSE)*. IEEE; 2019. p. 96–9609.

[23] Monga SK, Ramachandra SK, Simmhan Y, IEEE. ElfStore: A resilient data storage service for federated edge and fog resources. *2019 IEEE International Conference on Web Services (ICWS)*. 2019; p. 336–345.

[24] Satyanarayanan M. The emergence of edge computing. *Computer*. 2017;50(1):30–39.

[25] Google. Edge Computing Google Trend ; 2024. https://cloud.google.com/resources/2024-state-of-edge-computing-report (accessed: June 10, 2025).

[26] Sadri A, Rahmani A, Saberikamarposhti M, *et al.* Fog data management: A vision, challenges, and future directions. *Journal of Network and Computer Applications*. 2021;174:102882.

[27] European Union. General Data Protection Regulation; 2018. Available from: https://gdpr.eu/tag/gdpr/ (accessed: January 30, 2025).

[28] FR of Brazil. Lei Geral de Proteção de Dados Pessoais (LGPD); 2018. Available from: https://www.planalto.gov.br/ccivil_03/_ato2015-2018/2018/lei/l13709.htm (accessed: January 30, 2025).

[29] R. of South Africa. Protection of Personal Information Act; 2013. Available from: https://www.gov.za/sites/default/files/gcis_document/201409/3706726-11act4of2013protectionofpersonalinforcorrect.pdf (accessed: January 30, 2025).

[30] PIPCS Korea. Protection of Personal Information Act; 2011. Available from: https://www.privacy.go.kr/front/main/main.do (accessed: June 10, 2025).

[31] FMR of Austria. DSG; 2018. Available from: https://www.bmf.gv.at/en/data-protection.html (accessed: January 30, 2025).

[32] Wang H, Tan C, and Li Q. Snoogle: A search Engine for the Physical World. In: *27th IEEE International Conference on Computer Communications (INFOCOM)*. IEEE; 2008. p. 1382–1390.

[33] RI Systems. Retail Web Site Performance – Consumer Reaction to a Poor Online Shopping Experience; 2008. Available from: https://risnews.com/retail-web-site-performance-consumer-reaction-poor-online-shopping-experience (accessed: January 30, 2025).

[34] Brutlag J, Hutchinson H, and Stone M. User preference and search engine latency. In: *Quality and Productivity Research Conference*; 2008.

[35] Arapakis I, Bai X, and Cambazoglu B. Impact of response latency on user behavior in web search. In: *Proceedings of the 37th International ACM SIGIR Conference on Research & Development in Information Retrieval*. ACM; 2014. p. 103–112.

[36] Reinsel D, Gantz J, and Rydning J. Data Age 2025: The evolution of data to life-critical. Don't focus on big data; focus on data that's big. Int. Data Corporation (IDC); 2017.

[37] Alamouti S, Arjomandi F, and Burger M. Hybrid edge cloud: A pragmatic approach for decentralized cloud computing. *IEEE Communications Magazine*. 2022;60(9):16–29.

[38] Tsiftes N, and Dunkels A. A database in every sensor. In: *Proceedings of the 9th ACM Conference on Embedded Networked Sensor Systems*; 2011. p. 316–332.

[39] Bittencourt L, Immich R, Sakellariou R, *et al.* The internet of things, fog and cloud continuum: Integration and challenges. *Internet of Things*. 2018;3–4:134–155. Available from: https://www.sciencedirect.com/science/article/pii/S2542660518300635.

[40] Al-Dulaimy A, Jansen M, Johansson B, *et al.* The computing continuum: From IoT to the cloud. *Internet of Things*. 2024;27:101272. Available from: https://www.sciencedirect.com/science/article/pii/S2542660524002130.

[41] Taleb T, Samdanis K, Mada B, *et al.* On multi-access edge computing: A survey of the emerging 5G network edge cloud architecture and orchestration. *IEEE Communications Surveys & Tutorials*. 2017;19(3):1657–1681.

[42] Rosendo D, Costan A, Valduriez P, *et al.* Distributed intelligence on the edge-to-cloud continuum: A systematic literature review. *Journal of Parallel and Distributed Computing*. 2022;166:71–94. Available from: https://www.sciencedirect.com/science/article/pii/S0743731522000843.

[43] Liang B, Gregory MA, and Li S. Multi-access edge computing fundamentals, services, enablers and challenges: A complete survey. *Journal of Network and Computer Applications*. 2022;199:103308. Available from: https://www.sciencedirect.com/science/article/pii/S1084804521002976.

[44] E. Commission. InterConnect Project; 2019. Available from: https://cordis.europa.eu/project/id/857237 (accessed: January 30, 2025).

[45] IEC. SGAM Diagram; 2015. Available from: https://syc-se.iec.ch/deliveries/sgam-basics/ (accessed: January 30, 2025).

[46] Kumar NM, Chand AA, Malvoni M, *et al.* Distributed energy resources and the application of AI, IoT, and blockchain in smart grids. *Energies*. 2020;13(21):5739. Available from: https://www.mdpi.com/1996-1073/13/21/5739.

[47] Daki H, El Hannani A, Aqqal A, *et al.* Big data management in smart grid: Concepts, requirements and implementation. *Journal of Big Data*. 2017;4(1):1–19. Available from: https://doi.org/10.1186/s40537-017-0070-y.

[48] Zhang Y, Huang T, Bompard EF. Big data analytics in smart grids: A review. *Energy Informatics*. 2018;1(1):1–24. Available from: https://doi.org/10.1186/s42162-018-0007-5.

[49] Matera F, Di Bartolo S, and Nastri E. Nonlinear interaction length for transmissions based on subcarrier multiplexing. In: *20th Italian National Conference on Photonic Technologies (Fotonica 2018)*; 2018. p. 1–4.

[50] Deng S, Zhang L, and Yue D. Data-driven and privacy-preserving risk assessment method based on federated learning for smart grids. *Communications Engineering*. 2024;3(1):154. Available from: https://doi.org/10.1038/s44172-024-00300-6.

[51] Markovic DS, Zivkovic D, Branovic I, *et al.* Smart power grid and cloud computing. *Renewable and Sustainable Energy Reviews*. 2013;24:566–577. Available from: https://www.sciencedirect.com/science/article/pii/S136403211300227X.

[52] Bera S, Misra S, and Rodrigues JJPC. Cloud computing applications for smart grid: A survey. *IEEE Transactions on Parallel and Distributed Systems*. 2015;26(5):1477–1494.

[53] Sittón-Candanedo I, Alonso RS, García, *et al.* Edge computing, IoT and social computing in smart energy scenarios. *Sensors*. 2019;19(15):3353. Available from: https://www.mdpi.com/1424-8220/19/15/3353.

[54] Minh QN, Nguyen VH, Quy VK, *et al.* Edge computing for IoT-enabled smart grid: The future of energy. *Energies*. 2022;15(17):6140. Available from: https://www.mdpi.com/1996-1073/15/17/6140.

[55] Sena ASd, Ullah M, and Nardelli PHJ. In: Fathi M, Zio E, and Pardalos PM, editors. *Edge Computing in Smart Grids*. Cham: Springer; 2023. p. 1875–1885. Available from: https://doi.org/10.1007/978-3-030-97940-9_106.

[56] Saleem MU, Shakir M, Usman MR, *et al.* Integrating smart energy management system with internet of things and cloud computing for efficient demand side management in smart grids. *Energies*. 2023;16(12):4835. Available from: https://www.mdpi.com/1996-1073/16/12/4835.

[57] Fu W, Wan Y, Qin J, *et al.* Privacy-preserving optimal energy management for smart grid with cloud-edge computing. *IEEE Transactions on Industrial Informatics*. 2022;18(6):4029–4038.

[58] Bose SK, Kar B, Roy M, *et al. ADEPOS: Anomaly Detection based Power Saving for Predictive Maintenance using Edge Computing*; 2018. Available from: https://arxiv.org/abs/1811.00873 (accessed: January 30, 2025).

[59] Jasny M, Thostrup L, Ziegler T, *et al.* P4DB – The Case for In-Network Oltp. In: *Proceedings of the International Conference on Management of Data*. ACM; 2022. p. 1375–1389.

[60] Shanbhag A, Tatbul N, Cohen D, *et al.* Large-scale in-memory analytics on Intel optane DC persistent memory. In: *Proceedings of the 16th International Workshop on Data Management on New Hardware*. ACM; 2020. p. 1–8.

[61] Wu Y, Park K, Sen R, *et al.* Lessons learned from the early performance evaluation of Intel Optane DC persistent memory in DBMS. In: *Proceedings of the 16th International Workshop on Data Management on New Hardware.* ACM; 2020. p. 1–3.

[62] Jay M, Ostapenco V, Lefèvre L, *et al.* An experimental comparison of software-based power meters: focus on CPU and GPU. In: *2023 IEEE/ACM 23rd International Symposium on Cluster, Cloud and Internet Computing (CCGrid).* IEEE; 2023. p. 106–18.

[63] Khan KN, Hirki M, Niemi T, *et al.* RAPL in action: Experiences in using RAPL for power measurements. *ACM Transactions on Modeling and Performance Evaluation of Computing Systems (TOMPECS).* 2018;3(2):1–26.

[64] Murwantara IM, and Yugopuspito P. Evaluating energy consumption in a different virtualization within a cloud system. In: *2018 Fourth International Conference on Science and Technology (ICST).* IEEE; 2018. p. 1–6.

[65] Murwantara IM, and Bordbar B. A simplified method of measurement of energy consumption in cloud and virtualized environment. In: *2014 IEEE Fourth International Conference on Big Data and Cloud Computing.* IEEE; 2014, p. 654–661.

[66] I Corporation. Power Top Github; 2020. Available from: https://github.com/fenrus75/powertop (accessed: January 30, 2025).

[67] Bigerna S, Bollino CA, and Micheli S. Socio-economic acceptability for smart grid development – A comprehensive review. *Journal of Cleaner Production.* 2016;131:399–409. Available from: https://www.sciencedirect.com/science/article/pii/S0959652616304656.

Chapter 9

AI in energy management: the market view

Sa'd Shannak[1], Deeb Alashgar[1] and Abdalrahman Migdad[2]

This chapter examines the impact of artificial intelligence (AI) on energy management. A lot of attention was paid to the market and emerging technologies and their potential in addressing global problems in energy. In the course of such energy-related troubles as energy demand, environmental issues, and the urgency of switching to renewable resources, AI appears to be the spark in the radical and sustainable transformation of the energy sector.

A particular emphasis has been made on the most significant players in this industry, ranging from established energy firms to nimble new entrants; all of them apply AI in making energy use and its predictive maintenance and automation of the grid even smarter. We demonstrate these transformations through a number of practical examples, not only do the new technologies bring permanent changes in energy operations, but they are also transforming the way utilities do business with their customers.

There are however difficulties associated with the implementation of AI in the energy market such as data security, proper infrastructure, and human resources. However, the benefits of embarking on this journey are enormous. To be sure, a successful deployment of energy AI technologies can take us to a future world that is green and efficient for an individual, communities, and the society at large.

9.1 Introduction

The energy sector has pioneered technological advancements for a long time and has not yet changed. Meeting the demands of customers and industries has always been the driver for development and innovation, and it introduces changes and upgrades continuously. Artificial intelligence (AI) is one of the latest developments in the world of technology that has changed people's perceptions and management behavior toward energy systems [1]. Classical management techniques are usually described as manual and reactive. With AI and advancements in machine learning (ML), management techniques have shifted to become proactive and follow data-driven strategies.

[1]Qatar Environment and Energy Research Institute, Hamad Bin Khalifa University, Qatar
[2]Department of Islamic Economics and Finance, Istanbul Sabahatin Zaim University, Turkey

Such a shift from manual to data-driven strategies enhances both efficiency and sustainability. With AI involving machines and systems to perform tasks, AI develops human intelligence that may include learning, problem-solving, and reasoning [2]. The energy sector could utilize AI in varied and complex activities, including energy distribution and the best integration strategies of renewable energy sources [3]. For example, AI algorithms can process large volumes of big data from sensors, smart water or energy meters, and weather forecasts to present scientists and stakeholders with energy consumption patterns. Such knowledge enhances efficiency in resource allocation and waste reduction [4]. Powering energy management systems with AI enables an equilibrium point that best balances the supply and demand of energy in the market [5].

The influence of AI exceeds mere improvements in operational efficiency; AI contributes heavily to the presentation of innovations in business models, such as decentralizing energy sales. One new trend is households with solar panels possibly being on the supply side of the energy transaction, selling their excess energy back to the grid. In this example, AI algorithms optimize the timing of these energy sales transactions. With such advancements, communities have become stronger and more resilient.

9.1.1 Importance of AI in the market

The energy market faces multiple challenges and increasing concerns associated with climate change, an increasing demand for energy consumption, and rising resource scarcity. Due to this multi-layered and multidisciplinary growing problem, the role of AI in energy management is becoming more critical. The importance of AI makes up for shortcomings connecting to human shortcomings and the traditional methods that fail to manage the complexities of today's interconnectedness of today's energy market. AI solves human capabilities and provides swift real data processing and adaptable, reliable decision-making [4]. A critical problem for energy suppliers today is managing demand effectively, especially given the differing nature of renewable energy sources, which can cause fluctuations in supply, leading to potential grid instability. AI presents precision in forecasting demand fluctuations, helping managers ensure energy production, storage, and distribution are done efficiently. This approach enhances grid reliability and leads to cost efficiency for energy suppliers and consumers [4].

Additionally, AI can be critical in identifying causes and mitigating energy waste. AI advanced analytics systems can detect irregularities in consumption patterns, allowing businesses and households to address inefficiencies promptly. This proactive approach can yield economic benefits and environmental advantages, such as reducing energy consumption, lowering carbon emissions, and promoting sustainable practices [6]. As markets become increasingly competitive, companies that leverage AI technologies gain an edge. The ability to optimize operations, enhance customer experiences, and reduce costs is paramount in a landscape where consumers value efficiency and sustainability. Businesses that can effectively harness the potential of AI will position themselves advantageously in an evolving marketplace. Figure 9.1 shows the flowchart of the importance of AI in the market.

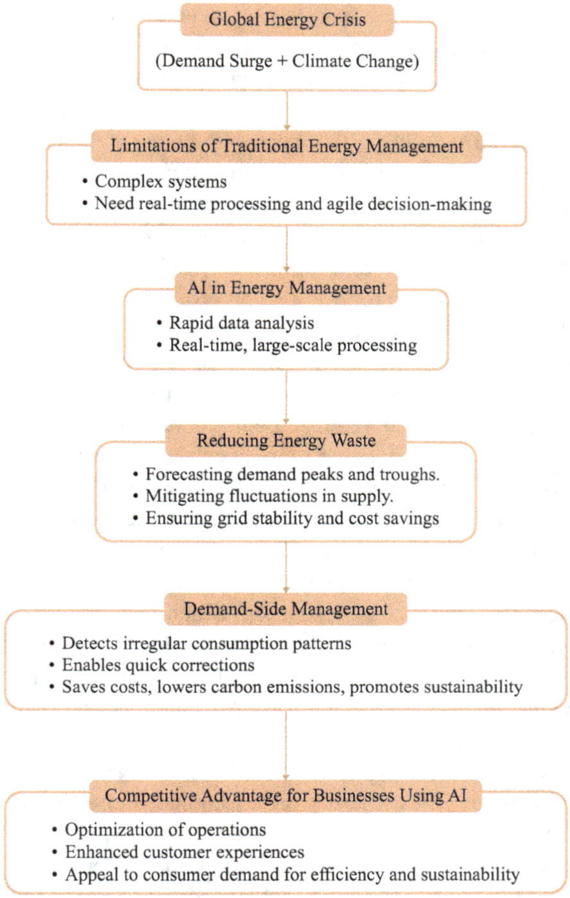

Figure 9.1 Importance of AI in the market

Figure 9.1 highlights the global energy crisis represented by a demand surge and climate change. The limitations of traditional energy management systems are mainly the inability to deal with complex systems and to process real-time data in high volumes to present stakeholders with up-to-date results that enable them to make prompt decisions. The flowchart explains how AI can manage tasks, learn faster, and innovate new techniques to minimize energy waste by identifying extravasation in energy use. Eventually, businesses using AI will be more competitive in the market by virtue of increased performance and operational efficiency.

9.1.2 Evolution of the market

Technological advancements and changing market dynamics coupled with AI shape a fascinating journey that captures the evolution of the energy management market.

In the late twentieth century, the energy sector generally operated within a framework characterized by centralized decision-making and predominantly defined by linear processes [3]. Before digitalization, the collected data was scattered and often manual, which hindered the prompt responsiveness of the energy systems. With the appearance of the digital revolution that began in the early twenty-first century, the landscape in different sectors started to shift dramatically. Many technological innovations, such as advanced metering infrastructure and the IoT, contributed to the development and efficiency of energy management. Innovations in such smart technology laid the groundwork for a better-connected energy ecosystem [3]. The energy ecosystem enhancement could be represented by promptly gathering large databases, which can now be easily channeled and analyzed using advanced AI technologies. Initially, AI technologies were primarily utilized for automated reporting, predictive analysis, maintenance scheduling, and demand forecasting [3]. With the maturity of ML algorithms, traditional energy management practices were transformed.

Recently, the market has witnessed more acceleration in AI adoption, driven by sophisticated algorithms, increasing computing power, and the accessibility of big data. Applications powered by AI algorithms have become more widely used, including everything from smart grids to electric vehicle management and energy storage solutions. The current energy landscape is shaped by a more decentralized model that accommodates decentralized energy resources, which has led to a structural change in how energy is generated and consumed [7]. Furthermore, the increased rise of regulatory frameworks designed to promote renewable penetration has further powered the need for efficient energy management systems. There is a recognition that AI can play a significant role in reaching energy transition targets, and policymakers further recognize the urge to support investment in AI technologies to facilitate the smart management of resources [8]. AI is not only a tool. It is becoming a fundamental component of energy infrastructure. AI allows for enhanced predictive analytics, real-time decision-making, and, most importantly, continuous optimization, all essential factors in managing future energy markets. Consequently, as more AI is incorporated into the energy market, we will achieve efficient energy production [9,10], consumption [11], and sustainability [12]. AI is not only helping to meet the demands of contemporary energy systems, it also paves the road for a greener and more efficient future [13].

9.2 Market drivers for AI in energy management

The drivers affecting AI integration in energy management can be summarized as follows: growing energy demand, regulatory pressures, the integration of renewable energy sources, and the need for cost optimization. We are going to discuss each driver in the following subsections.

9.2.1 Energy demand and efficiency

Rapid urbanization, population growth, and a growing reliance on energy-intensive technologies have increased the world's energy demand. There is an evident hindrance to the current energy infrastructure as more people require additional access to the grid. It is a formidable task for energy firms to meet the growing demand for electricity while simultaneously increasing energy efficiency [14]. Hence, AI becomes an obvious solution. It is a formidable ally in this conflict. For instance, understanding and predicting energy consumption requires examining enormous volumes of data more precisely from smart meters, weather trends, and customer behavior [15,16]. Consider a busy metropolis where AI systems can predict heatwave-related energy spikes, allowing utilities to modify their supply [17] proactively. Additionally, predicting energy spikes and adjusting supply would reduce waste and guarantee that homes and businesses stay powered [18,19].

Smart home appliances are another example where the data generated can be analyzed using AI to optimize energy utilization on a household level. AI-powered smart appliances can learn a pattern of use and optimize accordingly specific functionality [20], such as modifying lighting, heating, and cooling to maximize comfort and save energy use. This allows end users to be partners in energy saving and reducing their carbon footprint, which results in significant bill savings.

9.2.2 Regulatory pressures

Worldwide, there is momentum to promote sustainable energy solutions to curb the negative impact of climate change and reduce greenhouse gas emissions. Regulatory frameworks have been designed and revisited to increase reliance on green energy sources. These regulations and frameworks are not routine requirements; they reflect a global commitment and recognition of the urgent need for change. Within this landscape, AI emerges as a vital tool to meet the objectives of these frameworks and regulations [21]. On the one hand, governments are encouraging energy companies to utilize AI solutions that can streamline operations to enhance reporting capabilities and ensure adherence to environmental policies. One AI application is related to automating the collection and analysis of emission data, making it more convenient for operators to meet regulatory requirements and avoid any penalties. Moreover, several governments are enforcing policies that encourage AI in energy management [22,23] by providing grants, tax incentives, and research funding for projects that leverage AI to improve energy efficiency and reduce emissions. This creates a solid ground for innovation, where companies and operators can experiment with AI solutions that comply with regulations and position them as leaders in the sustainable energy space.

9.2.3 Renewable energy integration

Integrating variable renewable energy sources such as wind and solar causes a significant burden on governments and operators. The intermittent nature of these resources necessitates sophisticated management strategies to balance supply and demand effectively. This integration process can be eased by employing AI systems. AI systems can optimize the penetration of renewable energy into the existing grids by analyzing real-time data from weather forecasts and considering grid conditions and other factors such as energy consumption patterns. For instance, as solar radiations increase during a sunny day, AI models can adjust the grid to allow for more solar energy generation and accordingly store any excess energy, if any, or redirect it where it is needed most. Furthermore, the development of smart grids can benefit from AI systems for managing a decentralized energy landscape. To create a dynamic and responsive energy ecosystem, these grids, by employing AI algorithms, can enhance communication with various energy sources such as electric vehicles and household solar panels. This could significantly increase the reliability of renewable energy resources and empower consumers to contribute to energy generation and consumption decisions.

9.2.4 Cost optimization

A strategic goal of energy companies is to achieve profitability and sustainability targets, and here, cost optimization becomes necessary. Most operation costs associated with energy generation and distribution are significant capital investments and can be substantial; therefore, inefficiencies can cause reduced profit margins [24]. Employing AI offers a pathway to reduce significant operation costs through predictive maintenance and real-time analytics. AL algorithms can predict maintenance by analyzing equipment performance data and identifying potential failure before it occurs [25].

Take the example of wind turbines that can send operators an alert to investigate a specific mechanical issue before it leads to costly downtime. Addressing these maintenance requirements proactively could help energy companies extend the lifetime of their assets, reduce repair costs, and, most importantly, ensure continuous energy supply. Analyzing real-time data could further enhance the overall performance of equipment and optimize costs by offering insights into energy consumption patterns, operational efficiencies, and market dynamics. AI models provide insights into historical, current, and future trends to recommend cost-saving operational modifications. For example, during low-demand periods, AI systems can recommend temporarily shutting down less efficient power plants to save fuel, associated emissions, and overall costs.

In addition to savings associated with operational activities, AI systems can enhance financial forecasting and risk management. By analyzing market trends and consumer behavior, AI systems can inform energy policymakers with decisions related to pricing, investments, and resource allocation [26]. This approach improves profitability and helps companies to adapt swiftly to changing market conditions. The main factors in the AI energy management market are interconnected and represent a more significant cultural movement [27]. Accordingly, employing AI and integrating it in the energy sector is not only responding to market pressures, but also embracing a future where energy management solutions can be more responsive and aligned with the values of economic and environmental sustainability.

9.3 Fundamental AI technologies in energy management

Embracing the future of the energy sector, we find that innovative developments in solutions tailored to enhance efficiency, sustainability, and reliability remain fundamental to the sector's ability to meet growing demand. In today's rapidly growing economy, all sectors capitalize on AI technological advancements, and the energy sector is no different. This chapter explores critical AI technologies that contribute to facilitating energy management to allow for faster human decision-making. The chapter looks into ML and predictive analysis, AI for smart grids, IoT and AI in energy monitoring, and robotic process automation (RPA).

9.3.1 ML and predictive analytics

ML and predictive analysis are central to AI's impactful transformation of energy markets. By utilizing these technologies, energy providers can analyze large data sets, make informed predictions about energy usage, and identify general patterns in energy management behavior. Using AI ML, it is possible to provide daily energy forecasts that empower homeowners to adjust their energy consumption habits, for example, by shifting heavy energy usage to off-peak hours; such as washing machines and heaters. Such adjustment in consumption habits saves money while relieving pressure on the grid. A real-life clarifying example could be car monitors that show the driver instant fuel consumption patterns. Drivers will learn that driving uphill, overloading the car, fast acceleration, constant usage of breaks, and driving in traffic will consume more fuel. Realizing those patterns, a responsible driver will find alternative routes and become more efficient.

The advantage gained from utilizing AI ML is the instant compilation of real-time data streams, followed by instant learning and algorithm changes, leading to enhanced predictive power. Companies operating in the energy market can utilize the technology to anticipate periods of peak demand, which can be done instantly and accurately. With such capabilities, energy providers can allocate resources efficiently and minimize any possible outage that may disrupt people's lives. Predictive analytics can optimize overall power consumption by enabling utilities to observe and spot trends within large populations. This precision-based insight allows them to promote behavioral changes and support energy-saving efforts, benefiting consumers and the environment.

9.3.2 AI for smart grids

Energy demand grows with the expansion of cities and the urban sprawl into towns and villages close to main cities. To meet this growth, a higher supply is necessary. Reaching higher equilibrium points with a conventional business method will be very expensive. Smart grids are introduced as critical solutions in the field of energy management. The provision of solutions formulates by bringing AI ML and digital technology together to improve electricity distribution. The essential point about AI in this context is its ability to provide real-time data by monitoring electricity flow. The additional importance of AI surfaces when attempting to balance conventional power plants and decentralized power sources like wind turbines and rooftop solar panels. Besides balancing the energy sources entering the grid, advancement in AI capabilities improves load management, allowing energy providers to respond to real-time demand changes promptly. The swift response increases efficiency and facilitates higher utilization of decentralized energy already generated by solar panels while reducing the operations of conventional energy plants. By doing so, clean energy will transition from aspiration into a practical reality. Such outcomes may only result from quick responses. The lag in human-based responses will lead to the absence of such an opportunity and the reliance on conventional energy plants.

In sudden and unpredictable situations such as storms or heavy rain that causes floods, smart grids powered by AI can swiftly redirect electricity from less affected neighboring regions. AI-powered smart grids' instant ability to adapt to sudden changes minimizes power interruptions or blackouts for households and businesses. Investment in smart grids is significant and economically feasible. This is well known to business owners, especially in developing countries with occasional electricity interruptions and blackouts. In such cases, fuel-powered electricity generators must provide electricity during those regular blackouts. This means extra machinery, labor, and higher input costs.

9.3.3 IoT and AI in energy monitoring

The integration of AI with the IoT represents a significant development in the field of energy monitoring and management. Both technological developments can be allocated in smart sensors built to collect instantaneous data on energy consumption utilizing IoT while AI algorithms conduct the required analysis. Consider the analogy of a commercial building where AI-IoT-powered smart meters and sensors are used; building managers are empowered to reveal unusual patterns when and where energy is wasted, allowing them to take immediate corrective action. Many scenarios could cause waste: (1) a water leakage in an empty apartment, (2) excessive energy usage during moderate temperatures, and (3) wasteful behavior could also be the problem. Such scenarios cause energy waste and physical harm to the building structure, such as water leakage. Positive consequences, however, include reducing costs, minimizing risk, and reducing energy waste. All that represents positive outliers to the community, the energy sector, and the planet in the form of lower emissions.

Deterioration is an intended characteristic of machines and devices. Meters and sensors used by energy providers will wear out over time and require human maintenance intervention. While regular checks by physical human intervention reduce interruptions in the energy provision systems, IoT-powered devices further enhance predictive maintenance, which can be done by instant monitoring of the performance of the devices. AI complements the IoT by providing managers with highly accurate forecasts for machinery needed for maintenance, allowing power providers to avoid expensive fixes resulting from complete machine failures and ensuring optimality in machines' overall performance. The complementarity of IoT and AI denotes the importance of moving proactively toward energy management. This proactive movement should originate from human shortcomings, our awareness of the importance of technology in general, and the unique outcome that evolves from the synergy between AI and IoT.

9.3.4 RPA

Many jobs traditionally performed by humans, including repetitive and rule-based tasks, have now been automated. This was possible with RPA, an emerging and developing technology that uses software robots to make this automation possible [28]. RPA became a key player in enhancing administrative tasks, including data entry, report generation, and transaction processing [29,30]. When allowing for robotic automation, human administrators in the energy sector free up valuable human resources, such as time, allowing professional managers to focus their valuable time on strategic decision-making and focused involvement in innovation that further improves management in the energy sector [30].

Recently, increasing numbers of companies with thousands of customers, especially service providers, have been utilizing RPA. Instead of hiring hundreds of employees in call centers, that number can be significantly reduced with RPA. Tasks such as billing, processing invoices, and handling customer queries can all be efficiently done by RPA [31,32]. With such developments, operational efficiency and customer experience are optimized. The enhancement of customer experience increases their trust in energy providers. Part of that trust will also come from RPA's swift ability to ensure data accuracy and protect client privacy. Protecting client privacy and data accuracy is also crucial for meeting government standards and regulations along with environmental commitment.

9.4 Current trends in AI-driven energy management

Exploring current trends in AI-driven energy management takes us into the industry's future outlook. In this section, we address those trends that reflect technological advancements and commit to achieving goals related to cleaner environments and the establishment of more efficient energy systems. Those trends include decentralization of energy, AI for renewable energy forecasting, blockchain and AI in energy trading, and AI for demand response programs.

9.4.1 Decentralization of energy

Decentralized energy management systems (DEMS) usually refer to micro or small energy grids utilized in smaller communities by individuals or small businesses to provide adequate energy management and distribution solutions. The energy sector currently witnesses momentous movement toward DEMS. According to Delfino *et al.* [33], DEMS can optimize energy generation, distribution, and consumption in smaller localized energy grids. Some of the system's main advantages are scalability and robustness [34]. Microgrids enable communities to harness renewable energy sources, sell access to the leading energy providers during the day when most energy is generated, such as solar energy, and benefit from main energy grids when necessary. Such a possibility allows users to approach zero or negligible energy bills. DEMS, being powered by AI, can balance supply and demand [35].

Microgrids could be islanded, and in such cases, the owners benefit from renewable energy sources circulating within a private grid that is disconnected from the primary energy grids. However, if renewable energy sources are grid-connected, the benefit harvested from DEMS is more significant because such systems incorporate intelligent features such as evolutionary adaptive dynamic programming and reinforcement learning. Such intelligent features handle uncertainties associated with grid-connected operations of excess/shortage supply or demand [36]. Imagine a scenario of an independent small community with its own solar panels and storage systems. Suppose the community utilizes AI technologies to autonomously operating and manage energy production and consumption within a local microgrid. In that case, residents benefit from renewable energy sources circulating within the community's private network. However, according to Venayagamoorthy *et al.* [36], suppose these systems are interconnected. In that case, the advantages multiply, as AI-driven tools such as predictive analytics and dynamic load balancing can optimize energy flow, facilitating peer-to-peer trading and efficiently handling fluctuations in supply and demand. This enhances energy independence and transforms community members into active participants in their local energy ecosystem [37,38].

9.4.2 AI for renewable energy forecasting

Renewable energy forecasting is an essential current trend in AI-driven energy management. The importance stems from several vital reasons that include balancing supply and demand, optimizing resource allocation, optimal integration of renewable energy into the grid, reducing cost, market stability and pricing, infrastructure planning, improving demand response programs, and enhancing the resilience of the statuesque of the energy sector [39–41]. With accurate and precise forecasting, integrating available renewable energy into main grids becomes possible. It dramatically benefits private energy production and relieves the load off of conventional energy plants. Accurate forecasting may become feasible with equipped AI models to analyze weather data, historical energy generation patterns, and market trends to predict solar and wind peaks in electricity production [42]. With insights such as identifying clean electricity production peaks, energy providers can tailor their operations to fully consume clean energy sources during high demand before considering increasing production that utilizes fossil fuels. This energy mix is a proactive approach that favors sustainable energy production. Sustainable energy mixes become possible with accurate forecasting, making clean energy a reliable and growing constant in the energy system.

9.4.3 Blockchain and AI in energy trading

Blockchain technology has enhanced security measures in different sectors of the economy. The energy sector has also been revolutionized by introducing blockchain technology, enabling higher security measures in decentralized energy transactions, energy trading, and peer-to-peer energy trading systems [43]. While security is uplifted by blockchain, AI facilitates predictive analysis and automative decision-making [43,44]. Using blockchain for secure and transparent transactions, individual customers and institutions can directly trade excess energy generated from private solar panels to neighbors.

This combination provides an ecosystem that benefits the energy sector stakeholders. The AI-enhanced ecosystem provides optimized pricing strategies and guarantees safe and efficient transaction occurrence. This promotes a decentralized energy system that empowers individual customers to control their energy choices when buying and selling.

9.4.4 AI for demand response programs

AI-driven demand response programs are emerging as key technologies that optimize energy usage by dynamically adjusting consumption based on real-time data analysis. These systems analyze user behavior and market trends, enabling energy providers to suggest reductions during peak demand periods, thus enhancing grid stability [45]. For instance, such technological advancements allow users to receive notifications encouraging them to operate high-energy appliances like dishwashers or washing machines during off-peak hours [46]. Financial incentives for such participation help consumers save on energy costs and improve overall system efficiency. This collaborative model fosters an intelligent and responsive energy ecosystem that benefits stakeholders.

9.5 Challenges and barriers to AI adoption

While technological development in AI has widely contributed to efficiency, sector developments, speeding up processing, and presenting instant data analysis, barriers and challenges exist in AI adoption. Several challenges and barriers hinder the widespread adoption of AI in the energy sector. Such challenges include data availability and quality, infrastructure limitations, cyber security concerns, and the cost of AI implementation.

9.5.1 Data availability and quality

Since the quality of inputs determines the quality of outputs, one of the leading obstacles to AI is the quality and availability of data. Energy providers could face difficulties collecting high-quality, clean data to train AI algorithms. As such, the performance of the AI will depend on the provided inputs. Additionally, inconsistencies in data formats, legacy systems, and hardware limitations can all impede the effective use of AI [47]. Putting more effort into solving issues related to data collection is crucial to train AI based on reliable inputs. With high-quality and reliable data, AI systems can provide precise insights and drive much-needed efficiencies. Without this foundational element, the expected benefits of AI would remain out of reach.

9.5.2 Infrastructure limitations

The adoption of AI in infrastructure has limitations along with the opportunities available. Adopting AI could mean a complete structural makeover for some companies, while others would require significant changes to their infrastructure. As such, many energy markets still rely on old technologies and outdated systems that cannot support up-to-date AI applications [48]. Upgrading infrastructure is financially burdensome and technologically challenging, particularly for smaller energy companies. However, despite the extra financial burden in the short run, an updated infrastructure is a feasible investment in the long run. The utilization of technology is becoming a vital part of competition in the market. As such, investment in AI infrastructure today is essential for long-term survival. The transition from older and classic on-premise to contemporary and modern cloud-based systems is crucial for AI advancement and training AI algorithms, offering scalability and cost-efficiency [49].

Energy providers seeking a transition to AI-driven solutions must develop a roadmap for infrastructure upgrades. This is important because while AI can generate mechanisms of sustainable production and model climate change [50], AI's energy consumption paradoxes contribute to environmental degradation. Researchers are designing frameworks for energy-efficient resource allocation in ML training tasks to mitigate AI energy consumption paradoxes. These systems can dynamically shift workloads to locations with access to cleaner energy, all while ensuring performance standards are met [51]. Such advancements aim to promote a more environmentally sustainable AI infrastructure. Therefore, as energy provision systems evolve to remove infrastructure limitations to meet AI potential, companies must ensure alignment with energy demands and modern technological advancements.

9.5.3 Cybersecurity concerns

Security concerns are usually intriguing to those working with or through technology. Cybersecurity is a genuine concern in AI that requires high-level professionalism to overcome such risk [52,53]. AI techniques, particularly ML, have indeed advanced the security of renewable energy systems, with low false positive rates and detection rates of 85% [54].

In energy management, IoT devices, smart grids, and interconnected networks create susceptibilities that cybercriminals can exploit [55]. It is vital to secure AI-powered systems against potential attacks, protecting energy infrastructure and consumer data. To address these challenges, energy companies are implementing various measures, with 90% using cybersecurity measures and 50% employing blockchain technology [54]. Therefore, energy firms are encouraged to invest in advanced cybersecurity tools and spread security awareness in their workplace to ensure reliable operations and build trust in an increasingly interconnected world.

9.5.4 Cost of AI implementation

Challenges and barriers to AI adoption in the energy sector strongly connect to cost. The cost of AI adoption is high, especially for smaller-sized companies making the transition for the first time [56,57]. When small and medium-sized companies in the energy sector conduct a feasibility study comparing the benefits of AI adoption to the initial investment, the process may be discouraging.

On the bright side of the transaction is the position of the energy sector at the center of many countries' strategic plan objectives. Environmental protection and low-carbon emissions are usually targets for most governments, and governments will specialize funds and other support mechanisms to achieve their goals. As such, companies can utilize government grants, engage in partnerships, and innovative financing models that support AI initiatives. By showcasing the long-term value of

AI investments, which energy firms should have the ability to demonstrate, energy firms can position themselves for future success.

9.6 The competitive landscape of AI in energy markets

Energy markets operate under the principles of economic competition. In "markets," competition drives success or failure, growth or decline, and sparks innovation that leads to progress or stagnation. The energy sector stakeholders undergo transformative shifts driven by AI technologies to remain competitive players in the market. Such shifts include fundamental changes in energy production, distribution, and consumption. The competition and evolved innovation from the interaction of significant companies and innovative startups pave the way for a new era of energy management. This section introduces the major players in AI innovation that serves the energy sector, explores startups that may be considered disrupters to the major companies' classical ways of managing their operations, looks into investment trends in energy management, and examines the nature of partnerships and collaborations to reshape the energy sector.

9.6.1 Major players in AI innovation

Major players driving AI innovations in 2024 can be distributed into a few categories: technology giants, semiconductor leaders, AI-first companies, industry-specific innovators, and energy management-specific innovators. Technology giants include Microsoft, Google, Amazon, and Meta. Semiconductor leaders include NVIDIA, Intel, and AMD. AI-first companies include OpenAI, Anthropic, and Stability AI. Industry-specific innovators include Tesla in autonomous driving and Tempus in the healthcare sector. Finally, AI innovators in energy management include Siemens, General Electric (GE), IBM, Schneider Electric, Microsoft, and Google (Alphabet). The following section will address some of these companies.

Siemens, a global engineering and technology powerhouse and a leading company in AI energy management, has recognizable innovations and AI-based solutions for smart grid and energy systems. The company's services include ML tools and analytics designed to enhance energy use and optimize grid stability. With a strong focus on sustainability, Siemens remains a crucial force in shaping a more efficient energy future. General Electric is another major player that developed its Predix platform to enhance the efficiency of its energy operations. The company adopts AI and ML to analyze data for predictive maintenance and performance optimization. This approach helps energy firms minimize downtime and enhance reliability.

With its Watson platform, IBM, another leading company in AI-powered energy management, the company operates to empower energy providers to make informed decisions that improve efficiency and sustainability. On the other hand, Schneider Electric specializes more in energy automation and management. Their approach to utilizing AI in smart buildings helps their customers contribute to achieving sustainability targets.

9.6.2 *Startups and disruptors*

The relationship between established large firms and new startups in different industries is built on competition but on uneven grounds. Large companies dominating their sectors will continuously be challenged by new waves of innovative, vibrant startups that utilize unconventional mechanisms to prove their presence in the energy market. Startups in the energy market challenge the traditional energy models by adopting AI to create out-of-the-box solutions to pressing challenges and real problems facing the sector [58]. As such, observers in the energy markets will come to recognize the new entrants into the market, and with the increased levels of competition, there is a high likelihood for established companies to cooperate with startups [59].

Uplight is a notable startup in the energy sector that provides customer engagement solutions. The company's platform offers customer recommendations to ensure personalized energy-saving techniques. By incorporating AI, uplight is renovating how utilities interact with their customers to foster a culture of energy efficiency. Grid edge uses AI to optimize commercial energy usage in large establishments and buildings. Their platform provides actionable insights that help decrease costs and improve sustainable energy behavior in large commercial buildings. Another startup is Aurora Solar, which heavily invests in AI to rationalize the design and installation of solar systems. The company uses space images and other inputs in its analysis to enhance the time and cost associated with solar installations. Looking back at the operations and performance of the listed examples, those startups, despite facing liabilities such as being new entrants to the market and being small compared to larger existing companies, successfully presented solutions to existing problems in the market. Those startups adopted AI, mobilized resources, and aligned incentives to present new products and change traditional energy management models. By doing so, they proved their worthiness in the market and contributed to achieving goals related to environmental protection.

9.6.3 *Investment trends*

The surge in AI-driven innovation in the energy sector is fueled by significant investment from venture capital and other sources. Investors increasingly recognize AI technologies' potential to transform energy management and drive sustainability.

According to recent reports, investment in AI for energy management has seen exponential growth, with billions of dollars flowing into startups and established companies. Venture capital firms are particularly interested in funding innovative solutions that address the challenges of energy efficiency, renewable integration, and grid reliability. This influx of capital enables companies to scale their operations, develop new technologies, and expand their market reach.

Moreover, government initiatives and grants support AI innovation in the energy sector. Many governments invest in research and development to promote clean energy technologies and enhance grid resilience. This collaborative approach between the public and private sectors fosters an environment conducive to innovation, ultimately benefiting consumers and the planet.

9.6.4 Partnerships and collaborations

Strategic partnerships between AI firms, energy companies, and governments often drive the advancement of AI technology in the energy sector. These collaborations facilitate knowledge sharing, resource pooling, and the development of innovative solutions that address complex energy challenges.

For instance, partnerships between technology companies and energy providers enable AI integration into existing infrastructure. By collaborating with AI firms, energy companies can leverage cutting-edge technologies to enhance operations and improve service delivery. These partnerships are crucial for driving the adoption of AI solutions and ensuring that they align with the unique needs of the energy sector.

Governments are also playing a vital role in fostering collaboration. By supporting initiatives that promote research and development in AI and energy management, governments are helping to create a robust ecosystem for innovation. This collaborative approach accelerates the adoption of AI technologies and ensures that they are implemented in a manner that benefits society.

9.7 Case studies of AI applications in energy

9.7.1 Smart grids and AI in urban energy management

Cities worldwide increasingly turn to AI to manage their energy resources more effectively. For example, Barcelona has implemented an AI-driven energy management system that optimizes energy consumption across the city. The system can predict energy demand, adjust supply accordingly, and reduce waste by analyzing real-time data from smart meters and sensors. This initiative has led to significant energy savings and improved grid reliability, showcasing the potential of AI in urban energy management.

9.7.2 Renewable energy integration

AI is also critical in improving the integration of renewable energy sources into the grid. A notable case is the Hawaii Electric Company, which has implemented AI algorithms to manage solar and wind energy generation variability. The company can optimize its grid operations and ensure a stable electricity supply by predicting energy production based on weather patterns and historical data. This proactive approach enhances grid reliability and supports the state's ambitious renewable energy goals.

9.7.3 Energy storage optimization

As renewable energy sources become more prevalent, optimizing energy storage systems is essential for ensuring a reliable energy supply. Tesla's Powerwall, an advanced energy storage solution, utilizes AI to manage energy flow efficiently. The system learns from user consumption patterns and weather forecasts to optimize charging and discharging cycles. This intelligent management of energy storage not only maximizes the use of renewable energy but also provides homeowners with greater control over their energy consumption.

The competitive landscape of AI in energy markets is dynamic and rapidly evolving. Major players, innovative startups, and collaborative partnerships are driving the adoption of AI technologies that enhance energy management, improve efficiency, and support sustainability goals. As investment trends continue to fuel innovation, the potential for AI to transform the energy sector becomes increasingly evident.

Real-world applications of AI, from smart grids to renewable energy integration and energy storage optimization, demonstrate the tangible benefits of these technologies. As we move forward, the continued collaboration between technology firms, energy providers, and governments will be crucial in harnessing the full potential of AI, ultimately paving the way for a more sustainable and resilient energy future.

9.8 Recommendations and policy implications

As we delve into AI's promising yet intricate landscape in energy management, it is clear that achieving its full potential requires a coordinated approach. This should encompass innovation, education, ethical practices, consumer engagement, and robust policy support.

To begin with, fostering collaboration across various sectors is essential. We can drive significant advancements by cultivating partnerships among energy companies, technology firms, regulators, and academic institutions. Governments play a pivotal role by creating platforms that facilitate knowledge-sharing and collaboration. Public–private partnerships can act as incubators for innovative projects that tackle critical energy challenges on a larger scale.

In addition, we need to invest in education and training to address the growing demand for skills in AI and energy management. Policymakers should support initiatives that enhance STEM education and incorporate AI-focused curricula in universities and vocational programs. Furthermore, providing scholarships and incentives for students pursuing careers in this field will help create a diverse and skilled workforce ready to embrace the future of energy.

As AI becomes more integrated into energy management, championing ethical practices is another vital aspect. Organizations must uphold moral standards, prioritizing consumer trust and data privacy. Policymakers should establish regulations that govern the responsible use of AI, ensuring that transparency and accountability are integral to AI applications from the start. This will help alleviate concerns and foster public confidence in these technologies.

Encouraging consumer engagement is equally important. Energy providers should leverage AI to empower consumers, enabling them to make informed energy choices. Policymakers can promote this by incentivizing utilities to offer personalized energy-saving solutions. Clear communication strategies will be critical to demystifying AI technologies, making it easier for consumers to understand their benefits and engage actively in energy management.

Moreover, embracing innovative pilot programs can yield valuable insights. Government support for these initiatives – whether through funding, grants, or tax incentives – can encourage experimentation and lead to meaningful innovations.

These programs can serve as learning labs, showcasing best practices and helping to shape broader policy frameworks for adopting AI in the energy sector.

Finally, implementing supportive policies for integrating renewable energy sources is crucial. As AI plays a significant role in this transition, policymakers must develop regulations that facilitate the deployment of AI technologies in energy systems, provide incentives for renewable investments, and establish standards that enhance grid flexibility and resilience.

By comprehensively addressing these recommendations and considering their policy implications, we can truly unlock the transformative potential of AI in energy management. Together, we can pave the way for a future where sustainable energy is achievable and equitable, ensuring that all communities benefit from technological advancements. This collaborative effort will lead us toward a brighter, greener tomorrow, with energy systems designed to be as dynamic and resilient as the world they serve.

9.9 Conclusions

In this chapter, we have explored the transformative impact of AI technologies on energy management, stressing their essential role in shaping a more sustainable future. AI emerges as a powerful ally as we navigate the complexities of a rapidly changing world marked by fluctuating energy demands and the pressing need for environmental stewardship. It empowers us to rethink how we generate, distribute, and consume energy, offering innovative solutions that were once the stuff of dreams.

However, the path to fully integrating AI into energy systems is not without its hurdles. The challenges we face, from data privacy concerns to the need for robust infrastructure and skilled personnel—require careful consideration and collaborative efforts from all stakeholders. Yet, as daunting as these challenges seem, the potential rewards are substantial. By embracing AI, we can create a more sustainable, efficient, and resilient energy landscape that benefits industries, governments, communities, and individuals.

As we explore advancements in the energy sector, it becomes increasingly clear that AI technologies are not merely tools for optimization; they symbolize a pivotal shift in our approach to energy management. They represent a commitment to innovation and sustainability, a promise that we can harness the power of technology to address the urgent challenges of our time. Ultimately, the journey toward an AI-driven energy future is more than just efficiency or profitability; it is about creating a world where clean energy is accessible to all. It is about ensuring that future generations inherit a livable and thriving planet. Each step in integrating AI into our energy systems brings us closer to this vision: a brighter, greener future where energy is abundant, sustainable, and equitable.

As we look ahead, let us embrace this transformative journey with optimism and determination. Together, we can leverage the potential of AI to not only meet today's energy challenges but also to pioneer solutions for tomorrow. The future of energy is here, and with it comes the promise of a more sustainable world for everyone.

References

[1] J. A. Chatfield, "A Representation theorem for operators on a space of interval functions," *International Journal of Mathematics and Mathematical Sciences*, vol. 1, p. 589151, 1978.

[2] S. Raisch and K. Fomina, "Combining human and artificial intelligence: Hybrid problem-solving in organizations," *Academy of Management Review*, vol. 50, pp. 441–464, 2025.

[3] L. Zhang, J. Ling, and M. Lin, "Artificial intelligence in renewable energy: A comprehensive bibliometric analysis," Energy Reports, vol. 8, pp. 14072–14088, 2022, doi:10.1016/j.egyr.2022.10.347.

[4] M. A. Akbar, A. A. Khan, S. Mahmood, S. Rafi, and S. Demi, "Trustworthy artificial intelligence: A decision-making taxonomy of potential challenges," *Software: Practice and Experience*, vol. 54, no. 9, pp. 1621–1650, 2024, doi: 10.1002/spe.3216.

[5] B. Zeng, Y. Liu, F. Xu, Y. Liu, X. Sun, and X. Ye, "Optimal demand response resource exploitation for efficient accommodation of renewable energy sources in multi-energy systems considering correlated uncertainties," *Journal of Cleaner Production*, vol. 288, 2021, doi:10.1016/j.jclepro.2020.125666.

[6] A. N. Mackay, "From Algorithms to Arctic Ice: AI's Role in Climate Adaptation from Ottawa to Oslo," Master thesis, University of Stavanger, 2024.

[7] L. L. Kiesling, "Innovation and Decentralized Energy Markets: Technologies and Institutions for a Clean and Prosperous Energy Future," Center for Growth and Opportunity, Utah State University, 2020.

[8] S. Shannak and M. Contestabile, "Does Qatar Face a Trade-off Between Economic Growth and CO2 Emissions?" *Fronters in Environmental Science*, vol. 10, 2022, doi:10.3389/fenvs.2022.855971.

[9] S. Tiwari, "Artificial Intelligence (AI) in the Sustainable Energy Sector †," *Engineering Proceedings*, vol. 37, no. 1, 2023, doi:10.3390/ECP2023-14609.

[10] E. K. Iriakuma, O. N. Ukela, and A. P. Uche, "Forecasting Energy Consumption with AI: A Review for Sustainable Energy Management", *International Journal of Latest Technology in Engineering, Management & Applied Science*, vol. 13, no. 3, pp. 14–21, 2024.

[11] A. Raihan, "A comprehensive review of artificial intelligence and machine learning applications in energy consumption and production," *Journal of Technology Innovations and Energy*, vol. 2, no. 4, pp. 1–26, 2023, doi:10.56556/jtie.v2i4.608.

[12] S. Kaur, R. Kumar, K. Singh, and Y. Huang, "Leveraging artificial intelligence for enhanced sustainable energy management", *Journal of Sustainability for Energy*, vol. 3, no. 1, pp. 1–20, 2024, doi:10.51583/IJLTEMAS.

[13] A. Hamdan, K. Ifeanyi Ibekwe, V.I. Ilojianya, S. Sonko, and E. A. Etukudoh, "AI in renewable energy: A review of predictive maintenance and energy optimization," *International Journal of Science and Research Archive*, vol. 11, no. 1, pp. 718–729, 2024, doi:10.30574/ijsra.2024.11.1.0112.

[14] B. Jucker, P. Leupp, and T. Sjökvist, "Electrical energy: The challenge of the next decades," *ABB Review*, 2008. https://library.e.abb.com/public/ad4023256c4b27b5c12574270031e6be/08-13%201M801_ENG72dpi.pdf.

[15] E. Blasch, H. Li, Z. Ma, and Y. Weng, "The Powerful use of AI in the energy sector: Intelligent forecasting," 2021, https://doi.org/10.48550/arXiv.2111.02026.

[16] S. Nuthakki, S. Kumar, C. S. Kulkarni, and Y. Nuthakki, "Role of AI enabled smart meters to enhance customer satisfaction," *International Journal of Computer Science and Mobile Computing*, vol. 11, no. 12, pp. 99–107, 2022, doi:10.47760/ijcsmc.2022.v11i12.010.

[17] T. Michael Olatunde, A. Chukwudi Okwandu, D. Oluwajuwonlo Akande, and Z. Queen Sikhakhane, "Reviewing the role of artificial intelligence in energy efficiency optimization," *Engineering Science & Technology Journal*, vol. 5, no. 4, pp. 1243–1256, 2024, doi:10.51594/estj/v5i4.1015.

[18] B. Eck, F. Fusco, R. Gormally, M. Purcell, and S. Tirupathi, "AI modelling and time-series forecasting systems for trading energy flexibility in distribution grids," in *e-Energy 2019 – Proceedings of the 10th ACM International Conference on Future Energy Systems*, Association for Computing Machinery, Inc., 2019, pp. 381–382. doi:10.1145/3307772.3330158.

[19] S. Onwusinkwue, F. Osasona, I. A. Ibrahim Ahmad, *et al.*, "Artificial intelligence (AI) in renewable energy: A review of predictive maintenance and energy optimization," *World Journal of Advanced Research and Reviews*, vol. 21, no. 1, pp. 2487–2799, 2024, doi:10.30574/wjarr.2024.21.1.0347.

[20] J.-D. Ruvini and C. Dony, "APE: Learning user's habits to automate repetitive tasks," *International Conference on Intelligent User Interfaces*, New York, USA, 2000, pp. 229–232. doi:10.1145/325737.325854.

[21] R. Tillu, M. Muthusubramanian, and V. Periyasamy, "Transforming regulatory reporting with AI/ML: Strategies for compliance and efficiency," *Journal of Knowledge Learning and Science Technology*, vol. 2, no. 1, pp. 145–157, 2024, doi:10.60087/jklst.vol1.n1.P157.

[22] M. S. S. Danish and T. Senjyu, "AI-enabled energy policy for a sustainable future," *Sustainability (Switzerland)*, vol. 15, no. 9, 2023, doi:10.3390/su15097643.

[23] C. van Noordt, R. Medaglia, and L. Tangi, "Policy initiatives for Artificial Intelligence-enabled government: An analysis of national strategies in Europe," *Public Policy Administration*, 2023, doi:10.1177/09520767231198411.

[24] S. A. Shannak, J. I. Mikayilov, and R. Dua, "Energy price reform to mitigate transportation carbon emissions in oil-rich economies," *Environmental Economics and Policy Studies*, vol. 26, no. 2, pp. 263–283, 2024, doi:10.1007/s10018-024-00400-9.

[25] M. Shahin, F. F. Chen, A. Hosseinzadeh, and N. Zand, "Using machine learning and deep learning algorithms for downtime minimization in manufacturing systems: an early failure detection diagnostic service," *The International Journal of Advanced Manufacturing Technology*, vol. 128, pp. 3857–3883, 2023, https://doi.org/10.1007/s00170-023-12020-w.

[26] A. M. Rahmani, B. Rezazadeh, M. Haghparast, W. C. Chang, and S. G. Ting, "Applications of artificial intelligence in the economy, including applications in stock trading, market analysis, and risk management," *IEEE Access*, vol. 11, pp. 80769–80793, 2023, doi:10.1109/ACCESS.2023.3300036.

[27] T. Ahmad, D. Zhang, C. Huang, *et al.*, "Artificial intelligence in sustainable energy industry: Status Quo, challenges and opportunities," *Journal of Cleaner Production*, vol. 289, p. 125834, 2021, doi:10.1016/j.jclepro.2021.125834.

[28] R. R. Agrawal, B. Sharma, S. Gupta, and S. Shinde, *Fundamental of Robotic Process Automation (RPA)*. San International Scientific Publications, 2024. doi:10.59646/rpa/225.

[29] K. Kumar, R. Shah, N. Kumar, and R. P. Singh, "Application of Robotic Process Automation," in *Applications of Artificial Intelligence in Engineering*, X. Z. Gao, R. Kumar, S. Srivastava, and B. P. Soni (Eds.) Singapore: Springer, 2021, pp. 929–937. doi:10.1007/978-981-33-4604-8_75.

[30] W. A. Ansari, P. Diya, S. Patil, and S. Patil, "A review on robotic process automation – The future of business organizations," *SSRN Electronic Journal*, 2019, doi:10.2139/ssrn.3372171.

[31] S. Sahu, S. Salwekar, A. Pandit, and M. Patil, "Invoice processing using robotic process automation," *International Journal of Scientific Research in Computer Science, Engineering and Information Technology*, vol. 6, no. 2, pp. 216–223, 2020, doi:10.32628/CSEIT2062106.

[32] S. Kaushik, "Invoice processing using robotic process automation," *International Journal of Scientific Research and Engineering Trends*, vol. 9, no. 6, pp. 1–4, 2023, doi:10.61137/ijsret.vol.9.issue5.107.

[33] F. Delfino, L. Barillari, F. Pampararo, *et al.*, "Development and assessment of decentralized energy management system in a smart microgrid," in IISA 2014, *The 5th International Conference on Information, Intelligence, Systems and Applications*, IEEE, 2014, pp. 125–130. doi:10.1109/IISA.2014.6878746.

[34] C. Eksin, A. Hooshmand, and R. Sharma, "A decentralized energy management system," in *2015 European Control Conference (ECC)*, IEEE, 2015, pp. 2260–2267. doi:10.1109/ECC.2015.7330875.

[35] G. Sanchez-Guzman, W. Velasquez, and M. S. Alvarez-Alvarado, "Modeling a simulated Forest to get Burning Times of Tree Species using a Digital Twin," in *2022 IEEE 12th Annual Computing and Communication Workshop and Conference (CCWC)*, IEEE, 2022, pp. 639–643. doi:10.1109/CCWC54503.2022.9720768.

[36] G. K. Venayagamoorthy, R. K. Sharma, P. K. Gautam, and A. Ahmadi, "Dynamic energy management system for a smart microgrid," *IEEE Transactions of Neural Network Learning Systems*, vol. 27, no. 8, pp. 1643–1656, 2016, doi:10.1109/TNNLS.2016.2514358.

[37] S. Zhou, Z. Hu, W. Gu, M. Jiang, and X.-P. Zhang, "Artificial intelligence based smart energy community management: A reinforcement learning approach," *CSEE Journal of Power and Energy Systems*, vol. 5, no. 1, pp. 1–10, 2019, doi:10.17775/CSEEJPES.2018.00840.

[38] V. Atias, "Opportunities and challenges of using artificial intelligence in energy communities," in *2023 International Conference Automatics and Informatics (ICAI)*, Piscataway, NJ: IEEE, 2023, pp. 508–513. doi:10.1109/ICAI58806.2023.10339026.

[39] N. E. Benti, M. D. Chaka, and A. G. Semie, "Forecasting renewable energy generation with machine learning and deep learning: Current advances and future prospects," *Sustainability*, vol. 15, no. 9, p. 7087, 2023, doi:10.3390/su15097087.

[40] M. Taseen, WangYongli, and E. Ali, "Integration of Artificial Intelligence for Demand Forecasting and Resource Allocation in Renewable Energy Supply Chains," in *2023 25th International Multitopic Conference (INMIC)*, Piscataway, NJ: IEEE, 2023, pp. 1–7. doi:10.1109/INMIC60434.2023.10465757.

[41] K. Ukoba, K. O. Olatunji, E. Adeoye, T.-C. Jen, and D. M. Madyira, "Optimizing renewable energy systems through artificial intelligence: Review and future prospects," *Energy & Environment*, vol. 35, no. 7, pp. 3833–3879, 2024, doi:10.1177/0958305X241256293.

[42] M. Abdul Baseer, A. Almunif, I. Alsaduni, and N. Tazeen, "Electrical power generation forecasting from renewable energy systems using artificial intelligence techniques," *Energies (Basel)*, vol. 16, no. 18, 2023, doi:10.3390/en16186414.

[43] Syed Tanveer Alam, "The convergence of artificial intelligence, blockchain and fintech in energy, oil and gas trading: Increasing efficiency, transparency and automations," *World Journal of Advanced Research and Reviews*, vol. 22, no. 2, pp. 2064–2073, 2024, doi:10.30574/wjarr.2024.22.2.1625.

[44] S. Otoum and H. T. Mouftah, "Enabling Trustworthiness in Sustainable Energy Infrastructure Through Blockchain and AI-Assisted Solutions," *IEEE Wireless Communications*, vol. 28, no. 6, pp. 19–25, 2021, doi:10.1109/MWC.018.2100194.

[45] I. Antonopoulos, V. Robu, B. Couraud, *et al.*, "Artificial intelligence and machine learning approaches to energy demand-side response: A systematic review," *Renewable and Sustainable Energy Reviews*, vol. 130, p. 109899, 2020, doi:10.1016/j.rser.2020.109899.

[46] M. A. Khan, A. M. Saleh, M. Waseem, and I. A. Sajjad, "Artificial intelligence enabled demand response: Prospects and challenges in smart grid environment," *IEEE Access*, vol. 11, pp. 1477–1505, 2022, doi:10.1109/ACCESS.2022.3231444.

[47] A. Vergara-Romero, "Challenges and stakes of artificial intelligence in economic sciences," *Revista Amazonia Investiga*, vol. 12, no. 64, pp. 7–8, 2023, doi:10.34069/ai/2023.64.04.0.

[48] V. Franki, D. Majnarić, and A. Višković, "A comprehensive review of artificial intelligence (AI) companies in the power sector," *Energies (Basel)*, vol. 16, no. 3, p. 1077, 2023, doi:10.3390/en16031077.

[49] Maicon Roberto Martins, "From on-premise to cloud: Evolving IT infrastructure for the AI age," *World Journal of Advanced Research and Reviews*, vol. 20, no. 3, pp. 1898–1934, 2023, doi:10.30574/wjarr.2023.20.3.1590.

[50] A. Pachot and C. Patissier, "Towards sustainable artificial intelligence: An overview of environmental protection uses and issues," *Green and Low-Carbon Economy*, 2023, doi:10.47852/bonviewGLCE3202608.

[51] R. Sala, F. Filippini, D. Ardagna, D. Lezzi, F. Lordan, and P. Thiem, "Greening AI: A Framework for Energy-Aware Resource Allocation of ML Training Jobs with Performance Guarantees," in *Advanced Information Networking and Applications*, L. Barolli, (Ed.), Cham: Springer Nature, 2024, pp. 110–121. doi:https://doi.org/10.5281/zenodo.10592846.

[52] N. Mohamed, M. El-Guindy, A. Oubelaid, and S. K. Almazrouei, "Smart energy meets smart security: A Comprehensive review of AI applications in cybersecurity for renewable energy systems," *International Journal of Electrical and Electronics Research*, 11(3), 728–732, 2023, doi:10.37391/ijeer.110313.

[53] A. A. Bouramdane, "Cyberattacks in Smart Grids: Challenges and Solving the Multi-Criteria Decision-Making for Cybersecurity Options, Including Ones That Incorporate Artificial Intelligence, Using an Analytical Hierarchy Process," *Journal of Cybersecurity and Privacy*, vol. 3, no. 4, pp. 662–705, 2023, doi:10.3390/jcp3040031.

[54] N. Mohamed, M. E.-G. El-Guindy, A. Oubelaid, and S. khameis Almazrouei, "Smart energy meets smart security: A comprehensive review of ai applications in cybersecurity for renewable energy systems," *International Journal of Electrical and Electronics Research*, vol. 11, no. 3, pp. 728–732, 2023, doi:10.37391/ijeer.110313.

[55] N. Mengidis, T. Tsikrika, S. Vrochidis, and I. Kompatsiaris, "Blockchain and AI for the next generation energy grids: Cybersecurity challenges and opportunities," *Information & Security: An International Journal*, vol. 43, no. 1, pp. 21–33, 2019, doi:10.11610/isij.4302.

[56] A. Govori and Q. Sejdija, "Future prospects and challenges of integrating artificial intelligence within the business practices of small and medium enterprises," *Journal of Governance and Regulation*, vol. 12, no. 2, pp. 176–183, 2023, doi:10.22495/jgrv12i2art16.

[57] L. Oldemeyer, A. Jede, and F. Teuteberg, "Investigation of artificial intelligence in SMEs: a systematic review of the state of the art and the main implementation challenges," *Management Review Quarterly*, vol. 75, pp. 1185–1227, 2025, doi:10.1007/s11301-024-00405-4.

[58] K. Kotilainen, U. A. Saari, J. Valta, S. J. Makinen, and S. Kufeouglu, "Business model readiness of start-up driven energy innovations – An empirical review," in *2019 16th International Conference on the European Energy Market (EEM)*, Ljubljana, Slovenia: IEEE, 2019, pp. 1–6. doi:10.1109/EEM.2019.8916541.

[59] N. Hashai and S. Markovich, "Market entry by high technology startups: The effect of competition level and startup innovativeness," *Strategy Science*, vol. 2, no. 3, pp. 141–160, 2017, doi:10.1287/stsc.2017.0033.

Chapter 10
Federated learning for energy management applications

Rami Hamdi[1] *and Marwa Qaraqe*[2]

Remote monitoring systems play a crucial role in analyzing environmental dynamics across various smart industrial applications, including occupational health and safety, as well as environmental monitoring. In the context of industrial Internet of Things (IoT) systems, the vast proliferation of devices combined with the stringent performance requirements imposes significant pressure on resources such as computational power, network bandwidth, and device energy.

The challenge is particularly pronounced when it comes to the distributed training of machine learning (ML) and deep learning (DL) models, which are essential for enabling intelligent functionalities in industrial IoT applications. These tasks are especially difficult to execute on resource-constrained devices operating within heterogeneous wireless networks. Managing these limitations while ensuring the effective deployment of ML/DL models requires innovative approaches to optimize resource utilization, energy management, and maintain system performance. In this chapter, we develop a framework that enhances energy efficiency in Federated Learning (FL) over wireless networks considering various network architectures and assumptions, and integrating energy-harvesting technologies.

10.1 Introduction

ML is poised to be a transformative force in the design and optimization of 6G architectures, protocols, and operations. As next-generation networks evolve, ML will be essential for enabling intelligent management of wireless resources and devices. However, the centralized ML approaches traditionally employed are turned to be impractical due to the sheer volume of data generated at the edge, coupled with the constraints of limited power, bandwidth, and growing privacy concerns [1].

FL emerges as a promising alternative by allowing devices to train ML models locally, without requiring data transmission to a central server. Instead, each edge device trains its model independently and then shares only the model parameters with a central server for aggregation [2]. This process results in a shared ML model that

[1]AFG College, University of Aberdeen, Qatar
[2]College of Science and Engineering, Hamad Bin Khalifa University, Qatar

is refined iteratively over several communication rounds until the desired accuracy is achieved. By keeping data localized on the edge devices, FL not only enhances data privacy and security but also reduces the strain on network resources [3].

The potential of FL in edge computing has been demonstrated, particularly in scenarios like predicting content popularity based on user interactions. However, implementing FL in wireless networks is fraught with challenges. Multi-user interference, energy constraints, and the heterogeneity of devices – varying in data models and hardware capabilities – pose significant hurdles [4]. Additionally, the large number of edge devices involved and the limited bandwidth turn communication into a major bottleneck. To effectively deploy FL in wireless environments, it is imperative to address resource allocation challenges, including user scheduling, resource block allocation, energy management, and bandwidth partitioning [5,6].

Implementing FL across distributed users inevitably leads to an increase in the power consumption of these devices. To address this challenge, powering devices through energy-harvesting technologies presents a viable solution. Energy harvesting enables devices to continuously gather energy from natural sources [7]. This approach allows each device to utilize the harvested energy for updating and transmitting FL model parameters, thereby reducing the dependency on traditional power sources. By integrating energy harvesting into FL systems, a greater number of devices can take part in the training process without being constrained by battery limitations [5]. This not only broadens the scope of FL deployments but also enhances the efficiency of the training process. With more devices contributing to the model training, the convergence time of FL can be significantly improved, leading to faster and more accurate results [8]. Moreover, the ability to sustain longer participation in FL tasks through renewable energy sources can reduce training loss as the model benefits from the continuous input of diverse data sources. In essence, leveraging energy harvesting in FL networks not only mitigates power consumption concerns but also enhances the scalability and performance of FL implementations, making it a key enabler for the future of distributed AI systems.

In this chapter, we explore an energy-harvesting FL framework, where each user relies on an individual energy-harvesting source. In this context, a joint energy management and user scheduling problem is formulated to optimize the system's performance. The impact of energy management and wireless parameters on system performance is investigated. The framework is further extended to multiple base stations (BSs). Establishing the connection between FL's convergence and user association, the extended optimization problem is solved using branch and bound techniques. This approach aims to efficiently solve the user association challenge, thereby enhancing the system performance and scalability of the proposed energy-harvesting FL system.

10.2 Federated learning

10.2.1 Definition

FL is a distributed ML method that allows model training across multiple distributed devices or servers holding local data samples, without requiring the transfer of the data to a central location. This method is particularly beneficial in scenarios where data privacy, security, or bandwidth limitations make it impractical to centralize large amounts of data. Instead of sending raw data to a central server, each device involved in FL independently trains a local model on its own data and periodically shares only the model updates (e.g., gradients or weights) with a central server. The central server then aggregates these updates to form a global model, which is subsequently redistributed to the devices for further training in the next round. This iterative process continues until the global model converges, achieving a level of performance comparable to what would be expected if the data had been centralized.

10.2.2 System model

We investigate an FL-based wireless communication system [5], where a BS equipped with a massive multiple-input mutiple-output (MIMO) system, consisting of K antennas, serves N single-antenna users. Considering the condition $K \gg N$ holds, we ensure that the BS has far more antennas than the number of users it serves, as illustrated in Figure 10.1.

The communication process is divided into L time frames, each with a duration T_{fl}. Every user n in the system possesses its own dataset $\mathscr{D}_n = \{\mathbf{A}_n, \mathbf{b}_n\}$, where $\mathbf{A}_n = [\mathbf{a}_{n,1}, \mathbf{a}_{n,2}, \ldots, \mathbf{a}_{n,M_n}]$ represents the input data, and $\mathbf{b}_n = [b_{n,1}, b_{n,2}, \ldots, b_{n,M_n}]$ represents the corresponding outputs. Here, M_n denotes the number of samples collected by user n.

The total number of training samples across all users is given by $M = \sum_{n=1}^{N} M_n$. At each time frame t, the weight vector $\mathbf{w}_n(t)$ represents the parameters of the local FL model for user n. The BS's objective is to optimize the weight vectors of these

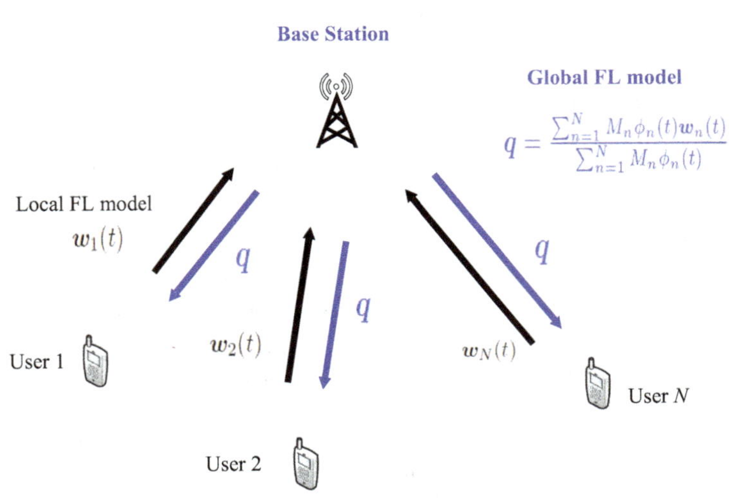

Figure 10.1 FL architecture for energy-harvesting wireless networks

local FL models to reduce a global loss function, leveraging the entire dataset from all participating users. The global loss function is mathematically expressed as

$$\frac{1}{N}\sum_{n=1}^{N}\sum_{m=1}^{M_n} f(\mathbf{w}_n(t), \mathbf{a}_{n,m}, b_{n,m}) \tag{10.1}$$

where $f(\mathbf{w}_n(t), \mathbf{a}_{n,m}, b_{n,m})$ indicates the loss corresponding to the data point $\mathbf{a}_{n,m}$. This formulation enables the system to aggregate and learn from distributed data, optimizing the performance of the overall FL model across the network.

After collecting the local FL models from users, the BS combines them to produce the global FL model [6]. We denote by $\phi_n(t)$ as an index that equals 1 if user n is scheduled at frame t, and 0 if not. Only the users whose signal-to-interference-plus-noise ratio (SINR) at frame t exceeds the minimum threshold γ_{th} can send their local FL models to the BS. Consequently, the global FL model update can be formulated as

$$\mathbf{q}(\phi(t)) = \frac{\sum_{n=1}^{N} M_n \phi_n(t) \mathbf{w}_n(t)}{\sum_{n=1}^{N} M_n \phi_n(t)} \tag{10.2}$$

The global loss function reflecting the efficiency of the FL model across all users is then given by

$$F(\mathbf{q}) = \frac{1}{N}\sum_{n=1}^{N}\sum_{m=1}^{M_n} f(\mathbf{q}(\phi(t)), \mathbf{a}_{n,m}, b_{n,m}) \tag{10.3}$$

To refine the local FL model, a gradient descent method is employed, updating the model parameters at each user as follows:

$$\mathbf{w}_n(t) = \mathbf{q}(\phi(t)) - \frac{\lambda}{M_n}\sum_{m=1}^{M_n} \nabla f(\mathbf{q}(\phi(t)), \mathbf{a}_{n,m}, b_{n,m}) \tag{10.4}$$

where λ is the learning rate and $\nabla f(\mathbf{q}(\phi(t)), \mathbf{a}_{n,m}, b_{n,m})$ denotes the gradient of the loss function $f(\mathbf{q}(\phi(t)), \mathbf{a}_{n,m}, b_{n,m})$ with respect to the global model $\mathbf{q}(\phi(t))$.

10.3 Wireless model

We describe in this section a typical massive MIMO wireless system. We denote by $\mathbf{h}_n(t) \in \mathbb{C}^{K\times 1}$ the channel for user n at frame t. We represent the path loss by $\beta_n = \zeta \frac{d_n^{-\nu}}{d_0^{-\nu}}$, where ν is the path loss exponent, d_n is the distance between the BS and user n, d_0 is the reference distance, and ζ is a constant related to the carrier frequency and the reference distance. Given these components, the overall channel vector $\mathbf{g}_n(t) \in \mathbb{C}^{N\times 1}$ for user n at frame t is defined as $\mathbf{g}_n(t) = \beta_n^{1/2}\mathbf{h}_n(t)$. The corresponding channel matrix that aggregates the channel vectors for all n users is given by $\mathbf{G}(t) = [\mathbf{g}_1(t), \mathbf{g}_2(t), \ldots, \mathbf{g}_n(t)] \in \mathbb{C}^{K\times N}$.

We use zero forcing (ZF) as receiver, which is shown to attain nearly optimal performance [9]. ZF is particularly effective for massive MIMO systems, offering a practical solution to manage interference and enhance signal quality. The uplink receiver $\mathbf{Z}(t)$ at frame t is defined as $\mathbf{Z}(t) = \mathbf{G}(t)\mathbf{J}(t)$, where $\mathbf{J}(t) = (\mathbf{G}^H(t)\mathbf{G}(t))^{-1}$. The SINR for user n in the uplink, assuming perfect channel state information (CSI), is given by

$$\gamma_n(t) = \frac{p_n(t)}{[\mathbf{J}(t)]_{n,n}\sigma^2} \tag{10.5}$$

where $p_n(t)$ represents the transmit power of user n at frame t and σ^2 is the noise variance.

This formulation highlights the role of ZF in mitigating multi-user interference by effectively inverting the channel matrix $\mathbf{G}(t)$, thereby enhancing the quality of the received signals. The use of ZF, combined with precise power allocation and accurate CSI, allows for significant improvements in uplink performance in massive MIMO systems.

10.4 Energy model

In our system, we assume that each user operates using a dedicated energy-harvesting source [7]. Each user harvests energy, modeled as a compound Poisson stochastic process, which is kept in a battery with a limited storage capacity of $B_{\max}$. We define $E_n(t)$ as the energy harvested by user n during frame t and $B_n(t)$ as the associated battery level. The consumed energy consumption by user n to train its model is denoted by $E_{\text{fl}} = \rho j_n$, where ρ is a fixed energy coefficient and j_n represents the input data size utilized by user n. The energy needed to transmit the FL model must remain within the available battery capacity [10], leading to the following constraint:

$$E_{\text{fl}} + p_n(t)T_{\text{fl}} \leq B_n(t) \tag{10.6}$$

where $p_n(t)$ represents the transmission power of user n at frame t and T_{fl} is the duration of the time frame. The battery level is updated at each frame according to the following equation:

$$B_n(i+1) = \min(B_{\max}, B_n(t) - p_n(t)T_{\text{fl}} + E_n(t)) \tag{10.7}$$

This update ensures that the battery level does not exceed its maximum capacity while accounting for both energy consumption and newly harvested energy.

Due to the limited availability of energy and the impact of multi-user interference, not all users may be able to perform the FL model update at every time frame. As a result, only a subset of users, determined by their energy levels and interference conditions, can contribute to the FL model training and transmission in each frame.

10.5 Energy management and user scheduling problem

The goal is to reduce the training loss of the FL model while efficiently utilizing the available harvested energy and guaranteeing an SINR threshold. Consequently, the problem of jointly managing energy and scheduling users in an energy-harvesting FL system is given by

$$\min_{\substack{\{\phi_n(t),p_n(t)\}\\ n=1,\ldots,N,t=1,\ldots,L}} \quad F(\mathbf{q}) \tag{10.8}$$

$$\text{s. t. } \gamma_n(t) \geq \phi_n(t)\gamma_{\text{th}}, \quad \forall n = 1, \ldots, N, t = 1, \ldots, L, \tag{10.8a}$$

$$\sum_{t=1}^{l} \phi_n(t)(p_n(t)T_{\text{fl}} + E_{\text{fl}}) \leq \sum_{t=1}^{l} E_n(t), \quad \forall n = 1, \ldots, N, l = 1, \ldots, L, \tag{10.8b}$$

$$\sum_{t=1}^{l} E_n(t) - \sum_{t=1}^{l-1} \phi_n(t)(p_n(t)T_{\text{fl}} + E_{\text{fl}}) \leq B_{\max}, \quad \forall n = 1, \ldots, N, l = 2, \ldots, L, \tag{10.8c}$$

$$p_n(t) \geq 0, \quad \forall n = 1, \ldots, N, t = 1, \ldots, L, \tag{10.8d}$$

$$\phi_n(t) \in \{0, 1\}, \quad \forall n = 1, \ldots, N, t = 1, \ldots, L. \tag{10.8e}$$

Constraint (10.8a) ensures that each user achieves a minimum received SINR, denoted as γ_{th}. Constraint (10.8b) addresses energy causality, meaning that the energy consumed by user n cannot exceed the energy currently available in their battery. Furthermore, constraint (10.8c) stipulates that the harvested energy during the current frame cannot surpass the battery's maximum capacity. Lastly, constraint (10.8d) guarantees that allocated power levels remain non-negative.

The optimization problem (10.8) is characterized as a mixed-integer non-linear problem due to non-linearity of the objective function and its combinatorial nature. Thus, it is essential to analyze the impact of both the transmit power $p_n(t)$ and the user scheduling $\phi_n(t)$ on the training loss of the FL model in order to simplify the problem and to propose an efficient solution.

10.6 Impact of wireless parameters on FL performance

To address the problem outlined in (10.8), we begin by establishing the connection between user scheduling and the performance of FL. Subsequently, we conduct a detailed analysis of the convergence rate of the FL process, as described below.

Theorem 1. *The expected convergence rate of the FL algorithm is given by*

$$\mathbb{E}\left[F\left(\mathbf{q}_{t+1}\right) - F\left(\mathbf{q}^*\right)\right] \leq \sum_{i=0}^{t} \left(\prod_{j=i+1}^{t} v_j \right) z_i + \mathbb{E}\left(F\left(\mathbf{q}_0\right) - F\left(\mathbf{q}^*\right)\right) \prod_{i=0}^{t} v_i \tag{10.9}$$

where

$$z_i = \frac{2\zeta_1}{VM} \sum_{n=1}^{N} M_n \left(1 - \phi_n(t)\right) \tag{10.10}$$

$$v_i = 1 - \frac{\mu}{V} + \frac{4\mu\zeta_2}{VM} \sum_{n=1}^{N} M_n \left(1 - \phi_n(t)\right) \tag{10.11}$$

The optimal FL model, denoted as $\mathbf{q}^*$*, represents the target to which the FL process aims to converge. The corresponding optimal training loss is defined as* $F(\mathbf{q}^*) = \frac{1}{N}\sum_{n=1}^{N}\sum_{m=1}^{M_n} f(\mathbf{q}^*, \mathbf{a}_{n,m}, b_{n,m})$*. Here, V represents the Lipschitz constant, ensuring the smoothness of the function. The coefficients* ζ_1*,* ζ_2*, and* μ *are derived from the second-order Taylor expansion of the loss function, providing a local approximation around* $\mathbf{q}^*$*.*

Proof. The proof is given in Appendix A. □

From (10.9), it is evident that a gap $\sum_{i=0}^{t-1}\left(\prod_{j=t+1}^{t-1} v_j\right) z_i$ exists between $\mathbb{E}\left[F(\mathbf{q}_t)\right]$ and $\mathbb{E}\left[F(\mathbf{q}^*)\right]$. This gap arises due to the user scheduling policy, which diminishes as the number of scheduled users increases. Consequently, optimizing the number of scheduled users enhances the convergence speed of the FL algorithm.

When every user is selected, i.e., $\phi_n(t) = 1$, the parameters simplify to $z_i = 0$ and $v_i = 1 - \frac{2\mu}{V}$. Under these conditions, the expected convergence rate of the FL algorithm can be expressed as

$$\mathbb{E}\left[F(\mathbf{q}_{t+1}) - F(\mathbf{q}^*)\right] \leq \mathbb{E}\left(F(\mathbf{q}_0) - F(\mathbf{q}^*)\right)\left(1 - \frac{2\mu}{V}\right)^t \tag{10.12}$$

The connection between wireless and FL parameters can be leveraged to simplify and ultimately solve the optimization problem defined in (10.8).

10.7 Optimal energy management and user scheduling

After identifying the connection between the FL convergence rate and the wireless parameters, the objective function of the primary problem (10.8) is reformulated and simplified to $\sum_{i=0}^{t}\left(\prod_{j=i+1}^{t} v_j\right) z_i$ based on the inequality in (10.9). Consequently, the main problem (10.8) can be decomposed into smaller sub-problems. The objective function at frame t is given by $z_t + v_t\mathbb{E}\left[F(\mathbf{q}_t) - F(\mathbf{q}^*)\right]$ using (A.3). Therefore, the optimization problem at frame t can be defined by

$$\min_{\substack{\{p_n(t),\phi_n(t)\}\\ n=1,\ldots,N}} z_t + v_t\mathbb{E}\left(F(\mathbf{q}_t) - F(\mathbf{q}^*)\right) \tag{10.13}$$

$$\text{s. t. } \gamma_n(t) \geq \phi_n(t)\gamma_{\text{th}}, \quad \forall n = 1,\ldots,N, \tag{10.13a}$$

$$\phi_n(t)(p_n(t)T_{\text{fl}} + E_{\text{fl}}) \leq B_n(t), \quad \forall n = 1,\ldots,N, \tag{10.13b}$$

$$B_n(t) + E_n(t) - \phi_n(t)(p_n(t)T_{\text{fl}} + E_{\text{fl}}) \leq B_{\max}, \quad \forall n = 1,\ldots,N, \tag{10.13c}$$

$$p_n(t) \geq 0, \quad \forall n = 1,\ldots,N, \tag{10.13d}$$

$$\phi_n(t) \in \{0,1\}, \quad \forall n = 1,\ldots,N. \tag{10.13e}$$

The impact of energy consumption on the training process can be thoroughly analyzed. Specifically, the energy required to train the model over a predefined duration.

In this context, the necessary transmit power to satisfy the quality of service (QoS) can be expressed as

$$p_n(t) = \phi_n(t)\gamma_{\mathrm{th}}[\mathbf{J}(t)]_{n,n}\sigma^2 \tag{10.14}$$

Therefore, the required energy required to transmit the weight $\mathbf{w}_n(t)$ for user n at frame t is determined by

$$e_n(t) = \phi_n(t)\left(\gamma_{\mathrm{th}}[\mathbf{J}(t)]_{n,n}\sigma^2 + E_{\mathrm{fl}}\right). \tag{10.15}$$

Consequently, the problem (10.13) can be simplified and expressed as

$$\min_{\substack{\{\phi_n(t)\}\\ n=1,\ldots,N}} z_t + v_t\mathbb{E}\left(F\left(\mathbf{q}_t\right) - F\left(\mathbf{q}^*\right)\right) \tag{10.16}$$

$$\text{s. t. } \phi_n(t)(\gamma_{\mathrm{th}}[\mathbf{J}(t)]_{n,n}\sigma^2 T_{\mathrm{fl}} + E_{\mathrm{fl}}) \leq B_n(t), \quad \forall n = 1,\ldots,N, \tag{10.16a}$$

$$B_n(t) + E_n(t) - \phi_n(t)(\gamma_{\mathrm{th}}[\mathbf{J}(t)]_{n,n}\sigma^2 T_{\mathrm{fl}} + E_{\mathrm{fl}}) \leq B_{\max}, \quad \forall n = 1,\ldots,N, \tag{10.16b}$$

$$\phi_n(t) \in \{0, 1\}, \quad \forall n = 1,\ldots,N. \tag{10.16c}$$

The objective function in (10.16) can be reformulated as

$$\begin{aligned} z_t + v_t u_t &= \frac{2\zeta_1}{VM}\sum_{n=1}^{N} M_n\left(1 - \phi_n(t)\right) \\ &+ \left(1 - \frac{\mu}{V} + \frac{4\mu\zeta_2}{VM}\sum_{n=1}^{N} M_n\left(1 - \phi_n(t)\right)\right) u_i \\ &= \left(\frac{2\zeta_1}{VM} + \frac{4\mu\zeta_2}{VM}u_t\right)\sum_{n=1}^{N} M_n\left(1 - \phi_n(t)\right) \\ &+ \left(1 - \frac{\mu}{V}\right) u_t \end{aligned} \tag{10.17}$$

As a result, the objective function can be simplified to the equivalent form $-\sum_{n=1}^{N} M_n\phi_n(t)$. Simultaneously, the constraints are also simplified, and the optimization problem is reformulated as

$$\min_{\substack{\{\phi_n(t)\}\\ n=1,\ldots,N}} -\sum_{n=1}^{N} M_n\phi_n(t) \tag{10.18}$$

$$\text{s. t. } \phi_n(t) \leq \frac{B_n(t)}{\gamma_{\mathrm{th}}[\mathbf{J}(t)]_{n,n}\sigma^2 T_{\mathrm{fl}} + E_{\mathrm{fl}}}, \quad \forall n = 1,\ldots,N, \tag{10.18a}$$

$$-\phi_n(t) \leq \frac{B_{\max} - B_n(t) - E_n(t)}{\gamma_{\mathrm{th}}[\mathbf{J}(t)]_{n,n}\sigma^2 T_{\mathrm{fl}} + E_{\mathrm{fl}}}, \quad \forall n = 1,\ldots,N, \tag{10.18b}$$

$$\phi_n(t) \in \{0, 1\}, \quad \forall n = 1,\ldots,N. \tag{10.18c}$$

Solving (10.18) corresponds to maximizing the number of selected users while meeting the requirements (10.18a)–(10.18c). Therefore, the optimal user scheduling can be determined by

$$\phi_n(t) = \begin{cases} 1, & B_n(t) + E_n(t) - B_{\max} \leq \gamma_{\text{th}}[\mathbf{J}(t)]_{n,n}\sigma^2 T_{\text{fl}} + E_{\text{fl}} \leq B_n(t), \\ 0, & \text{otherwise} \end{cases} \tag{10.19}$$

From (10.19), it is evident that user scheduling is heavily influenced by the channel coefficients, available energy, and the maximum battery capacity. These factors play a crucial role in determining the optimal scheduling strategy. To address this, we propose a user scheduling and energy management (USEM) algorithm, which efficiently balances these parameters. The developed algorithm is described in **Algorithm 1**.

The algorithm can be adapted for situations where users have multiple antennas by adopting block diagonalization (BD) as beamforming. This adaptation is particularly beneficial because, with multiple antennas, multi-user interference can be effectively mitigated using the BD approach. BD is recognized for achieving nearly optimal performance while maintaining computational efficiency, making it a suitable choice for enhancing system performance in multi-antenna environments.

We analyze now the implementation of the proposed USEM algorithm. To implement USEM, the BS must first compute two critical terms for each user: $B_n(t) + E_n(t) - B_{\max}$ and $\gamma_{\text{th}}[\mathbf{J}(t)]_{n,n}\sigma^2$. The BS station must be aware of both the current battery level $B_n(t)$ and the energy harvested $E_n(t)$ by each user to compute the first term. The battery level $B_n(t)$ at frame t is obtained after the BS calculates the energy consumed $e_n(t-1)$ and takes into account the harvested energy $E_n(t-1)$

Algorithm 1 Optimal user scheduling and energy management

```
B_n(1) ← E_n(1), n = 1...N
for t = 1 : L do
    for n = 1 : N do
        if B_n(t) + E_n(t) − B_max ≤ γ_th[J(t)]_{n,n} σ^2 then
            φ_n(t) ← 1
            p_n(t) ← γ_th[J(t)]_{n,n} σ^2
            e_n(t) ← p_n(t) T_fl + E_c
            weight vector w_n(t) sent to the BS
        else
            φ_n(t) ← 0
            e_n(t) ← 0
        end if
        if t < L then
            B_n(t+1) ← min(B_max, B_n(t) − e_n(t) + E_n(t+1))
        end if
    end for
    q ← (Σ_{n=1}^N M_n φ_n(t) w_n(t)) / (Σ_{n=1}^N M_n φ_n(t))
end for
```

from the previous frame for each user. The harvested energy $E_n(t)$ can either be directly reported by the users to the BS or predicted by the BS using prediction algorithms [7]. For the other term, the BS computes the matrix $\mathbf{J}(t)$, which is dependent on the channel $\mathbf{G}(t)$. Once these two terms are calculated, the BS proceeds with user scheduling and updates the battery levels by determining the energy consumption $e_n(t)$ for each user. Following the user scheduling process, the FL mechanism is executed: the BS transmits the FL model information to the scheduled users and collects their local updates. Finally, the BS aggregates these updates to refine and produce the global model.

The computational complexity of the USEM algorithm is investigated by examining its key operations. Calculating the matrix $\mathbf{J}(t)$ involves one matrix multiplication, which has a complexity of $O(KN^2)$, and one matrix inversion, which has a complexity of $O(N^3)$. Other terms, such as the consumed energy, are determined using straightforward arithmetic operations with negligible complexity. Therefore, the overall complexity for user scheduling across L frames is given by the polynomial complexity $O\left(L(KN^2 + N^3)\right)$. This indicates that the computational requirements of USEM scale polynomially with the number of users K, the number of antennas K, and the number of frames L, making it feasible for practical implementation in large-scale systems.

10.8 Multiple base stations

We extend the system model from a single BS to a network of multiple base stations, enabling the implementation of FL across a large-scale network with numerous BSs and users. In a multi-BS scenario, it is crucial to account for the interference not only between the BSs themselves but also among the users connected to them. Additionally, optimizing user association with different BSs becomes essential, particularly under the constraints of energy efficiency and QoS. The decision-making process for user association is guided by large-scale fading characteristics, with the user association index defined as follows:

$$\alpha_{n,c} = \begin{cases} 1, & \text{if user } k \text{ is associated to BS } c, \\ 0, & \text{otherwise} \end{cases} \tag{10.20}$$

Let C represent the total number of BSs. We consider that each user can be associated with no more than one BS at a time, as follows:

$$\sum_{c=1}^{C} \alpha_{n,c} \leq 1, \quad \forall n = 1, \ldots, N \tag{10.21}$$

Every BS possesses accurate CSI for all users. This allows each BS to employ a cooperative ZF to eliminate both intra-cell and inter-cell interference. The ZF receiver $\mathbf{J}_c(t)$ is expressed as
$\mathbf{J}_c(t) = \mathbf{G}_c(t)\left(\mathbf{G}_c(t)^H \mathbf{G}_c(t)\right)^{-1}$. Consequently, the SINR is formulated as

$$\gamma_{n,c}(t) = \frac{p_{n,c}(t)}{[\mathbf{J}_c(t)]_{n,n}\sigma^2} \tag{10.22}$$

where $p_{n,c}(t)$ represents the transmit power from user n to BS c.

With a particular user association setup, the BSs select users for the transmission of local FL models. Consequently, the problem of optimizing the joint tasks of user association, user scheduling, and power allocation in the FL context is given by

$$\min_{\substack{\{\alpha_{n,c},\phi_{n,c}(t),p_n(t)\} \\ n=1,\ldots,N,t=1,\ldots,L,c=1,\ldots,C}} F(\mathbf{q}) \tag{10.23}$$

$$\text{s. t. } \gamma_{n,c}(t) \geq \alpha_{n,c}\phi_{n,c}(t)\gamma_{\text{th}}, \quad \forall n = 1,\ldots,N, c = 1,\ldots,C, t = 1,\ldots,L, \tag{10.23a}$$

$$\sum_{c=1}^{C} \alpha_{n,c} \leq 1, \quad \forall n = 1,\ldots,N, \tag{10.23b}$$

$$\sum_{t=1}^{l} \phi_{n,c}(t)(p_{n,c}(t)T_{\text{fl}} + E_{\text{fl}}) \leq \alpha_{n,c}\sum_{t=1}^{l} E_n(t), \quad \forall n = 1,\ldots,N, \\ l = 1,\ldots,L, c = 1,\ldots,C, \tag{10.23c}$$

$$-\sum_{t=1}^{l-1} \phi_{n,c}(t)(p_{n,c}(t)T_{\text{fl}} + E_{\text{fl}}) \leq \alpha_{n,c}(B_{\max} - \sum_{t=1}^{l} E_n(t)), \\ \forall n = 1,\ldots,N, l = 2,\ldots,L, c = 1,\ldots,C, \tag{10.23d}$$

$$p_{n,c}(t) \geq 0, \quad \forall n = 1,\ldots,N, t = 1,\ldots,L, c = 1,\ldots,C, \tag{10.23e}$$

$$\alpha_{n,c} \in \{0,1\}, \quad \forall n = 1,\ldots,N, c = 1,\ldots,C. \tag{10.23f}$$

$$\phi_{n,c}(t) \in \{0,1\}, \quad \forall n = 1,\ldots,N, t = 1,\ldots,L, \\ c = 1,\ldots,C \tag{10.23g}$$

This formulation aims to balance the trade-offs between user association and resource allocation while ensuring efficient model updates and minimizing interference across the network. From (10.23b) and (10.23c), it is evident that when $\alpha_{n,c} = 0$, the user scheduling index $\phi_{n,c}(t)$ is automatically set to 0. However, when $\alpha_{n,c} = 1$, the user scheduling index $\phi_{n,c}(t)$ can take a value of either 0 or 1.

The inequality in (10.9), which depends on both the FL model and wireless parameters, is adapted for the multi-BS system model by incorporating the user association index $\sum_{c=1}^{C} \alpha_{n,c}$. Here, $\sum_{c=1}^{C} \alpha_{n,c} = 1$ indicates that user n is associated with exactly one BS, while $\sum_{c=1}^{C} \alpha_{n,c} = 0$ implies that user n is not associated with any BS. The expected convergence rate of the FL algorithm under these conditions is given by

$$\mathbb{E}\left[F(\mathbf{q}_{t+1}) - F(\mathbf{q}^*)\right] \leq \sum_{i=0}^{t} \left(\prod_{j=t+1}^{t} d_j\right) c_i + \mathbb{E}\left(F(\mathbf{q}_0) - F(\mathbf{q}^*)\right) \prod_{i=0}^{t} d_i \tag{10.24}$$

where

$$c_i = \frac{2\zeta_1}{VM}\sum_{n=1}^{N} M_n\left(1 - \sum_{c=1}^{C}\phi_{n,c}(t)\right) \tag{10.25}$$

and

$$d_i = 1 - \frac{\mu}{V} + \frac{4\mu\zeta_2}{VM}\sum_{n=1}^{N} M_n\left(1 - \sum_{c=1}^{C}\phi_{n,c}(t)\right) \tag{10.26}$$

The objective function can be simplified using the inequality in (10.24), resulting in the expression $\sum_{t=1}^{L}\left(\prod_{j=t+1}^{L} d_j\right) c_t$. The minimization of this expression can be further simplified by applying (10.25) and (10.26), leading to the minimization of $\sum_{t=1}^{L}\sum_{n=1}^{N} M_n\left(1-\sum_{c=1}^{C}\phi_{n,c}(t)\right)$. Additionally, the constraints can be simplified using (10.15), allowing the optimization problem initially presented in (10.23) to be reformulated as

$$\min_{\substack{\{\alpha_{n,c},\phi_{n,c}(t)\}\\ n=1,\ldots,N,t=1,\ldots,L,s=1,\ldots,S}} \sum_{t=1}^{L}\sum_{n=1}^{N} M_n\left(1-\sum_{c=1}^{C}\phi_{n,c}(t)\right) \tag{10.27}$$

$$\text{s. t. } \textstyle\sum_{c=1}^{C}\alpha_{n,c}\le 1,\quad \forall n=1,\ldots,N, \tag{10.27a}$$

$$\textstyle\sum_{t=1}^{l}\phi_{n,c}(t)(\gamma_{\text{th}}[\mathbf{J}_c(t)]_{n,n}\sigma^2 T_{\text{fl}}+E_{\text{fl}})\le\alpha_{n,c}\sum_{t=1}^{l}E_n(t),$$
$$\forall n=1,\ldots,N, l=1,\ldots,L, c=1,\ldots,C, \tag{10.27b}$$

$$-\textstyle\sum_{t=1}^{l-1}\phi_{n,c}(t)(\gamma_{\text{th}}[\mathbf{J}_c(t)]_{n,n}\sigma^2 T_{\text{fl}}+E_{\text{fl}})\le\alpha_{n,c}(B_{\max}-\sum_{t=1}^{l}E_n(t)),$$
$$\forall n=1,\ldots,N, l=2,\ldots,L, c=1,\ldots,C, \tag{10.27c}$$

$$\alpha_{n,c}\in\{0,1\},\quad \forall n=1,\ldots,N, c=1,\ldots,C. \tag{10.27d}$$

$$\phi_{n,c}(t)\in\{0,1\},\quad \forall n=1,\ldots,N, t=1,\ldots,L,$$
$$c=1,\ldots,C \tag{10.27e}$$

The optimization problem (10.27) is defined as an integer linear problem, which may be efficiently solved using branch and bound techniques. In the user association problem, similar parameter calculations as in the USEM algorithm are necessary, including matrix inversion and energy computation. The optimization problem involves $2KLC+K$ constraints and KC variables. The optimal user association is determined using the branch and bound technique, which can operate in polynomial time. Given that BSs typically possess sufficient computational resources, they are capable of executing this method efficiently.

10.9 Performance evaluation

We assess the performance of the developed FL and energy management framework via numerical simulations. The simulations consider users randomly distributed within a circular cell. We investigate two scenarios: a single BS (Scenario 1) and multiple BSs (Scenario 2). The simulation parameters utilized in this evaluation are detailed in Table 10.1.

In Figure 10.2, the effectiveness of the optimal user scheduling scheme is examined by analyzing the number of scheduled users as a function of the minimum received SINR. As anticipated, a more stringent QoS constraint results in a reduced average number of scheduled users. The figure clearly illustrates that the proposed scheduling algorithm significantly outperforms the round-robin approach, increasing the number of participating devices in FL by up to 100% at 5 dB and 50% at

Table 10.1 Simulation parameters

Symbol	Description	Value
ν	Path loss exponent	3.7
$B_{\max}$	Max battery capacity	50 J
K	Number of antennas	128
N	Number of users	10
L	Number of frames	300
	Noise PSD	−174 dBm/Hz
	Circuit power per RF chain	30 dBm
	Cell radius	500 m

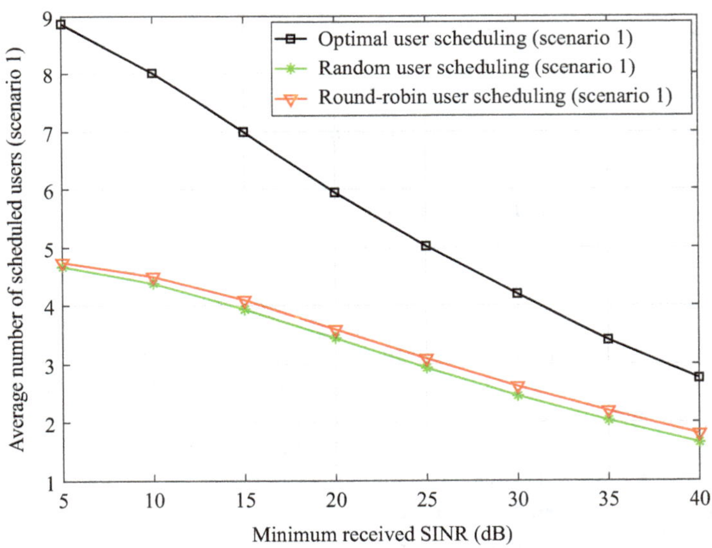

Figure 10.2 Impact of SINR on the average number of scheduled users across various scheduling schemes

40 dB. However, the performance gap narrows at higher SINR thresholds due to the dominant influence of interference. This improved performance can be attributed to the algorithm's efficient energy management strategy, which selectively deactivates users with higher energy demands, thereby optimizing the utilization of harvested energy.

Figure 10.3 illustrates the performance of the user scheduling algorithms in relation to the energy arrival rate. The figure demonstrates that as the energy arrival rate increases, both scheduling schemes can accommodate a higher number of scheduled devices. However, this improvement is not indefinite; the performance, particularly for the random scheduling algorithm, begins to plateau as the energy arrival rate continues to rise. This saturation indicates that, beyond a certain point, additional energy

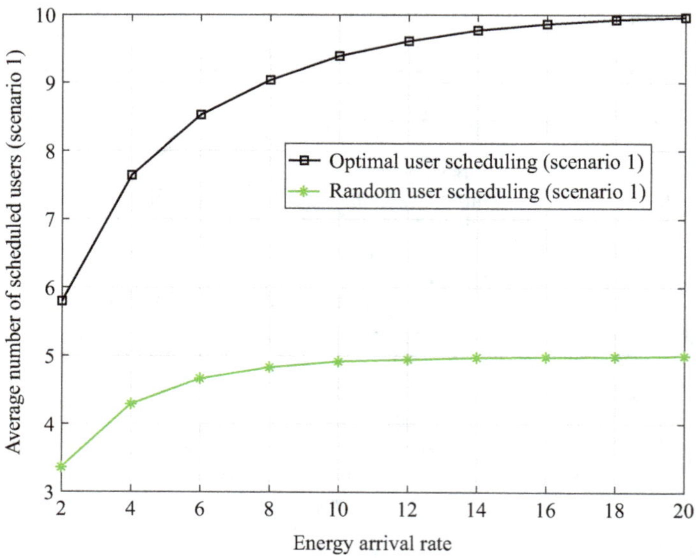

Figure 10.3 Effect of energy arrival rate on the number of scheduled users with various scheduling schemes

does not translate into a proportional increase in the number of scheduled devices, highlighting the limitations of the random scheduling approach in efficiently utilizing the available energy.

In Figure 10.4, the impact of different scheduling policies on FL performance is presented. The proposed USEM algorithm's effectiveness is evaluated in an FL setup over a wireless environment using a support vector machine (SVM) model. The study considers an SINR threshold of $\gamma_{\text{th}} = 20$ dB and focuses on a two-class classification task where each user is tasked with recognizing digits 0 and 8 from the MNIST dataset using two training samples per user. The model is evaluated every 10 communication rounds on a test set of 100 samples, with a learning rate of 0.00001. The performance results are averaged over five Monte Carlo simulations. A successful global aggregation in this context is defined as the scenario where a set of users effectively transmit their weights to the BS, satisfying both the SINR and energy constraints, allowing the BS to aggregate the received weights. The results clearly demonstrate that the optimal scheme improves system performance compared to other schemes. This improvement is primarily attributed to the algorithm's superior ability to manage interference and energy constraints, leading to a higher number of successful global aggregations. Despite the improvement, the proposed FL approach achieves a 93% accuracy rate, indicating that it does not perfectly identify all handwritten digits. This discrepancy is due to inherent limitations in the FL model, the training methodology, and the training datasets, which are not fully optimized by the proposed FL system. It is important to note that the goal of the study is to minimize training loss rather

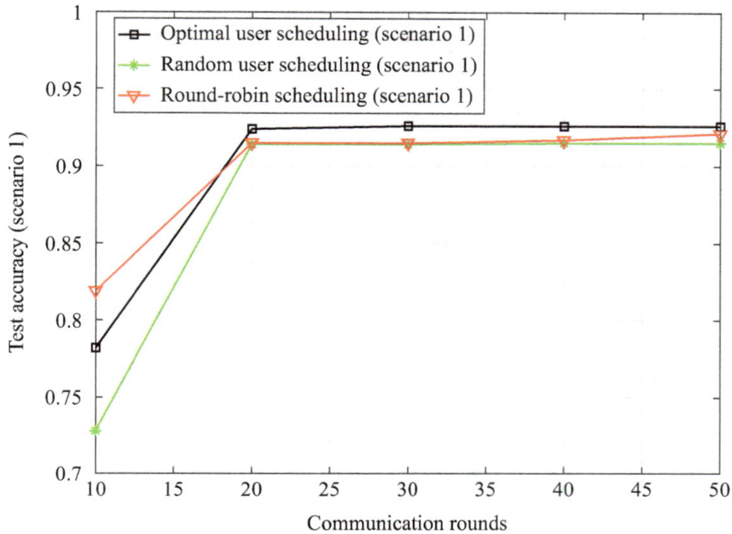

Figure 10.4 Performance of the trained SVM: Test accuracy across various scheduling schemes

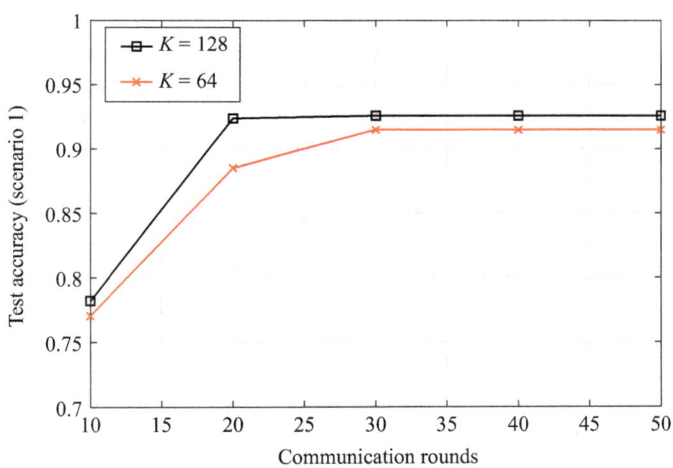

Figure 10.5 Effect of antenna number on the test accuracy of the trained SVM with optimal scheme

than completely eliminate it, which explains why the accuracy falls short of 100%. In practice, achieving 100% accuracy is exceedingly rare for any ML model.

Figure 10.5 illustrates the effect of increasing the number of antennas at the BS on the performance of FL during the training of an SVM model. The results indicate a clear trend: as the number of antennas at the BS increases, the number of successful global aggregations also rises. This boost in successful aggregations

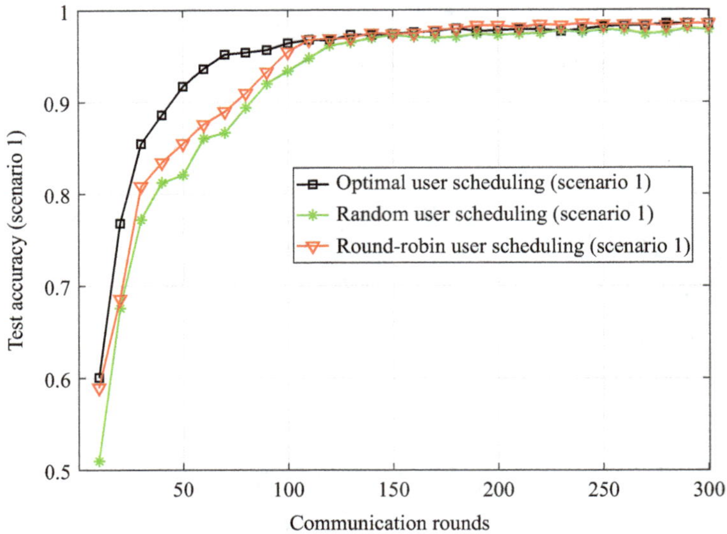

Figure 10.6 Performance of the trained CNNs: test accuracy across various scheduling schemes

directly translates into a significant improvement in the test accuracy of the trained SVM model. The additional antennas enhance the BS's ability to manage interference and maintain reliable communication with multiple users, thereby facilitating more effective and frequent aggregations of local model updates. This improvement underscores the critical role of BS antenna configuration in optimizing FL performance, particularly in scenarios where robust and efficient model training is required across a distributed network.

The performance of the optimal scheme is further assessed in Figure 10.6 using a convolutional neural network (CNN) model with an SINR threshold of $\gamma_{th} = 25$ dB. The CNN model consists of 3 layers, each with 8 filters, followed by a dense layer of 16 neurons. Similar to the previous experiment, the task involves a two-class classification to recognize digits 0 and 8 from the MNIST dataset, with each user having 10 training samples. The model is tested every 10 communication rounds using 100 test samples, and the learning rate is set to 0.00006.

The results demonstrate that the optimal scheme achieves faster convergence compared to other methods, primarily due to its efficient control of interference and energy management. The developed algorithm achieves a stable state after 100 iterations, whereas conventional FL systems require 150 training steps to achieve the same level of stability. This efficiency is reflected in the higher number of successful global aggregations facilitated by the optimal algorithm. However, it is important to note that the performance gap between the proposed and conventional schemes diminishes after a certain number of iterations. This is because, when the number of devices is small, the difference in successful aggregations becomes less significant, leading to a convergence in performance across the different scheduling methods.

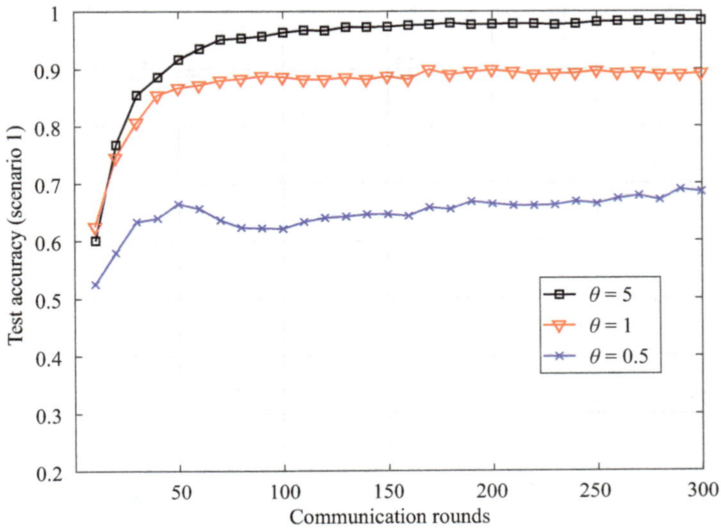

Figure 10.7 Effect of energy arrival rate on test accuracy of trained CNNs with optimal scheme

Figure 10.7 illustrates the effect of the energy arrival rate on system performance during the training of a CNN model. The results clearly indicate that as the energy arrival rate increases, there is a corresponding rise in the number of successful global aggregations. This increase in successful aggregations plays a crucial role in significantly enhancing the test accuracy of the trained CNN model. The improved energy availability enables more users to participate actively in the FL process, ensuring that their local model updates are effectively transmitted and aggregated at the BS. Consequently, this leads to a more accurate and robust global model, demonstrating the importance of energy management in optimizing FL performance.

In Figure 10.8, the performance of an optimal user association algorithm is examined within a multiple BSs system, assuming $C = 4$. The evaluation focuses on the number of scheduled users as a function of the minimum received SINR. The results indicate that the number of scheduled users decreases as the QoS constraints become more stringent. However, the optimal scheme is able to allow more connections compared other schemes, particularly under high SINR thresholds. This performance gap widens at higher SINR levels due to the optimal algorithm's enhanced control of harvested energy.

Figure 10.9 explores the performance of the various user association methods. The results clearly show that an increase in the energy arrival rate leads to a higher number of selected users. However, for the random association scheme, this performance improvement tends to plateau as interference becomes more prominent, limiting the scheme's effectiveness. This contrast highlights the advantages of the optimal user association approach, which better handles the challenges of interference and energy constraints, especially in scenarios with higher SINR thresholds.

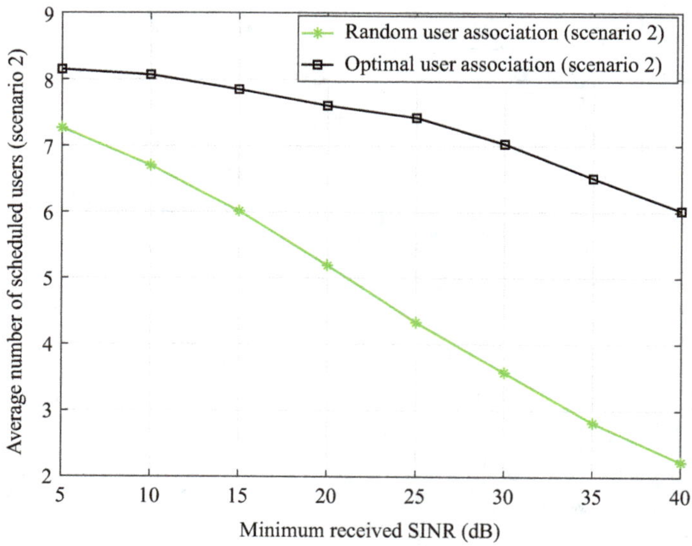

Figure 10.8 Relationship between SINR and the number of scheduled users across various user association schemes

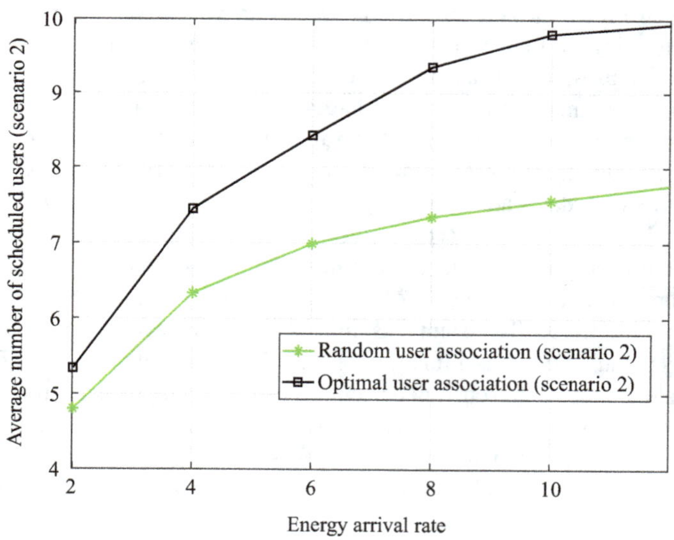

Figure 10.9 Effect of energy arrival rate on the number of scheduled users across different user association schemes

In Figure 10.10, the impact of different user association schemes on FL performance is presented. The evaluation focuses on the optimal user association in an FL scenario within a wireless system with $C = 4$ BSs and an SINR threshold of $\gamma_{\text{th}} = 25$ dB. The FL task involves training an SVM model for a two-class classification problem, where each user is tasked with recognizing digits 0 and 8 from the MNIST

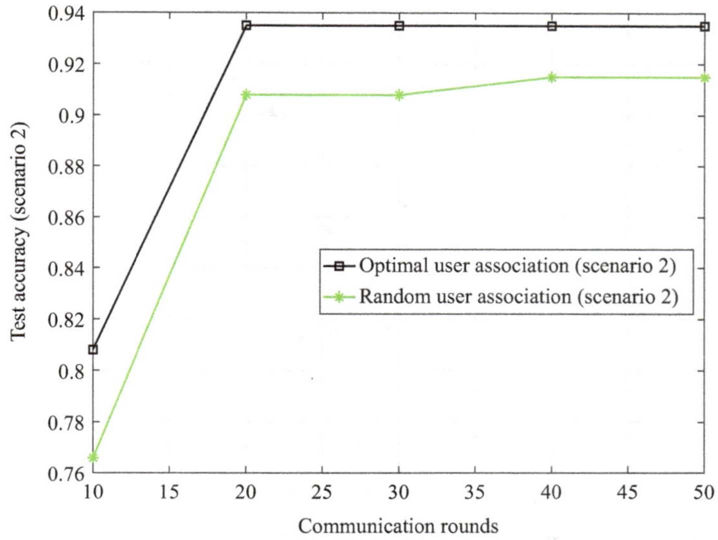

Figure 10.10 Performance of the trained SVM: test accuracy across various association schemes

dataset using two training samples. The model's performance is tested every 10 communication rounds using 100 test samples, with a learning rate of 0.00001. The results clearly demonstrate that the optimal user association scheme significantly enhances FL performance in terms of accuracy compared to random association. This improvement is primarily attributed to the optimal scheme's ability to more effectively manage interference and energy constraints, leading to a higher number of successful global aggregations. Consequently, the trained model achieves better accuracy, underscoring the importance of intelligent user association in maximizing FL efficiency in wireless environments.

In Figure 10.11, the performance of user association schemes in a multiple BSs system is evaluated using CNN. The study focuses on a two-class classification task where each user is tasked with recognizing digits 0 and 8 from the MNIST dataset, using 10 training samples. The model is tested every 10 communication rounds on 100 test samples, with a learning rate of 0.00006. The results clearly demonstrate that the proposed user association scheme significantly increases the total number of successful aggregations, which in turn enhances FL performance in terms of both accuracy and time to convergence. The optimal scheme outperforms the random scheme by converging faster, thanks to its efficient management of energy and interference.

10.10 Dynamic clustering in hierarchical federated learning (HFL)

The incorporation of dynamic clustering in HFL opens up numerous research opportunities, particularly when integrated with energy management strategies. As HFL

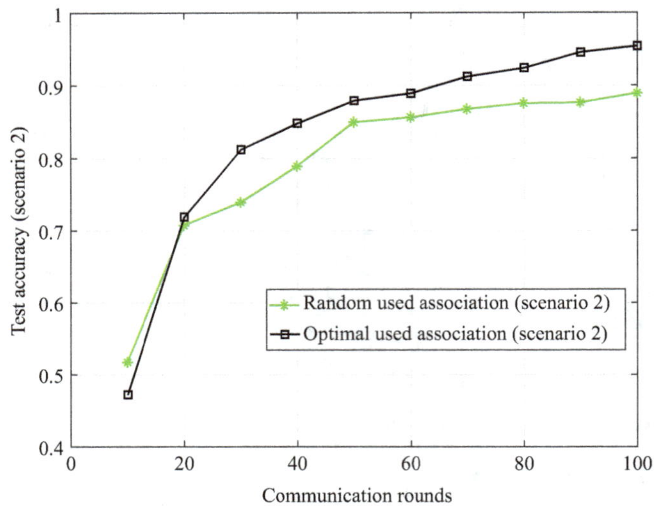

Figure 10.11 Performance of the trained CNNs: test accuracy across various association schemes

systems scale and diversify, addressing challenges related to energy efficiency, communication overhead, and data heterogeneity becomes crucial. Future research should explore the development of adaptive clustering algorithms that not only adjust cluster formations based on real-time data distribution and network conditions but also incorporate energy considerations. These algorithms could dynamically form clusters that optimize energy consumption across devices while maintaining model accuracy and communication efficiency. Techniques like reinforcement learning could be employed to balance the trade-offs between energy usage and other performance metrics in real-time.

Addressing the challenges of non-independent and identically distributed data in conjunction with energy efficiency is vital for HFL. Research could focus on clustering strategies that account for data heterogeneity while minimizing energy consumption. Investigating how dynamic clustering can reduce the energy costs associated with data skewness and imbalance within clusters could lead to more energy-efficient and equitable learning outcomes. As HFL systems scale, centralized management of clusters may become impractical, particularly in energy-constrained environments. Future research should investigate decentralized or fully distributed clustering methods that allow clusters to be formed and managed locally by devices or edge servers. These methods should prioritize energy efficiency, ensuring that devices with limited energy resources can still participate effectively in the learning process without excessive drain on their power reserves.

Applying dynamic clustering with energy management in real-world HFL applications, such as smart cities, healthcare, or autonomous vehicles, could provide valuable insights into practical challenges and benefits. Future research should focus on case studies that demonstrate the feasibility and impact of energy-aware dynamic clustering in these domains. Understanding the specific energy requirements and constraints of different applications will help tailor clustering strategies to achieve both performance and energy efficiency.

10.11 Conclusion

This chapter explores an energy-efficient FL framework over wireless systems by leveraging massive MIMO at BS and utilizing energy harvested by the users. We formulate a minimization problem that involves optimizing a global loss function while adhering to QoS constraints for each user and energy availability constraints. Our analysis delves into the influence of wireless parameters on FL performance, establishing a clear connection between FL convergence rate and user scheduling. We derive the optimal user scheduling strategy that reduces the FL training loss. Additionally, we extend our investigation to scenarios involving multiple BSs, where we optimize the user association in FL systems to achieve the best performance.

Appendix A

Proof of Theorem 1

The following inequality shows the relationship between the convergence rate of FL model and the wireless parameters in Chen *et al.* [11]. It is adapted to our system model as

$$\begin{aligned}
\mathbb{E}\left(\|\mathbf{o}\|^2\right) &= \mathbb{E}\left(\left\|\nabla F(\mathbf{q}_t) - \frac{\sum\limits_{n=1}^{N}\sum\limits_{m=1}^{M_n}\phi_n(t)\nabla f(\mathbf{q},\mathbf{x}_{n,m},y_{n,m})}{\sum\limits_{n=1}^{N}M_n\phi_n(t)}\right\|^2\right) \\
&= \mathbb{E}\left(\left\|-\frac{\left(K-\sum\limits_{n=1}^{N}M_n\phi_n(t)\right)\sum\limits_{n\in\mathcal{N}_2(t)}\sum\limits_{m=1}^{M_n}\nabla f(\mathbf{q},\mathbf{x}_{n,m},y_{n,m})}{N\sum\limits_{n=1}^{N}M_n\phi_n(t)}\right.\right. \\
&\qquad \left.\left. + \frac{\sum\limits_{n\in\mathcal{N}_1(t)}\sum\limits_{m=1}^{M_n}\nabla f(\mathbf{q},\mathbf{x}_{n,m},y_{n,m})}{N}\right\|^2\right) \qquad \text{(A.1)} \\
&\le \mathbb{E}\left(-\frac{\left(K-\sum\limits_{n=1}^{N}M_n\phi_n(t)\right)\sum\limits_{n\in\mathcal{N}_2(t)}\sum\limits_{m=1}^{M_n}\|\nabla f(\mathbf{q},\mathbf{x}_{n,m},y_{n,m})\|}{N\sum\limits_{n=1}^{N}M_n\phi_n(t)}\right. \\
&\qquad \left. + \frac{\sum\limits_{n\in\mathcal{N}_1(t)}\sum\limits_{m=1}^{M_n}\|\nabla f(\mathbf{q},\mathbf{x}_{n,m},y_{n,m})\|}{N}\right)^2
\end{aligned}$$

where $\mathcal{N}_1(t) = \{n \in \{1, \ldots, N\} \mid \phi_n(t) = 1\}$ is the set of scheduled users at frame t and $\mathcal{N}_2(t) = \{n \in \{1, \ldots, N\} \mid \phi_n(t) = 0\}$ is the set of unscheduled users at frame t. Hence, the following inequality holds:

$$\begin{aligned}\mathbb{E}\left[F(\mathbf{q}_{t+1}) - F(\mathbf{q}^*)\right] \leq \frac{2\zeta_1}{VM}\sum_{n=1}^{N} M_n\left(1-\phi_n(t)\right) + \\ \left(1 - \frac{\mu}{V} + \frac{4\mu\zeta_2}{VM}\sum_{n=1}^{N} M_n\left(1-\phi_n(t)\right)\right)\mathbb{E}\left(F(\mathbf{q}_t) - F(\mathbf{q}^*)\right).\end{aligned} \tag{A.2}$$

Let $u_t = \mathbb{E}[F(\mathbf{q}_t) - F(\mathbf{q}^*)]$, $z_t = \frac{2\zeta_1}{VM}\sum_{n=1}^{N} M_n(1-\phi_n(t))$, and $v_t = 1 - \frac{\mu}{V} + \frac{4\mu\zeta_2}{VM}\sum_{n=1}^{N} M_n\left(1-\phi_n(t)\right)$. Thus, the following inequality applies:

$$u_{t+1} \leq z_t + v_t u_t \tag{A.3}$$

The following inequality is now proven recursively:

$$u_{t+1} \leq \sum_{i=0}^{t}\left(\prod_{j=i+1}^{t} v_j\right) z_i + u_0 \prod_{i=0}^{t} v_i. \tag{A.4}$$

For $t = 0$, the preceding inequality holds true, as demonstrated by

$$u_1 \leq z_0 + v_0 u_0. \tag{A.5}$$

Assume the inequality holds for $t+1$; we will now prove it for $t+2$ as follows:

$$\begin{aligned} u_{t+2} &\leq z_{t+1} + v_{t+1}u_{t+1} \\ &\leq z_{t+1} + v_{t+1}\left(\sum_{i=0}^{t}\left(\prod_{j=i+1}^{t} v_j\right) z_i + u_0\prod_{i=0}^{t} v_i\right) \\ &\leq z_{t+1} + \sum_{i=0}^{t}\left(\prod_{j=i+1}^{t+1} v_j\right) z_i + u_0\prod_{i=0}^{t+1} v_i \\ &\leq \sum_{i=0}^{t+1}\left(\prod_{j=i+1}^{t+1} v_j\right) z_i + u_0\prod_{i=0}^{t+1} v_i \end{aligned} \tag{A.6}$$

This concludes the proof.

References

[1] Zhou Y, Tian L, Liu L, *et al.* Fog computing enabled future mobile communication networks: A convergence of communication and computing. *IEEE Communications Magazine*. 2019;57(3):20–27.

[2] Zhu G, Y Wang Y, and Huang K. Broadband analog aggregation for low-latency federated edge learning. *IEEE Transactions on Wireless Communications*. 2020;19(1):491–506.

[3] Li T, Sahu AK, Talwalkar A, *et al.* Federated learning: Challenges, methods, and future directions. *IEEE Signal Processing Magazine*. 2020;33(3):50–60.

[4] Li T, Sahu AK, Talwalkar A, *et al.* Federated learning over wireless fading channels. *IEEE Transactions on Wireless Communications*. 2020;19(5):3546–3557.

[5] Hamdi R, Chen M, Ben Said A, *et al.* Federated learning over energy harvesting wireless networks. *IEEE Internet of Things Journal*. 2022;9(1):663–675.

[6] Hamdi R, Ben Said A, Baccour E, *et al.* Optimal resource management for hierarchical federated learning over HetNets with wireless energy transfer. *IEEE Internet of Things Journal*. 2023;10(19):16945–16958.

[7] Hamdi R, Driouch E, and Ajib W. Energy management in large-scale MIMO systems with per-antenna energy harvesting. In: *IEEE International Conference on Communications (ICC)*; 2017. p. 1–6.

[8] Yang Z, Chen M, Saad W, *et al.* Energy efficient federated learning over wireless communication networks. *IEEE Transactions on Wireless Communications*. 2021;20(3):1935–1949.

[9] Yang Z, Chen M, Saad W, *et al.* Scaling up MIMO: opportunities and challenges with very large arrays. *IEEE Signal Processing Magazine*. 2013;30(1):40–60.

[10] Hamdi R and Qaraqe M. Resource management in energy harvesting powered LoRa wireless networks. In: *IEEE International Conference on Communications (ICC)*; 2021. p. 1–6.

[11] Chen M, Yang Z, Saad W, *et al.* A joint learning and communications framework for federated learning over wireless networks. *IEEE Transactions on Wireless Communications*. 2021;20(1):269–283.

Chapter 11

Quantum computing for energy management: a semi-non-technical guide for practitioners

Jirawat Tangpanitanon[1]

The pursuit of energy transition necessitates the coordination of several technologies, including more efficient and cost-effective distributed energy resources, smart grids, carbon capture, utilization, and storage, energy-efficient technologies, Internet of Things, edge computing, artificial intelligence (AI), and nuclear energy, among others. Quantum computing is an emerging paradigm for information processing at both hardware and software levels by exploiting quantum mechanical properties to solve certain computational tasks exponentially faster than classical computers. This chapter will explore the opportunities and challenges of using quantum computing for energy management applications, enabling the more efficient and economically optimal integration of DERs such as solar photovoltaic rooftops, energy storage systems, electric vehicles (EVs), and EV charging stations into the grid.

11.1 Introduction

Transitioning from fossil fuels due to climate change requires the adoption of green energy sources, leading to the rise of distributed energy resources (DERs) such as solar panels, wind turbines, energy storage systems, and electric vehicles (EVs) [1]. This energy transition is transforming the energy landscape and presenting new computational challenges, among others, to energy management due to the variability and decentralization of DERs [2,3]. These challenges include, for example, managing real-time optimal power flow to ensure grid stability and efficiency [4], developing effective DER control strategies for reliable virtual power plant operations [5], determining optimal locations for EV charging facilities [6], and devising trading strategies for peer-to-peer energy markets [7]. The underlying models frequently exhibit complex nonlinearities associated with power flow, memory effects from energy storage, a diverse array of DERs from local households to grid scales, and uncertainties stemming from the weather-dependent nature of renewable sources.

Quantum computing offers the potential to address these challenges with greater speed and efficiency, surpassing the capabilities of traditional computational

[1]Thailand Center of Excellence in Physics, Ministry of Higher Education, Science, Research and Innovation, Thailand

approaches [8]. Unlike a classical computer that uses bits as the fundamental unit of computation, a quantum computer uses quantum bits, or "qubits," which can exist in a superposition of both 0 and 1 simultaneously. Multiple qubits can be "entangled" to produce correlations that cannot be reproduced by a classical computer.

From a computational complexity perspective, these non-classical properties enable a class of decision problems, called bounded-error quantum polynomial time (BQP), that can be efficiently solved by a quantum computer with a certain error probability. BQP contains P, which is a class of decision problems efficiently solved by a classical computer or, more precisely, a deterministic Turing machine. In non-technical terms, this means that fundamentally, there exists certain computational problems that can be efficiently solved by a quantum computer but not by any imaginable classical computers – a capability referred to as "quantum advantage" [9].

In the four decades since its inception by Richard Feynman in 1982 [10], quantum computing has achieved several breakthroughs [11], including advancements in its theoretical foundations, Nobel Prize-winning experiments, and the demonstration of quantum advantage in 2019 [12]. Recently, quantum computing has expanded from purely academic research to commercial ventures [13], with global investments from industry and government exceeding 30 billion USD in 2022 [14]. Leading cloud providers such as AWS, Microsoft, and IBM offer access to quantum computers through a Software-as-a-Service (SaaS) model, enabling practitioners to run quantum experiments on actual quantum systems ranging from five to hundreds of qubits [15]. This "democratization" of quantum technology has spurred numerous studies on its real-world applications, ranging from artificial intelligence (AI) and optimization to simulations in various industry sectors, including finance [16–19], energy [20–29], materials science [30], logistics [31,32], aviation [33], supply chain management [34], telecommunications [35], insurance [36], and the pharmaceutical industry [37], among others.

Despite these advancements, it is important to note that there is currently no experimental demonstration of quantum advantage in solving real-world problems. The quantum advantage experiment conducted by the Google Quantum AI team in 2019 used specific mathematical problems that maximally favored quantum computation – a task of simulating quantum computation itself with randomized evolution [12]. Additionally, current quantum hardware is "noisy," limiting the number of logical operations that can be performed before quantum information is destroyed by a process called decoherence. These noisy qubits are referred to as physical qubits and "stable," which are error-corrected and more reliable for computations, qubits are called logical qubits [38]. A few tens of logical qubits have recently been realized experimentally [39]. However, depending on the applications, real-world use cases of quantum computing may require at least hundreds or thousands of high-quality logical qubits, which may still take years to realize [40,41].

It is a biased intention of this chapter not to leave the readers with an "interesting but not now" impression of quantum computing. First, although a fully fledged large-scale quantum computer is still far fetching, energy transition is also far fetching involving strategic planning over several decades [42]. Ignoring advances in emerging technologies like quantum computing could lead to over-investment in outdated strategies, resulting in higher electricity costs or even unnecessarily unfeasible business models. Second, a quantum computer is not an advanced digital computer. Integrating a quantum computer into existing digital infrastructures is a long-term process, which requires multi-discipline collaborations including quantum experts, system operators, power engineers, software engineers, network engineers, business developers, policy makers and regulators [43]. Third, there exists "quantum-inspired algorithms," such as tensor networks, that can be run on an existing classical computer by approximating quantum computation. Applications of these techniques for improving classical machine learning and optimization algorithms have been demonstrated [44,45]. Fourth, there are "quantum-inspired hardware" such as Fujitsu's digital annealer [46] and Toshiba's bifurcation machine [47] that are currently capable of solving large-scale optimization problems. These are classical hardware designed to mimic certain aspects of quantum optimization.

There is a wealth of material introducing quantum computing, along with an extensive list of prominent quantum algorithms [48–50]. Additionally, several reviews discuss the applications of quantum computing in energy management [20–29]. Instead of reiterating this existing content, this chapter intends to offer practical guidelines for energy practitioners interested in developing their own quantum computing use cases. We will begin by discussing the typical decision-making process involved in defining the research question and methodology for various quantum solutions. The chapter's layout is summarized in Figure 11.1. Note that the references mentioned serve as examples and do not necessarily encompass the entirety of the existing literature.

11.2 Choosing the "right" questions

An extensive literature on quantum computing use cases in the energy sector showcases a broad range of applications and involves various quantum methods at different stages of maturity. For first-time explorers, finding applications that align with their interests can be overwhelming. While there is no definitive right or wrong approach to selecting quantum computing use cases, this section aims to outline the pros and cons of various decision-making choices related to typical questions that may arise during the process.

11.2.1 Machine learning or mathematical optimization

Machine learning (ML) and mathematical optimization (MO) are two distinct but interconnected paradigms within the field of data analytics. ML leverages large datasets to train models that identify patterns and make predictions about unknown data. Conversely, MO utilizes relatively smaller datasets to define business logic and applies search algorithms to find optimal decisions that minimize a cost function, subject to a set of constraints.

There is an extensive body of literature on the applications of ML and MO in energy management [51,52], ranging from forecasting of variable renewable energy (VRE) production [53] to solving complex optimal power flow [54]. To understand

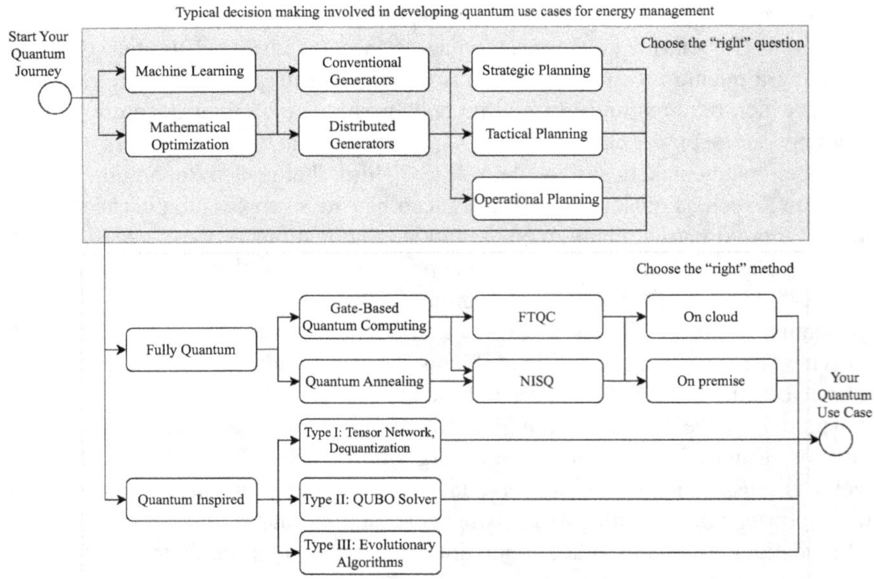

Figure 11.1 A diagram illustrating the typical decision-making process for developing quantum computing use cases in energy management. Each option presents various pros and cons concerning immediate impacts, provable quantum speedup, practicality, cost, and learning opportunities.

the new possibilities that quantum computing brings to these applications, it is beneficial to examine the theoretical foundations of quantum speedup in ML and MO.

11.2.1.1 Quantum machine learning

In the realm of machine learning, several quantum machine learning (QML) models [55–57] are expected to have exponential speedup compared to their classical counterparts, including least squares fitting [58], quantum Boltzmann machines [59], quantum principal component analysis [60], and quantum support vector machines [61]. Additionally, there are models that demonstrate quadratic speedup, such as Bayesian inference [62], online perception [63], and quantum reinforcement learning [64]. These speedups are achieved by substituting classical subroutines with more efficient quantum alternatives, such as matrix inversion, estimation of eigenvectors and eigenvalues for sparse or low-rank matrices, and fast Fourier transforms. These quantum subroutines frequently rely on the quantum Fourier transform (QFT) [65] and the Harrow–Hassidim–Lloyd (HHL) algorithm [66], the latter being a quantum method for numerically solving systems of linear equations.

In the energy domain, the HHL algorithm has been explored as a solution for the model predictive control (MPC) problem in nonlinear dynamical microgrids [67].

Additionally, several studies have applied the HHL algorithm to address power flow issues, which are common nonlinear challenges encountered in energy grids [68–72].

There are a few caveats regarding the QML models mentioned above. First, unless working with very small datasets, they require large-scale fault-tolerant quantum computers, which are not yet available. Second, these models necessitate the use of quantum random access memory (QRAM) [73], potentially requiring an exponential amount of quantum resources to load classical data into a quantum computer. Exploring end-to-end advantages of QML models over their classical counterparts remains an active area of research.

Additionally, there is a relatively new class of quantum algorithms called variational quantum algorithms (VQAs) that can be used for machine learning [74]. Unlike QFT and HHL, VQAs involve variational parameters. These parameters are optimized using a classical computer based on the results from multiple runs of a quantum computer. Consequently, VQAs are more robust to noise. This makes them more compatible with noisy intermediate-scale quantum (NISQ) devices [75], which refer to current quantum hardware consisting of several hundreds or thousands of physical qubits.

VQA has been used for photovoltaic (PV) power forecasting [76–78] as well as solar irradiance [79]. The latter uses the Folsom dataset, which spans from 2014 to 2016, containing minute-by-minute measurements of Global Horizontal Irradiance and Direct Normal Irradiance from Folsom, California, USA. Additionally, this method has been applied to forecast smart grid stability using a small dataset consisting of four nodes [80].

One of the drawbacks of this approach is the scalability of VQAs. Due to the nature of the quantum parameter space, the optimization landscape observed by a classical optimizer often becomes exponentially flat with an increasing number of variational parameters – a phenomenon known as barren plateaus [81,82]. This problem can be mitigated, to a certain extent, with good initial guess.

Lastly, we also want to mention a more modern quantum kernel method [83]. The idea behind quantum kernel methods is to use quantum computers to compute the kernel function for data embedded in a quantum feature space. The quantum feature space can represent data in a more complex and richer way than classical approaches, potentially leading to better performance in tasks like classification or regression.

Currently, there are theoretical foundations proving quantum advantage for QML, but this is limited to specific tasks, such as when the learning data originates from quantum systems, such as those in quantum chemistry [84,85] and cryptography problems [86]. However, practical applications related to these tasks remain unclear.

Based on this observation, there is currently no apparent, at least to the author, quantum speedup for machine learning applications in energy management. However, it is important to clarify that we refer to quantum speedup in a strong sense, where a QML model must outperform any known classical ML models. In practice, it is possible to achieve a "quantum-enhanced" ML model that outperforms a specific classical model in a particular use case. Much like the effectiveness of machine learning techniques such as deep learning, which often emerges from empirical results rather than purely theoretical proofs, QML might similarly show advantages in the future. Given that large-scale quantum computers do not yet exist, the current lack of theoretical foundation does not necessarily preclude future advantages over classical ML.

11.2.1.2 Quantum optimization

Mathematical optimization has been central to Operations Research (OR) since World War II [87]. Finding the best decision variables for real-world optimization problems often falls into the NP-hard class of decision problems. It is a strongly held conjecture in computer science that no classical algorithms can efficiently solve NP-hard problems. Unfortunately, the same limitation applies to quantum computers because the complexity class BQP does not include or overlap with NP-hard problems. Nevertheless, there are certain classes of optimization problems whose optimal solutions can be efficiently "approximated" by a quantum algorithm, but not by any known classical algorithms [88,89]. Identifying these problems and exploring their real-world applications is currently an active area of research [90].

One of the most common formulations used in quantum optimization is Quadratic Unconstrained Binary Optimization (QUBO) problems [91]. The task is to find a set of N binary variables $x_i \in \{0, 1\}$ that minimizes an objective function

$$\mathscr{C}_{\text{QUBO}} = \sum_{i,j=1}^{N} c_{i,j} x_i x_j \tag{11.1}$$

where $c_{i,j} \in \mathbb{R}$ is a QUBO coefficient. QUBO is computationally equivalent to the Ising model, which is commonly used to model spin systems in physics.

It is important to note that the widespread usage of QUBO in quantum optimization is primarily driven by the nature of quantum hardware itself, rather than its direct application to real-world problems. The QUBO formalism displays several distinct features compared to the commonly used Mixed-Integer Linear Programming (MILP). MILP involves finding a set of decision variables $u_i \in \mathbb{Z}^+$ and $u'_j \in \mathbb{R}^+$ to minimize a linear objective function

$$\mathscr{C}_{\text{MILP}} = \sum_{i=1}^{N_u} c_i u_i + \sum_{j=1}^{N_{u'}} c'_j u'_j \tag{11.2}$$

subject to linear constraints

$$\sum_{i=1}^{N_u} a_{k,i} u_i + \sum_{j=1}^{N_{u'}} a'_{k,j} u'_j \leq b_k \tag{11.3}$$

where $k \in \{1, 2, \ldots, K\}$, and $N_u, N_{u'}, K \in \mathbb{Z}^+$, with $c_i, c'_j, a_{k,i}, a'_{k,j}, b_k \in \mathbb{R}$

There exists a standardized protocol for transforming MILP into QUBO. First, it is necessary to represent u_i and u'_j in binary form by introducing additional binary variables, analogous to the floating-point representation of real numbers [92]. Next, all "hard" constraints in the MILP are to be converted into "soft" constraints, or equivalently, into a quadratic cost function within the QUBO framework, through the introduction of a Lagrange multiplier [93].

Example works that utilize the above transformation include

- The optimal placement of phasor measurement units (PMUs), aimed at minimizing the number of PMUs required while ensuring complete network observability according to established observability rules [94].
- The placement of EV charging stations, which focuses on identifying the combination of locations that requires the fewest charging stations [95–99].
- The optimal scheduling of battery energy storage systems (BESSs) to minimize total operating costs while adhering to charging and discharging constraints [100, 101].
- The formulation of optimal charging schedules for EVs in conjunction with local solar power generation, aiming to minimize reliance on the power grid [102].
- The optimal discount scheduling problems for demand response, a strategy that enables consumers to actively manage electricity demand [103].

There are a few caveats associated with the aforementioned transformation. First, converting hard constraints into soft constraints can lead to solutions that are optimal within the QUBO framework but infeasible in the original MILP framework. Second, the quadratic cost function related to the soft constraints can become highly complex, involving all-to-all connectivity among decision variables. This complexity can transform a relatively simple MILP problem into a significantly more complex QUBO problem. Lastly, the number of QUBO coefficients increases quadratically with the number of decision variables. Consequently, a medium-sized MILP with, for instance, 10^5 variables may transform into a gigantic QUBO instance with more than 10^{10} coefficients. This large number of coefficients could pose challenges related to communication bandwidth when attempting to send the data to a quantum computer.

Instead of transforming MILP to QUBO, it should be noted that it is possible and should be encouraged, to devise a problem that is fundamentally QUBO. A prominent example of such a problem is the quadratic assignment problem (QAP). In QAP, there are a collection of N facilities and N locations. For each pair of locations (i, i'), a distance $d_{i,i'}$ is specified and for each pair of facilities (j, j') a weight or flow $w_{j,j'}$ is specified (e.g., the amount of supplies transported between the two facilities). The problem is to assign all facilities to different locations with the goal of minimizing the sum of the distances multiplied by the corresponding flows. The objective function takes the form

$$\mathscr{C}_{\mathrm{QAP}} = \sum_{i,i',j,j'=1}^{N} d_{i,i'} w_{j,j'} x_{i,j} x_{i',j'} \tag{11.4}$$

where $x_{i,j}$ ($x_{i',j'}$) is equal to one if location i (i') is assigned with facility j (j') and zero otherwise. Facility location allocation problems of power plants have been studied with this approach [20]. A related problem has been studied for the tactical capacity problem to minimize loss in distribution networks [104,105].

Another standard QUBO problem is a maximum cut problem. Here, the objective is to partition the graph's vertices into two complementary sets S and T, such that the number of edges between S and T is as large as possible. The corresponding cost function is

$$\mathscr{C}_{\mathrm{MaxCut}} = \sum_{(u,v)\in E} \frac{1}{2}(1 - x_u x_v) \tag{11.5}$$

where E is a set of edges and x_u (x_v) denotes whether node u (v) is in subset S or T. This formulation has been used to analyze the stability of the power grid [106–109].

A typical quantum approach to solving QUBO problems involves using either a specific type of VQAs known as quantum approximate optimization algorithms (QAOA) [110], or another class of quantum methods known as quantum annealing [111]. Although these two methods are theoretically related and do not require a fault-tolerant quantum computer, they require different types of quantum hardware, which will not be discussed in this chapter. Similar to machine learning applications, in both cases, there is currently no theoretical guarantee of quantum speedup [90]. Again, this lack of theoretical foundation does not preclude the advantage of quantum optimization in solving particular real-world problems.

Lastly, it is worth highlighting recent proofs showing that quantum computers can achieve a super-polynomial advantage over classical computers for approximating certain combinatorial optimization problems. The problems addressed include the formula coloring problem [88], optimal polynomial intersection [89], and max-XORSAT [89], rather than QUBO problems. This advantage utilizes the well-known Shor's and QFT algorithms, which require a fault-tolerant quantum computer. Mapping these problems to energy management applications presents an interesting avenue for further research.

11.2.2 Conventional or distributed generators

When selecting use cases for quantum computing in energy management, it is essential to consider whether the focus is on conventional or distributed generators. Common optimization challenges for conventional generators include, for example, unit commitment (UC) [112] and economic dispatch (ED) [113]. UC involves determining which power generation units should be operational at any given time to meet anticipated electricity demand while adhering to operational constraints. It effectively decides the "commitment" of generators for the following day or operational period. ED, conversely, is concerned with optimizing the output levels of the committed generators to minimize overall generation costs, while also meeting demand and complying with operational constraints.

The application of quantum methods to UC and ED presents the advantage of addressing well-defined problems with clear business impacts [20,114–116]. However, there are significant challenges. System operators have often refined their current methods over decades, making it highly likely that these approaches are already near optimum, leaving little room for quantum computing to yield improvements. Additionally, regulatory frameworks in some regions impose specific planning guidelines that must be followed. For instance, "must-run" power plants are prioritized to ensure grid stability, followed by "must-take" plants based on long-term power purchase agreements. After these, remaining units are dispatched in merit order, giving preference to the most cost-effective options. Such regulations further streamline the computational challenges in optimizing generation strategies.

Distributed generators (DGs), such as battery stations, solar rooftops, and very small power producers, offer a distinct set of advantages and challenges for quantum computing applications compared to conventional generators. DG can be deployed in various contexts and introduces new computational challenges, particularly in managing distribution grid stability, which often involves nonlinear optimization [2,3]. As a result, DG presents a broader range of use cases for exploring quantum computing applications. However, in many countries, DG is still emerging and faces regulatory, management, and economic challenges. Consequently, the immediate impact of quantum computing on DG may not be readily apparent.

11.2.3 Strategic, tactical, or operational

Another important consideration is the phase of operations related to quantum computing use cases: strategic, tactical, and operational. Each phase presents its own set of potential challenges and benefits associated with the implementation of quantum computing.

11.2.3.1 Strategic phase

The strategic phase involves long-term planning for infrastructure investments, particularly in the energy sector, where plans often extend over several decades. On a national scale, these plans might include integrating more renewable energy sources such as wind, solar, and hydroelectric power to reduce carbon emissions and reliance on fossil fuels. They may also involve constructing and enhancing energy infrastructure, such as power plants, transmission lines, and smart grid technologies, to meet future demands and improve reliability [117]. On an enterprise scale, manufacturing companies might focus on strategically planning their renewable energy portfolio, BESSs, and EV fleets to align with global efforts to reduce carbon emissions [118].

The strategic phase presents an interesting opportunity for quantum optimization. First, strategic planning usually involves large MO models that may take standard MILP solvers hours or even days to solve to yield satisfactory solutions. Therefore, there is a gap that quantum optimization can attempt to fill. Second, slight improvements in strategic planning can have significant financial impacts or even make previously infeasible business models feasible, justifying the investment in quantum computing. Third, since strategic planning is typically a non-recurring activity, the cost of quantum computing can be viewed as a one-time capital expenditure (CAPEX) investment rather than an operational expense (OPEX), as it is in other phases.

However, a drawback of employing quantum computing in this phase is that the underlying MO models may not fully encapsulate all the decisions and concerns expressed by all relevant stakeholders. These may include subjective decision making, qualitative data sources, business negotiations, environmental complaints, and unforeseen regulatory restrictions. Such factors can complicate the quantification of the benefits of quantum optimization in terms of actual business impacts.

11.2.3.2 Tactical phase

The tactical phase in energy management typically involves day-ahead planning. Examples include

- For system operators, a unit commitment problem has to be solved in order to schedule and dispatch electric power generation resources [112,113].
- For power plants, operators must provide system operators with specific information about their plant's capabilities for the next day to ensure reliable grid operation. This includes data such as maximum output capacity, minimum operating level, ramp rates, and available ancillary services. While these tasks are routine for traditional power plants, determining optimal daily commitments can be complex for virtual power plants [119]. These facilities must coordinate their member DERs, which may be located in various locations and independently managed by different entities.
- In a peer-to-peer energy market, participants, including prosumers with their own generation capabilities, need to provide specific information for the next-day trading in a manner that optimizes their own objectives. This information typically includes energy generation forecasts, energy consumption forecasts, available excess energy, pricing information, trading preferences, and availability for ancillary services [120].
- For EV-fleet operators, tactical planning may involve optimizing routes and charging schedules [121]. This may include sourcing or bidding for green energy from DERs to minimize costs and environmental impact.
- For manufacturing operators, tactical planning may involve coordinating machine scheduling with battery usage to effectively utilize solar energy generated from PV rooftop installations. This approach can help optimize grid loads for peak shaving or participate in demand response programs, improving efficiency and reducing energy costs [122].

As seen above, the energy transition creates multiple optimization and forecasting opportunities for various players, from grid operators to prosumers. This makes it a compelling research area to explore quantum methods for addressing these challenges. Although tactical operations may require daily access to quantum resources, its OPEX remains relatively manageable compared to the demands of real-time operations in the operational phase.

Quantum computing use cases in the tactical phase face a similar drawback to those in the strategic phase. Since actual operations often differ from forecasts, the potential speedup provided by quantum computing in the tactical phase might be compromised by these discrepancies. Developing robust MO models can help mitigate this drawback.

11.2.3.3 Operational phase

The operational phase is where actual business value is realized. However, during real-time operations, decision-making options are often constrained by physical and managerial limitations of ongoing activities. It is not immediately clear, at least to the author, whether using a quantum computer to explore a vast solution space in that situation would outperform a simpler, more frequent feedback control loop running heuristic algorithms on a classical computer, such as those proposed in Dall'Anese *et al.* [123] for virtual power plant control. Moreover, the operational phase's demand for continuous use of quantum resources could significantly increase OPEX that outweigh potential benefits. Balancing the advantages of quantum computing with its cost implications and practical feasibility in real-time applications remains a largely unexplored territory.

In summary, each of the three phases – strategic, tactical, and operational – has its own pros and cons concerning quantum computing use cases. It is important to note that many of the challenges discussed arise not directly from quantum computing itself but from the fact that MO and ML models are only approximate representations of reality. Therefore, for quantum initiatives to succeed, it is crucial to incorporate expertise from various fields, including energy, quantum computing, AI, and operations research.

11.3 Choosing the "right" methods

In the previous section, we discussed the classifications of quantum computing use cases for energy management based on analytical techniques and operational phases. In this section, we will explore key considerations for selecting the appropriate quantum methods for specific use cases.

11.3.1 Quantum or quantum-inspired

Quantum methods, also known as fully quantum methods, utilize quantum hardware for execution. In contrast, quantum-inspired methods leverage digital hardware to run software influenced by quantum principles. Only fully quantum methods can achieve quantum advantage in a strong sense, as defined by computational complexity theory. However, due to the current unavailability of large-scale, fault-tolerant quantum computers, pursuing quantum-inspired methods presents a viable alternative when seeking tangible improvements over existing solutions.

The exploration of fully quantum methods offers benefits by deepening practitioners' understanding of quantum computing from both hardware and software perspectives. As noted previously, a quantum computer is not an advanced digital computer. For instance, while digital computers operate on deterministic principles, quantum computers are inherently probabilistic. This means that running the same software on a quantum computer may yield different results each time. Grasping these fundamental differences is essential for formulating a long-term quantum readiness strategy.

There are two primary types of fully quantum methods: gate-based quantum computing [8] and quantum annealing [111]. A gate-based quantum computer is universal, meaning that it can execute any quantum algorithm. In contrast, a quantum annealer is specifically designed to tackle a particular class of optimization problems, such as QUBO. Quantum advantage has only been demonstrated for gate-based quantum computers. Conversely, while a quantum annealer utilizes heuristic approaches, it does not require error correction, making it more suitable for near-term applications.

It is important to note that this classification is simplified. For instance, there are analog quantum computers that utilize continuous quantum states and employ continuous transformations instead of discrete qubits and gates. This type of quantum hardware is also commercially available. However, the details of analog quantum computing are beyond the scope of this chapter.

Quantum-inspired methods, on the other hand, can be categorized into three main types:

- Type I: This category employs tensor networks (TNs), numerical techniques originally developed to investigate the dynamics and ground state properties of quantum many-body systems [124]. Recent studies have noted similarities between certain tensor networks and artificial neural networks (ANNs) [125]. While most ANN models are nonlinear, TNs are linear, making them comparatively easier to analyze. For instance, singular value decomposition (SVD) can help clarify the role of long-range correlations in the prediction accuracy of TN models [126]. The explainability of machine learning models is crucial, as it enables users to understand decision-making processes, fostering trust and accountability in real-world applications. TNs have recently been explored across both ML [45] and MO [44] domains. Unrelated to TNs, we also like to mention dequantization techniques that can be used to classically approximate HHL-based algorithms [127,128]. However, the details of which may be challenging for beginners.
- Type II: This type encompasses software solutions such as Fujitsu's digital annealer [46] and Toshiba's simulated bifurcation machines [47]. These solutions address QUBO problems using heuristic techniques that draw loosely from the principles of quantum annealing. Additionally, it is worth noting that the standard Gurobi solver also offers a QUBO solver option. This class of quantum-inspired solutions is particularly effective for exploring large-scale QUBO applications involving between 10^4 and 10^5 variables.
- Type III: This category includes heuristic search algorithms [129], such as quantum-inspired genetic algorithms [130] and quantum-inspired particle swarm algorithms [131]. These methodologies introduce randomness to enhance exploitation through quantum superpositions. Although this type is simpler to implement, it typically neglects quantum entanglement, a crucial factor for achieving quantum advantage.

In summary, the "quantumness" of the aforementioned methods can be ranked from most to least quantum as follows: fully quantum, quantum-inspired Type I, Type II, and Type III. Methods that are more aligned with quantum principles tend to be more challenging to implement but offer closer proximity to quantum advantage, whereas those that are easier to implement may drift further from these advantages.

11.3.2 NISQ or FTQC

NISQ stands for near-term intermediate scale quantum computing [75], while FTQC refers to fault-tolerant quantum computing [132]. NISQ encompasses current or near-term quantum hardware with approximately a few hundred to a few thousand noisy qubits. In contrast, FTQC pertains to future large-scale quantum systems that utilize logical qubits.

Working with NISQ devices offers hands-on experience with fundamental quantum computing concepts, including superposition, entanglement, wave function collapse, and quantum gate decompositions. Practitioners can begin by using their laptops to simulate a small quantum computer with approximately 5–20 qubits. However, exactly simulating N qubit systems generally requires $\mathcal{O}(2^N)$ classical resources to store quantum states and $\mathcal{O}(4^N)$ for quantum gates. Consequently, as the number of qubits increases, laptop simulation quickly becomes impractical.

Next, practitioners can leverage advanced quantum simulators offered by cloud services to benchmark quantum algorithms with 20–60 qubits. These simulators typically utilize tensor networks for better utilization of classical resources [133]. Ultimately, these algorithms can be executed on NISQ devices, enabling exploration of the noise characteristics of quantum hardware and understanding the current limitations in qubit connectivity.

FTQC, in contrast, provides a solid theoretical framework for achieving quantum speedup, which can guide the exploration of relevant applications. However, studying FTQC often demands extensive mathematical analysis, which may seem abstract to non-experts. Despite this, most FTQC algorithms are built upon a few foundational "building-block" quantum algorithms, making it not impossible for enthusiastic practitioners to understand. These foundational algorithms include Shor's algorithm for factorization, Grover's search algorithm, and the QFT [8].

11.3.3 On-cloud or on-premise

Most quantum hardware providers offer cloud access to their quantum computers using pay-as-you-go models, which are popular due to their relatively accessible costs. However, certain energy management applications might be legally required to use on-premise facilities. For example, operations related to system operators are often connected to national security concerns, requiring cloud facilities to function within regulated areas.

Hosting quantum hardware demands a fundamentally different infrastructure than standard digital data centers. Therefore, practitioners should prepare a long-term plan for the installation and maintenance of quantum facilities. In addition, unlike most digital computers, which use complementary metal–oxide–semiconductor technology, quantum computing hardware is commercially available across various implementation platforms, including trapped ions [134], neutral atom arrays [135], photonic systems [136], and superconducting qubits [137]. Each platform has unique advantages and disadvantages and requires significantly different facilities. Detailed discussions on these platforms are beyond the scope of this chapter.

11.4 Some practical remarks

In this section, we address practical concerns that often arise when undertaking quantum projects. Readers primarily interested in an overview of quantum computing use cases for energy management may choose to skip this section.

11.4.1 Hardware availability

Unlike standard digital clouds, where resources are highly available, quantum resources are scarce. Submitting a job to quantum cloud services often results in long queue times, ranging from a few seconds to several hours. Some quantum cloud providers offer a dedicated session option for an additional cost to mitigate this issue. However, when integrating quantum clouds into a production environment, it is advisable to have backup digital servers capable of executing the same tasks in case the quantum server becomes unexpectedly unavailable. This ensures continuity and reliability in operations while leveraging quantum computing capabilities.

11.4.2 Hardware cost

In pay-as-you-go quantum clouds, providers typically charge based on the run time of quantum hardware. However, for beginners, estimating this run time prior to execution can be challenging. Users often write quantum programs using high-level quantum programming languages, which must then be compiled for compatibility with specific quantum hardware. This compilation process can involve breaking long-range quantum gates into smaller gates to align with the qubit connectivity constraints of the hardware.

Consequently, the same quantum program may result in a vastly different number of quantum gates across different problem instances after compilation, leading to cost unpredictability from a software deployment perspective. While most providers offer automation for the compilation process, it remains vital to cross-check the number of gates and relevant parameters before submitting to a quantum cloud to prevent unexpected expenses.

11.4.3 Communication overhead

As mentioned earlier, transmitting large data sets to quantum clouds can become a bottleneck in certain applications. For example, machine learning applications that utilize HHL algorithms require QRAM, which typically requires an exponentially large number of quantum resources to load classical data into a quantum computer.

In MO applications such as QUBO, the number of coefficients grows quadratically with the number of qubits, often resulting in several gigabytes of data in many real-world scenarios. Therefore, it is essential to be cautious when evaluating the end-to-end quantum speedup compared to classical methods, as the overhead associated with data transmission and loading can significantly impact performance assessments.

11.4.4 Open-source quantum software

It is generally advisable for software engineers not to modify software that is functioning well, as digital software does not degrade over time. However, this principle does not apply to quantum computing. As quantum hardware and quantum error correction protocols evolve rapidly, quantum software that is operational today may not work a year from now. In addition, some quantum hardware providers may deprecate compatibility with certain versions of open-source software on an annual basis. Therefore, it is crucial for practitioners to keep their quantum software up-to-date to ensure compatibility with ongoing developments in quantum hardware.

11.4.5 Debugging quantum program

Debugging quantum software can be a significant challenge for practitioners due to the probabilistic nature of quantum computing. It is helpful to categorize uncertainties into two classes.

The first class involves uncertainties that arise from the principles of quantum physics. These uncertainties are inherent and unavoidable, even with perfectly built quantum hardware. In fact, they are essential for achieving quantum speedup and should be embraced rather than eliminated. However, they can complicate the debugging process, especially for beginners. To mitigate this, it is advisable to start with small problem instances and utilize a state vector simulator. This type of simulator focuses on the probability function rather than sampling from it, effectively ignoring quantum uncertainties and providing clearer insights during debugging.

The second class of uncertainty originates from the imperfections of quantum hardware itself. NISQ devices particularly struggle with this issue as they lack robust error correction compared to FTQC. Debugging quantum software in the presence of this type of uncertainty can benefit from benchmarking NISQ devices with simulators that account for quantum uncertainties, allowing practitioners to better understand and manage the imperfections of the hardware while developing their algorithms.

11.5 Conclusion

In conclusion, this chapter has outlined the opportunities and challenges presented by quantum computing within the realm of energy management. By leveraging the unique properties of quantum systems, practitioners can potentially optimize complex problems associated with DERs, smart grids, and overall energy efficiency. We have discussed various use cases, from QML to optimization, and the implications for conventional and distributed energy generation.

Selecting appropriate methodologies, whether fully quantum or quantum-inspired, is critical in navigating this evolving landscape. While quantum advantage has not yet been universally proven in practical applications, especially in energy management, the groundwork laid by quantum-inspired algorithms and techniques demonstrates potential pathways for near-term applications.

Moreover, as our understanding of both quantum technology and its applications in energy systems continues to mature, strategic, tactical, and operational phases must be clearly defined to align quantum capabilities with real-world needs. This requires ongoing collaboration across disciplines, ensuring that quantum computing innovations are effectively integrated into existing energy infrastructure. The journey toward a quantum-enabled energy landscape is long and complex, but with the right approach, it can lead to transformative efficiencies and a more sustainable energy future.

References

[1] Akorede MF, Hizam H, and Pouresmaeil E. Distributed energy resources and benefits to the environment. *Renewable and Sustainable Energy Reviews*. 2010;14(2):724–734. Available from: https://www.sciencedirect.com/science/article/pii/S1364032109002561.

[2] Abdmouleh Z, Gastli A, Ben-Brahim L, *et al.* Review of optimization techniques applied for the integration of distributed generation from renewable energy sources. *Renewable Energy*. 2017;113:266–280. Available from: https://www.sciencedirect.com/science/article/pii/S0960148117304822.

[3] Alarcon-Rodriguez A, Ault G, and Galloway S. Multi-objective planning of distributed energy resources: A review of the state-of-the-art. *Renewable and Sustainable Energy Reviews*. 2010;14(5):1353–1366. Available from: https://www.sciencedirect.com/science/article/pii/S1364032110000146.

[4] Guerrero J, Gebbran D, Mhanna S, *et al.* Towards a transactive energy system for integration of distributed energy resources: Home energy management, distributed optimal power flow, and peer-to-peer energy trading. *Renewable and Sustainable Energy Reviews*. 2020;132:110000. Available from: https://www.sciencedirect.com/science/article/pii/S1364032120302914.

[5] Nosratabadi SM, Hooshmand RA, and Gholipour E. A comprehensive review on microgrid and virtual power plant concepts employed for distributed energy resources scheduling in power systems. *Renewable and Sustainable Energy Reviews*. 2017;67:341–363. Available from: https://www.sciencedirect.com/science/article/pii/S1364032116305160.

[6] Tao Y, Huang M, Chen Y, *et al.* Review of optimized layout of electric vehicle charging infrastructures. *Journal of Central South University*. 2021;28:3268–3278. Available from: https://doi.org/10.1007/s11771-021-4842-3.

[7] Zedan M, Nour M, Shabib G, *et al.* Review of peer-to-peer energy trading: Advances and challenges. *e-Prime – Advances in Electrical Engineering, Electronics and Energy*. 2024;10:100778. Available from: https://www.sciencedirect.com/science/article/pii/S2772671124003589.

[8] Nielsen MA and Chuang IL. *Quantum Computation and Quantum Information*. Cambridge University Press; 2010.

[9] Hangleiter D and Eisert J. Computational advantage of quantum random sampling. *Reviews of Modern Physics*. 2023;95:035001. Available from: https://link.aps.org/doi/10.1103/RevModPhys.95.035001.

[10] Feynman RP. Simulating physics with computers. *International Journal of Theoretical Physics*. 1982;21(6–7):467–488.

[11] Preskill J. *Quantum Computing 40 Years Later*. 2nd ed. Feynman Lectures on Computation. CRC Press; 2023.

[12] Arute F, Arya K, Babbush R, *et al.* Quantum supremacy using a programmable superconducting processor. *Nature*. 2019;574:505–510. Available from: https://www.nature.com/articles/s41586-019-1666-5#citeas.

[13] MacQuarrie ER, Simon C, Simmons S, *et al.* The emerging commercial landscape of quantum computing. *Nature Reviews Physics*. 2020;2:596–598. Available from: https://doi.org/10.1038/s42254-020-00247-5.

[14] Jaya Baloo ea. State of quantum computing: Building a quantum economy. *World Economic Forum*. 2022; Available from: https://www.weforum.org/publications/state-of-quantum-computing-building-a-quantum-economy/.

[15] Golec M, Hatay ES, Golec M, *et al.* Quantum cloud computing: Trends and challenges. *Journal of Economy and Technology*. 2024;2:190–199. Available from: https://www.sciencedirect.com/science/article/pii/S2949948824000271.

[16] Nguyen HT, Krishnan P, Krishnaswamy D, *et al.* Quantum Cloud Computing: A Review, Open Problems, and Future Directions; 2024. Available from: https://arxiv.org/abs/2404.11420.

[17] Herman D, Googin C, Liu X, *et al.* A Survey of Quantum Computing for Finance; 2022. Available from: https://arxiv.org/abs/2201.02773.

[18] Egger DJ, Gambella C, Marecek J, *et al.* Quantum computing for finance: State-of-the-art and future prospects. *IEEE Transactions on Quantum Engineering*. 2020;1:1–24.

[19] Orús R, Mugel S, and Lizaso E. Quantum computing for finance: Overview and prospects. *Reviews in Physics*. 2019;4:100028. Available from: https://www.sciencedirect.com/science/article/pii/S2405428318300571.

[20] Ajagekar A and You F. Quantum computing for energy systems optimization: Challenges and opportunities. *Energy*. 2019;179:76–89. Available from: https://www.sciencedirect.com/science/article/pii/S0360544219308254.

[21] Ajagekar A and You F. Quantum computing and quantum artificial intelligence for renewable and sustainable energy: A emerging prospect towards climate neutrality. *Renewable and Sustainable Energy Reviews*. 2022;165:112493. Available from: https://www.sciencedirect.com/science/article/pii/S1364032122003975.

[22] Giani A and Eldredge Z. Quantum computing opportunities in renewable energy. *SN Computer Science*. 2021;2:393.

[23] Paudel HP, Syamlal M, Crawford SE, *et al.* Quantum computing and simulations for energy applications: Review and perspective. *ACS Engineering Au*. 2022;2(3):151–196. Available from: https://doi.org/10.1021/acsengineeringau.1c00033.

[24] Golestan S, Habibi MR, Mousazadeh Mousavi SY, *et al.* Quantum computation in power systems: An overview of recent advances. *Energy Reports*. 2023;9:584–596. Available from: https://www.sciencedirect.com/science/article/pii/S2352484722025720.

[25] Morstyn T and Wang X. Opportunities for quantum computing within net-zero power system optimization. *Joule*. 2024;8(6):1619–1640. Available from: https://www.cell.com/joule/fulltext/S2542-4351(24)00155-7.

[26] Zhou Y, Tang Z, Nikmehr N, *et al.* Quantum computing in power systems. *iEnergy*. 2022;1(2):170–187.

[27] Ullah M, Eskandarpour R, Zheng H, *et al.* Quantum computing for smart grid applications. *IET Generation, Transmission & Distribution*. 2022;16(23):4239–4257.

[28] Kirihara K, Imai H, Kuroda E, *et al.* Exploring potential applications of ising machines for power system operations. *IEEE Access*. 2023;11:68004–68017.

[29] Eskandarpour R, Bahadur Ghosh KJ, Khodaei A, *et al.* Quantum-enhanced grid of the future: A primer. *IEEE Access*. 2020;8:188993–189002.

[30] Bauer B, Bravyi S, Motta M, *et al.* Quantum algorithms for quantum chemistry and quantum materials science. *Chemical Reviews*. 2020;120(22):12685–12717. Available from: https://doi.org/10.1021/acs.chemrev.9b00829.

[31] Harwood S, Gambella C, Trenev D, *et al.* Formulating and solving routing problems on quantum computers. *IEEE Transactions on Quantum Engineering*. 2021;2:1–17.

[32] Osaba E, Villar-Rodriguez E, and Oregi I. A systematic literature review of quantum computing for routing problems. *IEEE Access*. 2022;10:55805–55817.

[33] Vikstål P, Grönkvist M, Svensson M, *et al.* Applying the quantum approximate optimization algorithm to the tail-assignment problem. *Physical Review Applied*. 2020;14:034009. Available from: https://link.aps.org/doi/10.1103/PhysRevApplied.14.034009.

[34] Hassan A, Bhattacharya P, Dutta PK, *et al.*, editors. Quantum computing and supply chain management: A new era of optimization. *IGI Global*; 2024.

[35] Phillipson F. Quantum computing in telecommunication – A survey. *Mathematics*. 2023;11(15):3423. Available from: https://www.mdpi.com/2227-7390/11/15/3423.

[36] Tamturk M. Quantum Computing in Insurance Capital Modelling; 2023. Available from: https://www.mdpi.com/2227-7390/11/3/658.

[37] Zinner M, Dahlhausen F, Boehme P, *et al.* Toward the institutionalization of quantum computing in pharmaceutical research. *Drug Discovery Today*. 2022;27(2):378–383. Available from: https://www.sciencedirect.com/science/article/pii/S135964462100444X.

[38] Devitt SJ, Munro WJ, and Nemoto K. Quantum error correction for beginners. *Reports on Progress in Physics*. 2013;76(7):076001. Available from: https://dx.doi.org/10.1088/0034-4885/76/7/076001.

[39] Bluvstein D, Evered SJ, Geim AA, *et al.* Logical quantum processor based on reconfigurable atom arrays. *Nature*. 2024;626(7997):58–65.

[40] Katabarwa A, Gratsea K, Caesura A, *et al.* Early fault-tolerant quantum computing. *PRX Quantum*. 2024;5:020101. Available from: https://link.aps.org/doi/10.1103/PRXQuantum.5.020101.

[41] Kim IH, Liu YH, Pallister S, *et al.* Fault-tolerant resource estimate for quantum chemical simulations: Case study on Li-ion battery electrolyte molecules. *Physical Review Research.* 2022;4:023019. Available from: https://link.aps.org/doi/10.1103/PhysRevResearch.4.023019.

[42] Chen B, Xiong R, Li H, *et al.* Pathways for sustainable energy transition. *Journal of Cleaner Production.* 2019;228:1564–1571. Available from: https://www.sciencedirect.com/science/article/pii/S0959652619314738.

[43] Kaur M and Venegas-Gomez A. Defining the quantum workforce landscape: A review of global quantum education initiatives. *Optical Engineering.* 2022;61(8):081806.

[44] Cichocki A, Phan AH, Zhao Q, *et al.* Tensor networks for dimensionality reduction and large-scale optimization: Part 2 Applications and future perspectives. *Foundations and Trends® in Machine Learning.* 2017;9(6):431–673. Available from: http://dx.doi.org/10.1561/220000006.

[45] Huggins W, Patil P, Mitchell B, *et al.* Towards quantum machine learning with tensor networks. *Quantum Science and Technology.* 2019;4(2): 024001. Available from: https://iopscience.iop.org/article/10.1088/2058-9565/aaea94/meta.

[46] Matsubara S, Takatsu M, Miyazawa T, *et al.* Digital Annealer for High Speed Solving of Combinatorial Optimization Problems and Its Applications. In: *2020 25th Asia and South Pacific Design Automation Conference (ASP-DAC)*; 2020. p. 667–672.

[47] Goto H. Bifurcation-based adiabatic quantum computation with a nonlinear oscillator network: Toward quantum soft computing. *Scientific Reports.* 2016;6:21686.

[48] Abhijith J, Adedoyin A, Ambrosiano J, *et al.* Quantum algorithm implementations for beginners. *ACM Transactions on Quantum Computing.* 2022;3(4):1–92. Available from: http://dx.doi.org/10.1145/3517340.

[49] Montanaro A. Quantum algorithms: An overview. *npj Quantum Information.* 2016;2(1):15023. Available from: https://doi.org/10.1038/npjqi.2015.23.

[50] Dalzell AM, McArdle S, Berta M, *et al.* Quantum Algorithms: A Survey of Applications and End-to-End Complexities; 2023. Available from: https://arxiv.org/abs/2310.03011.

[51] Chang M, Thellufsen JZ, Zakeri B, *et al.* Trends in tools and approaches for modelling the energy transition. *Applied Energy.* 2021;290:116731. Available from: https://www.sciencedirect.com/science/article/pii/S0306261921002476.

[52] Perera ATD and Kamalaruban P. Applications of reinforcement learning in energy systems. *Renewable and Sustainable Energy Reviews.* 2021;137:110618. Available from: https://www.sciencedirect.com/science/article/pii/S1364032120309023.

[53] Abualigah L, Zitar R, Almotairi K, *et al.* Wind, solar, and photovoltaic renewable energy systems with and without energy storage optimization: A survey of advanced machine learning and deep learning techniques. *Energies.* 2022;15(2):578.

[54] Tang Y, Dvijotham K, and Low S. Real-time optimal power flow. *IEEE Transactions on Smart Grid*. 2017;8(6):2963–2973.

[55] Cerezo M, Verdon G, Huang H, *et al.* Challenges and opportunities in quantum machine learning. *Nature Computational Science*. 2022;2:567–576.

[56] Biamonte J, Wittek P, Pancotti N, *et al.* Quantum machine learning. *Nature*. 2017;549:195–202.

[57] Schuld M, Sinayskiy I, and Petruccione F. An introduction to quantum machine learning. *Contemporary Physics*. 2014;56(2):172–185.

[58] Wiebe N, Braun D, and Lloyd S. Quantum algorithm for data fitting. *Physical Review Letters*. 2012;109(5):050505.

[59] Amin MH, Andriyash E, Rolfe J, *et al.* Quantum Boltzmann machine. *Physical Review X*. 2018;8:021050. Available from: https://link.aps.org/doi/10.1103/PhysRevX.8.021050.

[60] Lloyd S, Mohseni M, and Rebentrost P. Quantum principal component analysis. *Nature Physics*. 2014;10:631–633.

[61] Rebentrost P, Mohseni M, and Lloyd S. Quantum support vector machine for big data classification. *Physical Review Letters*. 2014;113(13):130503.

[62] Low GH, Yoder TJ, and Chuang IL. Quantum inference on Bayesian networks. *Physical Review A*. 2014;89:062315. Available from: https://link.aps.org/doi/10.1103/PhysRevA.89.062315.

[63] Kapoor A, Wiebe N, and Svore K. Quantum Perceptron Models. In: Lee D, Sugiyama M, Luxburg U, *et al.*, editors. *Advances in Neural Information Processing Systems*. vol. 29. Curran Associates, Inc.; 2016. Available from: https://proceedings.neurips.cc/paper_files/paper/2016/file/d47268e9db2e9aa3827bba3afb7ff94a-Paper.pdf.

[64] Dunjko V, Taylor JM, and Briegel HJ. Quantum-enhanced machine learning. *Physical Review Letters*. 2016;117:130501. Available from: https://link.aps.org/doi/10.1103/PhysRevLett.117.130501.

[65] Coppersmith D. An Approximate Fourier Transform Useful in Quantum Factoring; 2002. Available from: https://arxiv.org/abs/quant-ph/0201067.

[66] Harrow AW, Hassidim A, and Lloyd S. Quantum algorithm for linear systems of equations. *Physical Review Letters*. 2009;103:150502. Available from: https://link.aps.org/doi/10.1103/PhysRevLett.103.150502.

[67] Jing H, Li Y, Brandsema MJ, *et al.* HHL algorithm with mapping function and enhanced sampling for model predictive control in microgrids. *Applied Energy*. 2024;361:122878. Available from: https://www.sciencedirect.com/science/article/pii/S0306261924002617.

[68] Feng F, Zhou YF, and Zhang P. Noise-resilient quantum power flow. *iEnergy*. 2023;2(1):63–70.

[69] Sævarsson B, Chatzivasileiadis S, Jóhannsson H, *et al.* Quantum Computing for Power Flow Algorithms: Testing on real Quantum Computers; 2022. Available from: https://arxiv.org/abs/2204.14028.

[70] Eskandarpour R, Ghosh K, Khodaei A, *et al.* Experimental Quantum Computing to Solve Network DC Power Flow Problem; 2021. Available from: https://arxiv.org/abs/2106.12032.

[71] Feng F, Zhou Y, and Zhang P. Quantum power flow. *IEEE Transactions on Power Systems*. 2021;36(4):3810–3812.

[72] Eskandarpour R, Ghosh K, Khodaei A, *et al.* Quantum Computing Solution of DC Power Flow; 2020. Available from: https://arxiv.org/abs/2010.02442.

[73] Giovannetti V, Lloyd S, and Maccone L. Quantum random access memory. *Physical Review Letters*. 2008;100:160501. Available from: https://link.aps.org/doi/10.1103/PhysRevLett.100.160501.

[74] Cerezo M, Arrasmith A, Babbush R, *et al.* Variational quantum algorithms. *Nature Reviews Physics*. 2021;3:625–644.

[75] Preskill J. Quantum computing in the NISQ era and beyond. *Quantum*. 2018;2:79.

[76] Jeong SG, Do QV, and Hwang WJ. Short-term photovoltaic power forecasting based on hybrid quantum gated recurrent unit. *ICT Express*. 2024;10(3):608–613. Available from: https://www.sciencedirect.com/science/article/pii/S2405959523001637.

[77] Sagingalieva A, Komornyik S, Senokosov A, *et al.* Photovoltaic power forecasting using quantum machine learning; 2023. Available from: https://arxiv.org/abs/2312.16379.

[78] Sushmit MM and Mahbubul IM. Forecasting solar irradiance with hybrid classical–quantum models: A comprehensive evaluation of deep learning and quantum-enhanced techniques. *Energy Conversion and Management*. 2023;294:117555. Available from: https://www.sciencedirect.com/science/article/pii/S0196890423009019.

[79] Oliveira Santos V, Marinho FP, Costa Rocha PA, *et al.* Application of quantum neural network for solar irradiance forecasting: A case study using the Folsom Dataset, California. *Energies*. 2024;17:3580.

[80] Hangun B, Eyecioglu O, and Altun O. Quantum Computing Approach to Smart Grid Stability Forecasting. In: *2024 12th International Conference on Smart Grid (icSmartGrid)*; 2024. p. 840–843.

[81] Bittel L and Kliesch M. Training variational quantum algorithms is NP-hard. *Physical Review Letters*. 2021;127:120502. Available from: https://link.aps.org/doi/10.1103/PhysRevLett.127.120502.

[82] McClean JR, Boixo S, Smelyanskiy VN, *et al.* Barren plateaus in quantum neural network training landscapes. *Nature Communications*. 2018;9:4812.

[83] Schuld M. Supervised Quantum Machine Learning Models Are Kernel Methods; 2021. Available from: https://arxiv.org/abs/2101.11020.

[84] Gyurik C and Dunjko V. Exponential Separations between Classical and Quantum Learners; 2023. Available from: https://arxiv.org/abs/2306.16028.

[85] Molteni R, Gyurik C, and Dunjko V. Exponential Quantum Advantages in Learning Quantum Observables From Classical Data; 2024. Available from: https://arxiv.org/abs/2405.02027.

[86] Liu Y, Arunachalam S, and Temme K. A rigorous and robust quantum speed-up in supervised machine learning. *Nature Physics*. 2021;17:1013–1017.

[87] Hillier FS and Lieberman GJ. *Introduction to Operations Research*. McGraw-Hill; 2015.

[88] Pirnay N, Ulitzsch V, Wilde F, Eisert J, and Seifert J-P. An in-principle super-polynomial quantum advantage for approximating combinatorial optimization problems via computational learning theory. *Science Advances*. 2024;10:eadj5170.

[89] Jordan SP, Shutty N, Wootters M, *et al.* Optimization by Decoded Quantum Interferometry; 2024. Available from: https://arxiv.org/abs/2408.08292.

[90] Abbas A, Ambainis A, Augustino B, *et al.* Challenges and opportunities in quantum optimization. *Nature Reviews Physics*. 2024;6:718–735.

[91] Glover F, Kochenberger G, and Du Y. A Tutorial on Formulating and Using QUBO Models; 2019. Available from: https://arxiv.org/abs/1811.11538.

[92] Braine L, Egger DJ, Glick J, *et al.* Quantum algorithms for mixed binary optimization applied to transaction settlement. *IEEE Transactions on Quantum Engineering*. 2021;2:1–8.

[93] Djidjev HN. Quantum Annealing with Inequality Constraints: The Set Cover Problem; 2023. Available from: https://arxiv.org/abs/2302.11185.

[94] Jones EB, Kapit E, Chang CY, *et al.* On the Computational Viability of Quantum Optimization for PMU Placement; 2020. Available from: https://arxiv.org/abs/2001.04489.

[95] Radvand T and Talebpour A. A Quantum Optimization Algorithm for Optimal Electric Vehicle Charging Station Placement for Intercity Trips; 2024. Available from: https://arxiv.org/abs/2410.16231.

[96] Rao PU and Sodhi B. Hybrid quantum-classical solution for electric vehicle charger placement problem. *Soft Computing*. 2023;27:13347–13363.

[97] Veshchezerova M, Somov M, Bertsche D, *et al.* A Hybrid Quantum-Classical Approach to the Electric Mobility Problem. In: *2023 IEEE International Conference on Quantum Computing and Engineering (QCE)*. vol. 01; 2023. p. 636–641.

[98] Różycki R, Józefowska J, Kurowski K, *et al.* A quantum approach to the problem of charging electric cars on a motorway. *Energies*. 2023;16:442.

[99] Chandra A, Lalwani J, and Jajodia B. Towards an Optimal Hybrid Algorithm for EV Charging Stations Placement Using Quantum Annealing and Genetic Algorithms. In: *2022 International Conference on Trends in Quantum Computing and Emerging Business Technologies (TQCEBT)*; 2022. p. 1–6.

[100] Ehsani D. Quantum-Powered Battery Scheduling in Modern Distribution Grids [Ph.D. thesis]. University of Denver; 2024. Electronic Theses and Dissertations, 2370. Available from: https://digitalcommons.du.edu/etd/2370.

[101] Kea K, Huot C, and Han Y. Leveraging knapsack QAOA approach for optimal electric vehicle charging. *IEEE Access*. 2023;11:109964–109973.

[102] Federer M, Müssig D, Klaiber S, *et al.* Application Benchmark for Quantum Optimization on Electromobility Use Case. In: *2022 IEEE Vehicle Power and Propulsion Conference (VPPC)*; 2022. p. 1–6.

[103] Bucher D, Nüßlein J, O'Meara C, *et al.* Incentivizing demand-side response through discount scheduling using hybrid quantum optimization. *IEEE Transactions on Quantum Engineering*. 2024;5:1–15. Available from: http://dx.doi.org/10.1109/TQE.2024.3407236.

[104] van der Linde SG, van der Schoot W, and Phillipson F. Efficient Quantum Solution for the Constrained Tactical Capacity Problem for Distributed Electricity Generation. In: Krieger UR, Eichler G, Erfurth C, *et al.*, editors. *Innovations for Community Services*. Cham: Springer Nature Switzerland; 2023. p. 203–221.

[105] Silva FFC, Carvalho PMS, and Ferreira LAFM. A quantum computing approach for minimum loss problems in electrical distribution networks. *Scientific Reports*. 2023;13:10777.

[106] Kaseb Z, Möller M, Vergara PP, *et al.* Power flow analysis using quantum and digital annealers: A discrete combinatorial optimization approach. *Scientific Reports*. 2024;14:23216.

[107] Bauer N, Yeter-Aydeniz K, Kokkas E, *et al.* Solving Power Grid Optimization Problems with Rydberg Atoms; 2024. Available from: https://arxiv.org/abs/2404.11440.

[108] Bucher D, Porawski D, Wimmer B, *et al.* Evaluating Quantum Optimization for Dynamic Self-Reliant Community Detection; 2024. Available from: https://arxiv.org/abs/2407.06773.

[109] Jing H, Wang Y, Li Y, *et al.* Quantum Approximate Optimization Algorithm-Enabled DER Disturbance Analysis of Networked Microgrids. In: *2022 IEEE Energy Conversion Congress and Exposition (ECCE)*; 2022. p. 1–5.

[110] Choi J and Kim J. A Tutorial on Quantum Approximate Optimization Algorithm (QAOA): Fundamentals and Applications. In: *2019 International Conference on Information and Communication Technology Convergence (ICTC)*; 2019. p. 138–142.

[111] Yarkoni S, Raponi E, Bäck T, *et al.* Quantum annealing for industry applications: Introduction and review. *Reports on Progress in Physics*. 2022;85(10):104001.

[112] Saravanan B, Das S, Sikri S, *et al.* A solution to the unit commitment problem – A review. *Frontiers in Energy*. 2013;7:223–236.

[113] Xia X and Elaiw AM. Optimal dynamic economic dispatch of generation: A review. *Electric Power Systems Research*. 2010;80(8):975–986. Available from: https://www.sciencedirect.com/science/article/pii/S0378779610000027.

[114] Koretsky S, Gokhale P, Baker JM, *et al.* Adapting Quantum Approximation Optimization Algorithm (QAOA) for Unit Commitment. In: *2021 IEEE International Conference on Quantum Computing and Engineering (QCE)*; 2021. p. 181–187.

[115] Nikmehr N, Zhang P, and Bragin MA. Quantum distributed unit commitment: An application in microgrids. *IEEE Transactions on Power Systems*. 2022;37(5):3592–3603.

[116] Mahroo R and Kargarian A. Learning infused quantum-classical distributed optimization technique for power generation scheduling. *IEEE Transactions on Quantum Engineering*. 2023;4:1–14.

[117] Terrados J, Almonacid G, and Hontoria L. Regional energy planning through SWOT analysis and strategic planning tools: Impact on renewables development. *Renewable and Sustainable Energy Reviews*. 2007;11(6):1275–1287. Available from: https://www.sciencedirect.com/science/article/pii/S1364032105001000.

[118] Seetharaman A, Sandanaraj LL, Moorthy MK, *et al.* Enterprise framework for renewable energy. *Renewable and Sustainable Energy Reviews*. 2016;54:1368–1381. Available from: https://www.sciencedirect.com/science/article/pii/S136403211501206X.

[119] Rahimiyan M and Baringo L. Strategic bidding for a virtual power plant in the day-ahead and real-time markets: A price-taker robust optimization approach. *IEEE Transactions on Power Systems*. 2016;31(4):2676–2687.

[120] Chen K, Lin J, and Song Y. Trading strategy optimization for a prosumer in continuous double auction-based peer-to-peer market: A prediction-integration model. *Applied Energy*. 2019;242:1121–1133. Available from: https://www.sciencedirect.com/science/article/pii/S0306261919305045.

[121] Yang H, Yang S, Xu Y, *et al.* Electric vehicle route optimization considering time-of-use electricity price by learnable partheno-genetic algorithm. *IEEE Transactions on Smart Grid*. 2015;6(2):657–666.

[122] Fang K, Uhan N, Zhao F, *et al.* A new approach to scheduling in manufacturing for power consumption and carbon footprint reduction. *Journal of Manufacturing Systems*. 2011;30(4):234–240. Selected Papers of 39th North American Manufacturing Research Conference. Available from: https://www.sciencedirect.com/science/article/pii/S0278612511000690.

[123] Dall'Anese E, Guggilam SS, Simonetto A, *et al.* Optimal regulation of virtual power plants. *IEEE Transactions on Power Systems*. 2018;33(2): 1868–1881.

[124] Biamonte J and Bergholm V. Tensor Networks in a Nutshell; 2017. Available from: https://arxiv.org/abs/1708.00006.

[125] Wang M, Pan Y, Xu Z, *et al.* Tensor Networks Meet Neural Networks: A Survey and Future Perspectives; 2023. Available from: https://arxiv.org/abs/2302.09019.

[126] Tangpanitanon J, Mangkang C, Bhadola P, *et al.* Explainable natural language processing with matrix product states. *New Journal of Physics*. 2022;24:053032.

[127] Tang E. A Quantum-inspired Classical Algorithm for Recommendation Systems. In: *Proceedings of the 51st Annual ACM SIGACT Symposium on Theory of Computing*. ACM; 2019. Available from: http://dx.doi.org/10.1145/3313276.3316310.

[128] Arrazola JM, Delgado A, Bardhan BR, *et al.* Quantum-inspired algorithms in practice. *Quantum*. 2020;4:307. Available from: https://doi.org/10.22331/q-2020-08-13-307.

[129] Gharehchopogh FS. Quantum-inspired metaheuristic algorithms: Comprehensive survey and classification. *Artificial Intelligence Review*. 2023;56:5479–5543.

[130] Narayanan A and Moore M. Quantum-inspired Genetic Algorithms. In: *Proceedings of IEEE International Conference on Evolutionary Computation*; 1996. p. 61–66.

[131] Meng K, Wang HG, Dong Z, *et al.* Quantum-inspired particle swarm optimization for valve-point economic load dispatch. *IEEE Transactions on Power Systems*. 2010;25(1):215–222.

[132] Shor PW. Fault-tolerant Quantum Computation. In: *Proceedings of 37th Conference on Foundations of Computer Science*; 1996. p. 56–65.

[133] Pan F, Gu H, Kuang L, *et al.* Efficient quantum circuit simulation by tensor network methods on modern GPUs. *ACM Transactions on Quantum Computing*. 2024;5(4):1–26.

[134] Monroe C, Campbell WC, Duan LM, *et al.* Programmable quantum simulations of spin systems with trapped ions. *Reviews of Modern Physics*. 2021;93:025001. Available from: https://link.aps.org/doi/10.1103/RevModPhys.93.025001.

[135] Wintersperger K, Dommert F, Ehmer T, *et al.* Neutral atom quantum computing hardware: Performance and end-user perspective. *EPJ Quantum Technology*. 2023;10:32. Available from: https://doi.org/10.1140/epjqt/s40507-023-00190-1.

[136] Zhong HS, Wang H, Deng YH, *et al.* Quantum computational advantage using photons. *Science*. 2020;370(6523):1460–1463. Available from: https://www.science.org/doi/abs/10.1126/science.abe8770.

[137] Kjaergaard M, Schwartz ME, Braumüller J, *et al.* Superconducting qubits: Current state of play. *Annual Review of Condensed Matter Physics*. 2020;11(Volume 11, 2020):369–395. Available from: https://www.annualreviews.org/content/journals/10.1146/annurev-conmatphys-031119-050605.

Part II

Applications

Chapter 12

Mapping all-sky images to GHI measurements for intra-hour solar forecasting

Nassma Mohandes[1], Daniel Perez-Astudillo[1], Dunia Bachour[1] and Antonio Sanfilippo[1]

Power grid performance and reliability depend on the balancing of continuously varying energy loads with equal amounts of energy generation. The integration of solar energy challenges grid performance and reliability because abrupt solar radiance changes due to meteorological events can cause mismatches between renewable energy generation and energy demand. Accurate solar forecasting is needed to anticipate these mismatches in order to operate grids that integrate solar energy reliably and regulate the resulting energy markets. Sky cameras have been increasingly recognized as a promising source of measurements to support intra-hour solar forecasting with reference to both cost and technical consideration. The goal of this chapter is to assess the utility of sky camera data as an alternative or additional data source for intra-hour solar forecasting.

12.1 Background

Solar irradiation displays significant variability throughout the year, as shown in Figure 12.1 with reference to Qatar. With higher solar energy penetration, this variability introduces significant complexity in the coordination of dispatchable (e.g., coal, nuclear, hydro, geothermal, and combined cycle power plants) and non-dispatchable (e.g., solar power plants) energy generation sources that are needed to balance energy supply and demand. This complexity arises from the limited capacity of conventional generators to accommodate the intermittency and variability of renewable energy generation by changing the rate at which generators can change output (ramp rate), reduce output (ramp range), or de-commit [1]. As discussed in Chapter 1, failure to match supply and demand gives rise to energy over- or under-supply, which threatens the stability of energy markets and the reliability of power grids whose system frequency must be managed within a tight band [2]. Solar forecasting can help mitigate or solve these adversities by providing power grid and energy market operators with advance warning about ebbs and flows in the amount

[1]Environment Centre, Qatar Environment and Energy Research Institute, Hamad Bin Khalifa University, Qatar.

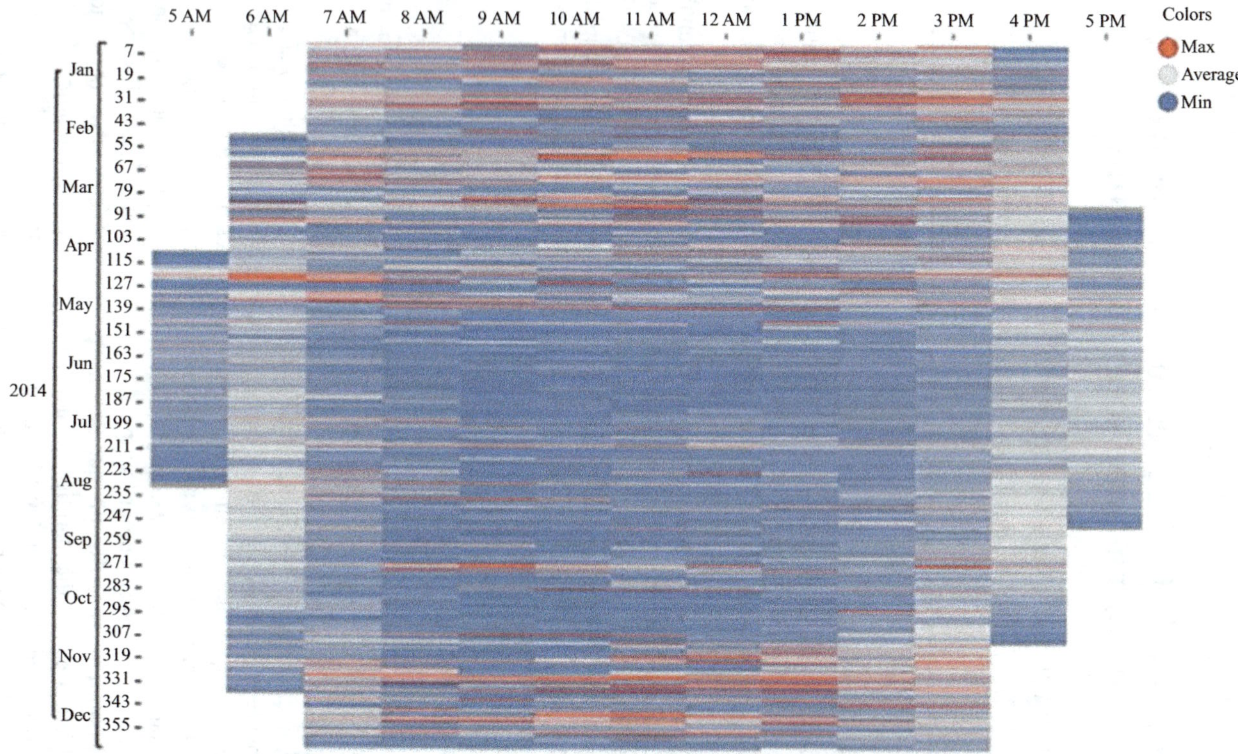

Figure 12.1 Yearly solar irradiation variability by hour in Qatar [5]

of solar radiation available to ensure grid stability and reduce dependency on fossil fuel generators to support a transition to net-zero emissions [3,4].

Specific use cases of solar forecasting include the following:

- *System Control*: Help maintain grid stability and optimize operating costs by committing appropriate amounts of solar energy through co-generation strategies.
- *Bridging Power*: Manage integration of solar energy and energy storage to optimize renewable energy use.
- *Demand Response*: Maximize the use of solar energy in times of peak consumption to reduce stress on the power grid and increase energy efficiency.
- *Avoidance of reverse flow of power*: Prevent the injection of excessive solar power into the grid during times of low demand and high photovoltaic (PV) productivity.
- *Protection of On Load Tap Changers (OLTC) that regulate voltage ratio in substation transformers*: Develop optimal OLTC control algorithms for electrical networks that include solar energy.
- *Management of Dynamic Electricity Pricing*: Fine-tune electricity sales in dynamic pricing scenarios that integrate solar energy by providing a reliable match between announced and delivered electricity prices and informing the pricing of ancillary services.
- *Optimization of PV plant performance*: Increase the productivity and longevity of solar PV plants.

The temporal resolution of solar forecasting needed for the use cases described above and others [6] can go from seconds or even microseconds up to years, depending on the context of application, as indicated in Table 12.1. The temporal resolution relevant to grid stability is typically within the scope of intra-hour operations.

Table 12.1 Characteristic time constant of power system components [6].

<10 s	Inertia response
	Protection system operations
	Switching of power electronics
	Battery switching between charge and discharge
1 min	Fast start of pumped hydropower plant
	Fast start of some combustion engines
15 min	Gas power plant from 1/3 to full power
1 h	Start-up and shutdown of most power plants
24 h	Commitment of generation units
1 year	Maintenance planning
10 years	Expanding transmission infrastructure
20+ years	Economic lifetime of PV systems
	Economic lifetime of grid assets

Many solar forecasting approaches have been developed during the last 25 years using statistical and machine learning (ML) methods or numerical weather prediction (NWP) models [7–11]. Statistical and ML approaches make use of time series forecasting algorithms based on regression techniques and are typically developed for near-real-time forecasting, AKA "nowcasting," e.g., $\leq$6 hours ahead. NWP uses physics-based models of the atmosphere and oceans to predict the weather based on current data from satellites and weather balloons. NPW models lack the temporal resolution needed for intra-hour forecasts and are typically used for longer horizons (e.g., days ahead) [12–14].

Global horizontal irradiance (GHI), which measures the solar irradiance received on a horizontal surface, is a key parameter in the estimation of solar radiation for solar PV [15] that is currently the fastest growing renewable energy sector [16]. The high variability of GHI (Figure 12.1), largely due to changes in cloud cover, requires precise, short-term forecasts, which traditional satellite and ground-based methods often do not provide due to their limited spatial and temporal resolution [17].

Sky imaging technology offers a powerful solution, with high-frequency, high-resolution data that captures real-time cloud movement, density, and type—factors critical for accurate GHI prediction [18]. Unlike satellite data, sky imaging provides localized updates that enhance forecast accuracy for applications that requires low temporal resolution. Advances in deep learning, particularly convolutional neural networks (CNNs), have enabled better feature extraction from sky images, facilitating GHI predictions [19]. However, challenges remain, such as adapting to variable sky conditions and the computational demands of real-time processing.

12.2 State-of-the-art in GHI prediction using all-sky images

Early GHI prediction methods based on cloud correlation model (CCM) and cloud motion vector (CMV) relied on analyzing cloud patterns and movements. CCM links cloud positions in sequential images to GHI by correlating coverage and position relative to the sun, assuming a predictable cloud trajectory. While effective under stable weather, this linear assumption limits CCM's adaptability to complex, changing weather [18]. Similarly, CMV uses vector analysis to predict irradiance based on cloud speed and direction, working best in uniform conditions like overcast or clear skies [8]. Both models struggle with GHI variability caused by non-linear cloud dynamics, particularly on partly cloudy days when cloud opacity, size, and velocity vary. Their static, rule-based approaches lack adaptability and do not learn from new data, making them ineffective for real-time predictions under diverse weather conditions [15,17].

To overcome the limitations of traditional methods based on CCM and CMV analysis, ML models have been adopted to handle non-linear data relationships and capture complex cloud patterns for GHI prediction. ML models can process high-dimensional sky image data and adapt more effectively to rapid weather changes [20] than CCM and CMV models. Early ML models such as support vector machines and simple neural networks improved accuracy through the classification of clear and cloudy skies and by capturing non-linear correlations between cloud features and GHI [21]. ML-based models, with their ability to learn from large datasets and adapt dynamically, mark a substantial improvement over traditional methods. Unlike models based on CCM and CMV analysis, ML models refine their predictions as they process new data, making them more adaptable to varying weather conditions and essential for short-term forecasting, which requires high temporal resolution [18]. By capturing non-linear patterns, ML models handle cloud variability more robustly, providing reliable irradiance predictions under partial cloud cover [20]. However, these models struggled with highly variable cloud dynamics, indicating the need for more sophisticated architectures [22].

CNNs have emerged as powerful tools for GHI prediction due to their ability to extract and interpret complex spatial features from high-dimensional sky image data. CNNs are particularly effective for predicting solar irradiance because they can identify detailed cloud patterns, textures, and edges that directly correlate with GHI [21]. Unlike previous models that rely on surface features, CNNs use multiple convolutional layers to capture hierarchical information within images. This allows them to accurately model the impact of cloud cover and density on irradiance. For example, Jiang *et al.* [23] developed a CNN model for GHI forecasting, focusing on performance during partly cloudy conditions. High-resolution sky images enabled the model to capture cloud nuances, resulting in more accurate predictions, particularly for scattered clouds where traditional models often falter. Zhao *et al.* [24] addressed noisy data in GHI forecasting by integrating regularization techniques, such as dropout and batch normalization, enhancing the model's robustness under variable cloud conditions. This approach improved accuracy by handling inconsistencies, such as partial cloud cover and uneven lighting, which are critical challenges in short-term forecasting. Ramadhan *et al.* [30] proposed a hybrid model that combines CNNs with physical data, aiming to address biases in ML models by introducing physical constraints. This hybrid approach leverages the predictive strengths of both CNNs and physics-based methods, enhancing model reliability by combining empirical and data-driven insights.

Further studies on the use of CNNs to predict solar radiation from sky images include [19,23–26]. In [24], the authors explore 3D CNN-based feature extraction techniques for analyzing ground-based cloud images, aimed at predicting direct normal irradiance. Zhen *et al.* [26] present a deep learning-based surface irradiance mapping model based on CNNs for solar PV power forecasting, utilizing sky images as input. SolarNet [19] provides a sky image-based deep CNN tailored for intra-hour solar forecasting. The study by Jiang *et al.* [23] focuses on solar irradiance capturing during cloudy sky conditions using a CNN regression approach. Their findings emphasize the challenges and potential of accurately predicting solar irradiance in less-than-ideal weather scenarios. By addressing similar concerns in our study, we aim to evaluate the capabilities of our CNN model in differing environmental conditions, particularly under the variability characterized by cloud cover. Song *et al.* [25] proposed a novel sky image-based solar irradiance nowcasting model that integrates a convolutional block attention mechanism.

To improve GHI prediction accuracy, researchers have integrated CNNs with long short-term memory (LSTM) networks and gated recurrent units (GRUs), which allow for capturing both spatial and temporal dependencies. While CNNs are adept at extracting spatial features, they do not inherently account for the temporal evolution of cloud cover. By adding LSTM or GRU layers, these hybrid models enhance prediction accuracy, especially during rapidly changing weather, by retaining information from previous frames and capturing sequential dependencies in cloud movement [22,27].

Recent advancements include infrared (IR) imaging and sensor fusion to improve GHI prediction. Traditional RGB sky images capture visible cloud features, but infrared imaging provides additional information on cloud temperature and density, significantly affecting irradiance. Combining IR data with RGB images allows models to predict solar irradiance more accurately under varied sky conditions [28]. Sensor fusion further enhances model robustness by integrating sky images with ground-based data, including temperature, humidity, and wind speed. These environmental factors influence irradiance, and combining them with visual data improves model generalization across diverse atmospheric conditions. Studies show that sensor fusion models outperform those relying solely on image data. However, the added complexity can present challenges in data handling and may limit applicability in environments with limited computational resources [30].

The quality and diversity of training data significantly influence the performance of CNN models for GHI prediction. Comprehensive datasets, such as those from the National Renewable Energy Laboratory (NREL), expose CNNs to various atmospheric conditions and seasonal variations, strengthening model robustness [20]. Like data augmentation, preprocessing techniques simulate varied cloud configurations, enhancing the model's generalizability. Additionally, anomaly detection methods help filter out sensor errors and outliers, ensuring a cleaner dataset for training [23]. Data balancing further aids in achieving equal representation of diverse sky conditions, preventing model bias toward certain weather patterns.

12.2.1 Advantages of CNN-based approaches

CNNs are adept at rapidly processing large volumes of image data and are particularly suited for the analysis of sky images for intra-hour solar forecasting. Their ability to autonomously learn feature hierarchies enables CNNs to model complex cloud structures impacting irradiance without manual feature engineering. This capacity for detailed feature extraction contributes to CNNs' high accuracy in GHI prediction, especially in capturing subtle cloud patterns that traditional methods might miss [24]. While CNN models can be computationally demanding, optimization techniques, such as layer reduction, help alleviate these requirements in resource-constrained settings [20].

The scalability of CNN models enhances their applicability across various solar energy contexts from centralized solar farms to decentralized microgrids. By adjusting CNN architectures, these models can operate on various hardware from high-performance servers in centralized operations to lower-power edge devices in remote areas. where CNN-based forecasts aid microgrid operators in balancing supply and demand, reducing reliance on backup power sources [21,28]. Model compression and transfer learning advances further enhance CNN deployment in low-power environments, making real-time, on-site GHI prediction more accessible [31,34].

Overall, CNN models based on sky images represent a significant advancement in solar irradiance forecasting, overcoming the limitations of traditional methods and early ML models. While traditional methods like the CCM and CMV rely on linear assumptions and struggle with dynamic weather conditions, CNNs excel at capturing intricate spatial patterns, such as cloud density and positioning, which directly impact GHI [8]. This advanced feature extraction capability positions CNNs as a promising solution for enhancing real-time GHI prediction accuracy and reliability [23].

12.3 Case study: A CNN model of GHI prediction with of sky images

The studies reviewed in the previous section showcase various methodologies for applying CNNs to sky images and related meteorological data that have provided significant advantages in solar forecasting. The study presented in this section complements and advances the insights provided in these works by incorporating GHI ground measurements with the CNN model of sky images to establish the extent to which GHI can be predicted directly from sky images. This dual approach enriches training datasets for solar forecasting and enables a more robust model performance. By evaluating the impact of visual inputs on GHI predictions from sky images, our investigation aims to contribute valuable insights into the optimization of solar forecasting models, ultimately advancing the accuracy and reliability of solar energy predictions for real-world applications.

The approach developed in this study focuses on the prediction of GHI values from sky images using the architecture of CNNs to capture complex patterns in the sky image data. The implementation of the approach is facilitated through the use of TensorFlow, an open-source machine learning framework developed by Google (https://www.tensorflow.org), which provides an efficient platform for building and optimizing neural networks. Using TensorFlow's Keras API, we developed a CNN model that predicts GHI values by training on a curated dataset of sky images paired with their corresponding GHI measurements where hyperparameters are optimized using Keras Tuner.

The sky images and GHI measurements used in this study were collected from equipment installed in Education City, Doha, Qatar. GHI measurements were collected as 1-min averages from 1-s sampling with a Kipp and Zonen CMP11 pyranometer [29]. Sky images was taken every minute at a resolution of 2272×1702 pixels with an EKO SRF-02 all-sky imager [33]. Figure 12.2 shows the sky imager and the solar radiation monitoring station that hosts the pyranometer used in this case study. Through systematic data preparation, model development, and performance evaluation, this study aims to demonstrate the effectiveness of CNNs in enhancing solar forecasting accuracy, ultimately contributing to more reliable renewable energy solutions.

Figure 12.2 Sky imager (left side) on its platform, and solar monitoring station (right side) hosting the pyranometer that measures GHI at QEERI's facilities in Education City, Doha

12.3.1 Data preparation and model development

The data used in developing the model consists of sky images and GHI measurements corresponding to these images. First, the data are filtered to remove bad GHI measurements using the data wrangling methodology as described in Chapter 5, *Data wrangling*, in this book and checked to ensure that all filtered GHI measurements are paired with a sky image. Then, sky images are preprocessed by resizing each image to a specified size and normalizing the pixel values to be within the range of [0, 1] to make them ready as input to the CNN algorithm. The final dataset includes GHI measurements and sky images from October to March 2023 at 1-minute intervals, for a total of about 140,000 images and corresponding GHI measurements.

The CNN model is trained on the reference dataset with a specified number of epochs and batch size and evaluated on selected test data. The number of epochs, i.e., the times the entire training dataset is processed by model, is initialized to be 100. Batch-size, i.e., the number of samples propagated through the model simultaneously, is set at 32. The model updates its weights after each batch, allowing for efficient learning. We use the RandomSearch Tuner in the Keras Tuner library (https://keras.io/keras_tuner/) to find the best hyperparameters (e.g., number of network layers, batch-size, learning rate, number of epochs) for the GHI prediction model.

12.3.2 Model validation

As mentioned above, the entire dataset includes a total of about 140,000 images and corresponding GHI measurements. 80% of this dataset is used for training and 20%

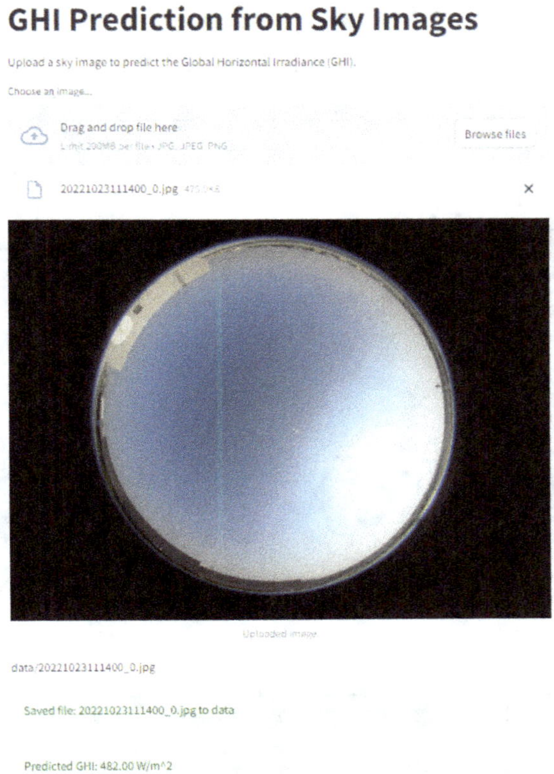

Figure 12.3 Exemplification of GHI predictions from sky images using an application developed in Streamlit-based (https://streamlit.io)

for testing. The objective of the model is to predict GHI from sky images. Consequently, the model results are validated by comparing the GHI values predicted from sky images with actual GHI measurements from solar radiation sensors. GHI values are generated by the model from input sky images, as exemplified in Figure 12.3.

The model's performance is evaluated on the test dataset using three key metrics: mean absolute error (MAE), mean squared error (MSE), and R^2 (R-squared). MAE measures the average absolute difference between predicted and actual values, as shown in (12.1). A lower MAE indicates better model performance, meaning that the predictions are closer to the actual values. MSE, shown in (12.2), calculates the average of the squares of the errors, which means that it penalizes larger errors more significantly than smaller ones. Similar to MAE, a lower MSE indicates a better fit for the model. R^2, the coefficient of determination, represents the proportion of the variance for a dependent variable explained by the model's independent variables and is calculated as the ratio of the sum of squares of residuals with respect to the linear regression to the total sum of squares, as shown in (12.3). It ranges from 0 to 1, with a value closer to 1 indicating a better fit. Table 12.2 summarizes the model evaluation

Table 12.2 Model evaluation values

Evaluation Metric	Value
MAE	0.027
MSE	0.0018
R^2	0.97

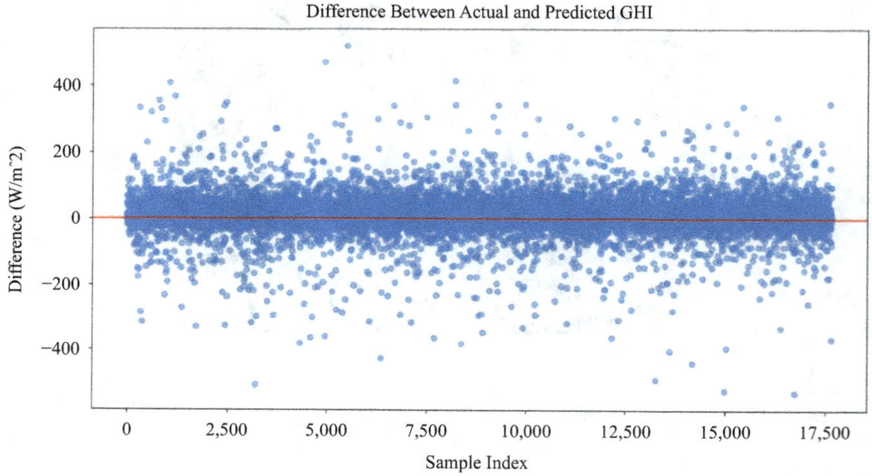

Figure 12.4 Difference between actual and predicted GHI

in terms of MAE, MSE, and R^2. These evaluation results indicate an excellent model performance, as graphically shown in Figure 12.4.

$$MAE : \frac{\sum_{i=1}^{n} |x_i - y_i|}{n} \tag{12.1}$$

$$MSE : \frac{\sum_{i=1}^{n} (x_i - y_i)^2}{n} \tag{12.2}$$

$$R^2 = 1 - \frac{\text{sum of squares of residuals}}{\text{total sum of squares}} \tag{12.3}$$

12.4 Conclusions

The high correlation between sky images and GHI values suggests that sky images can help improve multivariate forecasting models that include GHI measurements, as indicated in the literature reviewed in Section 2.2, and even predict GHI in the absence of solar radiation data. Both pursuits, especially the latter, need further exploration, and the present study corroborates the feasibility of this endeavor.

To enable the advancements that sky camera data can bring to solar forecasting, several challenges must be addressed. For example, variability in image quality and changing environmental conditions can impact the accuracy of cloud detection and forecasting models. Data wrangling can address data quality issues through the curation of sky image data. The integration of soiling dynamics and other meteorological and environmental factors using measured and modeled data from ground instruments and satellites can help improve the performance of cloud detection and forecasting models developed with sky images. Ultimately, sky camera data can be an important component of holistic, multifaceted predictive solar radiation models through the selective combination of ML and NWP models and the discerning integration of meteorological and environmental parameters.

Another challenge relative to the use of sky images for solar forecasting is the need for high-performance computational resources due to the need to process large volumes of high-resolution with ensuing limitations in time/cost efficiency and portability. Ongoing development in computing and ML offer promising avenues to address these issues. For example, the use of federated learning techniques [32] and edge computing can provide a way to reduce computational complexity and the requirement for high-end hardware. Promising contributions to reducing processing requirements can also be made available through the use of cellular neural networks [35,36] where computational complexity is reduced by allowing communication only between neighboring units, and novel neural network novel architecture for increasing the energy efficiency of deep learning [37,38].

References

[1] Denholm, P. and Mehos, M.S., 2011. Enabling greater penetration of solar power via the use of CSP with thermal energy storage. Golden, CO, USA: National Renewable Energy Laboratory.

[2] Utility Frequency, Wikipedia, https://en.wikipedia.org/wiki/Utility_frequency

[3] Wan, C., Zhao, J., Song, Y., Xu, Z., Lin, J. and Hu, Z., 2015. Photovoltaic and solar power forecasting for smart grid energy management. *CSEE Journal of Power and Energy Systems*, 1(4), pp. 38–46.

[4] Yang, D., Wang, W., Gueymard, C.A., *et al.*, 2022. A review of solar forecasting, its dependence on atmospheric sciences and implications for grid integration: Towards carbon neutrality. *Renewable and Sustainable Energy Reviews*, 161, p. 112348.

[5] Sanfilippo, A., Martin-Pomares, L., Mohandes, N., Perez-Astudillo, D. and Bachour, D., 2016. An adaptive multi-modeling approach to solar nowcasting. *Solar Energy*, 125, pp. 77–85.

[6] Alet, P.J., Efthymiou, V., Graditi, G., *et al.*, 2016. Forecasting and observability: Critical technologies for system operations with high PV penetration. In *Proceedings of the 32nd European Photovoltaic Solar Energy Conference and Exhibition (EUPVSEC)*, Munich, Germany (pp. 20–24).

[7] Diagne, M., David, M., Lauret, P., Boland, J. and Schmutz, N., 2013. Review of solar irradiance forecasting methods and a proposition for smallscale insular grids. *Renewable and Sustainable Energy Reviews* 27, 65–76.

[8] Inman, R.H., Pedro, H.T.C. and Coimbra, C.F.M., 2013. Solar forecasting methods for renewable energy integration. *Progress in Energy and Combustion Science*, *39*(6), pp. 535–576.

[9] Voyant, C., Notton, G., Kalogirou, S., *et al.*, 2017. Machine learning methods for solar radiation forecasting: A review. *Renewable Energy*, 105, pp. 569–582.

[10] Nespoli, A., Ogliari, E., Leva, S., *et al.*, 2019. Day-ahead photovoltaic forecasting: A comparison of the most effective techniques. *Energies*, 12(9), p. 1621.

[11] Mellit, A., Massi Pavan, A., Ogliari, E., Leva, S. and Lughi, V., 2020. Advanced methods for photovoltaic output power forecasting: A review. *Applied Sciences*, 10(2), p. 487.

[12] Heinemann, D., Lorenz, E. and Girodo, M., 2006. Forecasting of solar radiation. Solar energy resource management for electricity generation from local level to global scale. Nova Science Publishers, New York, pp. 83–94.

[13] Perez, R., Kivalov, S., Schlemmer, J., Hemker Jr, K., Renné, D. and Hoff, T.E., 2010. Validation of short and medium term operational solar radiation forecasts in the US. *Solar Energy*, 84(12), pp. 2161–2172.

[14] Fountoukis, C., Martín-Pomares, L., Perez-Astudillo, D., Bachour, D. and Gladich, I., 2018. Simulating global horizontal irradiance in the Arabian Peninsula: Sensitivity to explicit treatment of aerosols. *Solar Energy*, 163, pp. 347–355.

[15] Lorenz, E., Remund, J., Müller, S.C., *et al.*, 2009. Benchmarking of different approaches to forecast solar irradiance. *24th European Photovoltaic Solar Energy Conference*, pp. 21–25.

[16] REN21. 2024. Renewables 2024 Global Status Report collection, *Renewables in Energy Supply*. https://www.ren21.net/gsr-2024/

[17] Yang, D., Kleissl, J., Gueymard, C.A., *et al.*, 2018. History and trends in solar irradiance and PV power forecasting: A preliminary assessment and review using text mining. *Solar Energy*, *168*, pp. 60–101.

[18] Chu, Y., Pedro, H.T.C. and Coimbra, C.F.M., 2013. Hybrid intra-hour DNI forecasts with sky image processing enhanced by stochastic learning. *Solar Energy*, 98, pp. 592–603.

[19] Feng, C. and Zhang, J., 2020. SolarNet: A sky image-based deep convolutional neural network for intra-hour solar forecasting. *Solar Energy*, 204, pp. 71–78.

[20] Banalaxmi, B. and Rajesh, W. 2020. Solar irradiance forecasting based on deep learning methodologies and multi-site data. *Symmetry*, 12(11), p. 1830.

[21] Alonso-Montesinos, J. and Batlles, F.J. and Portillo, C., 2015. Solar irradiance forecasting at one-minute intervals for different sky conditions using sky camera images. *Energy Conversion and Management*, 105, pp. 1166–1177.

[22] Travieso-González, C.M., Cabrera-Quintero, F., Piñán-Roescher, A. and Celada-Bernal, S., 2024. A review and evaluation of the state of art in image-based solar energy forecasting: The methodology and technology used. *Applied Sciences*, 14(13), p. 5605.

[23] Jiang, H., Gu, Y., Xie, Y., Yang, R. and Zhang, Y., 2020. Solar irradiance capturing in cloudy sky days – A convolutional neural network based image regression approach. *IEEE Access*, 8, pp. 22235–22248.

[24] Zhao, X., Wei, H., Wang, H., Zhu, T. and Zhang, K., 2019. 3D-CNN-based feature extraction of ground-based cloud images for direct normal irradiance prediction. *Solar Energy*, 181, pp. 510–518.

[25] Song, S., Yang, Z., Goh, H., Huang, Q. and Li, G., 2022. A novel sky image-based solar irradiance nowcasting model with convolutional block attention mechanism. *Energy Reports*, 8, pp. 125–132.

[26] Zhen, Z., Liu, J., Zhang, Z., *et al.*, 2020. Deep learning based surface irradiance mapping model for solar PV power forecasting using sky image. *IEEE Transactions on Industry Applications*, 56(4), pp. 3385–3396.

[27] Shi, X., Chen, Z., Wang, H., Yeung, D.-Y., Wong, W.-K. and Woo, W.-c., 2015. Convolutional LSTM network: A machine learning approach for precipitation nowcasting. *Advances in Neural Information Processing Systems*, 28.

[28] Kim, B. and Suh, D., 2024. Solar PV generation prediction based on multisource data using ROI and surrounding area. *IEEE Transactions on Geoscience and Remote Sensing*, 62, p. 4704511.

[29] Perez-Astudillo, D. and Bachour, D., 2014. DNI, GHI and DHI Ground Measurements in Doha, Qatar. *Energy Procedia*, 49, pp. 2398–2404.

[30] Ramadhan, R.A.A., Heatubun, Y.R.J., Tan, S.F. and Lee, H.-J., 2021. Comparison of physical and machine learning models for estimating solar irradiance and photovoltaic power. *Renewable Energy*, 178, pp. 1006–1019.

[31] Pan, S. J. and Yang, Q., 2010. A survey on transfer learning. *IEEE Transactions on Knowledge and Data Engineering*, 22(10), pp. 1345–1359.

[32] Doe, J. and Smith, A., 2024. A review of federated learning in renewable energy applications: Potential, challenges, and future directions. *Renewable Energy Journal*, 52(3), pp. 123–135.

[33] Perez-Astudillo, D. and Bachour, D. All-sky Images for Solar Resource Assessment in Qatar. *EU PVSEC 2023 Conference Proceedings*, pp. 020477-001–020477-004. Available at: https://userarea.eupvsec.org/proceedings/EU-PVSEC-2023/4DV.4.49/

[34] Han, S., Mao, H. and Dally, W. J., 2016. Deep compression: Compressing deep neural networks with pruning, trained quantization and Huffman coding. In: *Proceedings of the International Conference on Learning Representations (ICLR)*.

[35] Chua, L.O. and Roska, T., 2002. *Cellular Neural Networks and Visual Computing: Foundations and Applications*. Cambridge University Press.

[36] Horváth, A., Hillmer, M., Lou, Q., Hu, X.S. and Niemier, M., 2017. Cellular neural network friendly convolutional neural networks—CNNs with CNNs.

In *Design, Automation & Test in Europe Conference & Exhibition (DATE)*, 2017 (pp. 145–150). IEEE.

[37] Shafique, M., Hafiz, R., Javed, M.U., *et al.*, 2017. Adaptive and energy-efficient architectures for machine learning: Challenges, opportunities, and research roadmap. In *2017 IEEE Computer society annual symposium on VLSI (ISVLSI)* (pp. 627–632). IEEE.

[38] Shafique, M., Theocharides, T., Bouganis, C.S., *et al.*, 2018. An overview of next-generation architectures for machine learning: Roadmap, opportunities and challenges in the IoT era. In *2018 Design, Automation & Test in Europe Conference & Exhibition* (pp. 827–832). IEEE.

Chapter 13

Real-time measurement of electrical signal in medium-voltage distribution network using synchrophasor technology

Mohd Zamri Che Wanik[1], *Abdullah Abdul Jabbar*[1], *Mohamed Gad*[1] *and Antonio Sanfilippo*[1]

The chapter presents and describes the synchrophasor technology for measuring the real-time electrical signal in the medium voltage distribution network. The motivation for using the technology is presented, the fundamental concept behind the technology is explained, and the infrastructure involved in applying it is described. The chapter also shares the experience of installing the PMU monitoring system in the Education City power network in Doha, Qatar, and delineates the characteristics of the data.

13.1 Introduction

Power system disturbances can result in considerable social and economic losses if not addressed in time. The typical window of opportunity for a system operator to do so varies from seconds to a few minutes. The ability to detect power system disturbances in real time and anticipate their occurrence is therefore crucial in maintaining energy security. The installation of phasor measurement units (PMUs) in electrical grids enables the collection of electrical wave measurements in real time and therefore provides data that can be effectively analyzed to detect disturbances promptly and anticipate their occurrence.

The chapter is organized in three sections. In the first section, the PMU technology is described and discussed. Then, a PMU data analytic case study is presented where machine learning algorithms for classification and time series forecasting are applied to PMU data to predict the location and type of power faults. In the last section, an industrial application is reviewed, and project experience with a PMU monitoring system is shared.

[1]Qatar Environment and Energy Research Institute, Hamad Bin Khalifa University, Qatar

13.2 Synchrophasor technology for measuring the real-time electrical signal in the medium voltage distribution network

The future distribution network is likely to have distributed generation and probably energy storage connected, as shown in Figure 13.1. The network is expected to have an advanced communication infrastructure, at least down to medium voltage. Advanced monitoring is required so that the operator may have up-to-date information about network status. One of the most promising power monitoring technologies is through synchrophasor measurements, utilizing PMU devices. PMUs provide measurements of voltage and current synchronized with a common time reference typically provided by UTC (coordinated universal time) or GPS (global positioning system) time.

PMU devices are normally distributed and located in the network at optimally selected locations [1]. The data from these devices are streamed to a phasor data concentrator (PDC), which is typically hosted in a server located normally close to the network being monitored or local monitoring and control center. The information is used to assess the network conditions so that the network operator can make well-informed decisions on how to operate available control devices to maintain the network operation according to standard operating conditions.

Data captured from PMU can be used by a network operator to monitor the network in real-time and also for protection and control purposes. An example of the

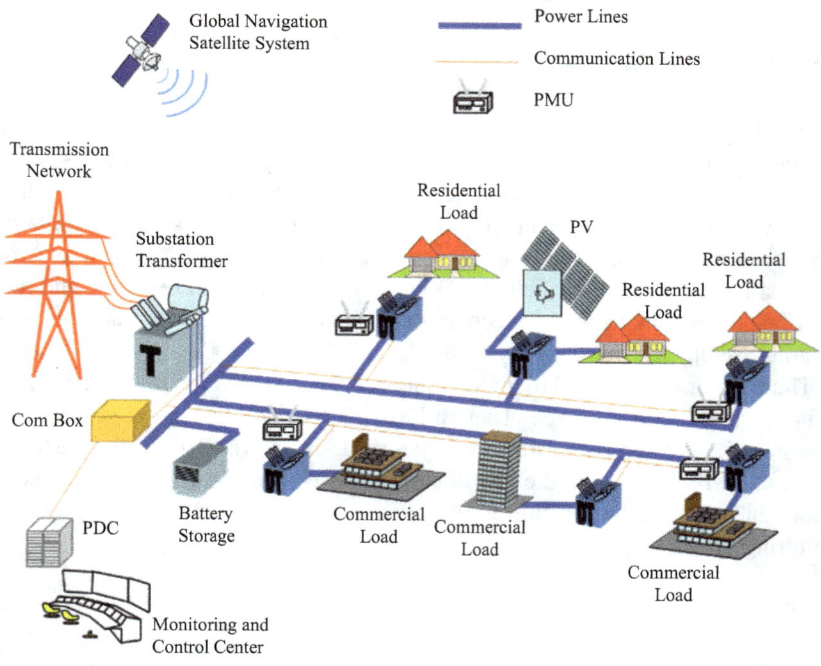

Figure 13.1 Advanced PMU-based monitoring, control, and protection [9]

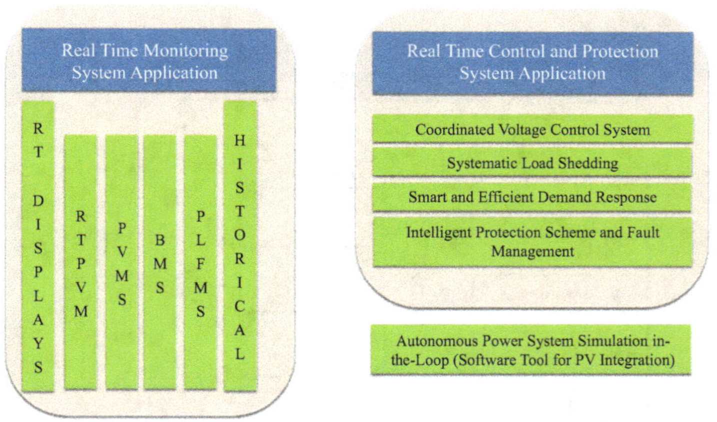

Figure 13.2 Real-time monitoring, control, and protection module [9]

network monitoring, protection, and control management module in the distribution network is depicted in Figure 13.2. PMU provides real-time information on the status of the network in terms of voltage profiles, power flow, to name a few. The data can also be used to coordinate the power system protection of the network. With some analytics, the data can be used to manage the network for more efficient operation.

13.2.1 Phasor measurement unit

A PMU is a device that provides synchrophasor and system frequency estimates, as well as other important information such as active power flow and reactive power flow. The PMU may provide synchrophasor estimates from one or more voltage or current waveforms. The PMU can be realized as a stand-alone physical device or as a part of a multifunction device such as a protective relay or meter. This information may be recorded locally or transmitted in real time to a central location, as illustrated in Figure 13.1. The IEEE Std. C37.118 standard addresses the real-time transfer of data from the PMU to the PDC or other devices [2].

Figure 13.3 outlines the architecture of the PMU. The PMU input is a digital signal of voltage and current waveform sampled by the simulation integration time-step T_s and the UTC timestamp provided by GPS. In our experiment, however, the UTC timestamp is taken from an internal clock. At every integration time-step, the input signal is processed by the modulated sliding discrete Fourier transform (MSDFT) algorithm to compute the five DFT bins surrounding the DFT maximum. The MSDFT outputs are given to the enhanced interpolated DFT (e-IpDFT) algorithm that estimates the signal parameters – frequency, rate of change of frequency (ROCOF), amplitude, and phase – for every newly acquired sample. These values are used together with the UTC timestamp to construct the C37.118 data frame that is then streamed out to PDC [3].

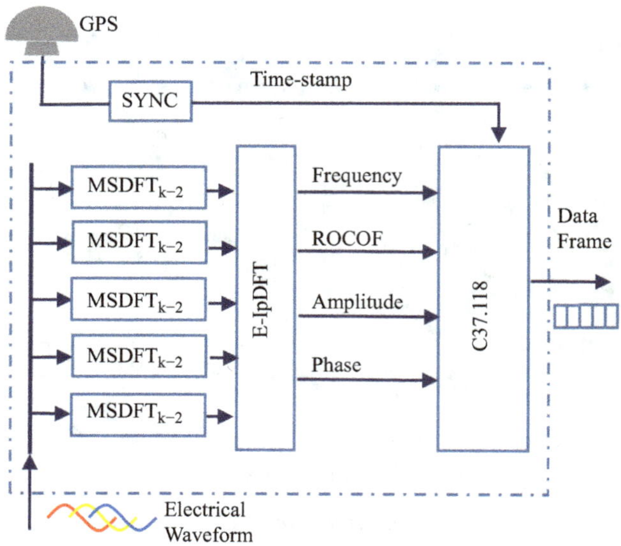

Figure 13.3 PMU graphical model architecture [9]

13.2.2 Synchrophasors

Phasor representation of the sinusoidal signal is commonly used in AC power system analysis. The sinusoidal waveform, which is input to the PMU, can be defined as in (1), commonly represented as the phasor shown in (2), where the magnitude X is the root-mean-square (rms) value, also known as the effective value of the sinusoidal waveform. The phasor representation of a sinusoidal signal can be represented in three forms, which is exponential, polar, and rectangular, as in (2)

$$x(t) = X_{\max} \cos(\omega t + \phi)$$

$$\begin{aligned} X &= Xe^{j\phi} \quad \text{(exponential)} \\ &= X\angle\phi \quad \text{(polar)} \\ &= X\cos\phi + jX\sin\phi \quad \text{(rectangular)} \end{aligned}$$

The synchrophasor representation of the signal $x(t)$ in (1) is the value X in (2) where ϕ is the instantaneous phase angle relative to a cosine function at the nominal system frequency synchronized to UTC. Under this definition, ϕ is the offset from a cosine function at the nominal system frequency synchronized to UTC. A cosine has a maximum at $t = 0$, so the synchrophasor angle is 0 degrees when the maximum of $x(t)$ occurs at the UTC second rollover (1 PPS time signal), and −90 degrees when the positive zero crossing occurs at the UTC second rollover (sin waveform). Figure 13.4 illustrates the phase angle/UTC time relationship [4].

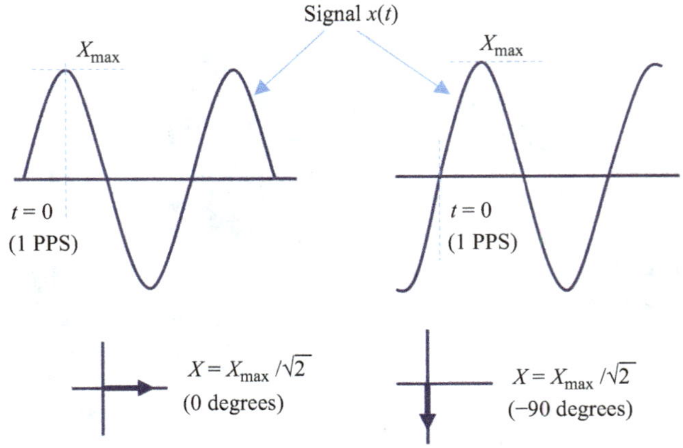

Figure 13.4 Phase angle/UTC time relationship [9]

13.3 PMU data analytic: Case study on power fault awareness and prediction

The goal of this section is to show how PMU data can be used as training material to develop situation awareness and forecasting models of power faults in electrical grids using machine-learning methods for classification and time series analysis.*

As shown in Figure 13.5, power system faults can be classified into two main categories: open- and short-circuit faults. Open-circuit faults are due to the failure of one or more conductors. When all conductors are involved, the fault is symmetrical, otherwise asymmetrical. Short-circuit faults result from abnormal connections of very low impedance between two points of different potential. These are the most common and serious power system faults, causing the flow of abnormally high currents through transmission lines, which can lead to extensive damage to power system equipment. When all phase conductors are involved either with or without the ground, the fault is symmetrical, and otherwise asymmetrical.

The use of PMU measurements to develop power-system fault diagnostics is still in its early stages. The previous work includes the adaptive protection technique for transmission systems based on fault detection, classification, and direction discrimination presented in [5,6], and the model for the location of faults for two-terminal multi-section compound transmission lines (the combination of overhead lines and underground power cables) provided in Rahideh *et al.* [7]. The work presented in this chapter adds a forecasting perspective to existing power-system fault diagnostics, which are primarily focused on detection.

Our initial focus is on two types of short-circuit faults: asymmetrical faults with specific reference to a single line to ground (SLG) and symmetrical faults with

*The study described in this section was initially published as [9].

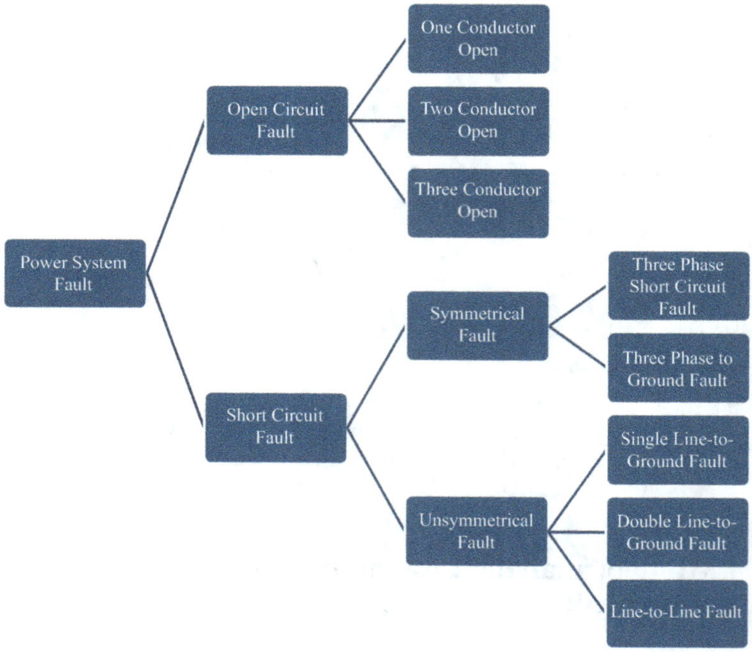

Figure 13.5 Classification of faults in power systems [9]

reference to three phases to ground (3PF). Successful application of this PMU analytics approach indicates that the approach can be extended to other power faults (Figure 13.5).

The data used in this study is generated from a lab demonstration setup of advanced PMU-based monitoring. The setup was developed for a combined hardware and software in the loop simulation. Both PMU and network are virtual and uploaded in the real-time simulator. For this demonstration setup, the OPAL-RT eMegasim PowerGrid Real-Time Digital simulator is utilized [8]. This simulation platform comprises a multi-core processor hardware able to perform simulations within an integration time-step up to tens of microseconds, as shown in Table 13.1.

The setup used for this simulation is shown in Figure 13.6. The electrical network and PMU models are converted to real-time models and uploaded to OPAL-RT real-time simulator (RTS). Both the network and PMU are virtual. The RTS and real-time monitoring station are connected via a network. For measurement visualization, the GUI is developed in LabVIEW. The voltage and current measurements from PMUs are displayed separately. At each measurement point, the magnitude and angle are displayed.

A CompactRIO-based PMU is used to model phasor estimation. The PMU is modeled utilizing an enhanced-interpolated MSDFT algorithm [9,10]. It provides a synchronized measurement of voltage and current phasors with accuracy up to TVE of 0.02%, subject to harmonics and frequency variation. The latency is low, with

Table 13.1 PMU simulation platform

Operating system	Redhat v2.6.29.6-opaltr-6.1
Chassis type	OP5600 Chassis, Rack mount
CPU	Intel Xeon Six-Core 3.46GHz 12M Cache
Total cores	12 Cores
Memory	8 GB
Additional features	External Clock Adapter, Spectracom Tsxhc-PCLe, Intel EXPI940PTL PT Quad Port Server Adapter

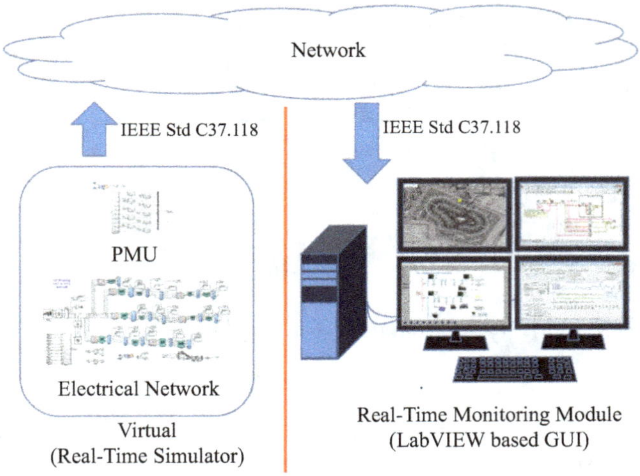

Figure 13.6 Software in the loop simulation setup [9]

an approximate value of 31 μs. The algorithm and PMU prototype satisfies every requirement defined in [4] for class-P PMUs and the majority of class-M PMUs. This PMU prototype has been validated in real electrical grids and it is currently installed in several pilot projects in Europe and also in Qatar.

Voltage and current from the network can be displayed in real-time and stored in the computer disc. Data that has been extracted is stored in the database to be used for data analysis and the development of the management module described in the following paragraphs.

In addition to displaying the phasor values in LabVIEW-based software, the PMU reading is also tested with open source PMU connection tester [11]. The PMU Connection Tester is a window GUI application used to validate, test, and trouble-shoot connections and data streams from phasor measurement devices, as well as visualize their synchrophasor data in real-time. A PMU connection Tester is used to verify that the data stream from modelled PMU devices is being received by the simulator.

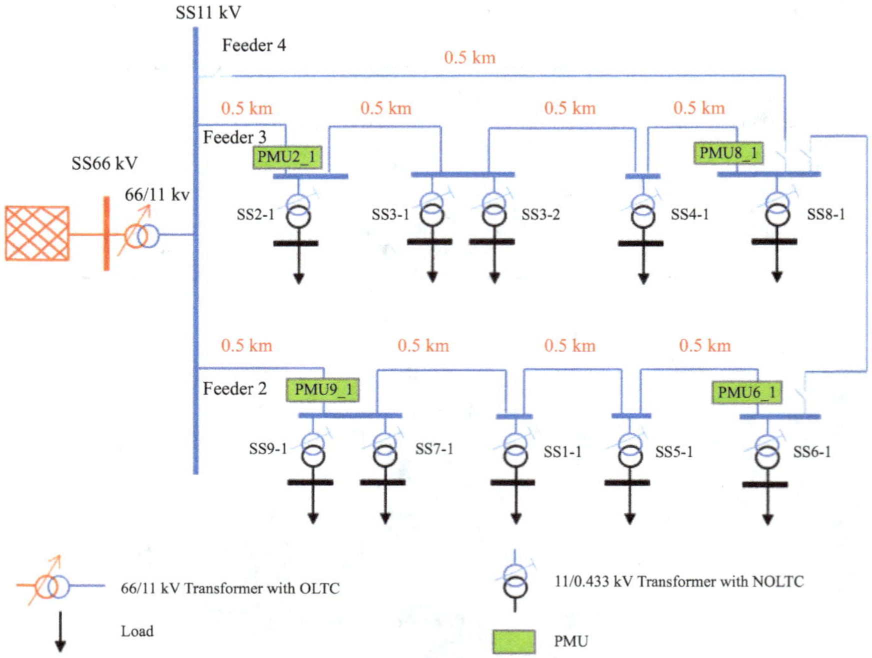

Figure 13.7 Single line diagram of ECCH Lot 2 power system with PMUs [9]

The IEEE C37.118.2-2011 standard was used to verify that the data from the virtual PMU is successfully received through the PMU connection tester. IEEE C37.118.2-2011 standardizes the method of synchronized phasor measurement data transfer between power system components. The standard specifies messaging that can be used with any suitable communication protocol for real-time communication between PMUs, PDC, and other applications. It defines message types, contents, and use, specifies data types and formats, and describes a typical measurement system as well as communication options and requirements.

For this case study, a part of the electrical network at Qatar Foundation Education City Housing Complex in Doha (Qatar) is used. The backbone voltage of the network is 11 kV, and there are eight 11/0.415 kV power transformers with a 1.6 MVA rating, each served by two feeders. The network single line diagram is shown in Figure 13.7. The network has two active feeders and is fed by a 20 MVA transformer of 66/11 kV. This transformer is equipped with an on-load tap changer, placed on the higher voltage side. The 11/0.415 kV transformer is equipped with an off-load tap changer where the position of the tap is already pre-set during installation. In the PMU simulation used in this study, four PMUs were placed at substations SS2-1, SS8-1, SS9-1, and SS6-1 at 11 kV side as depicted in Figure 13.7.

For the development of fault awareness and prediction models, phasor measurements due to two types of faults are generated: a three-phase balanced fault and an SLG fault. As most of the faults that occur in the power system is a SLG fault,

Table 13.2 PMU data generation schedule sample

Time (s)	Duration	Type of fault	Location of fault
0	15	Base	–
15	5	SLG	SS11kV, SS2-1
20	20	Base	–
40	5	3PF	SS11kV, SS2-1
45	25	Base	–
70	5	SLG	SS2-1, SS3-1
75	15	Base	–
90	5	SLG	SS3-2, SS4-1
95	30	Base	–
125	5	SLG	SS4-1, SS8-1
130	30	Base	–
160	5	3PF	SS2-1, SS3-1
165	35	Base	–
200	10	SLG	SS11kV, SS9-1
210	20	Base	–
...	...	...	...

the percentage of SLG faults generated is more compared to three-phase balanced faults, as shown in Table 13.2. Each fault is temporary, lasting only for 140 ms. The faults are applied in the middle of the line between substations and are assumed to be self-clearing.

For this kind of temporary fault, no tripping of the circuit breaker occurs, and the operation returns to normal once the fault clears itself. The different faults are applied alternately, as tabulated in Table 13.2. The simulation was performed continuously for 520 s. The phasor measurement data is captured every 100 ms.

The data generated in the simulation together with the UTC timestamp are streamed from the RTS every 100 ms and saved to the database of the real-time monitoring module. For each of the four PMUs in the simulated network system, the data is arranged as shown in Table 13.3, where

- "V" is volt (unit for voltage);
- "Rad" is radian (unit for angle);
- "Amp" is ampere (unit for current);
- "Va, Vb, Vc" stand for the magnitude of voltage for phase a, phase b, and phase c, respectively;
- "Ia, Ib, Ic" stand for the magnitude of current for phase a, phase b, and phase c, respectively.

  This data is used to train and test the fault classification and forecasting algorithms, as described in the following paragraphs:

The goal of PMU analytics is to detect and predict the type of electrical faults in a power network or lack thereof. As shown in Figure 13.8, the approach consists of three steps: PMU location detection, fault type detection, and fault type forecasting.

Table 13.3 Sample data

Parameter	PMU_21	PMU_91	...
Time	2018:09:18:07:31	2018:09:18:07:31	...
Va mag (V)	246.9323895	239.2371589	...
Va Ang (Rad)	−0.665130841	0.117671013	...
Vb mag (V)	246.932509	246.8667419	...
Vb Ang (Rad)	0.570432179	−0.651592174	...
Vc mag (V)	246.9324505	246.7729714	...
Vc Ang (Rad)	−2.759525556	1.444066999	...
Ia mag (Amp)	318.1647186	308.7791838	...
Ia Ang (Rad)	−0.910114373	−0.125862527	...
Ib mag (Amp)	318.1648679	318.0860291	...
Ib Ang (Rad)	1.18428075	−0.896523834	...
Ic mag (Amp)	318.1647833	317.9687065	...
Ic Ang (Rad)	2.41984418	1.199038302	...
Type of fault	no_fault	single_line	...
PMU location	PMU_21	PMU_91	...

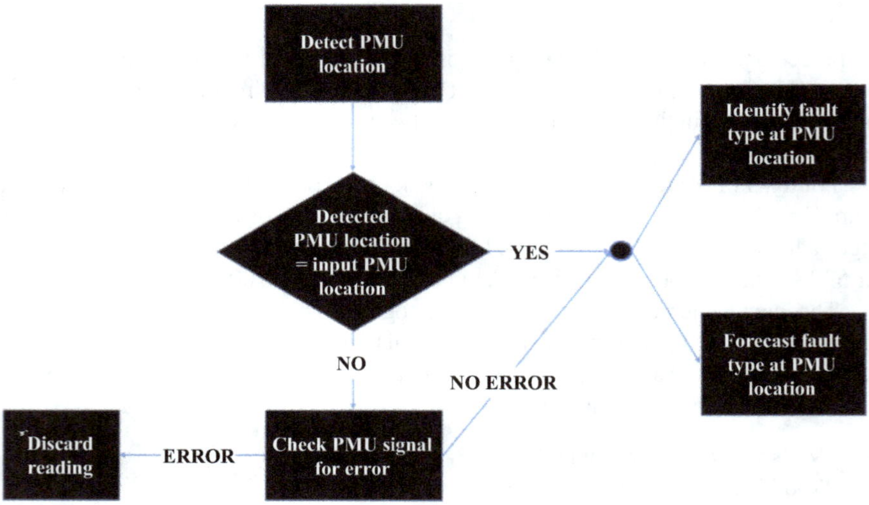

Figure 13.8 PMU analytics workflow [9]

The PMU location detection step provides a quality assurance check, as shown in Figure 13.8. PMU location detection is achieved through the development of a classifier that is capable of recognizing the location of each PMU in the network using as training material the data described in the following subsection (with the PMU location attribute removed). Given as input a PMU signal P containing information about voltage, angle, and current, if the location detected by the classifier is different from the location associated with P, then P is flagged for error checking. If the error is confirmed, then the signal is discarded. Otherwise, if the PMU location detected is the same as the location of the PMU input signal P, P is accepted, and classification and time series analysis are used to identify and forecast the type of fault at the detected location. Classification and forecasting models were developed using the WEKA machine-learning platform [12].

13.3.1 Classification for PMU location and fault type detection

The first goal of classification in the analysis of PMU data is to develop a function f that gives as input a list of attribute-value pairs $a_1 : v_1, a_2 : v_2, \ldots, a_n : v_n$ describing a PMU event (Table 13.3) with its location removed can identify the PMU location, e.g.

$$f(a_1 : v_1, a_2 : v_2, \ldots, a_n : v_n) \rightarrow \{PMU_{21} \mid PMU_{61} \mid PMU_{81} \mid PMU_{91}\}.$$

If the detected PMU location l from the input PMU signal is the same as the PMU location in the input signal, the PMU signal is accepted, and classification function g is developed to identify the occurrence of a given type of fault (SLG or 3PF), or lack thereof (Base) at the detected PMU location, e.g.

$$g(a_{l,1} : v_{l,1}, a_{l,2} : v_{l,2}, \ldots, a_{l,n} : v_{l,n}) \rightarrow \{SLG_l \mid 3PF_l \mid Base_l\}.$$

We run tests with several classification algorithms including Bayesian Networks [13], Support Vector Machine (SVM) [14], Multilayer Perceptrons [15], and Decision Trees [16]. The best results were obtained with Bayesian Networks and Decision Trees.

A Bayesian Network (BN) is a directed acyclic graph where each node is a domain variable, and each link between nodes represents a probability dependency. It represents a probability distribution $p(U) = \prod_{u \in U} p(u|\text{parents}(u))$, where U is the set of domain variables, and parents(u) is the set of parents of u. Within a BN approach to classification, the probability of a class variable C given a set of attribute variables $X \in U$ is calculated as $\arg\max_y p(C|X)$, where $p(C|X) = \frac{p(U)}{p(X)}$.

In the classification task at hand, the class variable refers to either PMU location ($\text{PMU}_{21}, \text{PMU}_{61}, \text{PMU}_{81}, \text{PMU}_{91}$) or power fault status at some PMU location l (SLG_l, 3PF_l, Base_l), and the attribute variables are the voltage, radian, and ampere parameters described in Table 13.3. In deriving a BN from a dataset such as the one described in Table 13.3, the network structure and its parameters need to be learned from the data. Several approaches have been proposed for learning the graphic structure of BNs from data [17,18]. The BN classifier in this study uses the K2 algorithm in [17] that recovers the underlying graphic structure based on a predetermined order of nodes in a greedy fashion according to a score-based procedure. At the outset, the algorithm assumes that a node has no parents. Parents are incrementally added to the node to maximize the probability of the resulting structure until the probability of the ensuing network structure can no longer be increased by the addition of yet another parent.

In decision-tree classification, the model identifies members of a class as the result of a sequence of decisions. A decision tree typically consists of a set test node linked to decision nodes. The test node is a feature describing the class (i.e. the voltage, angle, or current attribute that defines a PMU event in Table 13.3). The link

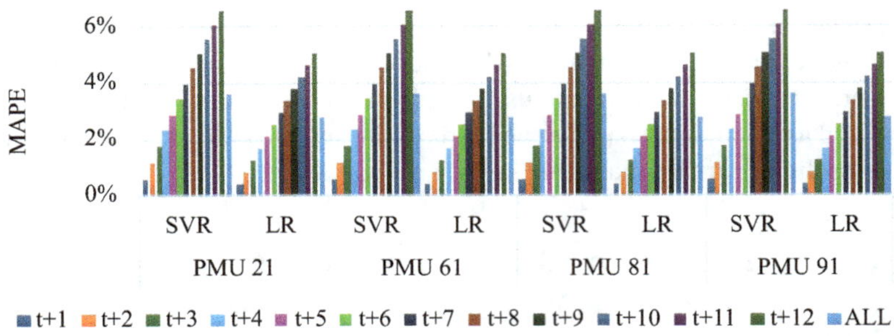

Figure 13.9 MAPE results for the forecasting of fault type ten steps ahead [9]

encodes the condition (feature value) that must be met in order to make a decision, and the decision node represents the outcome of the decision process, which can be another test node or the final class prediction, as shown in Figure 13.9 with reference to the PMU location class.

```
Vb mag (V) > 229.933235
|   Ib mag (Amp) <= 318.164989
|   |   Vb mag (V) <= 246.932142
|   |   |   Ia mag (Amp) <= 318.15497
|   |   |   |   Vc mag (V) <= 225.764914: 91PMU (11.75/1.75)
|   |   |   |   Vc mag (V) > 225.764914: 21PMU (106.5/47.75)
|   |   |   Ia mag (Amp) > 318.15497: 21PMU (140.5/27.0)
|   |   Vb mag (V) > 246.932142: 21PMU (7466.0/2.75)
|   Ib mag (Amp) > 318.164989: 91PMU (7547.25)
```

To learn a decision tree classifier from a training dataset such as the one exemplified in Table 13.3, the model establishes the sequential order of test nodes according to how informative the nodes' attributes are. The model determines the information content of an attribute by its information gain concerning the classification tasks [16,19]. The information gain of an attribute A concerning a class C is based on the reduction in entropy of the value for C when we know the value of the attribute, as shown below.

$$\text{gain}(C,A) = \text{entropy}(C) - \text{entropy}(C,X)$$

$$\text{entropy}(C) = \sum_{i=1}^{n} -p(C_i)\log_2 p(C_i)$$

$$\text{entropy}(C,A) = \sum_{x \in A} p(x)\,\text{entropy}(x)$$

Test nodes with attributes that have higher information gain occur earlier in the decision tree. The model creates test nodes using the available attributes until all data in the training dataset has been accounted for. Typically, not all attributes are used because decision tree learners use pruning strategies to reduce the number of nodes. The number of attributes depends on the specific implementation.

Table 13.4 Evaluation results for the first classifiers detecting location

PMU Location	Precision (%)		Recall (%)		F-Measure (%)	
	BN	DT	BN	DT	BN	DT
PMU21	98.9	98.8	99.7	99.9	99.3	99.4
PMU61	99.3	99.7	99.8	99.8	99.6	99.8
PMU81	99.8	99.9	99.6	99.7	99.7	99.8
PMU91	99.8	99.9	99.8	98.9	99.3	99.4
Weighted Average	**99.5**	**99.6**	**99.4**	**99.6**	**99.4**	**99.6**

Table 13.4 shows the evaluation results of the Bayesian network (BN) and decision tree (DT) classifiers that detect PMU locations. Each classifier is derived from the PMU data described in Table 13.3. We trained the classifier on 80 of the data and evaluated the remaining 20 using the precision, recall, and F-measure evaluation metrics, as shown in the equations below. For this experiment, we used a dataset of 40,804 simulated data records generated according to the schedule in Table 13.4, equally distributed across the 4 PMUs. These evaluation results show that PMU location can be reliably detected through classification from a list of PMU measurements, with an expected accuracy ranging from 99.8% to 99.4% when using decision tree and marginally lower when using Bayesian networks.

$$precision = \frac{True\ Positives}{True\ Positives + False\ Positives}$$

$$recall = \frac{True\ Positives}{True\ Positives + False\ Negatives}$$

$$F\text{-}Measure = \frac{2 \times (Precision \times Recall)}{Precision + Recall}$$

Table 13.5 shows the evaluation results for the second Bayesian net classifier, which identifies the type of power fault (or lack thereof) for the PMU location identified by the first classifier. We trained the classifier on 80 of the data and evaluated the remaining 20 using the precision-recall and the F-measure evaluation metrics. In this experiment, we used the same dataset used in the first classifier, where power faults and lack thereof were inserted in the data according to the schedule in Table 13.2. We developed one classifier for each of the four PMU locations by dividing the data into 4 equal subsets of 10,201 records each.

13.3.2 Forecasting

The aim of forecasting is to develop a function h that can predict the occurrence of an electrical fault (or lack thereof) and its type at some time $t >$ now for location l, given as input a list of located PMU measurements $m_{(t-1,l)}, \ldots, m_{(t-n,l)}$, where each

Table 13.5 Evaluation results for the second classifier detecting fault type

PMU	# records	Precision %				Recall %		PMU Fault
		BN	DT	BN	DT	BN	DT	
PMU 21	8,101	0.988	1	0.887	1	0.934	1	BSE
	1,400	0.752	1	0.953	1	0.840	1	SLG
	477	0.559	1	0.939	1	0.701	1	3PF
Weighted Average		**0.927**	**1**	0.899	**1**	**0.906**	**1**	
PMU 61	8,101	0.988	1	0.883	1	0.932	1	BSE
	1,400	0.738	1	0.953	1	0.832	1	SLG
	477	0.559	1	0.939	1	0.701	1	3PF
Weighted Average		**0.926**	**1**	**0.896**	1	**0.903**	**1**	
PMU 81	8,101	0.988	1	0.998	1	0.993	1	BSE
	2,100	0.992	1	0.950	1	0.971	1	SLG
	–	–	–	–	–	–	–	3PF
Weighted Average		**0.989**	**1**	**0.989**	**1**	**0.989**	**1**	
PMU 91	8,101	0.988	1	0.886	1	0.934	1	BSE
	1,400	0.749	1	0.953	1	0.839	1	SLG
	477	0.559	1	0.939	1	0.701	1	3PF
Weighted Average		**0.927**	**1**	**0.898**	1	**0.905**	**1**	

m is a set of attribute-value pairs $m = a_1 : v_1, \ldots, a_n : v_n$ describing a PMU event (see Table 13.3), as shown in (13.1).

$$h(m_{(t-1,l)}, \ldots, m_{(t-n,l)}) \rightarrow \{\mathrm{SLG}_{(t,l)} \mid \mathrm{PF}_{(t,l)} \mid \mathrm{Base}_{(t,l)}\} \tag{13.1}$$

The location l of the PMU is established by applying the classifier for PMU location detection (see 11.3.1) to the non-located PMU input prior to forecasting, as exemplified in (13.2).

$$f(m_{(t-1)}, \ldots, m_{(t-n)}) \rightarrow m_{(t-1,l)}, \ldots, m_{(t-n,l)} \tag{13.2}$$

We conducted tests with several classification algorithms, including linear regression (LR), support vector regression (SVR), and multilayer perceptrons. The best results were obtained with LR) and SVR. Due to space considerations, only the results for LR and SVR are reported in this section.

Within an LR approach to forecasting, the set of PMU measurements m_t of a time series $m_i, \ldots, m_n$, for $N \geq t \geq i \leq n$, can be estimated as the linear weighted sum of previous terms $m_{t-1}, \ldots, m_{t-n}$ in the series, as shown in (13.3), where

- c is a constant,
- ϵ_t is white noise,
- $a_i, \ldots, a_n$ are the coefficients that function as the weights in the sum.

$$m_t = c + \sum_{i=1}^{N} a_i m_{t-i} + \epsilon_t \tag{13.3}$$

These coefficients can be derived from the training data using Yule–Walker equations or regression techniques such as least squares estimation or maximum entropy [20].

The goal of forecasting within an SVR approach is to find a function that for each vector $\vec{x}_i \in \mathbb{R}^n$ representing a time series within a dataset with N training time series sequences approximates its value y_i $(i \geq 0 \leq N)$ as closely as possible, as shown in (13.4), where b is the bias. The objective of regression is to estimate the weight vector $\vec{w}$ with the smallest possible length to avoid over-fitting (see [21] for details)

$$y_i = \vec{w} \cdot \vec{x}_i + b \tag{13.4}$$

We used four datasets, one for each PMU containing 10,201 data records, generated according to the schedule shown in Table 13.2, as training and testing material for the development of four distinct forecasting models for each PMU. In operational use, the choice of forecasting model would be determined by the output of classification, whose function is to detect PMU location from a list of PMU measurements, as discussed in the previous session. We trained each model on 80% of the data and evaluated on the remaining 20% using the normalized root mean square error (NRMSE) and mean absolute percentage error (MAPE). The RMSE measures the standard deviation of the differences between predicted and observed value, as shown in (13.5). The NRMSE is obtained by dividing the RMSE by the mean of observed values (max – min). The MAPE expresses forecasting error as a percentage, as defined in (13.6). Forecasts were provided for five steps ahead at every 100 ms

$$RMSE = \sqrt{\frac{\sum_{t=1}^{n} \left(x_{\text{predicted}_t} - x_{\text{observed}_t}\right)^2}{n}} \tag{13.5}$$

$$MAPE = \frac{100}{n} \sum_{i=1}^{n} \left(\frac{\left| x_{\text{observed}_t} - x_{\text{predicted}_t} \right|}{x_{\text{observed}_t}} \right) \quad (8) \tag{13.6}$$

We performed various forecasting experiments, progressively removing the number of parameters in the model using information gain (IG) [16,19] as an attribute selection technique [22]. More precisely, we used IG to measure the bits of information obtained in recognizing the "Type of Fault" class (noFault, singleLineToGround, 3PhaseToGround) by removing one parameter at a time (Va mag (V), Va Ang (Rad), etc.). Using the WEKA implementation of IG [23], we ranked the 12 parameters in the data (Table 13.3). We then developed various forecasting models using a variety of parameter combinations, using the IG score as a parameter selection criterion. Overall, we found that the best results are obtained with univariate forecasting, using "Type of Fault" as the only parameter in the model.

Figures 13.9 and 13.10 provide the evaluation results relative to the prediction of electrical faults 12 steps ahead (1200 ms), or lack thereof, in each forecasting model. In addition to a low error rate, these results show that forecasting error grows linearly with greater numbers of step-ahead predictions. Trials with longer forecasting

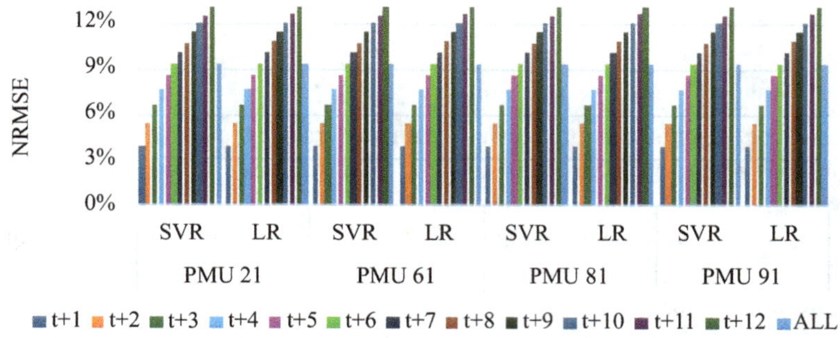

Figure 13.10 NRMSE results for the forecasting of fault type ten steps ahead [9]

horizons up to 40 steps ahead confirm this trend. These are very encouraging results, which we plan to corroborate with the full range of electrical faults in future work.

The collection of electrical wave measurements in power grids through PMU devices enables the development of computational models capable of detecting power faults in real time and anticipating their occurrence. The study presented in this section shows that machine learning algorithms for classification and time series forecasting can be successfully applied to PMU data to forecast the location and type of power faults, such as single line and triple line to ground faults. According to the approach developed, a classifier is first used to detect PMU location. Once the PMU location has been established, a second classifier detects and PMU-specific forecasting models predict the occurrence or lack thereof and type of fault at the established location. We recognize that the uncertainties of the first classification step used for PMU location detection will add to the uncertainties of the second classification step for fault type detection and the forecasting step for fault type prediction. However, given the strong performance of the classifier for PMU location detection (average accuracy 99.45%, see Table 13.4) we do not anticipate these added uncertainties to have a significant impact. Overall, the results obtained indicate that the same methodology may also be successfully applied to other types of power system faults.

13.4 PMU monitoring system: Industrial application and project experience

In this section, we discuss the industrial application of PMUs with reference to the installation carried out by the Qatar Environment and Energy Research Institute team with the collaboration of Qatar General Electricity and Water Company and Qatar Foundation Facility Management in Education City, Doha, Qatar.

13.4.1 Industrial application and devices

Distribution PMUs are devices designed for monitoring medium voltage distribution network. PMU monitoring systems typically include current and voltage sensors, GPS and 4G/5G antennaa, a PMU Box, and a server that functions as a PDC. Current sensors, voltage sensors, GPS antenna, and 4G/5G antenna are all connected to the PMU box as exemplified in Figure 13.11. These components are installed at the substation, as shown in Figure 13.12.

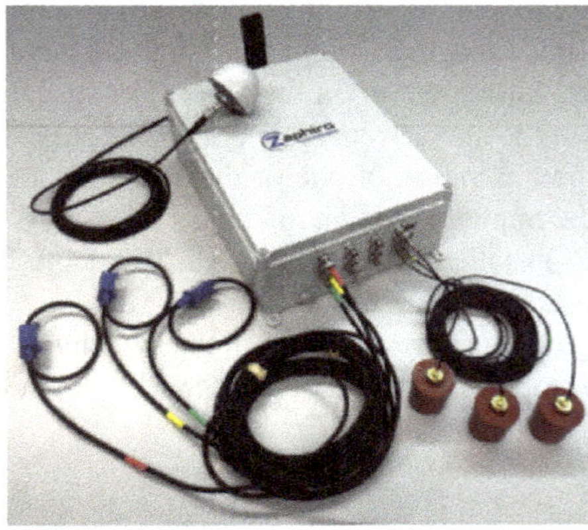

Figure 13.11 A commercial PMU devices [24]

Figure 13.12 PMU device Installation in the substation [25]

The current sensor (front right cable in Figure 13.11) is a flexible clip-around Rogowski coil for the electronic measurement of AC current with galvanic separation between the primary circuit and the secondary circuit (measurement). It is chosen so the sensor is thin, flexible, and light weight. It should have a very high accuracy and should not saturate with overcurrent and short-circuit currents. It should also have an excellent immunity to electromagnetic interferences and should be simple to installed. The base configuration of one of the PMU for medium voltage that is currently available in the market is shown in Table 13.6.

The PMU devices installed at Education City in Doha has the compliance characteristics listed in Table 13.7.

Current and voltage sensors are installed on the switchgear cable within substations, as shown in Figures 13.13 and 13.14. As the PMU monitoring system that is

Table 13.6 PMU configurations

Physical and environmental characteristics	
Dimensions	Width: 300 mm, depth: 400 mm, height: 120 mm
Weight	7.5 kg (max)
Operating temperature	Standard temperature range: −20°C to +55°C Extended temperature range: −40°C to +70°C
Power requirements	
Input voltage range	100–240 Vac
AC frequency range	45–65 Hz
Power consumption	35 W (max)
Power connector type	C14 plug
Low-level analogue inputs	
Number of channels	15
Connector type	RJ-45
ADC resolution	24 bits
ADC sampling rate	10 kHz
Input range	Voltage channels: 10 V, current channels: 1 V
Sensor compatibility	Voltage channels: resistive/capacitive voltage dividers Current channels: Rogowski coils
Sensor class	0.5 (voltage and current)
Rated primary voltage	$20/\sqrt{3}$ kV
Primary current range	1–20,000 A
Ethernet communication	
Number of ports	21
Connector type	RJ-45
Network interface	10/100/1000 Base-T
Supported PTP profiles	IEEE 1588-2008 (default end-to-end profile) IEEE 802.1AS-2011 (TSN profile)
Synchronization accuracy	100 ns typical, <1 μs maximum

Table 13.7 Synchrophasor compliance

Standard compliance	IEEE C37.118/IEC 60255-118
Performance class	P and M
Rated frequency	50/60 Hz
Typical accuracy	0.02% (TVE) 0.4 mHz (frequency error) 6 mHz/s (ROCOF error)
Reporting rate	Up to 100/120 Hz
Communication protocol	Default: IEEE C37.118 (2005 and 2011)
Reported phasors	Three-phase and symmetrical components (user selectable)
Reported analogues	Active, reactive and apparent power, power factor (user selectable)
Reported digitals	8
Numerical format	Integer or float (user selectable)
Phasor format	Rectangular or polar (user selectable)

Figure 13.13 Installation of PMU current sensor Education City [25]

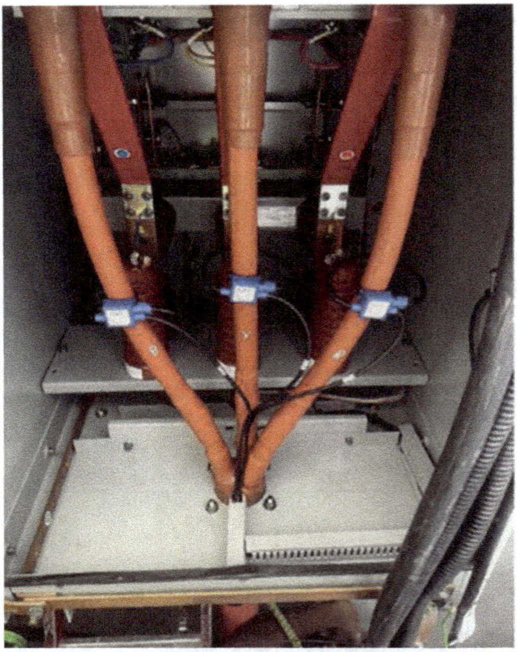

Figure 13.14 PMU devices voltage sensor installation in the substation [25]

Table 13.8 Voltage sensor characteristics

Rated insulation level	24/50/125 kV
Accuracy class*	0.5
Rated primary voltage, Upn*	$20/\sqrt{3}$ kV
Maximum voltage, Um	24 kV
Rated secondary voltage, Usn*	1 V
Rated voltage factor, Ku	1.9 for 8h
Nominal frequency	50/60 Hz
Nominal transformation ratio, Kn*	$20/\sqrt{3}$ kV / 1 V
Auxiliary supply voltage	±12 Vdc
Maximum supply current	15 mA
Creepage	375 mm
Primary terminal capacity	1 pF
Bandwidth (−1 dB)	30 Hz–20 kHz
Applicable standards	IEC 60044-7, IEC 60660
Accuracy on harmonics	According to IEC 60044-8
Weight	1.6 kg

described in this chapter is for medium-voltage applications, sensors are installed at the switchgear of the 11 kV substations.

The voltage sensor is also connected within the switchgear, as shown in Figure 13.14. Its main electrical and mechanical characteristics are listed in Table 13.8. Voltage sensors for PMU application are an innovative device that

Figure 13.15 GPS/4G antenna installation [26]

Figure 13.16 PMU monitoring system at distribution network [27]

incorporates two functions: voltage transducer (electric field sensing technology) and capacitive voltage detector for MV switchgear voltage indicating lamps.

GPS antennas are normally integrated with 4G/5G antenna in one unit. It is typically a compact heavy-duty antenna designed and suitable for a wide range of applications within the industry. The antenna is normally housed by a cover that is certified by standard of IP67 for water resistance, IK09 for high impact resistance, and IP69K standard from high pressure and hot water ingress. Figure 13.15 shows how the antenna is installed on the roof of the substation.

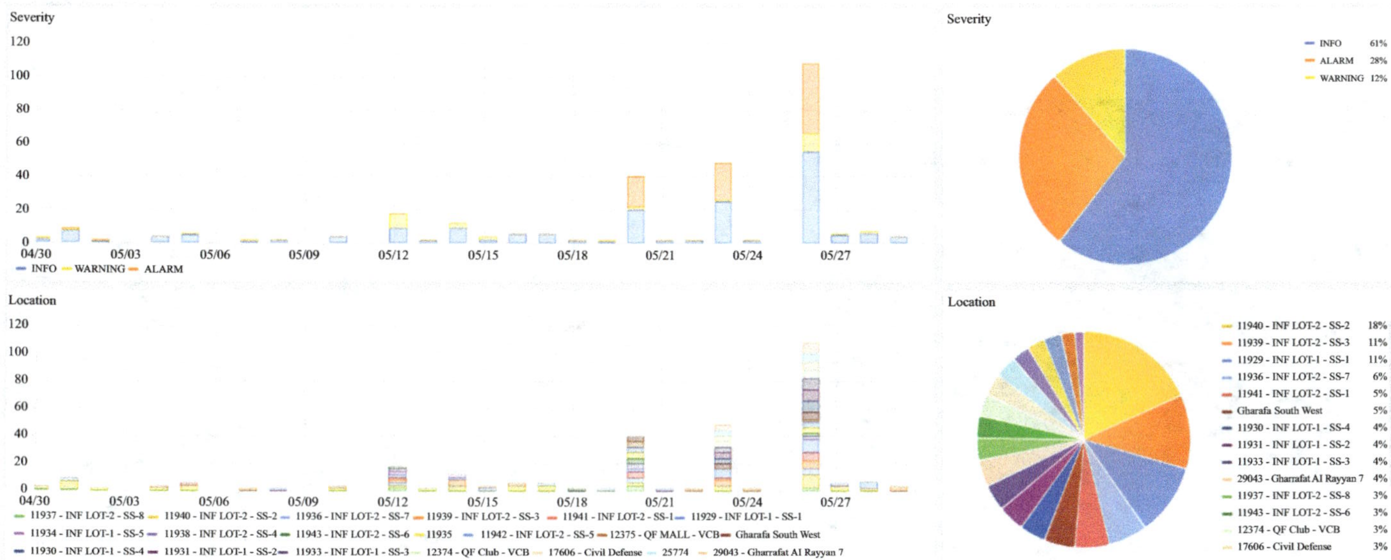

Figure 13.17 GUI for events detection alerts (courtesy: QEERI)

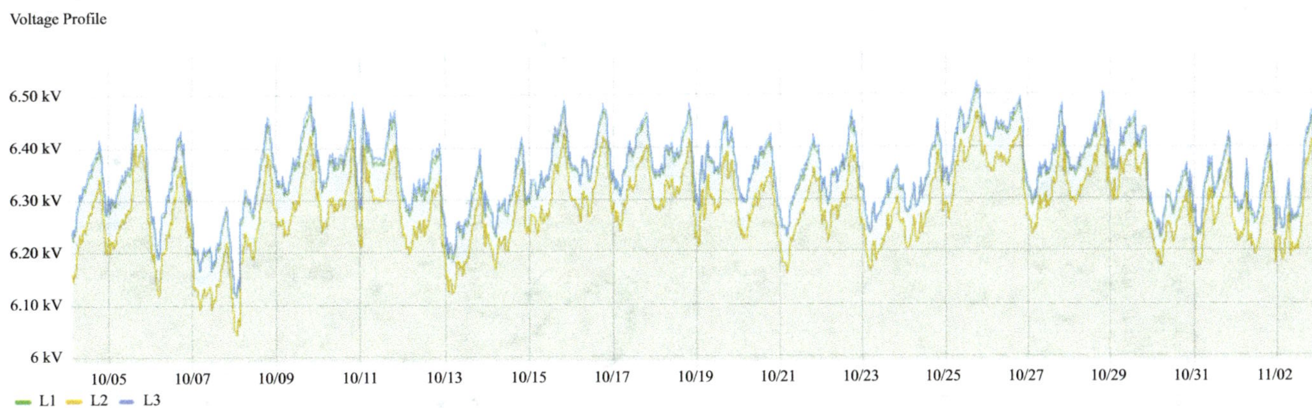

Figure 13.18 GUI for voltage profile (courtesy: QEERI)

Figure 13.19 GUI for substation location and installed PMU location (courtesy: QEERI)

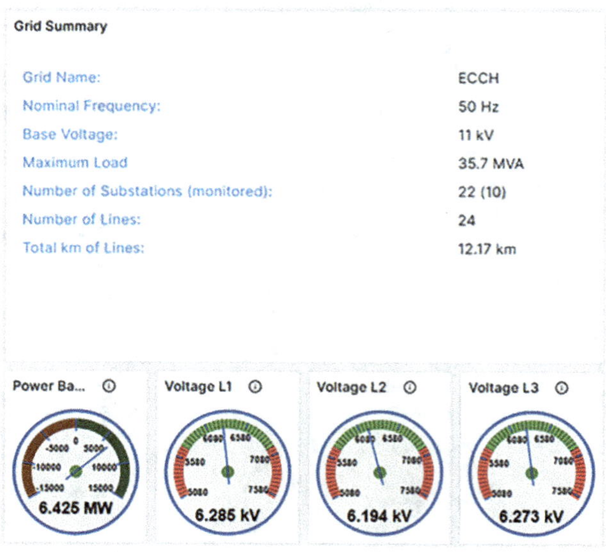

Figure 13.20 Grid summary page

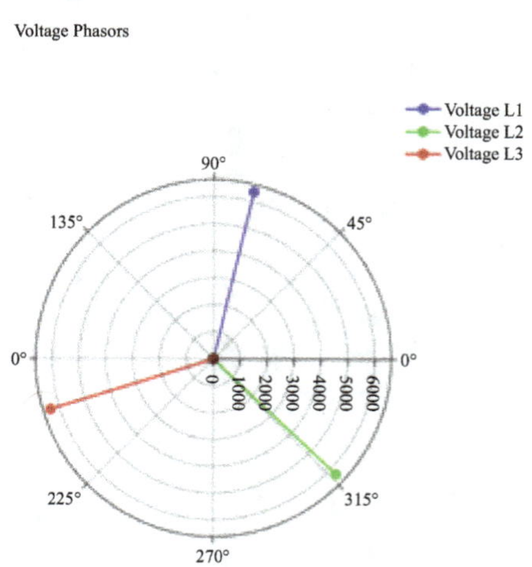

Figure 13.21 Voltage phasor information (courtesy: QEERI)

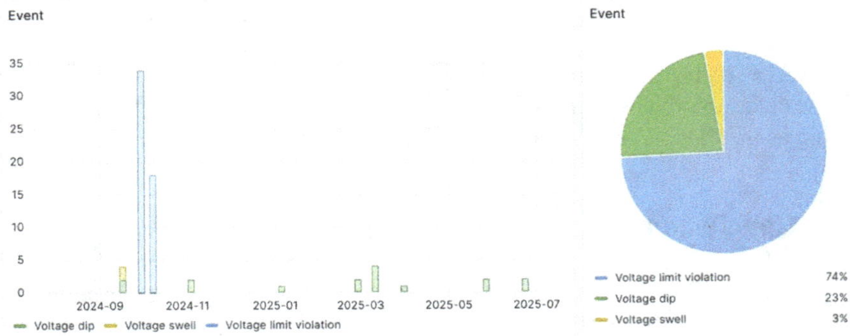

Figure 13.22 GUI for events detection details (courtesy: QEERI)

Table 13.9 PMU configurations

Characteristic	Value
Typical accuracy	0.02% (TVE)
Frequency error	0.4 mHz
ROCOF error	6 mHz/s
Reporting rate	Up to 100/120 Hz
Reported phasors	Three-phase and symmetrical components (user selectable)
Reported analogues	Active, reactive and apparent power, power factor (user selectable)
Numerical format	Integer or float (user selectable)
Phasor format	Rectangular or polar (user selectable)

13.4.2 Distribution PMU management and software application

The overview of the distribution network that deploys the PMU monitoring system in the Education City power network is shown in Figure 13.16. PMUs capture the electrical signal (voltage and current) from the selected substations. The PMU devices used in this installation are micro-PMUs, as they operate on microprocessors. These measured current and voltage are transformed into a phasor form, and it stamped with time from the GPS clock. This means that the measurements are synchronized to a common time reference. These data are then transmitted to a PDC that is placed at a selected location. From here the data is transmitted in real-time to a network operator control center. The utility will use the data from PMU to decide control action for distribution network control devices. This control action, which is currently mostly done manually. However, in the future, with more advanced ICT infrastructure, the control can be performed automatically in the closed loop.

13.4.3 The characteristics of the data captured by PMU

The software application that displays and provides the result of analytics of the captured data is called SynchroGuard. The user can see all the data captured with different resolutions. The data is also visualized depending on the information to be interpreted. The example visual graphic from the SynchroGuard is depicted in Figures 13.17, 13.18, 13.19, 13.20, 13.21, and 13.22.

The data captured by the PMU monitoring system are streamed to the server that is acting as the PDC. The characteristics of these data are detailed in Table 13.9.

13.5 Conclusion

This chapter presents and describes the synchrophasor technology for measuring the real-time electrical signal in the medium-voltage distribution network. The motivation for using the technology is presented, the fundamental concepts behind the technology were explained, and the infrastructure involved in applying this technology is described. The chapter also shared the experience in installing the PMU monitoring system at the Education City power network and the characteristics of the data captured were discussed. PMU, which is working based on synchrophasor technology, will enable many novel applications in managing the distribution network in the future. These applications can be categorized under system state awareness, event detection, operation auxiliary, and data mining.

References

[1] Manousakis, N. M., Korres, G. N., and Georgilakis, P. S., Taxonomy of PMU placement methodologies, *IEEE Trans. Power Syst.*, vol. 27, no. 2, pp. 1070–1077, 2012.

[2] *IEEE standard for synchrophasor measurements for power systems*, IEEE Std C37.118.1-2011, 2011.

[3] Romano, P., and Paolone, M., Enhanced interpolated-DFT for synchrophasor estimation in FPGAs: theory, implementation, and validation of a PMU prototype, *IEEE Trans. Instrum. Meas.*, vol. 63, no. 12, pp. 2824–2836, 2014.

[4] Pignati, M., Popovic, M., Barreto, S., *et al.*, Real-time state estimation of the EPFL-campus medium-voltage grid by using PMUs, *ISGT 2015*.

[5] Jiang, J.-A., Chen, C.-S., Fan, P.-L., Liu, C.-W., and Chang, R.-S., A composite index to adaptively perform fault detection, classification, and direction discrimination for transmission lines, *IEEE Conference Publications*, vol. 2, pp. 912–917, 2002.

[6] Jiang, J.-A., Chen, C.-S., and Liu, C.-W., A new protection scheme for fault detection, direction discrimination, classification, and location in transmission lines, *IEEE Trans. Power Deliv.*, vol. 18, no. 1, pp. 34–42, 2003.

[7] Rahideh, A., Gitizadeh, M., and Mohammadi, S., A fault location technique for transmission lines using phasor measurements, *Int. J. Eng. Adv. Technol.*, vol. 3, no. 1, pp. 241–248, 2013.

[8] Opal-RT, *eMEGAsim PowerGrid Real-Time Digital Hardware in the Loop Simulator*, [Online]. Available from: http://www.opal-rt.com/product/emegasim-powergrid-real-time-digital-hardware-in-the-loop-simulator.

[9] Wanik, M. Z. C., Sanfilippo, A., Singh, N., Jabbar, A., and Cen, Z., "PMU analytics for power fault awareness and prediction," *2019 International Conference on Smart Grid Synchronized Measurements and Analytics (SGSMA)*, College Station, TX, USA, 2019, pp. 1–8, doi:10.1109/SGSMA.2019.8784461.

[10] Romano, P., and Pignati, M., *Integration of an IEEE Std. C37.118 Compliant PMU into a Real-Time Simulator*, PowerTech 2015.

[11] PMU Connection Tester, Available from: https://github.com/GridProtectionAlliance/PMUConnectionTester.

[12] Frank, E., Hall, M. A., and Witten, I. H., *The WEKA Workbench*, Online Appendix for "Data Mining: Practical Machine Learning Tools and Techniques," Morgan Kaufmann, Fourth Edition, 2016.

[13] Friedman, N., Geiger, D., and Goldszmidt, M., Bayesian network classifiers, *Machine Learning*, vol. 29, no. 2–3, pp. 131–163, 1997.

[14] Platt, J. C., Fast training of support vector machines using sequential minimal optimization, in *Advances in Kernel Methods – Support Vector Learning*, B. Schoelkopf, C. Burges, A. Smola, Eds., pp. 185–208, 1999.

[15] Chaudhuri, B. B., and Bhattacharya, U., Efficient training and improved performance of multilayer perceptron in pattern classification, *Neurocomputing*, vol. 34, no. 1–4, pp. 11–27, 2000.

[16] Quinlan, J. R., Induction of decision trees, *Machine Learning*, vol. 1, no. 1, pp. 81–106, 1986.

[17] Cooper, G. F., and Herskovits, E., A Bayesian method for the induction of probabilistic networks from data, *Machine Learning*, vol. 9, pp. 309–347, 1992.

[18] Cheng, J., and Greiner, R., Comparing Bayesian network classifiers, in *Proc. of the Fifteenth Conference on Uncertainty in Artificial Intelligence* (UAI'99), K. B. Laskey and H. Prade, Eds., Morgan Kaufmann, pp. 101–108, 1999.

[19] Mitchell, T. M., *Machine Learning*, McGraw-Hill, New York, 1997.

[20] Brockwell, P. J., and Davis, R. A., *Introduction to Time Series and Forecasting*, Springer, Berlin, 2006.

[21] Smola, A. J., Ratsch, G., Scholkopf, B., Kohlgroen, J., and Vapnik, V., Predicting time series with support vector machines, in *Proc. of ICANN'97*, Springer LNCS 1327, pp. 999–1004, 1997.

[22] Yang, Y., and Pedersen, J., A comparative study on feature selection in text categorization, in *International Conference on Machine Learning (ICML)*, 1997.

[23] Frank, E., Hall, M. A., and Witten, I. H., *The WEKA Workbench. Online Appendix for Data Mining: Practical Machine Learning Tools and Techniques*, Morgan Kaufmann, Fourth Edition, 2016.

[24] Venture Kick, Portfolio-Zaphiro Technologies SA, https://www.venturekick.ch/zaphiro-technologies.

[25] Zaphiro Technologies SA, QEERI's SyncroGuard Commissioning Report, Dec 2023.

[26] Wanik, M. Z. C, Jabbar, A. A., Gad, M., and Sanfilippo, A., Phasor measurement unit (PMU) based monitoring and management for medium voltage network integrated with renewable energy resources, in *Proceeding of 2024 IEEE Sustainable Power and Energy Conference (iSPEC 2024)*, Malaysia, 24–27 Nov., 2024.
[27] Exoreti, SyncroGuard, https://exor-eti.com/software/zaphiro/.

Chapter 14
AI-powered power conversion

Ali Sharida[1], Naheel Kamal[2], Ugur Fesli[3], Sertac Bayhan[1] and Haitham Abu-Rub[4]

The convergence of artificial intelligence (AI) and power conversion systems has ushered in a new era of efficiency, adaptability, and reliability. This chapter explores the dynamic synergy between AI and power electronics, unraveling the transformative impact of advanced machine learning algorithms and data-driven techniques. Through a comprehensive overview, real-world use cases, and an in-depth analysis of benefits and challenges, we delve into the intricate relationship reshaping the landscape of power conversion.

14.1 Introduction

14.1.1 Evolution of power conversion systems

The evolution of power conversion systems has significantly influenced the efficiency, reliability, and flexibility of modern electrical and electronic systems. From their inception, power converters have progressed from simple rectifiers to sophisticated electronic devices capable of converting and regulating power in diverse applications [1]. This evolution has been driven by advancements in semiconductor technology, control methodologies, and real-time microcontroller capabilities, enabling more complex, efficient, and versatile power management solutions. The evolution of power electronic converters and their applications are demonstrated in Figure 14.1.

Initially, power conversion systems were rudimentary, relying on mechanical components like rotating machines for AC to DC conversion. With the advent of semiconductor devices such as diodes, thyristors, and later, transistors and metal-oxide-silicon field-effect transistors (MOSFETs), power converters became more compact, efficient, and reliable. These advancements allowed for the development of various converter topologies, including active rectifiers, inverters, and DC–DC converters [2].

[1]Qatar Environmental and Energy Research Institute, Hamad Bin Khalifa University, Qatar
[2]Department of Electrical and Computer Engineering, Texas A&M University at Qatar, Qatar
[3]Department of Energy Systems Engineering, Faculty of Technology, Gazi University, Turkey
[4]Department of Electrical and Computer Engineering, Hamad Bin Khalifa University, Qatar

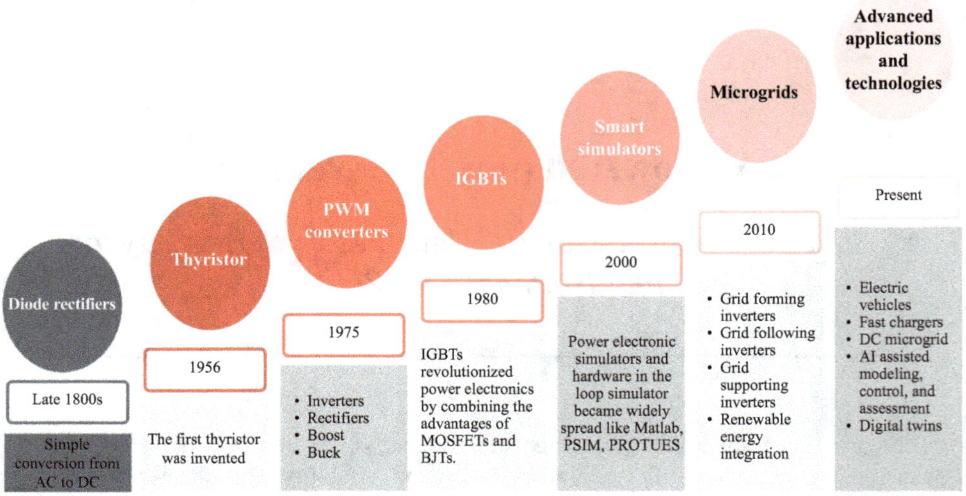

Figure 14.1 Evolution of power conversion systems

Today, power converters represent the backbone for various applications, ranging from consumer electronics to industrial systems and renewable energy solutions [2]. The current landscape of power conversion technology includes advanced power conversion functions and converter topologies including bidirectional converters, multilevel converters, and multifunctional converters.

Control techniques for these converters were also evolved in parallel [3]. Early control methods were mainly open-loop, relying on fixed parameters and lacking feedback mechanisms. As applications demanded more precision and stability, closed-loop control techniques emerged. Proportional-integral-derivative (PID) controllers became widely used [4], offering improved response and stability. Further advancements introduced modern control theories like state-space control [5], adaptive control techniques (ACT) [6], model predictive control (MPC) [7], and AI-based control (AIBC) [8], enabling finer adjustments and better performance under varying operating conditions. Table 14.1 shows a comparison between these converters and their application [3]. It can be noticed that the most modern control techniques like MPC and AIBC offer numerous advantages over other control techniques when it comes to control the entire power electronic systems. However, they require an intensive computational load. In addition, these techniques are inefficient in the context of complex systems that have many parameters to regulate and is exposed to a wide range of conditions changes.

Therefore, the advancements in power conversion systems and their control techniques were conditioned by the evolution of microcontrollers. Early microcontrollers were limited in processing power and memory, constraining their ability to handle complex control algorithms. As semiconductor technology advanced, microcontrollers became more powerful, featuring higher clock speeds, increased memory, and enhanced peripheral integration. Modern real-time microcontrollers are equipped

Table 14.1 Common control techniques for power electronic converters

Control method	Description	Advantages	Disadvantages	Applications
PID control	Uses PID components to control a system	Simple implementation, widely understood, and effective for many applications	Can struggle with highly nonlinear systems and may require tuning	Motor control, temperature control, and basic power converters
State-space control	Uses a mathematical model to control a system's state variables	Provides a detailed system description, can handle multivariable control	Requires a precise model of the system, complex to design	Advanced motor drives and simple power converters systems
ACT	Adjusts control parameters in real-time to cope with system changes	Can handle varying system dynamics and uncertainties	Complex to implement, may require significant computational resources	Automotive systems and adaptable industrial converters
SMC	Uses a discontinuous control action to drive system states to a sliding surface	Robust to parameter variations and disturbances, fast response	Can cause chattering, complex to design for nonlinear systems	Advanced power converters
MPC	Uses a model to predict future system behavior and optimize control actions	High performance, can handle constraints and multivariable systems	Computationally intensive, requires an accurate system model	Process control, energy management, and microgrids
AIBC	Utilizes AI techniques like neural networks and fuzzy logic	Can handle highly nonlinear and complex systems, adaptable	Requires extensive training data, high computational requirements	Autonomous systems, smart grids, and advanced power management operations

Table 14.2 Examples of modern microcontrollers

Microcontroller	Core	Clock speed	Memory	Key features
STM32H7 Series	ARM Cortex-M7	Up to 550 MHz	Up to 2 MB Flash, 1 MB RAM	Dual-core option (Cortex-M4), advanced peripherals, high-speed interfaces
ESP32-S3	Xtensa LX7 Dual-core	Up to 240 MHz	Up to 16 MB Flash, 512 kB RAM	Integrated Wi-Fi and Bluetooth, AI acceleration, low power consumption
SAM E70	ARM Cortex-M7	Up to 300 MHz	Up to 2 MB Flash, 384 kB RAM	High-speed connectivity, multiple communication interfaces, DSP capabilities
TMS 320F28388D	C28x + ARM Cortex-M4	Up to 200 MHz	Up to 1 MB Flash, 164 kB RAM	Real-time control, dual-core architecture, integrated FPU and CLA
Arduino Portenta H7	ARM Cortex-M7 + Cortex-M4	Up to 480 MHz (M7) + 240 MHz (M4)	16 MB Flash, 8 MB RAM	Dual-core architecture, advanced AI capabilities, industrial temperature range
Raspberry Pi Pico	ARM Cortex-M0+	Up to 133 MHz	2 MB Flash, 264 kB RAM	Dual-core processor, PIO (Programmable I/O), USB 1.1 host/device

with capabilities such as high-speed analog-to-digital converters, pulse-width modulation modules, specialized digital signal processing (DSP) units, high-speed clock, large memory, and many extra features, as shown in Table 14.2.

The advancements in microcontrollers, control algorithms, and the pressing needs to integrate AI techniques to address the complexity of smart grid systems led to a paradigm shift toward AI-powered solutions. These advancements enabled the implementation of sophisticated AI algorithms directly within embedded systems to enable real-time decision-making, optimizing energy flows, and ensuring stability and reliability of the power converters.

The integration of AI-powered solutions extended the range of control techniques from the converter level to system level including centralized control systems of grids, microgrids, and smart grids. Centralized control systems leverage AI to manage and optimize the operation of power electronic systems by analyzing vast amounts of data in real time, which enables coordination between generation units, storage systems, and loads to ensure optimal performance. In addition, energy management systems (EMSs) benefit from AI by optimizing energy production, storage, and consumption, with machine learning (ML) algorithms predicting energy demand based on historical data and user behavior. This capability is crucial for balancing supply and demand, which allows the utilities to manage resources more effectively and reduce energy wastage.

AI also plays a vital role in predictive maintenance [9], where data from sensors and historical records are used to anticipate equipment failures, reducing downtime and maintenance costs. Another crucial feature of AI-powered solutions is their ability to detect and isolate faults rapidly, which maintains the stability of the system even under fault conditions. Add to that, AI can significantly enhance monitoring systems and enable them to provide real-time analysis and decision-making capabilities.

One of the most exciting advancements in integrating AI with power electronics is digital twin (DT) technology [10]. DT is a virtual replica of a physical system that uses real-time data to simulate and predict system behavior. In power electronics, DTs can model and optimize power converter performance, predictive maintenance, and improve system design. Leveraging AI, DTs offer insights into system behavior under various conditions and enable more effective planning and operation of the power electronic systems. This paradigm shift towards AI solutions in power electronic systems promises to enhance the efficiency, reliability, and sustainability of modern power systems and smart grids.

14.1.2 Significance of AI in power conversion

The integration of AI into power conversion systems marks a transformative advancement in the field of power electronics. AI-driven techniques are revolutionizing the way power conversion systems operate, offering enhancements in efficiency, adaptability, and reliability that were previously unattainable with traditional methods.

Enhancing efficiency through intelligent algorithms: AI algorithms, particularly those involving ML and deep learning (DL), enable power conversion systems to optimize their performance dynamically. These intelligent algorithms analyze vast amounts of data in real-time, allowing systems to adjust their operations for maximum efficiency. For instance, AI can optimize switching sequences in power converters to reduce losses and improve overall energy efficiency. By learning from historical data and real-time inputs, AI systems can predict optimal operational states, leading to significant energy savings and reduced operational costs.

Adaptability to dynamic conditions: Power conversion systems often operate in environments with rapidly changing conditions. AI enhances the adaptability of these systems by enabling them to respond swiftly to fluctuations in load, input power, and other operational variables. Through continuous learning and real-time adjustments, AI algorithms ensure that power converters maintain optimal performance despite varying conditions. This adaptability is crucial for applications such as renewable energy integration, where power inputs can be highly variable.

Improving reliability and system longevity: AI plays a pivotal role in predictive maintenance and fault detection, which are essential for the reliability and longevity of power conversion systems. By analyzing sensor data and historical performance records, AI can predict potential failures before they occur, allowing for timely maintenance and reducing unexpected downtime. Additionally, AI can detect and diagnose faults in real-time, enabling quick corrective actions to prevent system failures. This proactive approach to maintenance not only extends the lifespan of power electronic components but also enhances the overall reliability of power systems.

The role of data-driven techniques: Data-driven techniques, powered by AI, are central to optimizing power electronics systems. These techniques involve the collection and analysis of large datasets to inform decision-making processes. For instance, AI can analyze patterns in energy consumption, load variations, and environmental conditions to optimize the operation of power converters. Data-driven insights enable more accurate forecasting of energy demand, improved load balancing, and efficient energy distribution, all of which contribute to the stability and efficiency of power systems.

In conclusion, the significance of AI in power conversion lies in its ability to enhance efficiency, adaptability, and reliability through intelligent algorithms and data-driven techniques. As AI continues to evolve, its integration into power electronics is expected to drive further innovations, leading to more efficient, adaptable, and reliable power conversion systems.

14.2 AI applications in power conversion

14.2.1 Intelligent control strategies

The rapid advancements in power electronics have necessitated the development of intelligent control methods to manage increasingly complex systems. The demand for higher efficiency, reliability, and adaptability in these systems has led to the evolution of advanced control strategies. ACT emerged to address the limitations of traditional fixed-parameter control systems. In conventional control systems, parameters are typically set based on the assumed operating conditions, which may not always accurately reflect real-world scenarios. As a result, the performance of the control system can degrade when the system operates outside the predefined conditions. ACT overcome this problem by dynamically adjusting control parameters in real-time based on the changes in system's behavior or operating environment.

One of the most common ACT is called self-tuning regulators (STR) [11]. STRs employ recursive parameter estimation algorithms to estimate the parameters of the dynamic model based on real-time data. The estimated parameters are then used to update the controller gains to ensure that the system remains stable and performs efficiently despite variations in operating conditions.

Parameter estimation aims to develop models that accurately represent the dynamics of physical systems based on observed data. One widely used method is recursive least squares (RLS) [12], which is particularly effective in adaptive filtering and real-time system identification. The RLS algorithm is an extension of the ordinary least squares method that updates parameter estimates recursively once the new data becomes available. This makes it highly efficient for applications where the system parameters may change over time. The RLS algorithm minimizes the sum of the squares of the differences between the observed and predicted values. Another advantage of RLS is its fast convergence properties, making it suitable for real-time adaptive control applications. RLS algorithm can be summarized as follows [12]:

$$\mathbf{K}(k) = \mathbf{P}(k-1)\varphi(k)\left[\mathbf{I} + \varphi(k)^T\mathbf{P}(k-1)\varphi(k)\right]^{-1} \tag{14.1}$$

$$\varepsilon(k) = \mathbf{Y}(k) - \varphi(k)^T\widehat{\boldsymbol{\theta}}(k-1) \tag{14.2}$$

$$\widehat{\boldsymbol{\theta}}(k) = \widehat{\boldsymbol{\theta}}(k-1) + \mathbf{K}(k)\varepsilon(k) \tag{14.3}$$

$$\mathbf{P}(k) = \left[\mathbf{I}_9 - \mathbf{K}(k)\varphi(k)^T\right]^{-1}\mathbf{P}(k-1) \tag{14.4}$$

where $\mathbf{K}(k)$ is the estimator gain matrix, $\mathbf{P}(k)$ is the error covariance matrix, $\varphi(k)$ is the measurements vector, $\widehat{\boldsymbol{\theta}}(k)$ is the estimated parameters matrix, $\varepsilon(k)$ is the estimation error vector at the instant k, $\mathbf{I}_9$is 9×9 identity matrix, $\mathbf{P}(k\text{-}1)$ is the previous value of the error covariance matrix, and $\widehat{\theta}(k-1)$ is the previous value of the estimated parameters matrix.

Another method is the Kalman filter, an algorithm that estimates the state of a linear dynamic system from a series of noisy measurements. The Kalman filter recursively updates the estimates of the system's state and the associated uncertainties. Due to its robustness in handling measurement noise and its ability to provide real-time estimates, it is widely used for adaptive and robust control techniques. The estimation process is done based on the error variance and covariance matrices. The main advantage of KFA is that it does not require parameter or gain tuning. Instead, only states initialization is required. KFA can be summarized as follows [13]:

1. Prediction step,

$$\widehat{\mathbf{x}}_{ek}^{-} = \mathbf{A}_{ed}\widehat{\mathbf{x}}_{ek-1} + \mathbf{B}_{ed}\mathbf{u}_{ek-1} \tag{14.5}$$

$$z\mathbf{P}_k^{-} = \mathbf{A}_{ed}\mathbf{P}_{k-1}\mathbf{A}_{ed}^T + \mathbf{Q}_k \tag{14.6}$$

2. Updating step,

$$\mathbf{K}_k = \mathbf{P}_k^{-}\mathbf{H}^T\left(\mathbf{H}\mathbf{P}_k^{-}\mathbf{H}^T + \mathbf{R}_k\right)^{-1} \tag{14.7}$$

$$\widehat{\mathbf{x}}_{ek} = \widehat{\mathbf{x}}_{ek}^{-} + \mathbf{K}_k\left(\mathbf{z}_k - \mathbf{H}\widehat{\mathbf{x}}_{ek}^{-}\right) \tag{14.8}$$

$$\mathbf{P}_k = \left(\mathbf{I} - \mathbf{K}_k\mathbf{H}\right)\mathbf{P}_k^{-} \tag{14.9}$$

where $\mathbf{A}_{ed}$ and $\mathbf{B}_{ed}$ are the discrete state matrix and the input matrix, $\widehat{\mathbf{x}}_{ek}$ is the vector of discrete estimated states, $\mathbf{u}_k$ is the vector of discrete inputs, $\mathbf{P}_k$ is the estimation error covariance matrix, $\mathbf{Q}_k$ is the process noise covariance matrix, $\mathbf{K}_k$is Kalman gain, $\mathbf{H}$is the output matrix, $\mathbf{R}_k$is the noise covariance error, $\mathbf{z}_k$ is the measurement vector, $\mathbf{I}$ is the identity matrix, and $(.^{-})$ denotes to the value before the estimation.

Once the parameters are estimated, they can be assigned to a control technique that depends on the dynamic model to update its rules or gains. Among the most common controllers used with adaptive gain or rules updating are the linear quadratic tracker (LQT), MPC, and PID. Based on the type of the adopted controller, a specific tuning algorithm can be selected. For the PID controller, it is widely used with the pole-placement method, where the estimated parameters and the predefined dynamic response are used to relocate the poles of the controlled system. For, LQT, gains

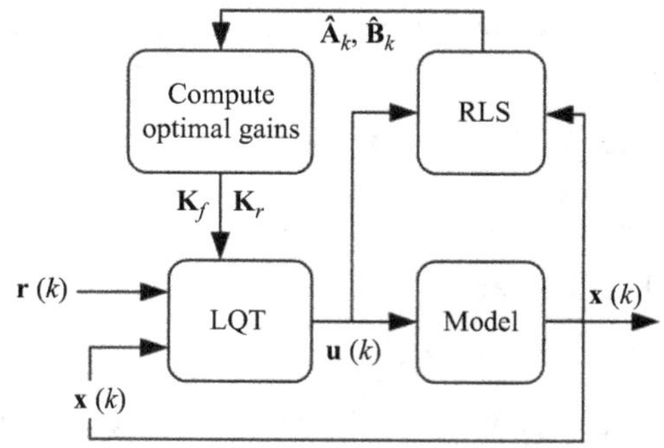

Figure 14.2 Conventional adaptive control and gains updating block diagram

updating is more systematic and more efficient than PID, as the LQT will always select the most optimal gains as a function of the estimated parameters, as shown in Figure 14.2. Finally, the MPC does not depend on gains updating as it contains constant weigh factors instead of gains. However, rely on predicting the response of the system based on the dynamic model. Therefore, the estimation process can be used effectively to update the dynamic model that is used in the prediction step. This enhances the accuracy of the prediction process and leads to a reliable optimization of the cost function.

Although conventional adaptive control methods were recognized for their efficient and reliable behavior during the last decades, they have several limitations with modern power electronic systems. One major limitation is their struggle with handling highly nonlinear systems. These traditional techniques often rely on linear approximations or models that may not capture the full complexity of a system's behavior. This leads to suboptimal performance in the presence of significant nonlinearities. Moreover, conventional adaptive control methods rely on the defining solid structure of the dynamic model in terms of inputs, outputs, states, and the nature of noise and disturbances. However, the modern power converter applications are very complex to be predicted with these methods, which will underestimate the effect of many variables such as the coupling effects, the nature of the connected loads, and the intermittent nature of RESs. Another significant limitation is the slow adaptation to changing conditions and the significant effect of parameters initialization. Conventional adaptive controllers may not adapt quickly enough to rapid changes in system dynamics or operating conditions, resulting in performance degradation or instability, particularly in power electronic systems that has high frequency and unpredictable variations.

Building on the foundations of adaptive control, the field has progressed towards more intelligent control methods, which incorporate elements of AI and ML. These techniques offer enhanced capabilities in dealing with the complexities and nonlinearities inherent in modern power electronic systems. One of the key benefits of AI-powered controllers is their ability to manage nonlinear relationships. AI-powered solutions are inherently capable of modeling and managing complex nonlinear dynamics. Another crucial advantage is the rapid adaptation and learning capabilities of AI algorithms. AI-powered techniques can quickly adapt to changing conditions and learn from new data. This maintains the optimal performance in power electronic converter applications, where load conditions and generation sources can vary rapidly. AI-based methods also perform effectively even with inaccurate system models by learning directly from the observed data. This reduces the reliance on precise mathematical models and improves robustness in real-world applications where perfect modeling is impractical.

Due to these advantages, AI becomes one of the most game-changers in the power electronic systems application, as shown in Table 14.3. Neural network-based control (NNC), for example, is employed in electric vehicle (EV) powertrain management and grid-connected converters, where it optimizes battery management and motor control while managing power quality and stability. Reinforcement learning (RL) algorithms learn optimal control policies through trial and error, making them ideal for EMSs and microgrid optimization. Therefore, RL becomes the best alternative for conventional optimal control techniques in power electronic applications, where they become widely used in balance supply and demand, improve efficiency, and reduce operational costs. Another example of AI-powered solutions is called support vector machines (SVM). SVM is widely used for fault detection and diagnosis in power electronics, where the main goal is to enhance the reliability of the system by classifying and diagnosing faults in power converters and other components. Moreover, genetic algorithms (GA) are among the most powerful tools to solve complex

Table 14.3 Common AI algorithms for power electronic converters

AI method	Applications	Description
NNC	EV powertrain management, grid-connected converters	Optimizes battery management and motor control, manages power quality and stability, and handles complex nonlinear relationships
RL	Energy management systems, microgrid optimization	Learns optimal control policies through trial and error, balances supply and demand, improves efficiency, and reduces operational costs
SVM	Fault detection and diagnosis	Classifies and diagnoses faults in converters and other components, enhancing system reliability
GA	Optimal design of components, control parameter tuning	Solves optimization problems by simulating natural selection, aids in the optimal design of power electronic components, and tunes control parameters
DLM	Predictive maintenance, load forecasting	Analyzes sensor data to predict equipment failures, forecasts future energy demands based on historical consumption data using CNNs and RNNs

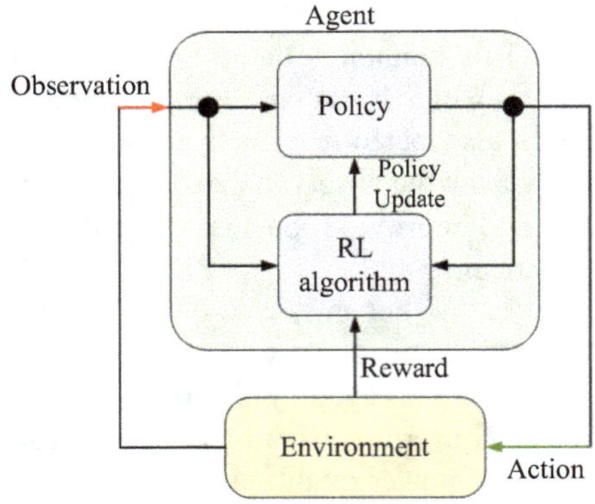

Figure 14.3 Block diagram of RL-based control

optimization problems. Deep learning models (DLM), such as convolutional neural networks (CNNs) and recurrent neural networks (RNNs), are widely utilized for predictive maintenance and load forecasting. They depend on analyzing the observed or measured data to predict equipment failures and future energy demands based on historical consumption data:

Among AIBC techniques, RL is increasingly popular in the primary control level of power electronic systems due to its adaptability, optimization capabilities, and robustness [8,14]. As the power electronic systems are highly dynamic, RL can continuously adapt to the changes such as loads and input variations. Moreover, RL improves its performance over time through the interaction with the environment. In addition, RL offers high flexibility to handle highly coupled and nonlinear systems without relying on the mathematical model. Additionally, RL-based controllers are robust to uncertainties and disturbances, which enhances the reliability of the system. Furthermore, RL's ability to learn from interaction and balance exploration and exploitation enables the discovery of optimal control actions.

RL-based control is a framework where an agent learns to make decisions by interacting with an environment. Figure 14.3 shows the block diagram of the RL process, where the agent comprises a policy and a reinforcement learning algorithm. The agent observes the current state of the environment, denoted as $O(t)$, and based on this observation, the policy generates an action $A(t)$. This action affects the environment, leading to a new state and generating a reward $R(t)$. Then, the RL algorithm uses this reward to update the policy, which improves the agent's future actions to maximize cumulative rewards. This iterative process continues, which allows the agent to adapt and optimize its performance in the environment through continuous learning and interaction.

To better understand the process of RL-based control, consider a simple boost converter. The main objective of the boost converter is to maintain a constant output voltage (V_o). For the boost controller, the agent represents the control algorithm, which consists of policy and RL algorithm. The policy determined the required duty cycle (D) of the switch while the RL algorithm adjusts the policy based on the feedback from the environment. The environment itself consists of the physical boost converter circuit, which steps up the input voltage to a higher output voltage. The agent interacts with this environment by observing the instant output voltage $Vo(t)$. Based on this observation, the agent's policy determines the duty cycle for the converter's switch. This duty cycle action affects the output voltage of the boost converter and leads to generate a new output voltage. The environment provides a reward to the agent, which is typically based on how close the actual output voltage is to the reference output voltage. This feedback loop allows the agent to learn and adjust its policy to maintain a stable output voltage despite the fluctuations in the input voltage or load conditions.

The process of RL-based control can be summarized as follows:

1. Initialization
   - The agent starts with an initial policy for the duty cycle (D0).
   - Set the output voltage reference r(t).
2. Observation
   - Observe the instant value of V_o from the boost converter.
3. Action
   - Based on the observation of V_o, the agent's policy determines the duty cycle D for the switch.
4. Environment response
   - As D is changed, the output voltage of the boost converter will change, and this will be observed by the RL algorithm.
5. Reward calculation
   - The environment provides a reward $R(t)$ to the agent. The reward could be a function of the error between the actual output voltage and the desired output voltage

     $$R(t) = |r(t) - V_o(t)| \tag{14.10}$$

   - The given reward pushes the agent to minimize voltage tracking error.
6. Policy update
   - The RL algorithm then uses the reward to update the policy, this can be done by numerous techniques including Q-learning.
7. Go to step 2.

14.2.2 Predictive maintenance and fault detection

The evolution of power electronic systems is pivotal to numerous industrial applications where systems' reliability and efficiency are critical. Failures in such systems can lead to significant operational disruptions and financial losses. Consequently, there is a growing emphasis on enhancing fault tolerance and implementing predictive maintenance strategies to ensure the continuous and optimal functioning of these systems.

Fault tolerance in power electronics involves designing systems that can continue operating, possibly at a reduced level, even when certain components fail. Predictive maintenance, conversely, is a proactive approach that uses data analytics to predict and preempt equipment failures before they occur, which allows system's operator to achieve timely maintenance and repairs.

AI has emerged as a transformative technology in this domain. AI provides advanced tools for early fault detection and predictive maintenance. Among the various AI-driven solutions, DT technology stands out as one of the most promising mechanism for enhancing fault tolerance and predictive maintenance capabilities in power electronic systems.

At its core, the DT comprises two interconnected components, a physical system model and a digital model [15]. The physical system model represents the real-world power electronic converter, capturing its physical attributes, electrical characteristics, and operational parameters. This physical model serves as the foundation upon which the virtual model is built. In contrast, the digital model is a numerical representation of the physical system, instantiated within a digital environment. This model encompasses various simulation tools, algorithms, and computational resources capable of accurately emulating the behavior of the physical system in real time. In general, the structure of power electronic system DT compromises five components as shown in Figure 14.4, dynamic models, simulated or physical power electronic system, data acquisition systems, control algorithms, and interfacing components [9].

Designing a DT involves collecting data from sensors and control signals of the physical system, then building and optimizing a deep neural network (DNN) model [10]. As shown in Figure 14.5, the DNN architecture typically includes one

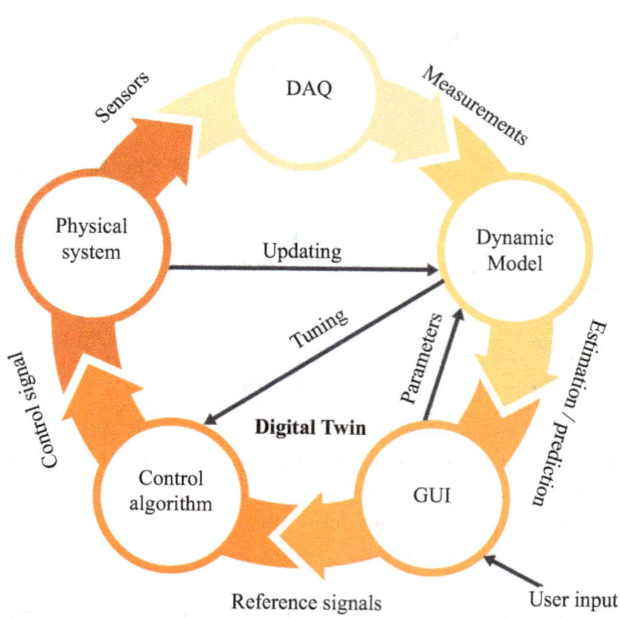

Figure 14.4 Digital twin structure for power electronic systems

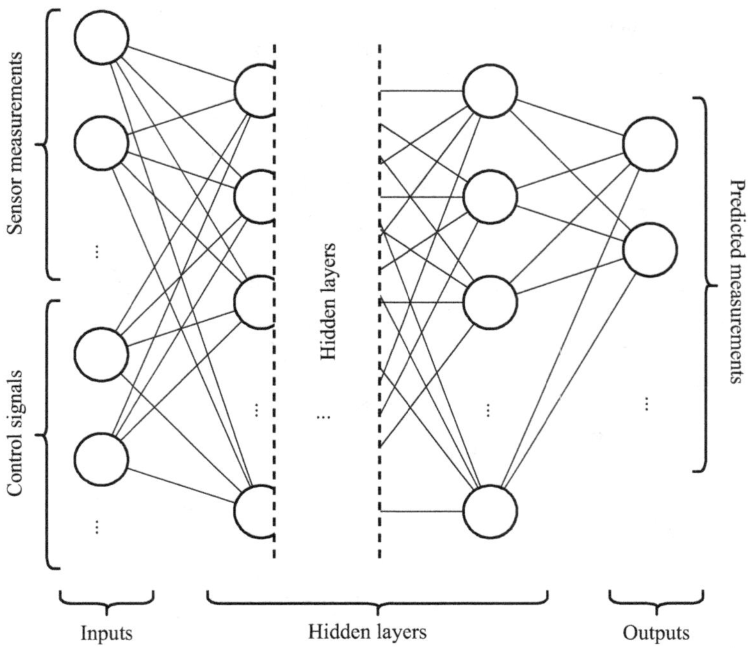

Figure 14.5 Generic DNN model of power converter's DT

or more hidden layers, with inputs and outputs based on the application's requirements. Typically, the inputs gathered from sensors' measurements and control signals and the outputs are the predicted measurements. The size and complexity of the DNN model must balance performance with the hardware constraints. Once trained and evaluated, the model is exported, compiled, and integrated into the embedded system's firmware for real-time online operation.

The next phase is to validate the obtained DT. The validation process is achieved online by comparing the outputs of the physical system with the predicted outputs from the DT. Figure 14.6 shows the validation process of an active front end rectifier, where the grid currents and the output DC voltage are the main states to observe and predict. In case of different converters like buck, boost, or buck-boost, the main states to observe are the inductor's current and the output voltage so they are used to validate the DT.

Once the DT is validated online, it can be used to tolerate numerous power electronic faults. The most common faults in power electronic systems applications are sensors and switches faults.

To understand the basic principle of utilizing DT for fault diagnosis, tolerant, and predictive maintenance, consider an inverter or active front end rectifier. The control process of these converters depends on measuring the grid voltages, grid currents, and the DC link voltage. If any measurement has a fault, the entire control process

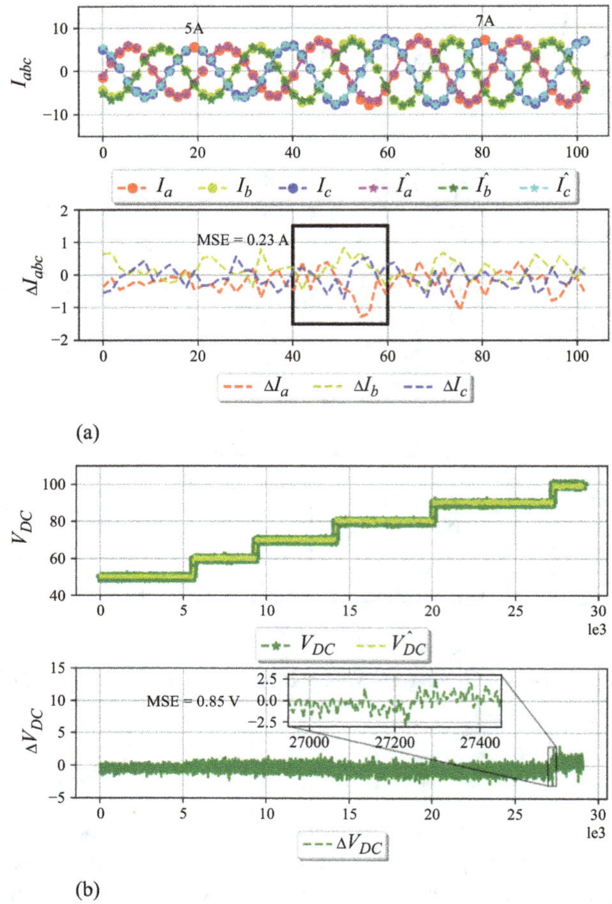

Figure 14.6 DT validation: (a) grid currents and (b) output voltage

will fail. To overcome this challenge, DT can be used and integrated with the physical sensors to provide the required feedback signals in case of sensors failure which increases the reliability of the controlled system.

Normally, the inputs of the physical system ($\mathbf{V}_{abc}$ and $\mathbf{U}_{abc}$) are applied to the DT, then DT provides an estimation of the physical system's outputs ($\mathbf{I}_{abc}$ and V_{dc}). Therefore, the outputs of the DT can be used for current-sensorless or fault tolerant control as shown in Figure 14.7a. In this case, the DT continuously receives the voltage measurements and the control signal, then computes the corresponding grid currents and feeds them directly to the control law. Conversely, if the current measurements are available and healthy, the inverse of the DT is activated. In this case, the DT receives the outputs of the physical system and provides the grid voltages as shown in Figure 14.7b.

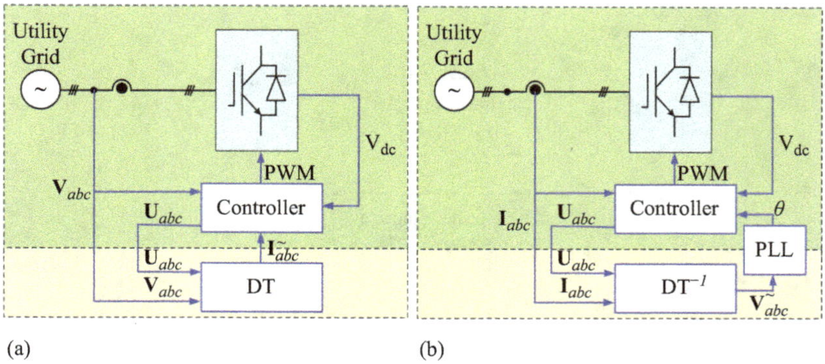

Figure 14.7 Sensor fault-tolerant control: (a) current sensor tolerant and (b) voltage sensor tolerant

The flow chart of DT-based sensors fault detection is shown in Figure 14.8. The measurements are collected from the sensors and compared with the DT outputs. If the voltage difference ($\Delta\mathbf{V}_{abc}$) between voltage measurement and DT estimated voltage is larger than a specific threshold value (T_{max}), then the voltage sensors contain false or high noise. Therefore, the estimated grid voltages vector will be considered as feedback instead of the physical measurements. Similarly, if the current difference ($\Delta\mathbf{I}_{abc}$) is larger than T_{max}, the system will consider the estimated currents vector as a feedback signal. Similarly, the DT can also be used to detect, locate, and tolerate other faults like open and short switches.

14.2.3 Energy harvesting and management

The growing integration of renewable energy sources and the increasing demand for efficient energy utilization have made energy harvesting and management critical areas in power conversion systems. AI-driven approaches provide significant advantages in optimizing energy harvesting from various sources and managing energy consumption to enhance sustainability and reliability.

AI-Driven Optimization of Energy Harvesting: Energy harvesting involves capturing energy from renewable sources such as solar, wind, and geothermal. The variability and unpredictability of these sources pose challenges in maximizing energy capture and ensuring stable supply. AI-driven techniques address these challenges by optimizing the processes involved in energy harvesting.

Solar Energy Harvesting: For solar energy systems, AI algorithms can optimize the angle and orientation of photovoltaic panels to maximize sunlight exposure throughout the day. ML models can predict solar irradiance based on weather forecasts and historical data, allowing for real-time adjustments to the panel positions.

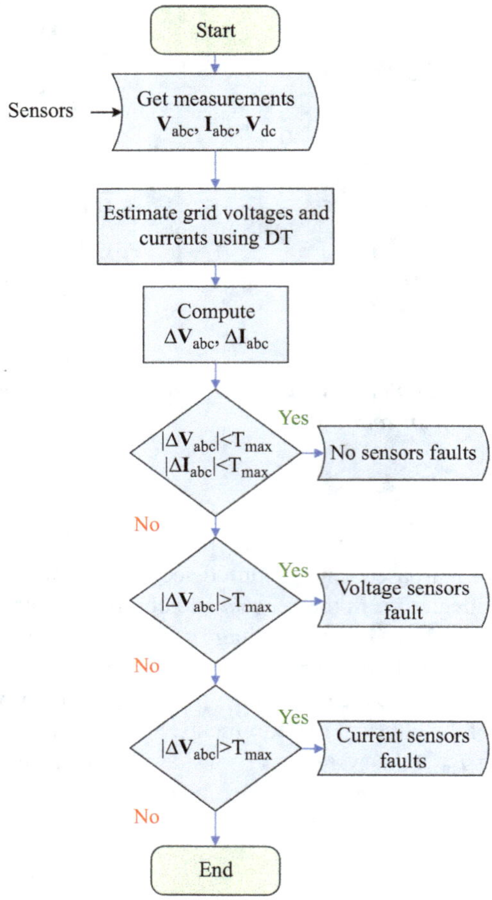

Figure 14.8 DT-based sensor fault diagnosis

Additionally, AI can manage the cleaning schedules of solar panels by predicting dust accumulation, ensuring maximum efficiency.

Wind Energy Harvesting: In wind energy systems, AI techniques can optimize the positioning and operation of wind turbines. ML models analyze wind patterns, speed, and direction to adjust the turbine blades' pitch and yaw angles for optimal energy capture. Predictive maintenance powered by AI ensures that turbines operate efficiently by anticipating mechanical issues and scheduling timely maintenance.

Energy Harvesting from Ambient Sources: AI can also enhance energy harvesting from ambient sources such as vibrations, thermal gradients, and radio frequency signals. By employing ML algorithms, systems can adapt to varying ambient conditions and optimize the energy conversion processes for maximum efficiency.

Smart Energy Management Strategies: Once energy is harvested, efficient management is crucial to ensure its optimal utilization. AI-driven smart EMS play a vital role in balancing supply and demand, reducing energy wastage, and enhancing the overall sustainability of power systems.

Load Forecasting and Demand Response: AI algorithms can predict energy demand based on historical data, user behavior, and external factors such as weather conditions. Accurate load forecasting enables the EMS to implement demand response strategies, where energy consumption is adjusted in real-time to match the available supply. For example, during peak demand periods, non-critical loads can be temporarily reduced or shifted to off-peak times, ensuring a balanced and efficient energy distribution.

Energy Storage Management: AI optimizes the use of energy storage systems, such as batteries and supercapacitors, by predicting energy demand and managing charge-discharge cycles. ML models can forecast periods of high-energy demand or low renewable generation, allowing the EMS to preemptively charge storage systems. This ensures that stored energy is available when needed, enhancing grid stability and reducing reliance on non-renewable backup sources.

Renewable Energy Integration: AI facilitates the seamless integration of multiple renewable energy sources into the grid. By analyzing real-time data from various sources, AI algorithms can dynamically allocate energy to different parts of the grid based on demand and generation levels. This improves the overall efficiency of the energy distribution network and supports the transition to a more sustainable energy mix.

Microgrid Management: In microgrids, AI-driven EMS can optimize the operation of local generation units, storage systems, and loads to achieve energy self-sufficiency. ML models predict energy production and consumption patterns within the microgrid, enabling efficient energy trading and ensuring that surplus energy is either stored or distributed to neighboring grids.

Enhancing Sustainability through AI: AI-driven EMSs contribute significantly to the sustainability of power systems. By optimizing energy harvesting and utilization, AI helps reduce greenhouse gas emissions, minimize energy wastage, and promote the use of renewable energy sources. The ability to predict and respond to energy demand fluctuations in real-time ensures that energy systems operate at peak efficiency, supporting global efforts to combat climate change.

Environmental Impact Reduction: AI techniques reduce the environmental impact of energy systems by improving the efficiency of renewable energy sources and minimizing the need for fossil fuel-based generation. By maximizing the use of clean energy, AI-driven systems help lower carbon footprints and promote sustainable development.

Economic Benefits: Efficient energy management driven by AI also offers economic benefits. By reducing energy wastage and optimizing the use of resources, both energy producers and consumers can achieve cost savings. Additionally, predictive maintenance powered by AI extends the lifespan of energy infrastructure, reducing the need for frequent replacements and associated costs.

Policy and Regulation Compliance: AI helps energy systems comply with regulatory requirements and sustainability targets. By continuously monitoring and optimizing energy usage, AI-driven systems ensure adherence to policies aimed at reducing emissions and promoting renewable energy adoption.

In summary, AI-driven energy harvesting and management strategies significantly enhance the efficiency, sustainability, and reliability of power systems. Through intelligent optimization of renewable energy sources and smart management of energy consumption, AI plays a crucial role in shaping the future of sustainable energy infrastructure.

14.3 Real-world use cases

14.3.1 Electric vehicle charging systems

Figure 14.9 illustrates the EV charging infrastructure pointing to AI applications within the infrastructure. In the central system management system (CSMS), AI can be employed for power forecasting and anomaly detection. Using the Open Charge Point Protocol, physical measurements like power, voltage, current, and state-of-charge (SoC) are transmitted from the EV supply equipment (EVSE) to the CSMS.

Given the chargers' data, AI algorithms can be used at the CSMS to apply anomaly detection mechanisms to identify unusual patterns in the metered values of the EV chargers, ensuring proper functioning and security [16]. Along with anomaly detection, the CSMS can apply AI methods to forecast the chargers' power consumption, which helps in demand management and grid efficiently [17].

In the EV side, a major application of AI is the SoH (State of Health) and SoC estimation [18]. SoH and SoC estimation involves AI algorithms that provide precise assessments of the battery's health and charge status, enhancing the reliability and longevity of the battery. Power converter modeling uses AI to optimize the power conversion processes within the EV, improving efficiency and performance.

Within the EVSE, particularly the Power Electronics (PE) part, AI can significantly help in power converter DT-based modeling and fault detection and tolerance [9,19,20]. Fault and anomaly detection and tolerance systems in the EVSE level

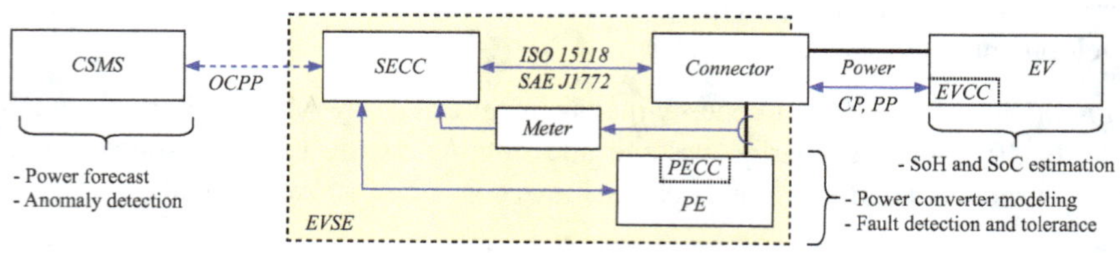

Figure 14.9 AI application in the EV charging infrastructure

can leverage AI algorithms to detect anomalies and manage faults, maintaining continuous operation and minimizing downtime [19,20]. Rule-based approaches have been used to detect anomalies in the communication channel [20]. This approach utilizes estimations and a threshold-based decision to detect anomalies. Anomaly detection systems can also be built as a classification model that predicts the confidence percentage of the current state being normal or anomalous [19]. This can be achieved by collecting sensors' data and communication channels' data in normal and anomalous conditions and then training a ML model to predict the difference. Another approach to detect anomalies is to model the charger's power converter as a DT model and then find patterns in the difference between the physical response and the DT model's response [9]. These patterns can be categorized to find the location of the anomaly and to tolerate the fault in the power converter.

Table 14.4 summarizes AI applications in EV charging systems by their location in the charging infrastructure and the type of the application. It can be seen that AI contributes to improved efficiency, reliability, and security by detecting anomalies, forecasting consumption, and optimizing various processes. These advancements significantly contribute to the overall performance and management of EV charging infrastructure.

14.4 Benefits of AI in power conversion

14.4.1 Enhanced efficiency and performance

The incorporation of AI into power conversion systems has dramatically transformed their efficiency and performance. AI algorithms, particularly those involving ML and DL, provide sophisticated methods for optimizing power conversion processes, reducing energy losses, and improving overall system performance. This section explores the various ways AI enhances efficiency and performance in power electronics.

Table 14.4 Categorization of AI applications in EV charging systems

Location	Application	Description	References
CSMS	Anomaly detection	Identifying unusual patterns in metered values	[16]
	Forecasting	Forecasting chargers' power consumption	[17]
EVSE	Power converter DT-based modeling	DT-based modeling for power converters	[9,10]
	Anomaly detection and tolerance	Detecting anomalies and managing faults in power converters	[19,20]
EV	SoH and SoC estimation	Precise assessments of battery's health and charge status	[18]

Optimization of Power Conversion Processes: AI-driven optimization techniques enable power conversion systems to operate more efficiently by fine-tuning various parameters in real-time. These techniques involve analyzing vast amounts of data collected from the system and its environment to make informed adjustments that maximize efficiency.

Real-Time Parameter Adjustment: AI algorithms can continuously monitor and adjust parameters such as switching frequencies, duty cycles, and voltage levels in power converters. By optimizing these parameters in real-time, AI ensures that the power conversion process operates at peak efficiency, reducing energy losses and improving overall performance. For instance, adaptive control algorithms can dynamically adjust the operation of DC–DC converters to match varying load conditions, ensuring optimal performance under different scenarios.

Loss Minimization: AI can identify and minimize energy losses in power conversion systems. ML models analyze patterns in energy consumption and loss data to pinpoint inefficiencies. For example, in inverters, AI algorithms can optimize the switching sequences and timing to minimize switching losses and reduce heat generation. This not only enhances efficiency but also extends the lifespan of the components by reducing thermal stress.

Predictive Control: MPC is an advanced control strategy that leverages AI to predict future system behavior and optimize control actions accordingly. MPC algorithms use real-time data to predict the future states of the power converter and determine the optimal control actions to achieve desired performance objectives. This predictive capability allows for more precise and efficient control, leading to improved system efficiency and reduced energy consumption.

Achieving Higher Efficiency and Reduced Losses: AI-driven techniques play a critical role in achieving higher efficiency and reducing losses in power electronic systems. By leveraging data and intelligent algorithms, these techniques enable more effective energy management and utilization.

Energy Harvesting Efficiency: AI enhances the efficiency of energy harvesting systems by optimizing the capture and conversion of energy from renewable sources. For example, in solar power systems, AI algorithms can predict optimal panel orientations and adjust the angle of photovoltaic panels to maximize sunlight exposure throughout the day. Similarly, in wind energy systems, AI can optimize the operation of wind turbines by adjusting blade angles to capture the maximum amount of wind energy.

Load Balancing and Energy Distribution: AI-driven EMSs can balance loads and distribute energy more efficiently across the grid. ML models can predict energy demand patterns and adjust the distribution of energy accordingly. This ensures that energy is delivered where it is needed most, reducing wastage and enhancing overall system efficiency. For example, AI algorithms can manage the charging and discharging cycles of battery storage systems to balance supply and demand, ensuring optimal energy utilization.

Smart Grid Integration: The integration of AI into smart grids enhances the efficiency of energy distribution and management. AI algorithms can analyze real-time data from various sources, such as smart meters and sensors, to optimize the operation of the grid. This includes demand forecasting, fault detection, and automated response to grid conditions. By optimizing the flow of energy and minimizing losses, AI contributes to a more efficient and resilient smart grid infrastructure.

Intelligent Control Strategies for Enhanced Performance: AI introduces intelligent control strategies that significantly enhance the performance of power conversion systems. These strategies leverage advanced algorithms to provide more precise and adaptive control, leading to improved system stability and responsiveness.

Adaptive Control Techniques: AI enables the implementation of ACT that adjust control parameters in real-time based on changing system conditions. These techniques, such as STR and adaptive PID controllers, use ML algorithms to continuously learn from the system's behavior and optimize control actions. Adaptive control ensures that power converters maintain optimal performance under varying load and environmental conditions.

Fault Detection and Diagnostics: AI enhances the reliability and performance of power conversion systems by enabling early fault detection and diagnostics. ML models can analyze sensor data and operational patterns to identify anomalies and potential faults before they lead to system failures. This proactive approach allows for timely maintenance and reduces downtime, ensuring continuous and efficient operation.

Neural Network-Based Control: Neural networks, a subset of AI, offer powerful tools for controlling complex and nonlinear power electronic systems. Neural NNCs can learn and adapt to the system's dynamics, providing precise control even in the presence of uncertainties and disturbances. These controllers can be used in applications such as motor drives, renewable energy systems, and grid-connected inverters to enhance performance and stability.

Reducing Operational Costs and Enhancing System Longevity AI-driven enhancements in efficiency and performance also lead to significant reductions in operational costs and improvements in system longevity. By optimizing energy use and minimizing losses, AI contributes to cost savings and sustainable operation.

Cost Savings Through Efficiency: Improved efficiency directly translates to cost savings in power conversion systems. By reducing energy losses and optimizing resource utilization, AI-driven systems lower operational expenses. For example, in industrial applications, AI algorithms can optimize the operation of power converters to reduce energy consumption during peak demand periods, leading to lower energy bills.

Prolonged Component Lifespan: AI-driven predictive maintenance and intelligent control strategies help extend the lifespan of power electronic components. By reducing thermal stress, minimizing switching losses, and ensuring optimal operating conditions, AI contributes to the longevity of components such as semiconductors, capacitors, and inductors. This not only enhances system reliability but also reduces the frequency of component replacements and associated costs.

Sustainable Energy Management: AI-driven EMSs support sustainable energy practices by optimizing the use of renewable energy sources and minimizing reliance on non-renewable resources. By efficiently integrating renewable energy into the grid and ensuring optimal energy utilization, AI contributes to the reduction of greenhouse gas emissions and promotes a more sustainable energy future.

In conclusion, AI-driven enhancements in efficiency and performance are revolutionizing power conversion systems. Through intelligent optimization of power conversion processes, loss minimization, adaptive control strategies, and proactive maintenance, AI significantly improves the efficiency, reliability, and sustainability of power electronic systems. These advancements not only reduce operational costs and enhance system longevity but also support the transition to a more efficient and sustainable energy infrastructure.

14.4.2 Adaptability to dynamic conditions

Power conversion systems operate in environments that are often subject to rapid and unpredictable changes. The integration of AI into these systems provides a robust solution to enhance their adaptability to dynamic conditions. AI-driven algorithms offer the flexibility and responsiveness needed to maintain optimal performance and stability, even in the face of varying loads, fluctuating input power, and changing environmental conditions. This section delves into the ways AI improves the adaptability of power conversion systems.

AI-Driven Algorithms for Dynamic Adaptation: AI algorithms, particularly those based on ML and DL, enable power conversion systems to dynamically adjust their operations in response to changing conditions. These algorithms analyze real-time data to make informed decisions and adjustments, ensuring that the system can adapt effectively.

Real-Time Monitoring and Adjustment: AI systems continuously monitor various parameters such as voltage, current, temperature, and load. By analyzing this data in real-time, AI algorithms can detect changes in the operating environment and adjust control parameters accordingly. For example, in a DC–DC converter, AI can dynamically adjust the duty cycle to maintain a stable output voltage despite fluctuations in input voltage or load conditions.

Predictive Adaptation: ML models can predict future system states based on historical and real-time data. This predictive capability allows AI to anticipate changes and adjust operations proactively. For instance, in renewable energy systems, AI can predict changes in solar irradiance or wind speed and adjust the operation of solar panels or wind turbines to maximize energy capture and maintain stability.

Adaptive Control Strategies: AI enables the implementation of adaptive control strategies that adjust control parameters in response to real-time changes. Techniques such as adaptive PID controllers and MPC use ML algorithms to continuously learn from the system's behavior and optimize control actions. These adaptive controllers ensure that power conversion systems can maintain optimal performance under varying conditions.

Improving System Response and Stability: The adaptability provided by AI significantly enhances the response and stability of power conversion systems. By quickly and accurately responding to dynamic conditions, AI ensures that these systems operate smoothly and efficiently.

Rapid Response to Load Variations: Power conversion systems often experience sudden changes in load, which can affect their performance and stability. AI-driven control algorithms can rapidly respond to these variations by adjusting operating parameters in real-time. For example, in an uninterruptible power supply (UPS) system, AI can quickly adjust the power output to match the load demand, ensuring continuous and stable power delivery.

Stabilizing Renewable Energy Sources: Renewable energy sources such as solar and wind are inherently variable and can cause instability in power systems. AI algorithms help stabilize these sources by dynamically adjusting their operation. In wind energy systems, for example, AI can optimize the pitch and yaw angles of turbine blades to maintain stable power output despite changes in wind speed and direction.

Fault Detection and Mitigation: AI enhances system stability by enabling early fault detection and mitigation. ML models can analyze sensor data to identify anomalies and potential faults before they escalate into serious issues. This proactive approach allows for timely interventions, reducing the risk of system instability and failures.

Enhancing Energy Management in Dynamic Environments: AI-driven EMSs play a crucial role in adapting to dynamic environments by optimizing energy distribution, storage, and utilization.

Dynamic Load Balancing: AI algorithms can predict energy demand patterns and dynamically balance loads across the grid. By adjusting the distribution of energy in real-time, AI ensures that energy is delivered where it is needed most, reducing wastage and improving efficiency. For example, AI can manage the charging and discharging cycles of battery storage systems to balance supply and demand, ensuring optimal energy utilization.

Smart Grid Integration: The integration of AI into smart grids enhances the adaptability of energy distribution and management. AI algorithms can analyze real-time data from various sources, such as smart meters and sensors, to optimize grid operations. This includes demand forecasting, fault detection, and automated response to grid conditions. By optimizing the flow of energy and minimizing losses, AI contributes to a more efficient and resilient smart grid infrastructure.

Microgrid Management: In microgrids, AI-driven EMS can optimize the operation of local generation units, storage systems, and loads to achieve energy self-sufficiency. ML models predict energy production and consumption patterns within the microgrid, enabling efficient energy trading and ensuring that surplus energy is either stored or distributed to neighboring grids.

Leveraging AI for Environmental Adaptation: AI's ability to adapt to environmental changes is particularly beneficial for power systems operating in diverse and challenging conditions.

Climate Adaptation: Power conversion systems operating in outdoor environments are subject to changing weather conditions. AI algorithms can optimize the performance of these systems by adapting to climate variations. For example, in solar power systems, AI can adjust the orientation and tilt of solar panels based on weather forecasts to maximize energy capture.

Environmental Monitoring: AI-driven systems can monitor environmental conditions such as temperature, humidity, and air quality. By integrating this data into power conversion operations, AI can optimize system performance to minimize the impact of environmental factors. For instance, AI can adjust cooling systems in power electronics to prevent overheating in high-temperature environments.

Resilience to Natural Disasters: AI enhances the resilience of power systems to natural disasters by enabling rapid adaptation and recovery. ML models can predict the impact of natural events such as storms or earthquakes on power infrastructure and adjust operations to mitigate damage. Additionally, AI-driven systems can coordinate emergency responses and ensure continuous power supply during and after disasters.

Benefits of AI-Driven Adaptability: The adaptability provided by AI-driven systems offers several key benefits for power conversion applications.

Increased Reliability: By continuously monitoring and adapting to changing conditions, AI enhances the reliability of power conversion systems. This reduces the risk of unexpected failures and ensures consistent performance.

Enhanced Efficiency: Adaptive AI algorithms optimize the use of resources, reducing energy wastage and improving overall system efficiency. This leads to cost savings and more sustainable operations.

Improved System Longevity: AI-driven predictive maintenance and fault detection extend the lifespan of power electronic components by preventing damage and reducing wear and tear.

Greater Flexibility: AI provides the flexibility needed to integrate diverse energy sources and adapt to varying operational conditions. This is particularly important for renewable energy systems and smart grids.

In conclusion, AI-driven adaptability is a transformative feature for power conversion systems, enabling them to respond effectively to dynamic conditions. Through real-time monitoring, predictive adaptation, and intelligent control strategies, AI enhances the performance, stability, and reliability of these systems. This adaptability not only improves efficiency and reduces costs but also supports the integration of renewable energy sources and the development of sustainable energy infrastructure.

14.4.3 Improved reliability and system longevity

The integration of AI into power conversion systems significantly enhances their reliability and extends the lifespan of their components. By leveraging advanced AI algorithms for predictive maintenance, fault detection, and proactive monitoring, these systems can operate more efficiently and with fewer interruptions. This section explores the various ways AI improves reliability and system longevity in power electronics.

Predictive Maintenance Strategies Preventing Unexpected Failures: AI-driven predictive maintenance is a proactive approach that anticipates potential issues before they result in system failures. By continuously monitoring the condition of power electronic components and analyzing data patterns, AI can predict when maintenance is required, thereby preventing unexpected downtimes.

Real-Time Condition Monitoring: AI algorithms can continuously monitor the health of power electronic components by analyzing data from sensors that measure temperature, voltage, current, and vibration. ML models can detect anomalies and trends that indicate wear and tear or impending failures. For example, an increase in temperature or vibration in a motor drive could signal bearing wear, prompting maintenance before a breakdown occurs.

Failure Prediction: ML techniques such as regression analysis, neural networks, and time-series forecasting can predict the remaining useful life (RUL) of components. These models use historical data and real-time inputs to estimate when a component is likely to fail, allowing for timely replacement or repair. For instance, in a photovoltaic inverter, AI can predict the degradation rate of capacitors, enabling their replacement before they cause system failure.

Maintenance Scheduling: AI can optimize maintenance schedules based on predictive analytics, reducing both the frequency and cost of maintenance activities. By accurately predicting when maintenance is needed, AI helps avoid unnecessary maintenance while ensuring that critical interventions are performed in a timely manner. This leads to more efficient use of resources and minimizes downtime.

Fault Detection and Diagnosis: AI enhances the reliability of power conversion systems by enabling early detection and diagnosis of faults. By quickly identifying and addressing issues, AI reduces the risk of prolonged outages and minimizes the impact of faults on system performance.

Anomaly Detection: AI algorithms can detect anomalies in power conversion systems by analyzing real-time data and comparing it to normal operating patterns. ML models can identify deviations from expected behavior, such as unusual voltage fluctuations or irregular current spikes, which may indicate a fault. This allows for rapid intervention and corrective action before the fault escalates.

Fault Classification: Advanced AI techniques, including neural networks and SVMs, can classify different types of faults based on their signatures in the data. For example, an AI system can differentiate between faults caused by overheating, electrical surges, or mechanical failures. This classification helps technicians diagnose the root cause of the problem more accurately and efficiently.

Root Cause Analysis: AI can assist in root cause analysis by correlating data from various sensors and identifying the underlying factors contributing to a fault. By understanding the root cause, maintenance teams can implement more effective corrective actions and prevent recurrence. For instance, if a power inverter frequently trips due to overcurrent, AI can analyze the load patterns and suggest adjustments to prevent future occurrences.

Proactive Monitoring and Control: AI enables proactive monitoring and control of power conversion systems, ensuring that they operate within optimal parameters and respond effectively to changing conditions.

Dynamic Performance Monitoring: AI algorithms can continuously monitor the performance of power converters, ensuring they operate within specified limits. By analyzing real-time data, AI can detect performance degradation and initiate corrective measures. For example, in a DC–DC converter, AI can monitor the output voltage and adjust the duty cycle to maintain stability.

Environmental Adaptation: Power conversion systems often operate in environments with varying conditions. AI can adapt the operation of these systems based on environmental factors such as temperature, humidity, and load variations. For instance, AI can adjust the cooling mechanisms of a power electronic device based on ambient temperature to prevent overheating and ensure reliable operation.

Autonomous Decision-Making: AI-driven systems can make autonomous decisions to optimize performance and prevent failures. For example, an AI system controlling a microgrid can autonomously switch between different energy sources based on availability and demand, ensuring continuous power supply and preventing overloads.

Extending the Lifespan of Power Electronic Components: AI contributes to extending the lifespan of power electronic components by optimizing their operation and reducing stress.

Thermal Management: Excessive heat is a major factor in the degradation of power electronic components. AI can optimize thermal management by adjusting the operating parameters to minimize heat generation. For instance, AI can modulate the switching frequency of power transistors to reduce thermal stress and extend their lifespan.

Load Management: AI can balance the load distribution among multiple components to prevent overloading and uneven wear. By distributing the load more evenly, AI reduces the stress on individual components, thereby extending their operational life. In a battery storage system, for example, AI can manage the charging and discharging cycles to ensure that no single battery cell is overburdened.

Component Degradation Monitoring: AI can monitor the degradation of components such as capacitors, inductors, and semiconductors by analyzing changes in their electrical characteristics. By identifying signs of degradation early, AI enables timely interventions that prevent catastrophic failures and extend component life. For instance, AI can detect changes in the equivalent series resistance of capacitors, indicating the need for replacement.

Benefits of Improved Reliability and Longevity: The improvements in reliability and longevity provided by AI-driven systems offer several significant benefits for power conversion applications.

Reduced Downtime: By predicting and preventing failures, AI minimizes unexpected downtimes, ensuring continuous operation and reducing the impact on productivity and service delivery.

Cost Savings: AI-driven predictive maintenance and fault detection reduce maintenance costs by avoiding unnecessary interventions and preventing costly failures. This leads to significant cost savings over the lifespan of the system.

Enhanced Performance: Reliable and well-maintained power conversion systems perform more efficiently, leading to better energy utilization and reduced operational costs.

Sustainability: Extending the lifespan of components reduces the need for frequent replacements, contributing to sustainability by minimizing electronic waste and conserving resources.

Increased Safety: Proactive monitoring and fault detection enhance the safety of power conversion systems by preventing hazardous conditions and ensuring safe operation.

In conclusion, AI significantly improves the reliability and longevity of power conversion systems through predictive maintenance, fault detection, and proactive monitoring. These enhancements lead to reduced downtime, cost savings, enhanced performance, and increased safety, contributing to the overall efficiency and sustainability of power electronic systems. As AI technology continues to advance, its role in ensuring the reliable and long-lasting operation of power conversion systems will become increasingly important.

14.5 Challenges in implementing AI in power conversion

14.5.1 Data quality and availability

The successful implementation of AI in power conversion systems heavily relies on the quality and availability of data. High-quality data ensures the accuracy and reliability of AI models, while data availability ensures that these models are well-trained and capable of making informed decisions. However, there are several challenges associated with obtaining and maintaining high-quality data in power conversion applications.

Challenges in Data Quality Noise and Interference: Power conversion systems often operate in environments with significant electrical noise and interference. This noise can corrupt sensor data, leading to inaccuracies that can affect the performance of AI models. Ensuring data integrity in such environments requires robust filtering and noise reduction techniques.

Sensor Accuracy and Calibration: The accuracy of data collected from sensors depends on their precision and calibration. Inaccurate sensors can lead to erroneous data, which in turn can degrade the performance of AI algorithms. Regular calibration and maintenance of sensors are essential to maintain data quality.

Data Completeness: Incomplete data can pose significant challenges for AI models. Missing data points can result from sensor failures, communication issues, or data logging errors. Handling missing data requires sophisticated imputation techniques to ensure that AI models can still function accurately despite gaps in the data.

Data Consistency: Ensuring data consistency across different sensors and systems is crucial. Inconsistent data formats, units, and sampling rates can lead to integration challenges and affect the performance of AI models. Standardizing data collection protocols and formats can help mitigate these issues.

Challenges in Data Availability Data Scarcity: In many power conversion applications, collecting sufficient data for training AI models can be challenging. This scarcity is especially prevalent in new or unique systems where historical data may be limited. Data augmentation and synthetic data generation techniques can help address this challenge by creating additional training data.

Real-Time Data Collection: For AI models to make accurate real-time decisions, they require access to up-to-date data. Ensuring real-time data availability involves implementing robust data acquisition and communication systems that can handle the high-speed data transfer requirements of power conversion systems.

Data Security and Privacy: Protecting the data collected from power conversion systems is essential to prevent unauthorized access and ensure privacy. Implementing secure data transmission and storage protocols, as well as compliance with data protection regulations, is crucial for maintaining data availability and integrity.

Strategies for Addressing Data Quality and Availability Data Preprocessing and Cleaning: Implementing data preprocessing and cleaning techniques is essential to ensure high-quality data. This includes filtering out noise, handling missing values, and standardizing data formats. Advanced algorithms can automatically detect and correct anomalies, improving the reliability of the data used by AI models.

Sensor Fusion: Combining data from multiple sensors can enhance data quality and availability. Sensor fusion techniques leverage the strengths of different sensors to provide a more accurate and comprehensive dataset. For example, combining temperature, vibration, and electrical measurements can provide a more holistic view of the system's condition.

Advanced Calibration Techniques: Regular calibration of sensors using advanced techniques ensures data accuracy. Automated calibration systems that use AI to detect and correct sensor drift can maintain data quality over extended periods without manual intervention.

Synthetic Data Generation: In situations where real data is scarce, synthetic data generation can provide additional training datasets. Techniques such as generative adversarial networks can create realistic synthetic data that mimics the characteristics of real-world data, enhancing the training of AI models.

Cloud and Edge Computing: Leveraging cloud and edge computing solutions can enhance data availability and processing capabilities. Cloud platforms can store vast amounts of data and provide powerful computing resources for AI model training. Edge computing can enable real-time data processing and decision-making closer to the source, reducing latency and improving data availability.

14.5.2 Explainability and Trustworthiness

As AI systems become increasingly integrated into power conversion applications, ensuring their explainability and trustworthiness is crucial. Stakeholders need to understand how AI models make decisions to trust their outputs and to ensure that these models align with regulatory and ethical standards. Explainability and trustworthiness are essential for gaining user confidence and ensuring the reliable and safe operation of AI-driven systems.

Challenges in Explainability Complexity of AI Models: Many AI models, particularly DL algorithms, operate as "black boxes," making it difficult to understand how they arrive at their decisions. The complexity of these models can obscure the relationship between input data and outputs, hindering transparency.

Nonlinear Relationships: Power conversion systems often involve nonlinear relationships between variables. AI models that capture these relationships can be inherently complex and challenging to interpret. Explaining the decisions of such models requires sophisticated techniques that can unravel these nonlinearities.

Feature Importance: Determining which features (input variables) are most influential in an AI model's decision-making process can be difficult. Understanding feature importance is crucial for validating the model's behavior and ensuring it aligns with domain knowledge.

Challenges in Trustworthiness Model Robustness: Trustworthiness requires that AI models are robust and perform reliably under various conditions, including edge cases and unexpected scenarios. Ensuring robustness involves extensive testing and validation to confirm that the model can handle real-world variability.

Bias and Fairness: AI models can inadvertently learn and propagate biases present in the training data. Ensuring fairness and avoiding biased decisions is critical, particularly in applications where decisions impact safety and operational efficiency. Identifying and mitigating biases in AI models is a significant challenge.

Regulatory Compliance: Power conversion systems often operate in regulated environments where compliance with standards is mandatory. Ensuring that AI models adhere to these regulations and can be audited for compliance is essential for their deployment.

Strategies for Enhancing Explainability and Trustworthiness Interpretable AI Models: Developing interpretable AI models that provide clear and understandable explanations for their decisions is key. Techniques such as decision trees, rule-based systems, and linear models are inherently interpretable and can be used where explainability is a priority.

Model-Agnostic Explainability Tools: Tools such as LIME (Local Interpretable Model-agnostic Explanations) and SHAP (SHapley Additive exPlanations) can provide post hoc explanations for complex models. These tools analyze the model's behavior by perturbing input data and observing the impact on outputs, helping to identify important features and explain decisions.

Transparency in Model Development: Documenting the development process of AI models, including data sources, feature selection, and training procedures, enhances transparency. Providing detailed documentation helps stakeholders understand how the model was created and ensures that it can be audited and validated.

Bias Detection and Mitigation: Implementing techniques to detect and mitigate biases in AI models is crucial for ensuring fairness. This includes analyzing training data for bias, using fairness-aware algorithms, and continuously monitoring model outputs to identify and address biased decisions.

Robustness Testing: Conducting extensive robustness testing to evaluate the model's performance under various conditions is essential for building trust. This involves testing the model on diverse datasets, including edge cases, to ensure it can handle real-world variability.

Regulatory Audits and Compliance: Ensuring that AI models comply with regulatory standards involves regular audits and validation. Implementing processes for continuous monitoring and updating models to maintain compliance is crucial for their long-term deployment.

User Feedback and Iterative Improvement: Engaging with users and stakeholders to gather feedback on AI model performance and explainability is important for continuous improvement. Iterative development processes that incorporate user feedback help build trust and ensure that the models meet operational needs.

In conclusion, addressing the challenges of data quality, availability, explainability, and trustworthiness is critical for the successful integration of AI in power conversion systems. Implementing robust strategies to enhance data quality, ensure model transparency, and build trustworthy AI systems will lead to more reliable, efficient, and accepted AI-driven solutions in the power conversion domain.

14.5.3 Computational requirements and real-time processing

Balancing the computational demands of AI algorithms with real-time processing constraints is a critical challenge in the deployment of AI-based models in power electronics applications. If not crafted carefully, an AI model can become large and resource intensive.

DNNs and other traditional ML models can replicate the behavior of physical systems like power electronics converters [10,21]. These models can predict system outputs, monitor performance, and detect faults, which are essential for applications such as smart grids and real-time system monitoring [9]. However, the complexity of ML models in terms of their structure, hyper-parameters, and the nature of their computations can results in significant computational and memory demands. This complexity often necessitates advanced hardware resources, such as high-performance GPUs or FPGAs [22,23], which may not be feasible for cost-sensitive industrial applications that require low-cost microcontrollers [10].

Power electronics applications often demand rapid responses to ensure system stability and performance. The delay introduced by processing complex AI models can be intolerable for real-time operation, making it crucial to optimize these models. The key challenge is to minimize the time and space complexity (i.e. execution time and memory consumption) of the AI algorithms while maintaining their predictive error as low as possible [10]. Building efficient AI models that can be deployed in power electronics systems can be successfully accomplished by following some strategies such as AI algorithm selection, DNN model simplification and optimization, use of physics-informed DNN, and efficient use of computing hardware.

Based on the application, the AI algorithm can be chosen. Traditional ML techniques such as **SVM and K-nearest neighbors (KNN)** can be effective for classification tasks, such as fault detection and diagnosis in power converters [24]. **Decision trees** and similar approaches like random forest and eXtreme Gradient Boosting (XGBoost) can be used for both, classification and regression tasks [21,24]. Execution time in such tree-based models is dependent on the depth of the trees and the branch instructions taken to reach the predicted value. This can result into lower performance if the application requires complex decisions [21]. **DNN-based models** are typically superior when compared to other techniques. One limitation of DNNs is that it requires more data to train the models. However, data can be relatively easy to collect when measurements are directly taken from a physical power converter [9]. Other more complex **DNN-based models for time series** prediction and forecasting can be used such as RNNs and long short-term memory models [25].

One effective approach in the context of DNN models is to simplify the DNN architecture and calculations without significantly compromising their accuracy. This can be achieved by reducing the number of the DNN's network layers and/or neurons [10]. Additionally, quantization can be used to lower the precision of the weights and activations, which reduces both memory usage and computational load. Quantization, for instance, can convert 32-bit floating-point weights to 8-bit integers, substantially decreasing the model size and inference time [28].

Physics-informed neural networks (PINNs) can significantly improve the reliability and efficiency of power converter components [17]. Using automatic differentiation, PINNs can effectively model embed mathematical models describing complex power conversion processes into the DNN model's training. This enables fast convergence of the model with lower prediction error of power converter systems' behavior. Additionally, PINNs can help detect anomalies and predict RUL for power converters, reducing maintenance costs and improving overall system reliability [27].

Leveraging the capabilities of low-cost microcontrollers through specialized libraries like TensorFlow Lite Micro (TFLM) is another strategy [28]. TFLM is designed to run ML models on microcontrollers with limited computational resources. By converting the trained DNN models into a format compatible with TFLM and optimizing them for the specific architecture of the microcontroller, it is possible to achieve efficient real-time performance. For example, the STM32 microcontrollers, which feature multi-core ARM-based architectures, can run the AI model on one core while managing other tasks on another, effectively distributing the computational load. Different AI software toolchains can be also used for microcontroller implementations, each offering its own advantages and limitations [26]. The designer of power converters must carefully consider the intended outcomes and hardware requirements to develop an effective model.

Table 14.5 summarizes techniques used for AI models to match the requirements of power converters. By applying these strategies and considering their limitations, it is possible to design AI-based models that are both computationally efficient and capable of operating within the stringent real-time constraints required by power electronics applications. This balance ensures that the benefits of advanced AI techniques can be realized in practical, cost-effective deployments.

Table 14.5 Categorization of AI strategies

Strategy	**Description**	**References**
AI algorithm selection	Choosing AI algorithms based on application needs, covering traditional ML techniques and DNN-based models	[9,21,24,25]
DNN model simplification and optimization	Simplifying DNN architectures and using quantization to reduce memory and computational load	[10,26]
Use of physics-informed DNN	Using PINNs to improve model reliability and efficiency	[17,27]
Efficient use of computing hardware	Running ML models on low-cost microcontrollers	[10,26,28]

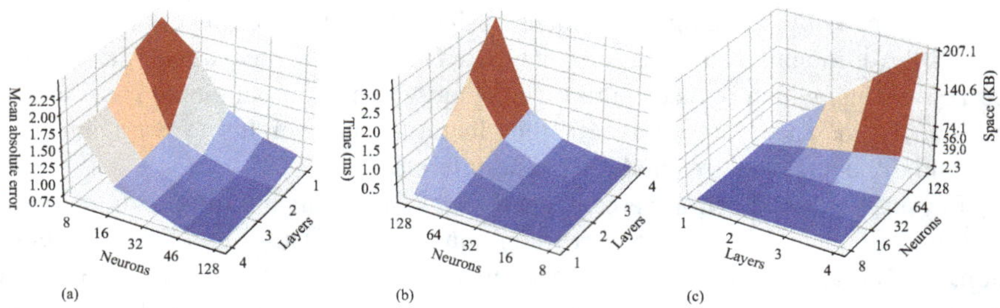

Figure 14.10 Model size with respect to (a) MAE, (b) computational time, and (c) memory usage [10]

As a use-case of balancing model performance and efficiency in the context of predicting the dynamic behavior of a power electronics inverter, multiple models were trained and evaluated with different shapes in [10] as presented in Figure 14.10. To evaluate the models, 20 different neural network architectures examined. Each model had a varying number of hidden layers ranging from 1 to 4 and with neuron sizes ranging from 8 to 128. The training process took place over 50 epochs, with MAE values averaged over 10 rounds for smoothing purposes.

Figure 14.10a shows that larger models with more neurons and hidden layers tend to have lower mean absolute errors. This is due to the fact that deeper models can better generalize training data. However, this increased performance comes at the cost of a larger memory footprint and longer processing time. In terms of space complexity, Figure 14.10c shows that the model's size increases rapidly with the number of hidden layers and neurons. For instance, a single hidden layer with 8 neurons consumes approximately 2.3 kB, while a model with 4 hidden layers and 128 neurons each uses about 207 kB. That is because every neuron shares connections with every other neuron in the next layer, leading to more calculations and longer execution time as can be seen in Figure 14.10b.

14.6 Future directions and emerging trends

14.6.1 Integration of edge AI in power electronics

Edge computing refers to the practice of processing data closer to where it is generated, rather than relying on a centralized data-processing warehouse or cloud. This approach reduces latency, saves bandwidth, enhances security, and improves response times for real-time applications [29]. Edge computing plays a crucial role in integrating AI closer to power electronic systems by enabling real-time data processing and decision-making closer to the source of data generation [30]. This integration offers several benefits. By processing data on an edge device, the time taken to analyze data and infer decisions is minimized when compared to a cloud-based solution. When compared to running an AI model locally on a limited microcontroller, edge devices offer more powerful computing capabilities allowing for more advanced and complex AI algorithms to be utilized. Minimizing time is critical for power electronic systems as it requires real-time monitoring and immediate responses to anomalies or changes in operational conditions. Additionally, processing data on-site enhances reliability and security by reducing the amount of data transmitted over networks, decreasing the risk of data breaches, and ensuring higher data integrity [31]. This is particularly important for critical power infrastructure where data security is paramount. Furthermore, edge computing allows for distributed data processing, which can be scaled across numerous devices and systems. This decentralization reduces the burden on centralized cloud servers and improves overall system efficiency, leading to cost savings by lowering bandwidth requirements and associated costs, as well as reducing the need for extensive cloud storage and computational resources.

In the context of power electronics, edge computing enables real-time anomaly detection and predictive maintenance [32]. Edge devices equipped with AI algorithms can continuously monitor the health of power electronic components, such as inverters, converters, and drives, detecting thermal anomalies, vibration patterns, and electrical signatures indicative of potential failures. By processing this data locally, these systems can predict maintenance needs and prevent costly downtimes. Additionally, AI at the edge can enable power electronic systems to adapt to changing conditions in real-time.

Emerging trends in edge AI for real-time decision-making include the deployment of advanced ML models, such as neural networks and DL algorithms, on edge devices. These models can handle complex data patterns and provide accurate predictions and insights, making them ideal for applications like anomaly detection and predictive maintenance [23,30,32]. Another significant trend is federated learning, which involves training ML models across multiple edge devices without transferring raw data to a central server [23]. Each edge device processes its own data and shares only model updates, enhancing privacy and reducing communication overhead. Federated learning is particularly useful in power electronics where data security and minimal latency are crucial.

The development of specialized hardware, such as AI accelerators and tensor processing units (TPUs) chips, designed for edge computing is also enabling more efficient and powerful AI processing at the edge [33]. These hardware advancements allow AI algorithms to execute with lower power consumption and greater speeds. Additionally, edge-orchestrated systems, which coordinate multiple edge devices to work together seamlessly, are improving overall decision-making processes. For power electronic systems, this could mean better coordination of distributed energy resources (DERs), more robust fault detection, and improved system resilience.

Overall, edge computing brings AI closer to power electronic systems and enables real-time, on-site data processing and decision-making at the cost of some network latency. As illustrated in Figure 14.11, this trade-off must be considered based on the application while integrating edge devices in the power electronics system. Emerging trends in edge AI, such as advanced ML models, federated learning, specialized hardware, are further enhancing the capabilities and applications of AI in power electronics. By leveraging these advancements, power electronic systems can achieve higher efficiency, reliability, and sustainability, paving the way for smarter and more resilient power infrastructure.

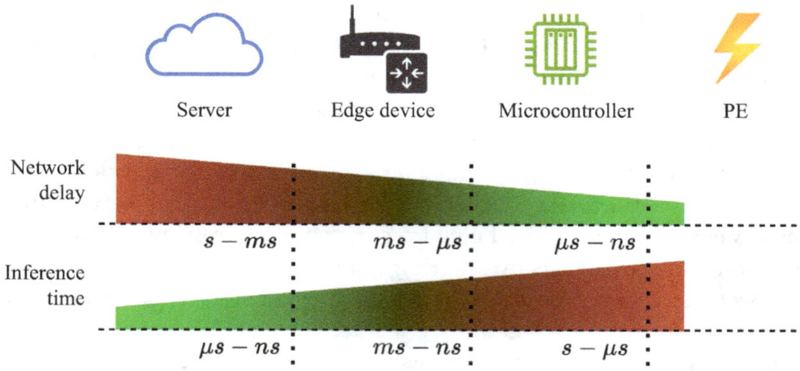

Figure 14.11 Overview of edge computing network delay-inference time trade-off

14.6.2 Collaborative AI-powered energy systems

The integration of AI in power conversion systems is transforming the energy sector, enabling more intelligent, interconnected, and collaborative energy networks. Collaborative AI-powered energy systems leverage the strengths of various AI technologies to optimize the generation, distribution, and consumption of energy across multiple stakeholders and infrastructures. This section explores the potential and impact of collaborative AI-powered energy systems, focusing on their key components, benefits, and future directions.

Integration of AI in a broader energy ecosystem collaborative AI-powered energy systems encompass a range of technologies and practices aimed at improving the efficiency, reliability, and sustainability of energy networks. These systems integrate AI at various levels, from local generation units to large-scale grids, enabling seamless coordination and optimization.

Smart Grids: Smart grids are a cornerstone of collaborative AI-powered energy systems. They use AI to manage and optimize the flow of electricity from producers to consumers. AI algorithms analyze data from smart meters, sensors, and other grid components to predict demand, detect faults, and optimize energy distribution. This real-time data analysis enables more efficient and reliable grid operations.

DERs: AI facilitates the integration and management of DERs such as solar panels, wind turbines, and battery storage systems. By predicting energy generation and consumption patterns, AI algorithms can optimize the use of DERs, ensuring that energy is available when and where it is needed. This enhances the resilience and sustainability of the energy network.

EMS: AI-powered EMS coordinate the generation, storage, and consumption of energy within buildings, microgrids, and entire communities. These systems use ML models to predict energy demand, optimize energy usage, and manage storage systems. By balancing supply and demand, AI-driven EMS improve energy efficiency and reduce operational costs.

Vehicle-to-Grid (V2G) Systems: Electric vehicles (EVs) equipped with V2G technology can act as mobile energy storage units. AI algorithms manage the bidirectional flow of energy between EVs and the grid, enabling EVs to supply energy during peak demand periods and recharge during off-peak times. This collaborative approach enhances grid stability and maximizes the utilization of renewable energy.

Collaborative Efforts for Intelligent Energy Networks: Creating intelligent, interconnected energy networks requires collaboration among various stakeholders, including utilities, technology providers, policymakers, and consumers. Collaborative AI-powered energy systems foster such cooperation by enabling seamless communication and coordination across the energy ecosystem.

Data Sharing and Interoperability: Effective collaboration in AI-powered energy systems relies on data sharing and interoperability. Standardized data formats and communication protocols enable different systems and devices to exchange information seamlessly. This interoperability allows AI algorithms to access and analyze data from diverse sources, providing a holistic view of the energy network.

Collaborative Decision-Making: AI facilitates collaborative decision-making by providing real-time insights and predictive analytics to all stakeholders. For example, utilities can use AI to forecast energy demand and coordinate with DER operators to ensure sufficient supply. Consumers can use AI-driven applications to manage their energy consumption and participate in demand response programs. This collaborative approach leads to more efficient and resilient energy networks.

Decentralized Energy Markets: AI enables the development of decentralized energy markets where multiple stakeholders can buy and sell energy in real-time. ML models predict energy prices based on supply and demand, allowing participants to make informed trading decisions. This decentralized approach promotes competition, reduces energy costs, and encourages the use of renewable energy sources.

Policy and Regulation Support: Policymakers can leverage AI to design and implement regulations that promote sustainable energy practices. AI algorithms can analyze the impact of different policies on energy consumption and generation, helping policymakers create regulations that support the integration of renewable energy and the development of smart grids. Collaborative efforts between policymakers, utilities, and technology providers ensure that regulations are effective and adaptable to evolving technologies.

Benefits of Collaborative AI-Powered Energy Systems: The integration of AI in collaborative energy systems offers numerous benefits, enhancing the efficiency, reliability, and sustainability of energy networks.

Enhanced Efficiency: AI optimizes the generation, distribution, and consumption of energy, reducing waste and improving overall efficiency. By predicting demand and managing supply in real-time, AI ensures that energy is used more effectively, lowering operational costs and reducing greenhouse gas emissions.

Increased Reliability: Collaborative AI-powered energy systems enhance the reliability of energy networks by enabling real-time monitoring and fault detection. AI algorithms can quickly identify and address issues, minimizing downtime and ensuring continuous power supply. This reliability is crucial for critical infrastructure and essential services.

Scalability and Flexibility: AI-driven energy systems are highly scalable and adaptable to changing conditions. They can integrate new energy sources, technologies, and participants without disrupting existing operations. This flexibility allows energy networks to evolve and expand, accommodating future energy demands and technological advancements.

Sustainability: Collaborative AI-powered energy systems promote the use of renewable energy sources and support sustainable energy practices. By optimizing the integration and management of DERs, AI reduces reliance on fossil fuels and lowers carbon emissions. This contributes to global efforts to combat climate change and promote environmental sustainability.

Consumer Empowerment: AI empowers consumers to actively participate in energy management. AI-driven applications provide consumers with insights into their energy consumption, enabling them to make informed decisions and reduce their energy usage. Demand response programs incentivize consumers to adjust their consumption patterns, contributing to grid stability and efficiency.

Future Directions and Emerging Trends: The future of collaborative AI-powered energy systems is shaped by emerging technologies and innovative approaches that further enhance the capabilities and benefits of these systems.

Edge AI and Internet of Things (IoT) Integration: The integration of edge AI and the IoT in energy systems enables real-time data processing and decision-making at the source. Edge AI devices can analyze data locally, reducing latency and enhancing the responsiveness of energy networks. IoT devices provide granular data on energy consumption and generation, supporting more precise and effective AI algorithms.

Blockchain Technology: Blockchain technology offers a secure and transparent way to manage decentralized energy markets and data sharing. Smart contracts can automate transactions between energy producers and consumers, ensuring trust and reducing administrative overhead. Blockchain can also enhance data security and integrity, supporting the collaborative nature of AI-powered energy systems.

Artificial Intelligence of Things (AIoT): AIoT combines AI and IoT to create intelligent and connected energy systems. AI algorithms analyze data from IoT devices to optimize energy management, predict maintenance needs, and enhance system performance. AIoT enables seamless communication and coordination across the energy network, supporting collaborative efforts and improving overall efficiency.

Federated Learning: Federated learning allows AI models to be trained across multiple decentralized devices without sharing raw data. This approach enhances privacy and security while enabling the development of robust AI models that leverage data from diverse sources. Federated learning supports collaborative AI efforts by enabling data-driven insights without compromising data ownership and confidentiality.

Energy Storage Innovations: Advances in energy storage technologies, such as high-capacity batteries and supercapacitors, enhance the capabilities of AI-powered energy systems. Improved energy storage supports the integration of renewable energy sources, provides backup power during outages, and enables more effective demand response programs. AI algorithms optimize the use and management of these storage systems, enhancing their performance and longevity.

In conclusion, collaborative AI-powered energy systems represent a transformative approach to managing the generation, distribution, and consumption of energy. By leveraging AI technologies, these systems enhance efficiency, reliability, and sustainability, supporting the development of intelligent, interconnected energy networks. Future advancements in edge AI, IoT, blockchain, AIoT, federated learning, and energy storage will further enhance the capabilities of collaborative AI-powered energy systems, driving innovation and promoting sustainable energy practices.

14.7 Conclusion

14.7.1 Recapitulation of key findings

The exploration of AI-powered power conversion systems has unveiled significant insights into how artificial intelligence can revolutionize the efficiency, adaptability, and reliability of power electronics. This section summarizes the key findings discussed throughout the chapter, highlighting the transformative impact of AI on power conversion systems.

Enhancing Efficiency and Performance: AI-driven techniques optimize power conversion processes, leading to substantial improvements in efficiency and performance. ML and DL algorithms enable real-time parameter adjustments, loss minimization, and predictive control. These advancements reduce energy losses, improve operational efficiency, and enhance the overall performance of power electronic systems.

Adaptability to Dynamic Conditions: AI provides the adaptability required to manage the dynamic and often unpredictable conditions in power conversion systems. Through real-time monitoring, predictive adaptation, and adaptive control strategies, AI ensures that power converters can respond effectively to varying loads, fluctuating input power, and changing environmental conditions. This adaptability enhances system stability and reliability.

Improved Reliability and System Longevity: AI significantly improves the reliability and longevity of power electronic components by enabling predictive maintenance, early fault detection, and proactive monitoring. These capabilities prevent unexpected failures, reduce downtime, and extend the lifespan of critical components. AI-driven systems ensure continuous and efficient operation, enhancing the overall reliability of power conversion systems.

Data Quality and Availability: The quality and availability of data are crucial for the successful implementation of AI in power conversion systems. Addressing challenges related to noise, sensor accuracy, data completeness, and data consistency ensures that AI models can operate effectively. Strategies such as data preprocessing, sensor fusion, and synthetic data generation improve data quality and availability, supporting the development of robust AI models.

Explainability and Trustworthiness: Ensuring the explainability and trustworthiness of AI models is essential for gaining user confidence and meeting regulatory standards. Techniques for enhancing explainability include developing interpretable AI models, using model-agnostic explainability tools, and maintaining transparency in model development. Ensuring trustworthiness involves addressing biases, testing for robustness, and ensuring compliance with regulatory requirements.

Collaborative AI-Powered Energy Systems: Collaborative AI-powered energy systems integrate AI at various levels to create intelligent, interconnected energy networks. These systems leverage AI to optimize energy generation, distribution, and consumption, enhancing efficiency, reliability, and sustainability. Collaboration among stakeholders, data sharing, and the development of decentralized energy markets are key components of these systems.

14.7.2 Implications for the future

The integration of AI into power conversion systems has profound implications for the future of the energy sector. These advancements not only improve the performance and reliability of power electronic systems but also support the transition to more sustainable and resilient energy infrastructures. This section discusses the anticipated advancements and the role of interdisciplinary collaboration in shaping the future of AI-powered power conversion.

Anticipated Advancements Edge AI and IoT Integration: The continued integration of edge AI and IoT devices will enable real-time data processing and decision-making closer to the source. This will enhance the responsiveness and efficiency of power conversion systems, supporting more effective energy management and fault detection.

Federated Learning and AIoT: Federated learning and the combination of AI with IoT (AIoT) will play a significant role in future energy systems. Federated learning enables collaborative AI model training without sharing raw data, enhancing privacy and security. AIoT will create intelligent and connected energy systems that optimize energy management, predict maintenance needs, and improve system performance.

Blockchain Technology: Blockchain will enhance the transparency, security, and efficiency of decentralized energy markets and data sharing. Smart contracts and blockchain-based transactions will automate and secure energy trading, supporting the development of more competitive and reliable energy networks.

Advanced Energy Storage: Innovations in energy storage technologies will enhance the capabilities of AI-powered energy systems. Improved storage solutions will support the integration of renewable energy sources, provide backup power during outages, and enable more effective demand response programs. AI will optimize the use and management of these storage systems, enhancing their performance and longevity.

The Role of Interdisciplinary Collaboration: The successful integration of AI into power conversion systems requires collaboration across multiple disciplines, including electrical engineering, computer science, data science, and policy making.

Interdisciplinary collaboration will drive innovation and ensure that AI technologies are effectively applied to address the challenges and opportunities in the energy sector.

Engineering and Computer Science: Collaboration between electrical engineers and computer scientists will lead to the development of advanced AI algorithms and their integration into power electronic systems. This partnership will enable the creation of more efficient, reliable, and adaptable power conversion technologies.

Data Science and Analytics: Data scientists will play a crucial role in developing models that can analyze vast amounts of data generated by power conversion systems. Their expertise in ML, data preprocessing, and analytics will enhance the accuracy and reliability of AI models.

Policy Making and Regulation: Policymakers will need to work closely with technologists to develop regulations that support the deployment of AI-powered energy systems. Ensuring that AI models comply with regulatory standards and addressing ethical considerations will be critical for the widespread adoption of these technologies.

Industry and Academia: Collaboration between industry and academia will drive research and development in AI-powered power conversion. Academic institutions can provide the theoretical foundations and conduct experimental research, while industry partners can focus on practical applications and commercialization.

In conclusion, the integration of AI into power conversion systems offers significant benefits in terms of efficiency, adaptability, reliability, and sustainability. The anticipated advancements in AI technologies and the ongoing interdisciplinary collaboration will shape the future of energy systems, supporting the development of intelligent, resilient, and sustainable energy infrastructures. As AI continues to evolve, its impact on power conversion and the broader energy sector will become increasingly profound, driving innovation and promoting sustainable development.

Acknowledgments

This publication was made possible by ARG01-0428-230023 from the Qatar Research, Development, and Innovation (QRDI). The statements made herein are solely the responsibility of the authors.

References

[1] Gomis-Bellmunt O. Guest Editorial Special Section Advances in research and applications of power electronics in T&D systems. *IEEE Transactions on Power Delivery*. 2024;39(1):339–340.

[2] Refaat SS, Ellabban O, Bayhan S, *et al. Smart grid and enabling technologies*. 1st ed. Hoboken, NJ: Wiley; 2021.

[3] Komurcugil H, Bayhan S, Guzman R, *et al. Advanced control of power converters: Techniques and Matlab/Simulink implementation*. Hoboken, NJ: Wiley; 2023.

[4] Ahmed KY, Bin Yahaya NZ, Asirvadam VS, *et al.* Development of power electronic distribution transformer based on adaptive PI controller. *IEEE Access*. 2018;6:44970–44980.

[5] Sharida A, Bayhan S, Abu-Rub H, *et al.* Voltage-sensorless open-switch fault-tolerant control of three-phase T-type rectifier. *IEEE Transactions on Power Electronics*. 2023;38(12):15365–15376.

[6] Sharida A, Bayhan S, and Abu-Rub H. Adaptive control strategy for three-phase three-level T-type rectifier based on online disturbance estimation and compensation. *IEEE Access*. 2023;11:40967–40977.

[7] Sharida A, Bayhan S, Abu-Rub H, *et al.* Inverse model predictive control for power electronic converters. *IEEE Access*. 2024;12:68485–68496.

[8] Mahazabeen M, Abianeh AJ, Ebrahimi S, *et al.* Enhancing EV charger resilience with reinforcement learning aided control. *e-Prime-Advances in Electrical Engineering, Electronics and Energy*. 2023;5:100276.

[9] Sharida A, Kamal NF, Alnuweiri H, *et al.* Digital-twin-based diagnosis and tolerant control of T-type three-level rectifiers. *IEEE Open Journal of the Industrial Electronics Society*. 2023;4:230–241.

[10] Kamal NF, Sharida A, Bayhan S, *et al.* Low-cost digital twin design for power electronics using deep neural networks. In: *2024 4th International Conference on Smart Grid and Renewable Energy (SGRE)*. IEEE; 2024. p. 1–6.

[11] Sharida A, Bayhan S, and Abu-Rub H. Online self-tuning current-controller for three-phase three-level T-type rectifier. In: *IECON 2022–48th Annual Conference of the IEEE Industrial Electronics Society*. IEEE; 2022. p. 1–6.

[12] Sharida A, Bayhan S, and Abu-Rub H. Fault-tolerant self-tuning control for three-phase three-level t-type rectifier. *IEEE Transactions on Power Electronics*. 2023.

[13] Hashlamon I and Erbatur K. An improved real-time adaptive Kalman filter with recursive noise covariance updating rules. *Turkish Journal of Electrical Engineering and Computer Sciences*. 2016;24(2):524–540.

[14] Dorokhova M, Martinson Y, Ballif C, *et al.* Deep reinforcement learning control of electric vehicle charging in the presence of photovoltaic generation. *Applied Energy*. 2021;301:117504.

[15] Chen H, Zhang Z, Karamanakos P, *et al.* Digital twin techniques for power electronics-based energy conversion systems: A survey of concepts, application scenarios, future challenges, and trends. *IEEE Industrial Electronics Magazine*. 2022;17(2):20–36.

[16] Kern D, Krauß C, and Hollick M. Detection of anomalies in electric vehicle charging sessions. In: *Proceedings of the 39th Annual Computer Security Applications Conference*; 2023. p. 298–309.

[17] Kamal NF, Sharida A, Bayhan S, *et al.* Enhancing electric vehicle charging predictions: A physics-informed neural network approach. In: *IECON 2024 – 50th Annual Conference of the IEEE Industrial Electronics Society*. IEEE; 2024. p. 1–6.

[18] Kumar RR, Bharatiraja C, Udhayakumar K, *et al.* Advances in batteries, battery modeling, battery management system, battery thermal management,

SOC, SOH, and charge/discharge characteristics in EV applications. *IEEE Access*. 2023;11:105761–105809.

[19] Mavikumbure HS, Cobilean V, Wickramasinghe CS, *et al.* Cy-Phy ADS: Cyber physical anomaly detection framework for EV charging systems. *IEEE Transactions on Transportation Electrification*. 2024;10(4):9904–9917.

[20] Abu-Rub O, Sharida A, Saeedifard M, et al. Real-time anomaly detection and remedy actions for resilient operation of DC-coupled fast EV chargers. In: *2024 IEEE Energy Conversion Congress and Exposition (ECCE)*; 2024. p. 1748–1752.

[21] Bayindir AB, Al-Khateeb A, Sharida A, *et al.* Design and analysis of digital twin models for dual active bridge. In: *IECON 2024 – 50th Annual Conference of the IEEE Industrial Electronics Society*. IEEE; 2024. p. 1–6.

[22] Liu X, Qiu L, Rodríguez J, *et al.* Data-driven neural predictors-based robust MPC for power converters. *IEEE Transactions on Power Electronics*. 2022;37(10):11650–11661.

[23] Zhao L, Li J, Li Q, *et al.* A federated learning framework for detecting false data injection attacks in solar farms. *IEEE Transactions on Power Electronics*. 2021;37(3):2496–2501.

[24] Moosavi S, Kazemi A, and Akbari H. A comparison of various open-circuit fault detection methods in the IGBT-based DC/AC inverter used in electric vehicle. *Engineering Failure Analysis*. 2019;96:223–235.

[25] Li X, Lin F, Wang H, *et al.* Temporal modeling for power converters with physics-in-architecture recurrent neural network. *IEEE Transactions on Industrial Electronics*. 2024;71(11):14111–14123.

[26] Saha SS, Sandha SS, and Srivastava M. Machine learning for microcontroller-class hardware: A review. *IEEE Sensors Journal*. 2022;22(22):21362–21390.

[27] Fassi Y, Heiries V, Boutet J, *et al.* Towards physics-informed machine learning-based predictive maintenance for power converters – A review. *IEEE Transactions on Power Electronics*. 2023;39(2):2692–2720.

[28] David R, Duke J, Jain A, *et al.* Tensorflow lite micro: Embedded machine learning for tinyml systems. *Proceedings of Machine Learning and Systems*. 2021;3:800–811.

[29] Syu JH, Lin JCW, Srivastava G, *et al.* A comprehensive survey on artificial intelligence empowered edge computing on consumer electronics. *IEEE Transactions on Consumer Electronics*. 2023;69(4):1023–1034.

[30] Bhoi SK, Chakraborty S, Verbrugge B, *et al.* Intelligent data-driven condition monitoring of power electronics systems using smart edge–cloud framework. *Internet of Things*. 2024;26:101158. Available from: https://www.sciencedirect.com/science/article/pii/S2542660524000994.

[31] Zheng R, Sumper A, Aragüés-Peñalba M, *et al.* Advancing power system services with privacy-preserving federated learning techniques: A review. *IEEE Access*. 2024;12:76753–76780.

[32] Gómez PI, Gajardo MEL, Mijatovic N, *et al.* A self-commissioning edge computing method for data-driven anomaly detection in power electronic systems. *IEEE Transactions on Industrial Electronics*. 2024;71(10):13319–13330.

[33] Hosseininoorbin S, Layeghy S, Kusy B, *et al.* Exploring edge TPU for deep feed-forward neural networks. *Internet of Things*. 2023;22:100749.

Chapter 15
Empowering resilience: AI and the future of microgrids

Ali Sharida[1], Anas Karaki[1], Sertac Bayhan[1], Ugur Fesli[2], and Haitham Abu-Rub[3]

15.1 Introduction to microgrids and their importance

15.1.1 Definition and components of microgrids

Microgrids are small-scale, localized energy systems that can operate independently or in conjunction with the main power grid. They consist of various distributed energy resources (DERs), energy storage systems (ESSs), and loads within a defined geographical area. Microgrids offer a flexible and reliable solution for integrating renewable energy sources (RES) into the energy system and can be designed for different purposes. These purposes include enhancing energy resilience, improving energy efficiency, or providing power to remote areas. These systems are especially beneficial in applications that require high reliability, like critical infrastructure, hospitals, and industrial facilities.

Microgrids generally function in two modes: grid-connected and islanded (off-grid). In grid-connected mode, the microgrid synchronizes with the main grid and allows bidirectional energy flow. In islanded mode, the microgrid disconnects from the main grid and operates independently. This can be beneficial during grid outages, where the microgrid offers continuous power to local loads. This capability enables microgrids to contribute to grid resilience, as they can continue to supply energy to critical loads even if the main grid is down [1].

Microgrids are composed of multiple interconnected components that work together to ensure a stable and efficient power supply. These components include DERs, ESSs, loads, and a microgrid controller, as illustrated in Figure 15.1. Each of these components has a specific function that contributes to the overall performance and functionality of the microgrid [2].

[1]Qatar Environment and Energy Research Institute, Hamad Bin Khalifa University, Qatar
[2]Department of Energy Systems Engineering, Faculty of Technology, Gazi University, Turkey
[3]College of Science and Engineering, Hamad Bin Khalifa University, Qatar

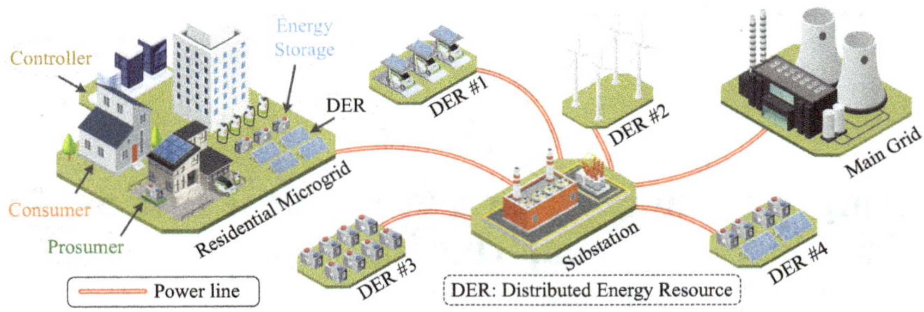

Figure 15.1 General block diagram of a microgrid connected to the upstream grid with different types of DERs

DERs are small-scale power generation sources within the microgrid that produce energy locally. This local energy production reduces the dependency on centralized power plants and transmission infrastructure, helps reduce transmission losses, and enhances overall energy efficiency.

ESSs are critical components of microgrids, they help to balance supply and demand, mitigate power fluctuations, and enhance system reliability. ESS can store excess energy produced during high-generation periods and release it during high-demand or low-generation periods.

The load component of a microgrid represents the power demand within the microgrid. Loads can be classified as critical, non-critical, or deferrable loads. Critical Loads are essential loads that must be supplied continuously, even during power outages such as hospitals, data centers, and emergency systems. Conversely, the non-critical loads are the loads that can be interrupted or shed temporarily without significantly impacting operations. In some microgrids, non-critical loads are disconnected during high-demand or emergency situations to preserve power for critical loads. Finally, deferrable loads are loads that do not require immediate power and can be shifted to off-peak times or periods of high renewable generation such as electric vehicle charging.

The microgrid controller is the central control system that manages all components within the microgrid. It plays a crucial role in maintaining grid stability, optimizing energy usage, and ensuring reliable power supply. The controller can monitor, control, and dispatch power from DERs and ESS, as well as manage load demand, as presented in Figure 15.2. The microgrid controller is responsible for achieving many tasks such as (i) energy management, (ii) grid synchronization and seamless transition, (iii) load shedding and demand response, and (iv) fault detection and protection [1].

Finally, a robust communication and control infrastructure is essential for microgrids, as it enables effective coordination and monitoring of all components. Communication networks connect sensors, controllers, DERs, ESS, and other devices. Modern microgrids often use advanced communication protocols like Internet of Things (IoT) technology, which allows remote monitoring, predictive maintenance, and online data acquisition, monitoring, and analytics [1,3].

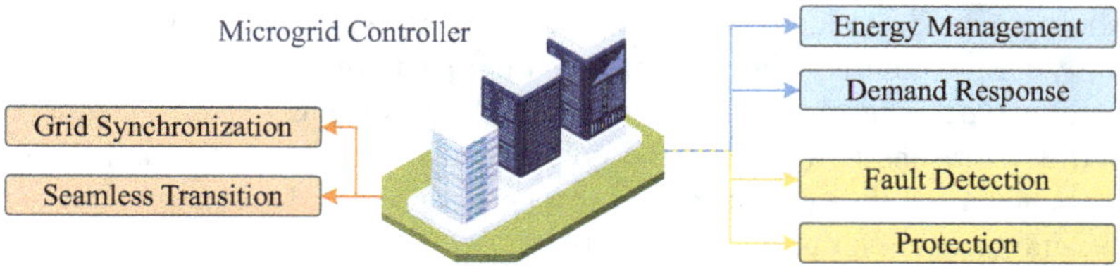

Figure 15.2 Microgrid controller tasks

15.1.2 The role of microgrids in modern energy systems

Microgrids are recognized as critical components of modern energy systems. As society faces challenges like rising energy demand, grid vulnerabilities, the need for resilience, and a shift toward cleaner energy sources, microgrids offer a versatile and sustainable solution. Microgrids contribute to energy decentralization, efficiency, and reliability by utilizing DERs and ESSs. These systems can operate independently or in concert with the main power grid. Microgrids can contribute to the transformation and modernization of energy systems in several ways [2],

- Enhancing energy resilience and reliability.
- Supporting the integration of RESs.
- Promoting decentralization and grid modernization.
- Enabling energy access in remote and underserved areas.
- Supporting grid stability and reliability.
- Enabling EV integration and smart charging solutions.
- Economic benefits and cost savings.

Energy resilience and reliability have become a priority as extreme weather events, natural disasters, and cyber threats increasingly jeopardize the stability of the main power grid. As microgrids have the ability to operate independently of the main grid, they offer continuous power to critical loads or essential services. In addition, by incorporating both renewable and conventional power sources, microgrids can maintain power during grid outages or even intentionally operate in islanded mode if the power quality of the main grid deteriorates. The resilience of microgrids is further enhanced by their ability to adapt to changing conditions and manage localized energy sources and storage. They can switch from grid-connected to islanded mode quickly when disruptions occur, a feature that allows uninterrupted operation in various scenarios. As a result, microgrids are increasingly being used in regions prone to power outages and where grid restoration may take time. Their flexibility and autonomy help reduce dependency on centralized grid infrastructure, which enhances the resilience of energy supply chains.

Conversely, traditional grids often face challenges when incorporating large amounts of variable renewable energy due to the intermittency and variability of RES output. Microgrids, however, are equipped with ESSs and advanced energy management technologies that can balance the generation and consumption of renewable energy locally. These technologies also reduce stress and strain on the main grid and improve overall efficiency.

Microgrids also reduce the transmission losses associated with long-distance electricity transportation. This happens by the means of on-site generation where the power can be generated close to where it will be consumed. Such feature makes microgrids more energy-efficient and cost-effective. Furthermore, microgrids can help maintain grid stability by storing excess energy produced during peak generation periods and releasing it during periods of high demand.

In addition, microgrids represent a shift away from the traditional centralized power grid model and offer a distributed modular approach to the energy supply. In a decentralized grid, energy is generated and consumed locally, which can significantly reduce the need for extensive transmission and distribution infrastructure. Decentralization also allows for a more flexible and resilient energy system, as failures in one part of the grid are less likely to affect other areas. Microgrids thus provide a scalable solution to energy supply, which allows consumer communities to manage their own energy needs while reducing reliance on centralized utilities.

Moreover, microgrids provide a practical solution for electrification in regions where access to a centralized power grid is limited or non-existent. For remote areas and islands, extending traditional grid infrastructure can be financially and logistically challenging. Microgrids offer a decentralized approach that can bring reliable, clean energy to these areas at a lower cost and in a shorter time frame. In addition, renewable energy-powered microgrids can reduce or eliminate the need for costly fuel imports in remote locations. For example, microgrids that rely on solar, wind, or small hydropower systems can provide sustainable energy without the operational expenses associated with fossil fuel generation. This approach is not only more sustainable but also enhances energy independence and security, and empowers the local communities to manage their own energy resources.

In terms of grid stability, microgrids can support overall grid stability by providing ancillary services, such as frequency regulation, voltage support, and reactive power management. When microgrids operate in grid-connected mode, they can participate in demand response programs, shifting load demand away from peak hours and helping to balance the grid in real-time. This load-shifting capability is particularly valuable in reducing the stress on the grid during high-demand periods. This prevents blackouts, enhances the stability of the grid, and enhances the overall power quality. Additionally, microgrids with energy storage systems can act as virtual power plants (VPPs), aggregating and dispatching power from multiple DERs. These VPPs can participate in energy markets, supplying additional capacity when required and enhancing the grid's flexibility. In times of sudden grid fluctuations, the distributed nature of microgrids and their rapid response capabilities contribute to maintaining grid frequency and reducing the risk of cascading failures.

Furthermore, microgrids provide an effective platform for EV integration. Microgrids can manage EV charging loads intelligently, minimizing peak demand on the grid, and optimizing charging based on renewable generation availability. Through vehicle-to-grid (V2G) technology, EVs can even serve as distributed energy storage resources within microgrids, and discharge the energy back to the grid or to local loads when needed. Microgrids equipped with EV infrastructure can play a crucial role in supporting the electrification of transportation by offering smart charging solutions. This helps to manage the additional load from EVs while supporting the decarbonization of both the power and transportation sectors. Moreover, by coordinating EV charging within microgrids, operators can balance supply and demand more effectively [4].

In terms of economic benefits, microgrids offer economic advantages to consumers, operators, and local communities. Microgrids can reduce energy costs associated with transmission and distribution losses by generating power on-site. Moreover, they reduce the need for expensive fuel imports or large infrastructure investments by enabling the use of local renewable resources. Businesses and industries can benefit from reduced energy costs, greater control over their energy usage, and lower emissions. Additionally, microgrids can take advantage of various market mechanisms, such as demand response programs and time-of-use pricing, to further optimize energy costs. In cases of high electricity prices from the main grid, microgrids can disconnect and operate in islanded mode [5].

15.1.3 Challenges faced by traditional grid systems and microgrid opportunities

Traditional grid systems have served as the primary means of power distribution for over a century. They were designed to supply electricity to large areas, from urban centers to rural communities based on centralized power generation, extensive transmission networks, and distribution systems. However, as energy demand rises and environmental threats intensify, traditional grids are becoming increasingly vulnerable to various challenges. These challenges range from natural disasters and cyber threats to energy inefficiency and the complexities of integrating RESs. As microgrids are decentralized, flexible, and have resilient design, they offer innovative solutions to these challenges and enable modern energy systems to be more adaptable, reliable, and sustainable.

Traditional grids are highly susceptible to extreme weather events, including hurricanes, wildfires, floods, and winter storms. Severe weather can damage power lines, substations, and transformers, which leads to extensive outages that disrupt entire regions and may take days or weeks to repair. For instance, hurricanes have caused widespread blackouts in coastal areas, while wildfires have caused preemptive shutdowns to prevent fire hazards. With climate change increasing the frequency and intensity of weather events, the grid's resilience is constantly under strain. As microgrids can operate in islanded mode, they can disconnect from the main grid to avoid widespread outages, and keep supplying power to local loads using their own energy resources and storage systems. For example, a hospital or community with a microgrid powered by solar panels and battery storage can remain operational during a hurricane or wildfire, even if the main grid is down. Microgrids can also be deployed in modular configurations, allowing them to be easily scaled or relocated based on changing environmental conditions [2].

Another challenge that faces the traditional grid is its centralized nature. The centralized nature makes them vulnerable to single points of failure. A failure at a large power plant or substation can cause cascading failures throughout the grid, which leads to blackouts over vast areas. Additionally, centralized grids are limited in their ability to adapt to sudden changes in power supply and demand. Therefore, it is difficult to accommodate localized disruptions or varying energy needs by the traditional grid. On the other hand, microgrids reduce the dependency on single, centralized sources. This reduces the risk and possibility of widespread outages. If a component within a microgrid fails, it is more likely to affect only a localized area, while disruptions from cascading across a larger region can be prevented. Furthermore, microgrids can incorporate multiple sources of power, such as solar, wind, and backup generators, which enhances their ability to supply power reliably even if one source fails.

Another critical challenge that threatens the traditional grids is cyberattacks. Cybersecurity threats can compromise critical infrastructure, disrupt grid operations, or even cause widespread outages. Hackers may target grid control systems, substations, or communication networks. This leads to potential severe disruptions that could impact power availability and public safety. Conversely, microgrids can be designed with modern cybersecurity measures to protect critical assets from cyber threats. As they are smaller and more localized than traditional grids, microgrids can implement targeted, robust cybersecurity protocols to secure their systems and prevent unauthorized access. Moreover, if a cyberattack occurs within a microgrid, its limited scope reduces the potential impact on the larger grid. Therefore, it is easier to isolate, contain, and address the disruption.

However, to unlock the full potential of microgrids, advanced algorithms are essential. These algorithms enable microgrids to manage complex operations, such as balancing distributed energy resources, optimizing energy storage, and dynamically responding to shifts in demand and supply. With the ability to make real-time decisions, forecast energy production, and adjust for peak usage patterns, microgrids can achieve the adaptability, efficiency, and resilience that modern energy systems require. Artificial intelligence (AI) plays a critical role in this process by providing the computational power and analytical capabilities needed to handle vast data streams, predict energy trends, and enable autonomous decision-making within microgrids. Through machine learning (ML), predictive analytics, and intelligent control, AI enhances the efficiency, reliability, and flexibility of microgrids, making them indispensable for their successful operation in dynamic environments.

15.2 Introduction to AI applications in energy systems

15.2.1 Challenges faced by traditional grid systems and microgrid opportunities

AI has emerged as a transformative force in modern energy systems. AI offers advanced tools for managing and optimizing the increasingly complex demands of power generation, distribution, and consumption. AI has become crucial in numerous aspects of energy systems, especially within microgrids, due to its capability to analyze vast amounts of data, make predictive assessments, and support decision-making in real-time. Moreover, AI technologies play an essential role in achieving crucial goals in microgrids including intelligent, adaptable, and efficient control strategies to operate autonomously or in conjunction with larger grids.

In the context of microgrid operations, AI serves as a cornerstone technology that enables multiple crucial functions, as illustrated in Figure 15.3. The diagram

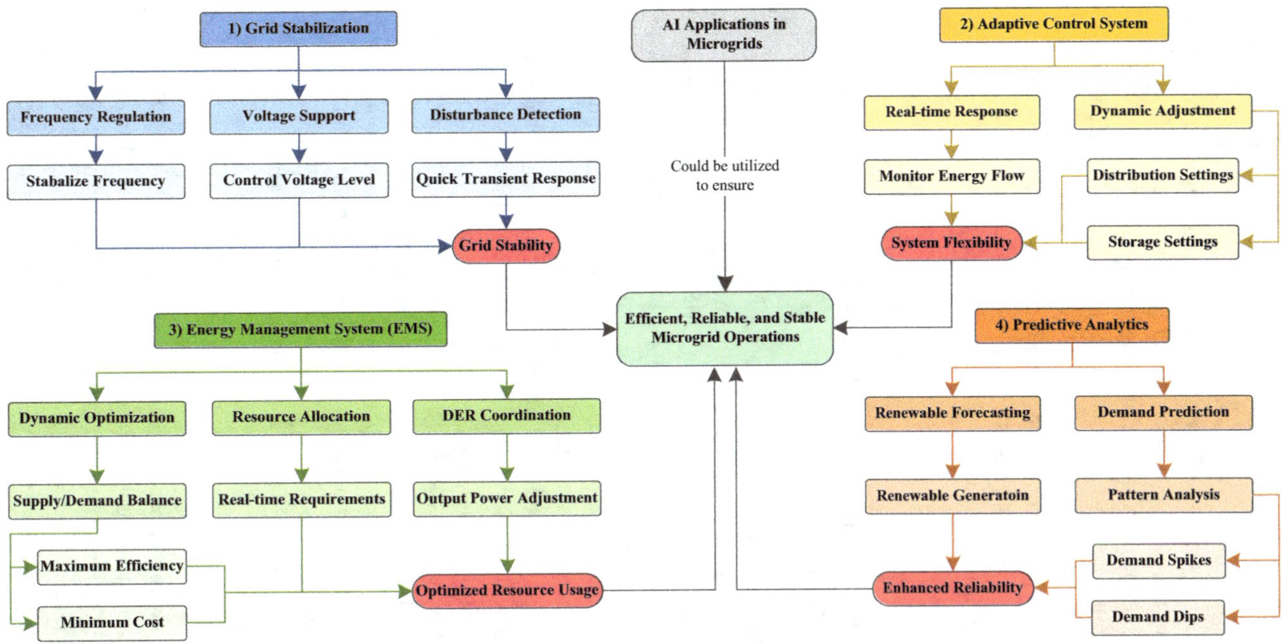

Figure 15.3 Potential AI integrations within microgrids

showcases four primary applications of AI in microgrids: EMS, Predictive Analytics, Adaptive Control Systems, and Grid Stabilization. Through EMS, AI orchestrates dynamic resource allocation and coordinates DERs to optimize supply–demand balance, while predictive analytics capabilities forecast demand patterns and renewable energy availability to enhance system reliability [6]. The adaptive control systems provide real-time responsiveness by dynamically adjusting distribution and storage settings, ensuring operational efficiency and load balance [7]. Finally, AI's role in grid stabilization is paramount, maintaining consistent frequency and voltage levels despite intermittent loads and renewable inputs [8]. All these functions work in concert to achieve the goal of efficient, reliable, and stable microgrid operations, significantly reducing dependency on conventional backup power sources while minimizing the risk of system overloads through sophisticated algorithms and real-time monitoring capabilities [9].

15.3 AI-driven energy management systems (EMSs) for microgrids

AI-driven EMSs have emerged as transformative solutions for managing complex and dynamic microgrid environments. Unlike traditional EMS, which relies on predefined, static control strategies, AI-driven EMS leverages advanced computational intelligence to optimize energy distribution, handle unpredictable RESs, and respond in real-time to fluctuating demands. These systems apply machine ML [10], deep learning (DL) [11], and reinforcement learning (RL) [12] techniques to coordinate DERs, optimize storage, and maintain grid stability.

AI-driven EMSs use intelligent algorithms to dynamically balance energy supply and demand within microgrids. Microgrids often incorporate diverse energy sources, including solar, wind, and battery storage, which require flexible and adaptive management to ensure consistent power availability. AI enables microgrids to make data-driven decisions that improve energy efficiency, reduce costs, and support sustainability. By continuously learning from real-time data, AI-driven EMS can enhance the microgrid's operational resilience, economic efficiency, and ability to integrate renewables [13].

Therefore, AI-driven EMS is designed to achieve several key objectives within microgrids, including [14],

- Optimizing energy allocation and dispatch: AI-driven EMS focuses on determining the optimal schedule for energy generation and distribution, balancing different DERs to meet demand in the most cost-effective manner. This includes minimizing reliance on fossil fuels and prioritizing renewable energies.
- Minimizing operational costs: One primary goal of an AI-driven EMS is to reduce operational costs by optimizing energy harvesting, storage utilization, and load management. AI-driven EMS can dynamically shift loads based on energy pricing and demand response programs, reducing peak-time costs.

- Enhancing reliability and resilience: AI-driven EMS provides reliable power by predicting energy demand and supply accurately. This ensures consistent delivery even during unexpected disruptions. This feature is especially critical for microgrids operating in areas prone to outages or in islanded operations.
- Supporting demand-side management (DSM) and load forecasting: AI-driven EMS can implement DSM strategies to shift or reduce loads, which supports stabilizing the grid during peak periods.
- Maximizing renewable energy utilization: AI-driven EMS optimizes the use of RES by forecasting energy production and aligning it with load requirements. This minimizes dependency on conventional backup systems.

Based on the aforementioned objectives, the EMS objective functions are formulated to quantitatively express the goals of the system. These objective functions are then minimized or maximized to achieve optimal performance. One of the most common objectives is to minimize the cost of the system, and this can be represented as follows:

$$Minimize\ C_{total} = C_{generation} + C_{storage} + C_{operation} \tag{15.1}$$

where C_{total} is the total cost of the system, $C_{storage}$ is the cost of storage, $C_{generation}$ is the cost of generating the energy, and $C_{operation}$ is the operational cost.

The generation cost represents the cost associated with generating electricity from various DERs, including both renewable and non-renewable sources, and can be represented by

$$C_{generation} = \sum_{i=1}^{N} P_i G_i + F_i G_i \tag{15.2}$$

where P_i is the price per unit energy, G_i is the power generated by the $i^t h$ DER, F_i is the fuel cost, and N is the number of DERs.

The storage cost represents the expenses related to energy storage activities, including charging, discharging, and maintenance costs associated with battery or other storage systems. This can be represented mathematically as

$$C_{storage} = \sum_{j=1}^{k} [(S_c Cc)^j (S_d C_d)^j] + M_{storage} \tag{15.3}$$

where S_c and S_d represent the energy charged to and discharged from storage unit j, C_c and C_d are the charging and discharging costs per unit of storage, and $M_{storage}$ is the maintenance cost for all storage units, and M is the total number of storage units.

Finally, operational costs include routine maintenance, inspections, and operational wear on the microgrid's components, and labor costs. This can be represented mathematically as

$$C_{operation} = \sum_{k=1}^{K} C_{maintenance,k} + C_{labor} \tag{15.4}$$

Another common cost function is related to minimize the fluctuation in both voltage and frequency profiles to enhance the stability of the microgrid. This objective can be represented as

$$Minimize\ C_{ripple} = W_v C_{vr} + W_f C_{fr} + C_{adj} \tag{15.5}$$

where C_{vr} and C_{fr} are the individual costs associated with voltage and frequency ripples, respectively, W_v and W_f are weighting factors that represent the importance of minimizing voltage and frequency ripples, which can be adjusted based on system requirements, and C_{adj} accounts for costs related to control adjustments needed to minimize ripples, such as component adjustments or switching operations.

Voltage and frequency ripple costs are calculated based on the deviation of the microgrid's voltage and frequency from their nominal values (V_{nom}, f_{nom}) as follows:

$$C_{vr} = \int_0^T \left(\frac{V(t) - V_{nom}}{V_{nom}}\right)^2 dt \tag{15.6}$$

$$C_{fr} = \int_0^T \left(\frac{f(t) - f_{nom}}{f_{nom}}\right)^2 dt \tag{15.7}$$

where $V(t)$ and $f(t)$ denotes the instantaneous measurements of grid voltage and frequency, respectively.

The adjustments costs include the costs adjust settings related to DERs, energy storage systems, and load distribution. This can be represented as

$$C_{adj} = \sum_{a=1}^{N} C_{switch,i} + C_{ramp,i} \tag{15.8}$$

where $C_{switch,i}$ represents the cost of switching on/off DERs, switching loads, or storage systems, and $C_{ramp,i}$ is the cost associated with ramping up or down power from sources or storage units to be injected to the grid.

15.3.1 Predictive analytics and forecasting of energy demand and RES supply

The integration of RESs into microgrids introduces variability and uncertainty in power generation, as the output of any RES depends on environmental conditions like sunlight and wind. Similarly, predicting energy demand in real-time is complex due to fluctuating usage patterns, particularly in dynamic environments. AI-based EMSs address these challenges through predictive analytics and forecasting, to enable the microgrids to optimize energy supply, improve stability, and manage costs effectively. AI-based EMS can accurately forecast both energy demand and renewable generation by employing advanced ML, DL, and other AI-driven approaches. With high-quality predictions, microgrids can optimize energy dispatch, enhance load balancing, improve storage management, and reduce operational costs [7].

Predictive analytics and forecasting AI-based EMS typically involve several core components and methodologies to manage both energy demand and RES supply predictions effectively. These components include data collection, preprocessing, model

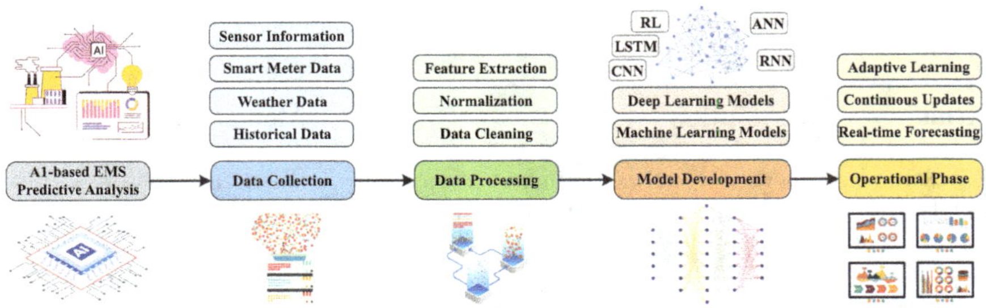

Figure 15.4 Areas for AI applications in the EMS of mirogrids

training, and real-time forecasting. Data collection and preprocessing predictive analytics require a significant amount of data, including historical load profiles, weather conditions, solar irradiance, wind speed, and occupancy patterns. Data from sensors, smart meters, and weather databases are aggregated, cleaned, and normalized to ensure consistency, as presented in Figure 15.4. Preprocessing techniques handle outliers, fill missing values, and transform data into formats suitable for training AI models [15].

Relevant features are then identified to improve prediction accuracy. For instance, in demand forecasting, features like temperature, time of day, and day of the week can influence consumption patterns, while for RES forecasting, variables such as cloud cover, wind speed, and humidity play a role. Feature engineering techniques further transform raw data into useful attributes for predictive models. Then, ML models like linear regression, decision trees, support vector machines, and ensemble methods (e.g., random forests and gradient boosting) are commonly used in demand forecasting. These models learn patterns in historical data, which helps to predict demand based on time, weather, and other variables. AI-driven EMS can also apply clustering techniques to segment similar usage patterns, which enhances the accuracy of demand predictions across different types of loads [16].

DL models, particularly recurrent neural networks (RNN) and long short-term memory (LSTM) networks, excel in time-series forecasting and are widely used for RES forecasting [17]. LSTM networks are effective for capturing temporal dependencies, which is useful for modeling fluctuating renewable energy generation based on past trends and weather conditions. Convolutional neural networks (CNNs) are also applied in cases where image data, such as cloud cover patterns from satellite imagery, are used to predict solar generation.

Once the training phase is done and the operation model is exported, AI-based EMS starts the forecasting process while continuously updating its predictions as new data becomes available, which allows it to refine forecasts in real time. Adaptive learning techniques enable models to adjust to evolving patterns in demand and supply, particularly useful in environments with seasonal variations. For example, RL allows the EMS to learn from past forecasting errors and to improve the accuracy iteratively. Table 15.1 summarizes the most common techniques used in AI-based demand and supply forecasting.

AI-based predictive analytics in EMS offers significant benefits, including enhanced efficiency through accurate dispatch of DERs, which reduces both energy

Table 15.1 Common techniques used in demand and supply forecasting

Method	Description	Algorithms	Ref.
Regression models	Used for baseline demand forecasting and identifying trends in past usage data.	Linear regression and support vector regression	[18]
Time-series analysis	Useful in forecasting demand and generation, especially at the short-term level.	ARIMA, ML models for nonlinear patterns	[19]
Recurrent neural networks	Frequently used in forecasting applications due to their ability to handle sequential data. LSTMs are effective for RES supply forecasting by capturing time-dependent variations in renewable output, such as the daily solar irradiance cycle.	RNN and LSTM	[16]
RL	Adjust predictions based on feedback. RL enables EMS to optimize energy dispatch based on updated forecasts, learning from errors in previous forecasts and improving over time.	Q-Learning, Deep Q-Network, and SARSA	[12]
Ensemble learning methods	Combine multiple models to improve forecasting accuracy and reduce the impact of individual model errors. Ensemble methods are effective for capturing diverse demand patterns across different user types.	Random forests, gradient boosting, and ensemble learning	[20]

waste and operational costs. The capability to forecast RESs availability enables the EMS to increase renewable integration to align with sustainability goals. Predictive analytics also supports grid stability by balancing load and managing frequency, thereby reduces outage risks. Additionally, accurate demand and supply forecasting decreases reliance on fossil fuels, as it reduces the need for backup generators, which leads to a smaller carbon footprint.

However, AI-based predictive analytics in EMS faces multiple challenges. High-quality data is crucial, and poor data quality can lower forecasting accuracy, which limits the system's effectiveness. Some AI models, particularly those using DL, require substantial computational power, which may pose difficulties in real-time deployment. Furthermore, RES predictions are highly sensitive to environmental changes; this makes unexpected weather conditions a potential disruptor to forecast accuracy.

15.3.2 Adaptive control systems for efficient energy distribution

Microgrids typically integrate a diverse mix of energy sources, such as solar, wind, batteries, and backup generators, alongside varying load demands. This diversity and variability introduce complexities in managing energy flows efficiently, especially as renewable energy sources are inherently intermittent. Traditional control systems often lack the flexibility to adjust to rapid changes in RES availability or fluctuating demand. Furthermore, conventional systems may require manual intervention to accommodate these changes, leading to inefficiencies and potentially suboptimal energy distribution.

AI-based adaptive control systems address these challenges by learning and adapting in real-time. They can analyze large amounts of operational data, predict fluctuations in demand and supply, and adjust energy dispatch to maintain stability and efficiency. These systems become increasingly accurate at responding to microgrid conditions through continuous learning. This allows them to autonomously adapt to seasonal changes, unexpected load spikes, or weather-related variability in RES. This adaptability is particularly valuable in microgrids operating in isolated or off-grid environments, where efficient energy distribution is critical to prevent power shortages or overloading.

Many AI-based adaptive control techniques were recently proposed in the literature as shown in Table 15.2. These methods include RL [21], fuzzy logic [22], model predictive control (MPC) [23], multi-agent systems (MAS) [24], and DL [25]. Among them, MPC is the most widely used control strategy in microgrids due to its ability to handle multi-variable systems with constraints, predict system behavior, and optimize control actions over a future horizon. This makes MPC particularly suitable for the complex, dynamic environment of microgrids, which involve multiple energy sources, variable demand, and the integration of RESs. In this section, a detailed case study for utilizing AI-based MPC approach for frequency and voltage regulation in microgrids will be illustrated.

The microgrid system's dynamics for frequency and voltage regulation can be represented in a state-space model as follows:

$$\begin{aligned} x(t+1) &= Ax(t) + Bu(t) + w(t) \\ y(t) &= Cx(t) + v(t) \end{aligned} \tag{15.9}$$

where $x(t)$ represents the state vector, $u(t)$ represents the control input vector, $w(t)$ represents external disturbances like changes in demand or renewable generation, and $y(t)$ is the output vector.

Table 15.2 Parameter and initial values

Parameter	Value	Parameter	Value
EV rated power	30 kW*3	AFR rated power	30 kW*3
PV rated power	50 kW	Grid voltage	220 V
Storage power	50 kW	DC link voltage	600 V^3

As the microgrid is mainly consisting of many DERs interfaced through DC–AC converters, most of the dynamic model of a microgrid is related to the model of the attached AC–DC converters. Therefore, a detailed model for LCL three-phase inverter is described in this section.The output voltage vector (v_{iabc}) of each inverter is described as

$$v_{i,abc} = SV_{dc} \tag{15.10}$$

where $S \in [1, 0, -1]$, and V_{dc} is the inverter input voltage. The input current (to the LCL filter) dynamics can be described as follows:

$$\frac{di_{i,\alpha\beta}}{dt} = -\frac{R_i}{L_i}i_{i,\alpha\beta} - \frac{1}{L_i}v_c + \frac{1}{L_i}v_{i,\alpha\beta} \tag{15.11}$$

and the capacitor's voltages as

$$\frac{dv_{c,\alpha\beta}}{dt} = -\frac{1}{C}i_{i,\alpha\beta} - \frac{1}{C}i_{o,\alpha\beta} \tag{15.12}$$

while the output current (from the LCL filter) as

$$\frac{di_{o,\alpha\beta}}{dt} = -\frac{R_o}{L_o}i_{o,\alpha\beta} + \frac{1}{L_o}v_c - \frac{1}{L_i}v_{o,\alpha\beta} \tag{15.13}$$

where $i_{i,\alpha\beta}$ is the LCL filter input current, $v_{c,\alpha\beta}$ is the voltage across the capacitors of the filter, $i_{o,\alpha\beta}$ the is output current of the filter, and $v_{o,\alpha\beta}$ is the output voltage of the filter. Based on (15.11) - (15.13), the dynamic model can be represented in state space representation as follows:

$$\begin{bmatrix} \frac{di_{i,\alpha\beta}}{dt} \\ \frac{dv_{c,\alpha\beta}}{dt} \\ \frac{di_{o,\alpha\beta}}{dt} \end{bmatrix} = A \begin{bmatrix} i_{i,\alpha\beta}dt \\ v_{c,\alpha\beta}dt \\ i_{o,\alpha\beta}dt \end{bmatrix} + Bv_{i,\alpha\beta} + Cv_{o,\alpha\beta} \tag{15.14}$$

where

$$A = \begin{bmatrix} \frac{-R_i}{L_i} & \frac{-1}{L_i} & 0 \\ \frac{1}{C} & 0 & \frac{-1}{C} \\ 0 & \frac{1}{L_o} & \frac{-1}{L-O} \end{bmatrix}, \; B = \begin{bmatrix} \frac{1}{L_i} \\ 0 \\ 0 \end{bmatrix}, \; C = \begin{bmatrix} 0 \\ 0 \\ \frac{-1}{L_o} \end{bmatrix} \tag{15.15}$$

For control objectives, 15.16 can be discretized as follows:

$$\begin{bmatrix} i_{i,\alpha\beta}(k+1) \\ v_{c,\alpha\beta}(k+1) \\ i_{o,\alpha\beta}(k+1) \end{bmatrix} = A_k \begin{bmatrix} i_{i,\alpha\beta}(k) \\ v_{c,\alpha\beta}(k) \\ i_{o,\alpha\beta}(k) \end{bmatrix} + B_k v_{i,\alpha\beta}(k) + Cv_{o,\alpha\beta}(k) \tag{15.16}$$

where

$$\begin{aligned} A_k &= I_{3x3} + T_s A \\ B_k &= T_s B \end{aligned} \tag{15.17}$$

After defining the model, a comprehensive cost function that includes voltage and frequency regulation objectives should be defined,

$$Minimize\ J = \sum_{k=0}^{T} W_f(f(k) - f_{nom})^2 + W_v(V(k) - V_{nom})^2 + W_u(u_k)^2 \quad (15.18)$$

where W_f, W_v, and W_u are the weighting factor that prioritize frequency and voltage stability while minimizing control effort.

The control of a microgrid is usually exposed to multiple constraints to ensure operation feasibility, such as power balance constraints, voltage and frequency limits, and batteries state of charge (SoC) limits. These constraints can be represented mathematically as

$$\begin{aligned} &\sum_i G_i(t) + S_{discharge}(t) = D(t) + S_{charge}(t) \\ &V_{min} \leq V_{max}\ and\ f_{min} \leq f(k) \leq f_{max} \\ &SoC_{min} \leq SoC(t) \leq SoC_{max} \end{aligned} \quad (15.19)$$

where $G_{i(t)}$ is the generation from *ith* DER, S_{charge} and $S_{discharge}$ are the battery charge and discharge rates, respectively, and $D(t)$ is the instant demand.

Although the basic dynamic model can be easily established, the mathematical mode is highly exposed to uncertainties and external disturbances like changes in demand or renewable generation. Therefore, AI models (e.g., LSTM networks for time-series data) can be trained to predict these disturbances with high accuracy based on past data, helping to improve the MPC's forecasting capability. Finally, AI-based algorithms (such as deep RL or neural network-based solvers) can be used to handle the objective function (15.18) while adhering to the constraints (15.19).

15.4 Enhancing microgrid resilience through AI

Enhancing microgrid resilience through AI involves advanced techniques for maintaining stability, security, and functionality, especially under challenging conditions. AI enables rapid fault detection and system restoration, identifying anomalies in voltage, frequency, or load patterns that signal potential faults. By predicting and identifying these issues early, AI-driven systems can initiate preemptive adjustments or repairs and minimize disruptions. During outages, AI facilitates autonomous decision-making for grid reconfiguration, which enables microgrids to disconnect from the main grid and optimize local resources to maintain power.

Additionally, AI-powered security measures protect against cyberattacks by monitoring network traffic, identifying abnormal patterns, and blocking unauthorized access, which is crucial as microgrids become more digitally connected. In disaster scenarios, AI supports response and recovery by prioritizing power restoration to critical infrastructure, such as hospitals and emergency facilities, ensuring that essential services remain operational. Through these capabilities, AI significantly enhances microgrid resilience, ensuring reliable and secure energy supply during both routine operations and emergencies. In this section, a detailed case study is illustrated

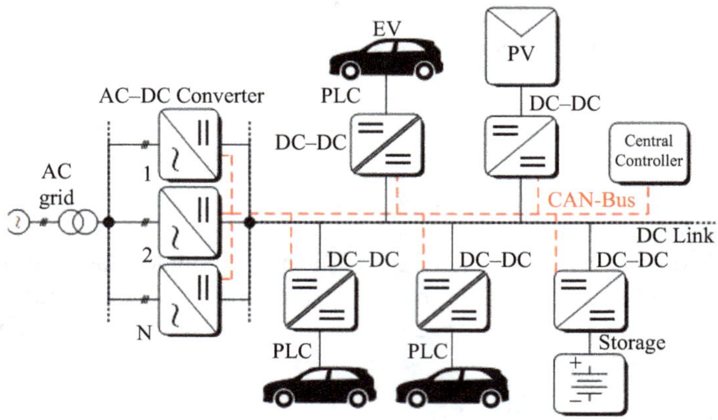

Figure 15.5 Block diagram of the adopted microgrid

about how AI and forecasting can be utilized to enhance the security and reliability of a microgrid consisting of EVs, PVs, storage systems, and multiple converters as shown in Figure 15.5 [26].

15.4.1 Communication-based control

To better understand the role of communication in stability and reliability of the charging system, it is worth describing the communication-based control technique. In this setup, the PV is always acting as a source, while each EV and the storage system may act as a source or a load. When serving as a source, the available power is positive, whereas it becomes negative when the battery is being charged. Therefore, the total required and available DC power is computed in the central controller as follows:

$$\begin{aligned} P_{av} = {} & sign(P_{ev1})P_{ev1} + sign(P_{ev2})P_{ev2} \\ & + sign(P_{ev3})P_{ev3} + sign(P_s)P_s + P_{pv} \end{aligned} \tag{15.20}$$

$$\begin{aligned} P_{re} = {} & (1 - sign(P_{ev1}))P_{ev1} + (1 - sign(P_{ev2}))P_{ev2} \\ & + (1 - sign(P_{ev3}))P_{ev3} + (1 - sign(P_s))P_s + P_{pv} \end{aligned} \tag{15.21}$$

where P_{av} is the total available power, P_{re} is the total required power, P_{ev} is the available power in the i-th EV, P_{pv} is the available power from PV, P_s is the available power in the storage system, $sign(x) = 1 ifx > 0, sign(x) = 0 ifx < 0$.

Each controller periodically sends its power to the central controller, which in turn computes the power-sharing ratio for the contributing devices (i.e. devices with positive power) based on their power ratings. Moreover, if $P_{av} < P_{re}$, the central controller will send a reference signal to the active front-end rectifiers (AFRs) to compensate the required power, or to consume the extra power in case of $P_{av} > P_{re}$.

If any device is exposed to attack, the required and available powers will not be computed correctly. This deteriorates the response of the entire system. In the first hand, it deteriorates the regulation of the DC link voltage due to imbalance between injected and consumed power. Conversely, it will increase the communication and computation time due to receiving more packets than expected leading to critical instability issues. Moreover, this problem leads to computing wrong power-sharing ratios and reference signals, which may lead some devices to inject power more than their ratings, leading to faster degradation, and exposing these converters to possible fatigue failures.

15.5 Types of attacks and their detection method

In general, there are numerous possible types of cyber-physical attacks on this infrastructure. One of the most common attacks is the reply attack. In this kind, the attacker captures a packet and keeps sending it over the communication link. One simple solution to detect such attacks is to check the periodic sampling of transmission (PST). If the PST of one of the devices is less than the adjusted periodic interval (API), this means there is an attacker who is performing replay attacks. This solution can be easily implemented by adding a unique ID for each device before sending its frame. Similarly, each device is expected to receive a periodic reference signal from the centralized controller. If the PST of the centralized controller is less than the API, all devices will recognize that there is a reply-attack occurring on the centralized controller level.

False data injection is another critical attack that deteriorates the response of the charging infrastructure. In this attack, the attacker sends false reference signals or false measurements leading to deteriorate the process of DC link voltage regulation and the control process efficiency. As the charging infrastructure is equipped with three AFRs, which are considered the most powerful converters in regulating the DC link voltage, it is expected to have an almost constant DC link voltage all the time. Therefore, if the DC link voltage has an error greater than a specific threshold (V_{dcTH}), the central controller will include that there is an attack. If the DC link voltage is not affected, the central controller estimates the available power based on the SoC of the EVs and the storage system periodically. Then, compares them with the received SoC from each charging controller. SoC can be estimated using Coulomb counting as follows [27].

Once the EV is connected to the charger, the EV sends all rated parameters including the rated power, and rated voltage. Therefore, even if the charging/discharging current rate is different than $2C_r$, it can be updated online using estimation techniques like recursive least square method [28] based on the provided data through the communication and the physical measurements. Then, the error in SoC estimation (ΔSoC), and power, estimation error ΔP_{av} is given by

$$\Delta SoC = |S\hat{o}C - SoC| \tag{15.22}$$

$$\Delta P_{av} = |\hat{P} - P| \tag{15.23}$$

Finally, (15.22) and (15.23) can be compared with a specific threshold (SoC_{TH} and P_{avTH}) to detect the presence of the attack.

Conversely, as the PV has a constant power rating, the available power from PV can be easily predicted based on the time of the day. For example, the peak generation occurs at noon, while there is no generation during night. Moreover, the power of the PV is always positive. If any received signal from the PV controller does not fit with these facts, it will be considered an attack. For more accurate and reliable diagnosis, ML approaches can be used to accurately predict the available power of the PV based on the current time and the day of the year, as discussed in the previous sections.

Sensor faults and cyber-physical attacks are also common in applications that include user interaction. Attacker may manipulate the sensor measurements to reduce the charging cost or to increase the profits of the injected energy. This kind of attacks can be identified and tolerated using state observers such as Kalman filter, state feedback observer, and sliding mode observer [8]. State observers can be used to compare the physical measurements from the sensors with estimated measurements. If the difference between the measured and estimated values is large, it can be concluded that there is an attack fault on the sensors [29].

Furthermore, another feature can be used to detect such attacks, which is the rate of change in batteries' SoC ($RoC-SoC$). It is well known that EVs have relatively high-capacity batteries, which means that the batteries require a long time to significantly change the SoC (compared to control sampling time). Therefore, if the received $RoC-SoC$ is higher than a specific threshold (usually zero), the central controller recognizes it as an attack.

Once an attack is detected, the system will switch its operation into communication-less control mode. In this case, the converters will not depend on communication for generating their reference signals. For AC–DC converters, a detailed and comprehensive technique for communication and communication-less control approaches were described in [3] with a seamless transition between each mode. For DC–DC converters, a droop control technique is adopted to achieve communication-less control.

15.5.1 Simulation results

The adopted system is implemented using MATLAB®/Simulink® to validate the proposed technique. The rated parameters of all devices are shown in Table 15.2. For page limit constraints, only false data injection results are shown in this section. It is assumed that as long as the PV and the AC side can provide the required charging power, the EVs and the storage system will not contribute to regulating the DC link voltage. Therefore, EVs and storage system still contribute when there is a drop in the DC link, or the required power is larger than 140 kW. The first test is conducted under normal operations conditions, where there are no faults or attacks.

Figure 15.6 shows the power of each device, while Figure 15.7 shows the AC side and DC link voltage response. At the beginning, there are no EVs connected, and the storage system is fully charged. Therefore, all available power comes from the PV, and it is injected into the grid. When the first EV with 30 kW rated power is connected, the available power becomes 20 kW. Therefore, the AC–DC converters

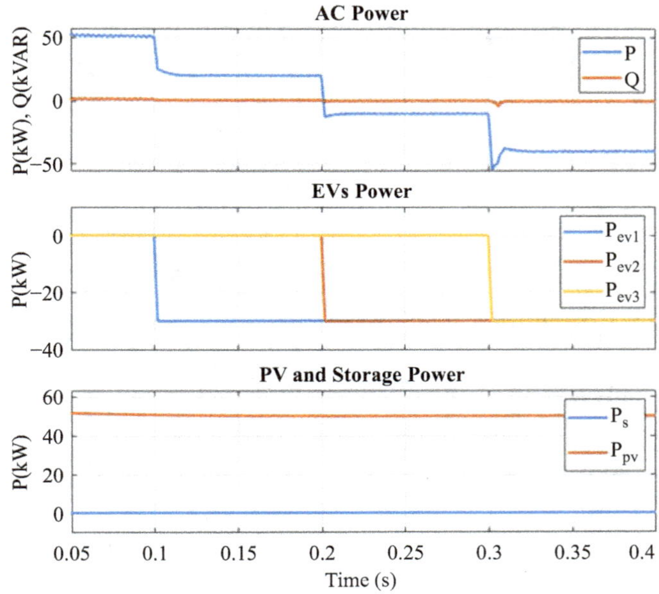

Figure 15.6 Power response during normal operation

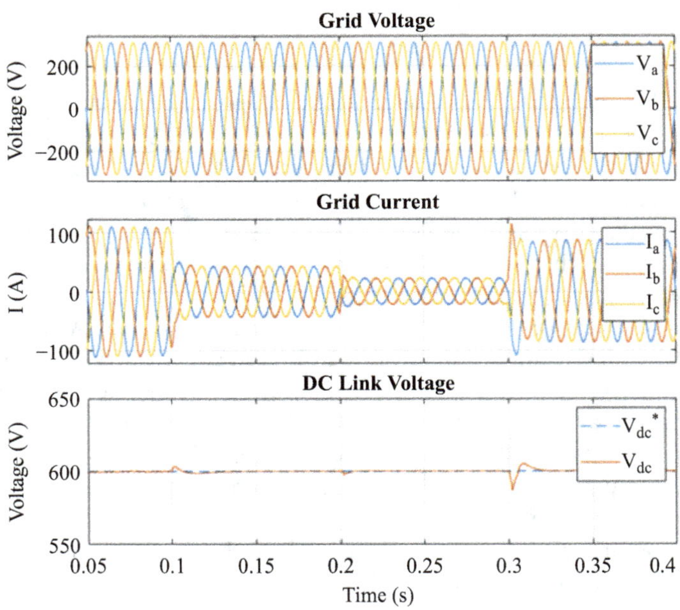

Figure 15.7 AC side and DC link voltage response during normal conditions

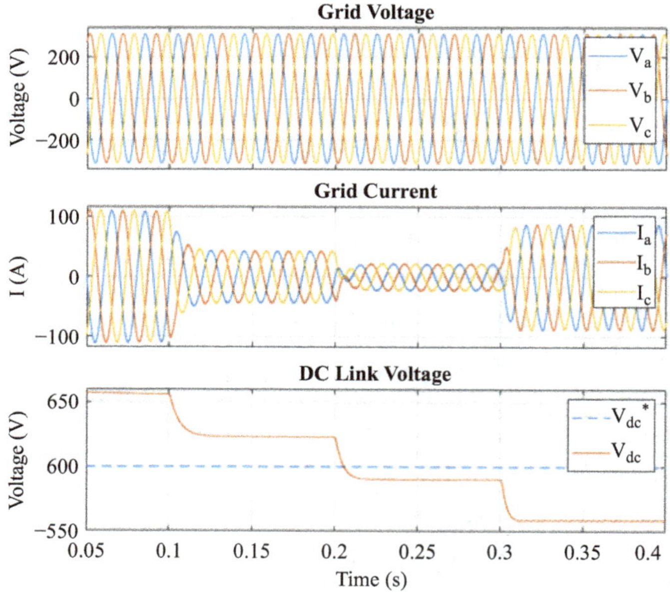

Figure 15.8 AC side and DC link voltage response during reply

still work as inverters. When the second EV is connected, the total required power becomes 60 kW, while the available is 50 kW. In this case, the AC–DC converters are switched into rectifiers to provide the remaining 10 kW. Similarly, when the third EV is connected, the required power is increased to 90 kW, and therefore, 40 kW is consumed from the grid. It can be noticed that the communication-based controller shows an excellent response, the DC link voltage is well regulated, and the voltage sag is less than 2% when a new EV is connected, the grid currents are well regulated and sinusoidal, and the THD is around than 1%.

In the next test, false-data injection is applied to the PV system. A false PV packet is sent to the central controller showing that the available PV power is 40 kW, while the correct power is 50 kW. As the storage is fully charged and there are no EVs, the only possible solution is to inject the available power into the grid. However, the injected power is less than the generated, and therefore, the DC link increased. Based on this, the computation of the available and required power will be wrong all the time, as shown in Figure 15.8.

Activation of the proposed technique leads to the detection mechanism being triggered by the substantial error and rapid rate of change of the DC link voltage, subsequently enabling communication-less control. Therefore, AC–DC converters and the storage system start injecting/ consuming power based on the droop characteristic and the deviation of the DC link voltage. Thus, the DC link voltage is restored and becomes well regulated. However, the voltage dips are increased to around 5%, as the maximum voltage deviation is set to 5% as shown in Figure 15.9.

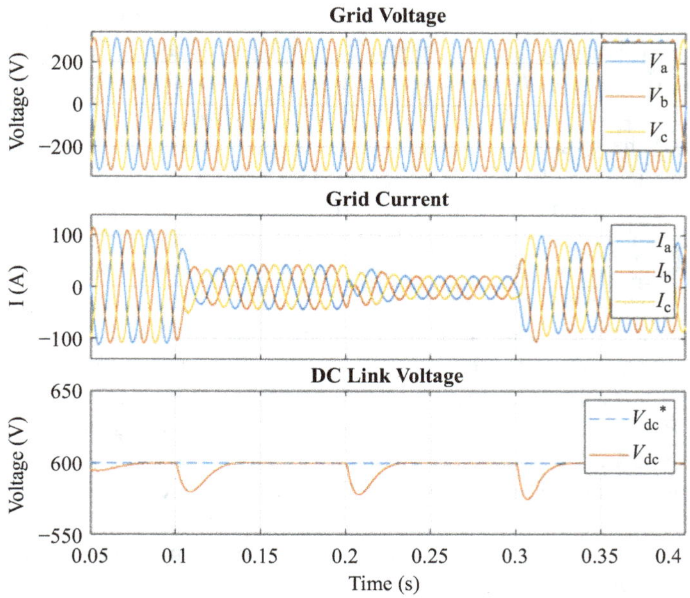

Figure 15.9 AC side and DC link voltage response during tolerant control

15.6 AI for energy storage management in microgrids

One of the primary roles of AI in energy storage management is to maximize the utilization of battery storage systems within a microgrid. AI algorithms analyze real-time data such as energy production, consumption patterns, and weather forecasts to determine the optimal charging and discharging schedules for batteries. This reduces energy losses, extends battery life, and improves overall microgrid efficiency [30].

Machine learning models, for example, can predict periods of peak energy demand and ensure that batteries are sufficiently charged to meet those needs. Conversely, during periods of low demand or high renewable energy generation, AI can prioritize charging batteries to store excess energy that would otherwise be curtailed. These adaptive strategies help balance energy flows, while reducing the dependency on grid energy and minimizing operational costs. Additionally, AI-driven optimization techniques, such as RL and metaheuristic algorithms, enable microgrids to dynamically adjust storage strategies according to the operating conditions. These systems consider variables like energy tariffs, battery degradation rates, and future load predictions to maximize the value extracted from energy storage.

Another application of AI in storage management is to balance storage and supply–demand gaps [31]. Balancing supply and demand is a cornerstone of efficient microgrid operation, and AI plays a pivotal role in mitigating the challenges posed by intermittency in renewable energy sources like solar and wind. AI systems continuously monitor energy supply and demand to identify potential imbalances and take proactive measures to bridge gaps using storage systems. Predictive analytics powered by AI can anticipate supply–demand mismatches hours or even days in advance.

For example, during a predicted drop in solar generation due to cloudy weather, AI can ensure that batteries are preemptively charged from available energy sources. Similarly, during periods of surplus generation, AI can divert excess power into storage, which prevents curtailment and ensures that the energy is available for later use.

AI-based energy storage management algorithms are also widely used to coordinate charging and discharging schedules for EVs based on real-time grid conditions, user preferences, and mobility needs [32]. For example, during peak load times, AI can discharge energy from connected EVs to supplement the grid. This reduces strain on other storage systems. Conversely, during periods of low demand or excess generation, AI ensures that EVs are charged efficiently without overloading the grid. Furthermore, AI facilitates the aggregation of multiple EVs into VPPs, allowing microgrids to treat these vehicles as a unified energy resource. Moreover, AI systems can optimize the participation of EVs in grid support activities by analyzing patterns of EV availability and charging behavior. This ensures seamless integration of V2G systems into microgrid operations.

Dynamic pricing and energy trading are also among the most common applications for AI-based energy storage management [33]. Dynamic pricing and peer-to-peer (P2P) energy trading are emerging trends in modern microgrids, and AI serves as a critical enabler of these concepts. AI algorithms can dynamically adjust energy prices to incentivize desired behaviors among consumers and prosumers by analyzing market conditions, energy availability, and demand fluctuations. For instance, during periods of high demand and limited storage, AI can increase energy prices to encourage users to reduce consumption or utilize the stored energy. Conversely, prices can be lowered during low demand and high renewable generation to promote energy usage or storage. Therefore, AI ensures an economically efficient and balanced microgrid ecosystem.

In P2P energy trading, AI facilitates the transactions between participants by determining the optimal trade matches based on energy availability, demand, and pricing preferences. Machine learning models evaluate historical trading data to predict future patterns, which enables more effective market participation. Moreover, AI-powered forecasting models enable microgrids to anticipate market trends, optimize trading strategies, and maximize financial benefits from stored energy. These capabilities allow microgrid operators to leverage energy storage as a revenue-generating asset while maintaining grid stability.

15.7 Case study: AI-driven microgrid project

Clemson University has established a cutting-edge microgrid powered by a 1 MW solar PV plant located on its campus in South Carolina. The project serves as a testbed for advanced energy management strategies, leveraging AI to optimize renewable energy integration, enhance grid resilience, and provide valuable insights into future energy systems. The microgrid employs AI algorithms to manage the intermittent nature of solar energy. These algorithms analyze real-time weather data, solar irradiance forecasts, and load patterns to dynamically adjust energy storage and

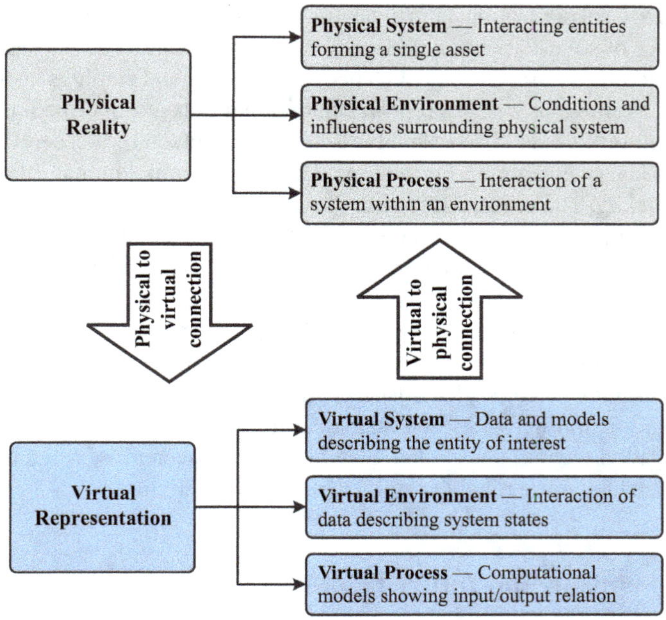

Figure 15.10 General block diagram of Clemson University's 1 MW solar PV plant DT

distribution. This ensures the efficient utilization of solar power while minimizing reliance on the external grid [34].

This project employs a digital twin (DT) framework enhanced by AI to optimize energy forecasting and management. At the core of this project is a robust methodology that integrates an ensemble of Elman ERNNs to replicate the behavior of the solar PV plant. As shown in Figure 15.10 the DT integrates

- Physical reality: The 1 MW solar PV plant with IoT-enabled sensors that collect data such as solar irradiance, temperature, and power output.
- Virtual representation: A computational model housed in Clemson's Real-Time Power Systems Laboratory, simulating the plant's performance under varying weather conditions.
- Bidirectional communication: A seamless data exchange system that ensures the DT stays updated while providing actionable insights for control and management.

The methodology begins with the collection of comprehensive data from IoT sensors deployed at the PV site. These sensors monitor critical variables such as solar irradiance, ambient temperature, and power output, generating a dataset of approximately 371,520 entries over 86 days. To improve prediction accuracy, the dataset is categorized into weather-based subgroups: clear, partially cloudy, moderately cloudy, and mostly cloudy days. This classification enables specialized training of the ERNN ensemble, with each neural network focusing on a specific weather condition. This specialization is critical for capturing the nonlinear and dynamic characteristics of solar PV output under diverse meteorological scenarios.

Each network is trained using batch backpropagation and employs historical data inputs, such as previous time-step power output and delayed intermediate weights, to predict future outputs. The ensemble size and architecture are carefully tuned, with hidden layers containing 60 neurons, balancing computational efficiency and model complexity. The model aggregates the outputs of individual ERNNs using two techniques: unweighted averaging and the "previous best" method, which selects the network with the lowest recent prediction error. These approaches allow the system to adapt dynamically to changing conditions.

After implementation, the testing of the DT under real-world conditions revealed significant accuracy, with mean absolute percentage errors as low as 1.11% for clear days. The ensemble method demonstrated superior performance compared to traditional single neural networks and statistical models like ARIMA. The "previous best" aggregation technique excelled in stable weather scenarios, while unweighted averaging provided more reliable results under high variability.

15.8 Conclusion and future trends

The integration of AI into microgrid operations is rapidly transforming the energy sector, promising a future where energy systems are increasingly efficient, decentralized, and resilient. AI and ML technologies are at the forefront of innovations aimed at optimizing energy distribution, managing renewable energy intermittencies, and facilitating the shift toward sustainable energy ecosystems. However, the path forward also presents challenges and ethical considerations that must be addressed to ensure equitable and secure energy solutions.

15.8.1 Emerging trends in AI and machine learning for energy systems

AI is revolutionizing energy systems through several emerging trends. ML models are increasingly employed to forecast energy demand, optimize storage utilization, and predict renewable energy generation based on weather patterns. Advanced techniques like RL are being adopted to develop self-learning systems capable of autonomously managing energy flows in complex microgrids.

One significant trend is the use of DTs, which are virtual replicas of physical microgrid systems. These AI-driven platforms simulate various operational scenarios in real time, which allows the operators to test and implement strategies that maximize efficiency and resilience. Additionally, federated learning is gaining traction as a decentralized approach for training AI models across distributed energy resources, which enhances data privacy while improving system intelligence.

The integration of IoT devices with AI further enhances the granularity and accuracy of data collection, which enables more precise energy management. For instance, sensors embedded within microgrids provide real-time data on load, generation, and storage conditions, which is employed by AI algorithms to optimize energy flows dynamically.

15.8.2 AI's role in the transition to decentralized energy systems

AI is a key enabler of the transition from centralized energy grids to decentralized systems, such as microgrids. AI ensures that energy generation and consumption are balanced locally by managing DERs. This reduces the dependence on large, centralized power plants and enhances energy resilience at the community level.

P2P energy trading is another area where AI plays a pivotal role. Blockchain-integrated AI platforms facilitate secure, automated energy transactions among prosumers—individuals or entities that produce and consume energy. These systems use smart contracts to match supply with demand in real-time, which fosters localized energy markets and empowers consumers to participate in the energy economy.

Moreover, AI aids in grid stability by predicting and mitigating potential disruptions. For example, during extreme weather events, AI can forecast outages and preemptively reconfigure microgrid operations to maintain supply for critical loads. This proactive approach minimizes downtime and enhances the overall resilience of decentralized energy networks.

15.8.3 Potential challenges and ethical considerations in AI-driven microgrids

The future of energy systems lies in the widespread adoption of resilient, AI-enabled microgrids. These systems will not only support the integration of renewable energy but also redefine the way energy is produced, distributed, and consumed. AI will enable communities to achieve energy self-sufficiency by dynamically optimizing resource allocation and seamlessly integrating new technologies, such as EV charging stations and smart appliances.

Moreover, AI-powered microgrids will serve as the backbone of smart cities, providing reliable and sustainable energy to power urban infrastructure. By leveraging advanced analytics and real-time control, these microgrids will enhance quality of life through reduced energy costs, lower carbon footprints, and increased resilience to environmental disruptions.

As these systems evolve, the collaboration between policymakers, technologists, and stakeholders will be essential. Clear regulatory frameworks, public–private partnerships, and community engagement will play pivotal roles in scaling AI-driven microgrids and ensuring their benefits are equitably distributed.

References

[1] Poonahela I, Krama A, Bayhan S, *et al.* Hierarchical model-predictive droop control for voltage and frequency restoration in AC microgrids. *IEEE Open Journal of the Industrial Electronics Society*. 2023;4:85–97.

[2] Refaat SS, Ellabban O, Bayhan S, *et al. Smart Grid and Enabling Technologies*. John Wiley & Sons; 2021.

[3] Sharida A, Kamal N, Bayhan S, and Abu-Rub H. Adaptive fault-tolerant communication based-control for parallel connected rectifiers. *IEEE Open Journal of the Industrial Electronics Society*. 2023;4:709–719.

[4] Shen Z, Wu C, Wang L, *et al.* Real-time energy management for microgrid with EV station and CHP generation. *IEEE Transactions on Network Science and Engineering*. 2021;8(2):1492–1501.

[5] Aljohani TM, Ebrahim AF, and Mohammed OA. Dynamic real-time pricing mechanism for electric vehicles charging considering optimal microgrids energy management system. *IEEE Transactions on Industry Applications*. 2021;57(5):5372–5381.

[6] Mohammadi E, Alizadeh M, Asgarimoghaddam M, *et al.* A review on application of artificial intelligence techniques in microgrids. *IEEE Journal of Emerging and Selected Topics in Industrial Electronics*. 2022;3(4):878–890.

[7] Lv L, Wu Z, Zhang L, *et al.* An edge-AI based forecasting approach for improving smart microgrid efficiency. *IEEE Transactions on Industrial Informatics*. 2022;18(11):7946–7954.

[8] Safiullah S, Rahman A, Lone SA, *et al.* Robust frequency–voltage stabilization scheme for multi-area power systems incorporated with EVs and renewable generations using AI based modified disturbance rejection controller. *Energy Reports*. 2022;8:12186–12202.

[9] Akpolat AN, Habibi MR, Dursun E, *et al.* Sensorless control of DC microgrid based on artificial intelligence. *IEEE Transactions on Energy Conversion*. 2020;36(3):2319–2329.

[10] Joshi A, Capezza S, Alhaji A, *et al.* Survey on AI and machine learning techniques for microgrid energy management systems. *IEEE/CAA Journal of Automatica Sinica*. 2023;10(7):1513–1529.

[11] Suresh V, Janik P, Guerrero JM, *et al.* Microgrid energy management system with embedded deep learning forecaster and combined optimizer. *IEEE Access*. 2020;8:202225–202239.

[12] Xiong L, Tang Y, Mao S, *et al.* A two-level energy management strategy for multi-microgrid systems with interval prediction and reinforcement learning. *IEEE Transactions on Circuits and Systems I: Regular Papers*. 2022;69(4):1788–1799.

[13] Bustos R, Marín LG, Navas-Fonseca A, *et al.* Hierarchical energy management system for multi-microgrid coordination with demand-side management. *Applied Energy*. 2023;342:121145.

[14] Kumar A, Alaraj M, Rizwan M, *et al.* Novel AI based energy management system for smart grid with RES integration. *IEEE Access*. 2021;9:162530–162542.

[15] Arumugham V, Ghanimi HM, Pustokhin DA, *et al.* An artificial-intelligence-based renewable energy prediction program for demand-side management in smart grids. *Sustainability*. 2023;15(6):5453.

[16] Massaoud M, Saleh MA, Eddin ME, *et al.* Harnessing recurrent-based deep learning models for time series photovoltaic power forecasting. In: *IECON*

2023-49th Annual Conference of the IEEE Industrial Electronics Society. IEEE; 2023. p. 1–6.

[17] Wang H, Lei Z, Zhang X, *et al.* A review of deep learning for renewable energy forecasting. *Energy Conversion and Management*. 2019;198:111799.

[18] Aktepe A, Yanık E, and Ersöz S. Demand forecasting application with regression and artificial intelligence methods in a construction machinery company. *Journal of Intelligent Manufacturing*. 2021;32(6):1587–1604.

[19] Parol M, Piotrowski P, Kapler P, *et al.* Forecasting of 10-second power demand of highly variable loads for microgrid operation control. *Energies*. 2021;14(5):1290.

[20] Keskin T, İnce G. An ensemble learning approach for energy demand forecasting in microgrids using fog computing. In: *Intelligent and Fuzzy Techniques for Emerging Conditions and Digital Transformation: Proceedings of the INFUS 2021 Conference*, August 24–26, 2021. Volume 2. Springer; 2022. p. 170–178.

[21] She B, Li F, Cui H, *et al.* Fusion of microgrid control with model-free reinforcement learning: Review and vision. *IEEE Transactions on Smart Grid*. 2022;14(4):3232–3245.

[22] Al Sumarmad KA, Sulaiman N, Wahab NIA, *et al.* Energy management and voltage control in microgrids using artificial neural networks, PID, and fuzzy logic controllers. *Energies*. 2022;15(1):303.

[23] Joshal KS and Gupta N. Microgrids with model predictive control: A critical review. *Energies*. 2023;16(13):4851.

[24] Tazi K, Abbou FM, and Abdi F. Multi-agent system for microgrids: Design, optimization and performance. *Artificial Intelligence Review*. 2020;53(2):1233–1292.

[25] Chandrasekaran K, Kandasamy P, and Ramanathan S. Deep learning and reinforcement learning approach on microgrid. *International Transactions on Electrical Energy Systems*. 2020;30(10):e12531.

[26] Abu-Rub O, Sharida A, Bayhan S, *et al.* Real-time anomaly detection and remedy actions for resilient operation of DC-coupled fast EV chargers. In: *2024 IEEE Energy Conversion Congress and Exposition (ECCE)*; 2024, IEEE. p. 6595–6599.

[27] Ko Y, Cho K, Kim M, *et al.* A novel capacity estimation method for the lithium batteries using the enhanced coulomb counting method with Kalman filtering. *IEEE Access*. 2022;10:38793–38801.

[28] Sharida A, Bayhan S, Abu-Rub H, and Fesli U. Voltage-sensorless open-switch fault tolerant control of three-phase T-type rectifier. *IEEE Transactions on Power Electronics*. 2023;38(12):15365–15376.

[29] Sharida A, Bayhan S, and Abu-Rub H. Fault-tolerant self-tuning control for three-phase three-level T-type rectifier. *IEEE Transactions on Power Electronics*. 2023;38(6):7049–7058.

[30] Al-Saadi M, Al-Greer M, and Short M. Strategies for controlling microgrid networks with energy storage systems: A review. *Energies*. 2021;14(21):7234.

[31] Kanakadhurga D and Prabaharan N. Demand side management in microgrid: A critical review of key issues and recent trends. *Renewable and Sustainable Energy Reviews*. 2022;156:111915.

[32] Gholami M, Muyeen S, and Lin S. Optimizing microgrid efficiency: Coordinating commercial and residential demand patterns with shared battery energy storage. *Journal of Energy Storage*. 2024;88:111485.

[33] Bui VH, Hussain A, and Su W. A dynamic internal trading price strategy for networked microgrids: A deep reinforcement learning-based game-theoretic approach. *IEEE Transactions on Smart Grid*. 2022;13(5):3408–3421.

[34] Walters M and Venayagamoorthy GK. Digital twin for solar photovoltaic power estimations based on an ensemble of recurrent neural networks. In: *2024 IEEE 19th Conference on Industrial Electronics and Applications (ICIEA)*. IEEE; 2024. p. 1–6.

Chapter 16

Building trust by design through explainable AI for resilient and cognitive smart grids

Mohamed Massaoudi[1], Haitham Abu-Rub[2] and Ali Ghrayeb[2]

Traditional electrical power grids have historically faced challenges with operational reliability, stability, flexibility, and efficiency. Machine learning (ML) aids in enhancing smart grids (SGs) by enabling predictive maintenance, optimizing energy distribution, and ensuring efficient and stable grid operations. Despite the ubiquitousness of ML methods, these methods lack transparency and explainability for SG applications due to their non-linear multilayer structures. Thus, explainable artificial intelligence (XAI) emerged as the missing piece toward human-centered AI that prioritizes the humanistic design perspective. This study aims to explore the potential of XAI for SGs. A background review of the explainability paradigm and its urgent need for SGs is provided. In addition, multiple XAI models are discussed while citing their advantages and drawbacks. Next, the chapter discusses a wide spectrum of SG-based XAI applications. This chapter promotes the exploration of the emerging XAI paradigms, specifically for SG applications. With this study, the authors hope that this work will serve as a touchstone for theorists and practitioners alike, promoting trust in AI systems through enhanced transparency and ultimately enhancing the resilience of SGs.

16.1 Introduction

The smart grid (SG) is a modernized version of the traditional electrical grid that uses advanced digital technology to improve the reliability, efficiency, and sustainability of power systems (PSs). The SG proliferation plays a pivotal role in the integration of renewable energy sources (RES) and the enhancement of energy efficiency. Among the different techniques deployed to safeguard these modern grid systems, artificial intelligence (AI), and specifically machine learning (ML), have emerged [1]. Although AI models can yield strong analytical results, they often lack transparency in revealing crucial details and explanations regarding how the models function. The deployment of these AI techniques has been hampered due to their

[1]Department of Electrical and Computer Engineering, Texas A&M University, USA
[2]College of Science and Engineering, Hamad Bin Khalifa University, Qatar

"black-box" nature, which hinders interpretability and accountability [2]. Due to their lack of interpretability, ML methodologies remain impractical for direct application in the majority of industrial decision-making processes. Consequently, there is now a pressing need to develop more scientifically grounded, human-interpretable methods. These methods should integrate with ML models, maintain high predictive performance, and ensure transparency, interpretability, and robustness [2].

To instill confidence and trust in AI, XAI has emerged as a promising alternative, offering a nuanced human-centered AI perspective that prioritizes intelligibility, accountability, and ethical considerations [3]. Recognizing the need for such explanatory insight to users in AI, the US Defense Advanced Research Projects Agency (DARPA) initiated a 5-year research program, with an investment of 50 million USD, to develop XAI models that function as "glass-box" (aka transparent) systems, providing explanations to human users without significantly compromising AI performance [4]. XAI has the potential to demystify the decision-making processes of AI models, providing users with an understanding of how and why specific decisions are made [5]. With the increasing integration of AI into SGs, ensuring the explainability of AI systems has become not just an academic interest but a practical necessity. Particularly, explainable DL (XDL) techniques can be used to understand the behavior of DL models and to identify the input features that are most important for a given prediction [6]. The annual increase in published research illustrates the fast-paced advancements in the fields of interpretable AI (IAI), interpretable DL (IDL), XAI, and XDL, as illustrated in Figure 16.1. Table 16.1 captures recent advancements in XAI as showcased in state-of-the-art review papers [7–11]. These works highlight the significant impact of XAI in the IT sector. However, the focus remains largely within this sector, leaving the scope for reviewing the potential of XAI in fortifying SGs relatively unexplored. More precisely, yet comparatively little attention has been paid to XDI in the electrical PSs realm.

To the best of our knowledge, this is the first review article that discusses the applicability of XAI to electrical PSs. Considering the existing research gaps, this study delves into the recent developments in the field of XAI for SGs and provides a comprehensive overview of the state-of-the-art methods, applications, and challenges in this area. The paper begins with a delineation of the need for XAI in SGs, followed by a comprehensive background on XAI. Further, a deeper dive is taken into the intricate world of XAI models, dissecting them into model-specific,

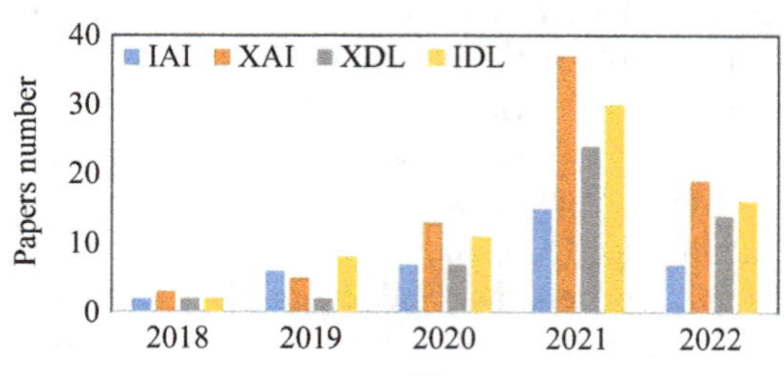

Figure 16.1 Yearly publications for IAI, IDL, XAI, and XDL. The data is derived from the Web of Science (January 2018 to March 2022)

Table 16.1 Scope of different literature on the XAI subject

Reference		[7]	[8]	[9]	[11]	[10]	Ours
	Year	2022	2023	2023	2023	2023	2023
Scope	XDL	○	◐	○	◐	○	●
	XAI	○	●	●	●	●	●
	SG	●	○	◐	○	◐	●

Note: ● Fully addressed; ◐ Partially addressed; ○ Not addressed.

model-agnostic, and domain-specific interpretability methods. Subsequently, pioneering application scenarios for XAI in SGs are examined. The aim is to illuminate future works by delineating current challenges and potential research directions in this domain. Finally, the findings and observations are synthesized in the conclusion.

16.2 The need for XAI in SG

SGs represent an evolution of the traditional electricity grid, utilizing digital communication technology to manage and distribute electricity more efficiently and sustainably. SGs integrate various technological components, such as smart meters, renewable energy resources, and advanced analytics, to optimize the production, distribution, and consumption of electricity. The proliferation of AI in SGs presents novel opportunities for system optimization, reliability enhancement, and risk mitigation. However, the opacity inherent in many AI models can hinder their acceptance and widespread deployment due to difficulties in understanding their underlying reasoning mechanisms [12]. As a response, XAI has emerged as an essential tool to provide interpretable, transparent, and human-understandable insights into AI model decisions. According to the theoretical perspectives on explanations from philosophy and psychology, Salmon [13] demonstrated that explanations can be conceptualized as sequences of causative events leading to a particular decision. These sequences, or causal chains, can be divided into five distinct types: temporal, coincidental, unfolding, opportunity, and preemptive chains, each corresponding to a different kind of explanation. The majority of AI/ML methods provide a function $g : E^m \rightarrow [0, 1]$, for some $m \in \mathbb{N}$, that measures only the correlation rather than causation between events in E and decision D.

In SGs, this transparency is pivotal, where safety, regulatory compliance, and stakeholder trust are paramount [4]. Figure 16.2 illustrates the timeline of the XAI for PSs. Starting in 2010, the first XAI models were introduced to reach wide adoption in 2020. Hence, XAI's role in enhancing decision-making transparency, traceability, and ultimately the credibility of AI applications in SGs, underscores its crucial importance. This AI transparency not only builds trust but also allows for more effective management of the grid, as operators can understand and anticipate the system's behavior and decisions [14].

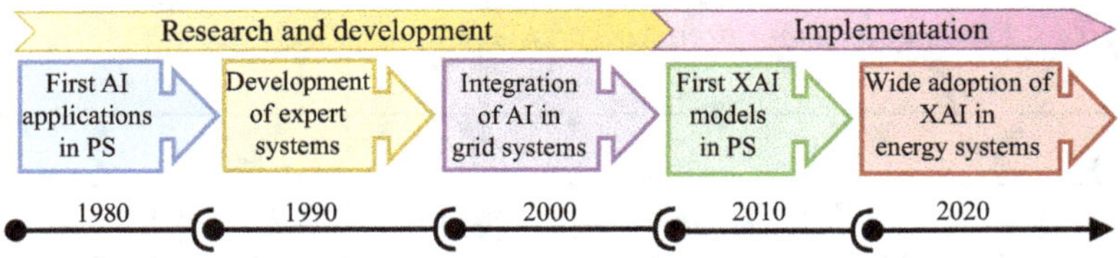

Figure 16.2 Path and milestones toward XAI for PS

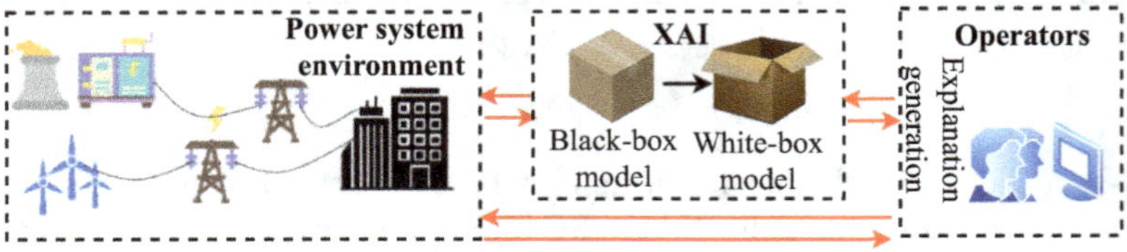

Figure 16.3 Flowchart of power system operations with XAI

Building on the foundational role of XAI in SGs, recent research continues to underscore its significance in enhancing AI-based applications. This shift from opaque, black-box models to transparent and interpretable controls aligns with the broader trend toward accountability and comprehensibility in AI applications within SGs, a sector where complexity and national significance are paramount [15]. Kabir *et al.* [16] proposed the adoption of XAI in smart city solutions, focusing on cybersecurity. XAI can significantly enhance detection capabilities in SGs, providing control and transparency. This is crucial in smart cities, where understanding the reasoning behind AI decisions can lead to more efficient energy generation and consumption. The paper also presents several commercial XAI platforms such as Kyndi and Factmata. Figure 16.3 outlines the integration of XAI into power system operations. As can be seen from the figure, XAI acts as an intermediary between the power system environment and the operators, indicating its role in translating complex AI decisions into actionable intelligence for human operators. For new AI technologies to be adopted within established PSs, they must be understood by the existing workforce. XAI facilitates this integration by making new AI technologies accessible and understandable to engineers and operators.

16.3 Background of XAI

The National Institute of Standards and Technology (NIST) adopts the definition of XAI, which emphasizes that XAI systems should adhere to certain core principles [17]. It specifies XAI as "a system delivers or contains accompanying evidence or reason for outputs and/or processes." The explanations given by these systems must be understandable to the intended audience, ensuring that the reason behind the generation of the output or the functioning of the system is conveyed accurately. Furthermore, the system should acknowledge its operational boundaries, functioning only within the conditions it was designed for and when there is sufficient certainty in its output. Explainability is acknowledged alongside a multitude of desirable features that contribute to the trustworthiness of AI systems, such as accuracy and privacy, among others, with responsibility also being an essential component of AI dependability [18].

XAI is a field of AI that seeks to make ML models and their decision-making processes more understandable to end users [19]. To present the basics of XAI, the assumption is that the ML model is already trained, encapsulating the input-output relationship through a designated function $f : \mathbb{R}^d \rightarrow \mathbb{R}$. f receives a vector of real-valued features $\mathbf{x} = (x_1, \ldots, x_d)$, typically associated with a variety of sensor data within the SG infrastructure. The function f computes an output as a real-valued score, which lays the groundwork for critical decision-making processes in the management and operation of SGs. In the SG domain, where ML classification plays a pivotal role in system monitoring and fault detection, the output score generated by f is interpreted as an indicator of support for or against a certain operational class. For instance, a stable/unstable class might represent a state of stability or an anomaly in the grid. To reach a classification conclusion, the output score is evaluated against a predetermined threshold. In the case of multi-class problems, such as differentiating between various types of grid anomalies, the classification is determined by comparing whether the output score for a given class outweighs the scores for all other possible classes.

Figure 16.4 presents a taxonomy of ML interpretable model categories. Practitioners must consistently consider these aspects to accurately identify the method that best aligns with their requirements. Seven principal categories have been delineated for interpretability methods: applicability scope (encompassing model-agnostic and model-specific methods), interpretation Scope (global versus local), interpretability type (intrinsic versus post hoc), interpretation fidelity (ranging from high to low), interpretation dynamics (either static or interactive), and interpretation objectives (including trust-building, debugging, and knowledge discovery). These categories help to structure the approach to understanding and explaining ML models, which is critical for validation, debugging, and ethical considerations in AI.

The field of XAI is characterized by several key terminologies, including intelligibility, explainability, transparency, and interoperability (Figure16.5(a)). Intelligibility (*Int*) refers to the ratio of correctly predicted instances (*C*) to the total number of instances (*Ins*) [8]. An intelligible model presents its logic in an understandable way, making it easier for practitioners to follow how the model arrives at its decisions [20]. Explainability (*E*) is about the capacity of an AI model to offer insights into its internal workings, allowing humans to understand the logic behind its decisions. It can be formulated as the proportion of output variance (*V*) that can be accounted for by the

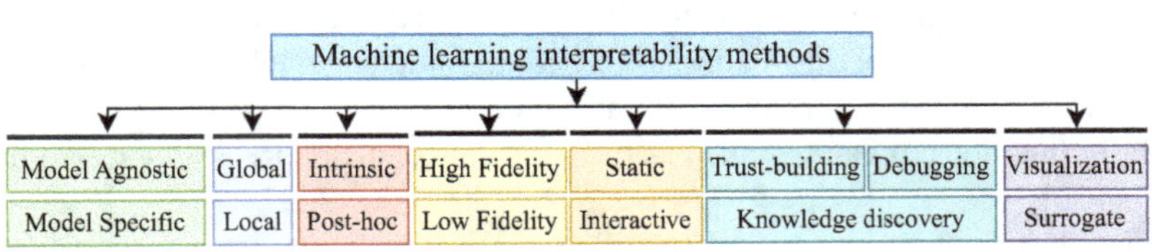

Figure 16.4 Interpretable model categories

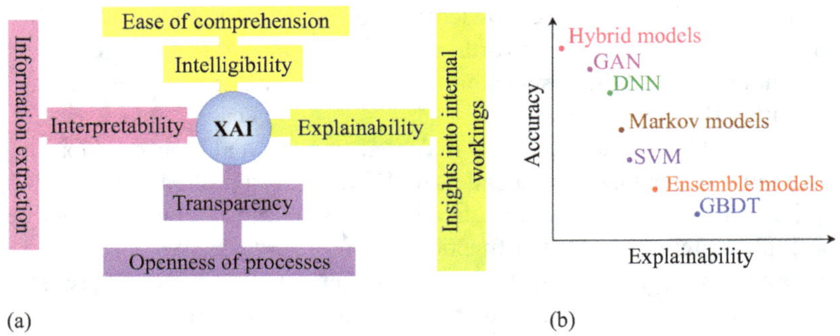

Figure 16.5 (a) XAI terminologies and (b) explainability vs accuracy (as presented by DARPA [4])

features (*F*) [7]. For instance, Figure 16.5(b) showcases the explainability-accuracy tradeoff. It is worth noting that hybrid methods and generative models possess the least explainability and the highest accuracy. In contrast, gradient boosting and decision trees represent the most explainable methods [4]. Transparency (*Tr*) relates to the openness of an AI model's processes to its internal mechanisms. A transparent model could be defined as the proportion of the inner workings of a model that are understood by a human (*U*) from the total operations in a model [7]. Interpretability (*I*) refers to the proportion between the total count of masked features, which have negligible relevance on the final categorization outcome result, and the entire set of dataset features [21]. Additionally, fidelity (*Fid*) is a set of processes or methods that quantify the congruence in accuracy between the original model (A_O) and its fully interpretable counterpart (A_i). The simplified equations of these concepts are given as follows [7]:

$$I = \frac{N}{F}, E = \frac{V}{F}, Tr = \frac{U}{T}, Int = \frac{C}{Ins}, Fid = \frac{A_i}{A_O} \tag{16.1}$$

XAI is categorized from several perspectives, accounting for the drive to integrate XAI into cyber security, the classifications of XAI, and the challenges XAI faces. XAI categories are multifaceted, and a specific XAI technique may fall under more than one category. Thus, categorizing one XAI technique from multiple perspectives enables a more comprehensive understanding of the technique. Intrinsic and post hoc classification targets achieving explainability either by restricting the AI model's complexity (intrinsic) or by analyzing the model's methodology after training (post hoc) [22]. Intrinsic XAI provides explanations concurrently with predictions, whereas post hoc explanations are interpretation methods employed after models have been trained to facilitate providing understandable information about how an already developed model generates its estimations. From the intrinsic interpretability methods, it can be taken as examples the decision trees and sparse linear models [23].

The classification into model-specific or model-agnostic depends on the models to which the XAI methods are applied. Model-specific methods are unique to a particular model or model group. A model-specific XAI method, represented as $x_{ms} : M_{spec} \times In \rightarrow Ex$, is applicable to a subset of AI models, $M_{spec} \subset M$, which are characterized by specific features such as architecture. It takes an input *in* from the set *In* and produces an explanation *ex* in the set *Ex*. Conversely, a model-agnostic XAI method, denoted as $x_{ma} : M \times In \rightarrow Ex$, is designed to work with any model in the comprehensive set *M* of AI models, regardless of their distinct characteristics. Both x_{ms} and x_{ma} generate explanations *ex* for inputs *in*, but their scope differs based on the range of models they can effectively interpret. In neural networks, layer-wise relevance propagation (LRP) is a technique used to trace the prediction of the network back to its input features, attributing relevance scores to each neuron in each layer that contributed to the final decision. The relevance R_j of the *j*-th neuron in a given layer is computed as [24]

$$R_j = \sum_k \frac{a_j(w_{jk} + \gamma v_{jk}^+)}{\sum_{0j} a_j(w_{jk} + \gamma v_{jk}^+)} R_k \tag{16.2}$$

where R_k denotes the relevance of the *k*-th neuron in the subsequent layer, a_j is the activation of the *j*-th neuron, w_{jk} represents the weight connecting the *j*-th neuron to the *k*-th neuron, and γv_{jk}^+ is a term introduced to consider the positive contributions of the weights, with γ being a scaling factor. The notation $\sum_{0j}$ indicates that the sum is taken over all neurons *j* in the lower layer, incorporating a bias term w_{0k} with $a_0 = 1$. This equation is a generalization of the standard LRP computation rule, which is typically given by $R_j = \sum_k \frac{a_j w_{jk}}{\sum_{0,j} a_i w_{ik}} R_k$, allowing for an adaptive distribution of relevance scores according to both positive and negative contributions of each neuron, thus providing a more nuanced understanding of the network's decision-making process.

Model-agnostic tools can theoretically be used with any ML model, often by analyzing feature inputs and outputs. The taxonomy of global and local interpretability depends on the scope of the model [25]. Local interpretability sheds light on the rationale behind specific model predictions. To approximate the decision-making process of a black-box model at a local level, a linear model can be employed as follows:

$$\hat{y} = \beta_0 + \beta_1 x_1 + \ldots + \beta_n x_n \tag{16.3}$$

where $\hat{y}$ represents the predicted outcome and β_i are coefficients that quantify the influence of each feature x_i on the prediction. These coefficients are determined by training the model on perturbed instances close to the instance being explained. Expanding on this, the decision function's behavior in the proximity of a specific point $\hat{x}$ is further elucidated using the gradient of the function at that point to provide a more detailed linear approximation as

$$\hat{y} \approx \sum_{i=1}^{d} [\nabla f(\hat{x})]_i \cdot (x_i - \hat{x}_i) \tag{16.4}$$

where the prediction $\hat{y}$ is now approximated by a weighted sum of the deviations of feature values from $\hat{x}$, their reference point. In this context, the weights $[\nabla f(\hat{x})]_i$ correspond to the coefficients β_i in the first equation, explaining the impact of each feature x_i on the prediction. A feature's relevance is emphasized when it distinctly manifests in the data, diverging from the reference point $\hat{x}_i$, and when the model's output is sensitive to it, indicated by a non-zero component of the gradient, $[\nabla f(\hat{x})]_i \neq 0$.

Global interpretability takes into account the entire learning algorithm, including training data, proper applications, and any potential flaws or misapplication [25]. One approach is to use feature importance scores, which can be represented as

$$I(f_j) = \frac{1}{N}\sum_{i=1}^{N} |f_j(x^{(i)})|, \tag{16.5}$$

where $I(f_j)$ is the importance of feature f_j, N is the number of samples, and $x^{(i)}$ is the i-th sample. Nevertheless, global interpretability can be inaccurate, struggle with data continuity, and may not explain the relationship between features. It is also challenging to achieve in cases with numerous relevant input variables, which is a common scenario in the PS applications [25]. Local interpretability, while useful for individual forecasts, may not provide a comprehensive understanding of the model's overall logic [25]. Explanation output is another vital aspect of XAI classification because the output format greatly influences users. Argument-based explanations and model-based explanation approaches aim to outline the black-box model behavior using a more interpretable model.

16.4 Existing XAI models

This section depicts a novel taxonomy of XAI methods based on the level of specificity, dissecting them into model-specific, model-agnostic, and domain-specific interpretability methods. These categories are explored, juxtaposing their strengths and limitations as introduced in Table 16.2.

1) Model-specific interpretability methods: These methods primarily capitalize on the unique structures and properties of specific model types to offer insights into their decision-making processes. For instance, the decision rules method [26] delivers easily interpretable outputs in the form of rules. This method is particularly suited to models that are inherently rule-based, utilizing the structure of these models to deliver comprehensive explanations. Similarly, the gain and cover technique [44] is designed specifically for use with models like XGBoost, providing permutation feature importance (PFI) based on improvement in model accuracy or coverage of data samples. The attention mechanism [37] underscores influential parts of the input, hence revealing the focus of the model during decision-making. Each of these techniques uniquely leverages the structure and properties of their corresponding model types to facilitate interpretability [45].

2) Model-agnostic interpretability methods: These methods offer a universal approach to understanding AI models, shedding light on model predictions rather than the intricacies of their internal operations. Among these techniques exists the local linear interpreter, which provides a balance between local fidelity and interpretability [27]. Furthermore, Shapley additive explanations (SHAP) offer a unified measure of feature importance across a variety of models [46]. SHAP conceptualizes

Table 16.2 Comparison of XAI techniques

Technique	Advantages	Limitations
Decision rules [26]	Easily interpretable, handles categorical variables	Can be sensitive to small changes in data
Local linear interpreter [27]	Provides local explanations	Requires careful selection of neighborhood
SHAP [28–33]	Handles interactions between features	High Computing, less interpretable for complex models
LIME [34]	Provides local explanations, works with any model	Local fidelity may not represent overall model behavior
ELI5 [35]	Simple, Useful for feature importance ranking	Simplified explanations might lack depth
Knowledge-based approach [36]	Incorporates domain knowledge and human expertise	Requires expert input, scalability may be challenging
Attention mechanism [37]	Highlights important parts of the input features	Difficult to interpret for complex models
Conditional entropy [38]	Uncovers dependencies between features	Not always intuitive, depends on data distribution
PFI [39]	Simple, works with any model	Biased if features are correlated
Surrogate model [40]	Provides a simplified model for interpretation	Model's fidelity can vary with its complexity
Rule-based approach [41]	Provides human-understandable rules	Can be hard to create for complex models
Partial dependence plots [42]	Visualizes feature influence, easy to interpret	Doesn't capture feature interactions
Occlusion sensitivity [5]	Good visualization, useful for image data	Does not always accurately represent model's reasoning
GRAD-CAM [5]	Visual explanations for CNN, useful for image data	Difficult to apply to non-visual data
DeepLIFT [43]	Effectively handles feature interactions	High computation, may not work with all model types
Gain and cover [44]	Quantifies the contribution of each input feature	Specific to XGBoost, lacks instance-level explanations

a model's prediction for a given instance in the context of cooperative game theory, with model features acting as players. This framework interprets the original model through a conditional expectation function. Simplified model inputs, denoted by x', are reverted to their original form using a mapping function h_x, ensuring $x = h_x(x')$. This guarantees that $g(z')$ approximates $g(h_x(z'))$ as closely as possible when $z' \approx x'$, with z' representing a binary vector in $\{0, 1\}^M$, where M is the number of simplified input features. Each feature is assigned an impact value ϕ_i, with the sum of these values aiming to approximate the output of the model's function $f(x)$ as

$$g(z') = \phi_0 + \sum_{i=1}^{M} \phi_i z'_i \tag{16.6}$$

In parallel, the Shapley value is employed to quantify the individual contributions of each feature within a coalition P. It examines the effect of including or excluding each feature i across all possible subsets S of P, in terms of the total predictive payoff $v(S)$. The Shapley value for feature i is then calculated as

$$\phi_i = \sum_{S \subseteq P \setminus \{i\}} \alpha_S \cdot (v(S \cup \{i\}) - v(S)) \tag{16.7}$$

where α_S is the weight accorded to each subset S, given as

$$\alpha_S = \frac{|S|! \cdot (|P| - |S| - 1)!}{|P|!} \tag{16.8}$$

The axioms of efficiency, symmetry, linearity, and null contribution are fundamental to Shapley values, ensuring that the sum of the individual contributions matches the total predicted outcome; features with identical impacts receive equivalent shares; the approach is additive for various predictive scenarios; and features that do not affect the prediction are assigned a zero value.

The system's explainability is enhanced through the use of SHAP decision graphs, which illustrate the decision-making process of complex models [47]. These graphs include the decision plot, waterfall plot, and summary plot, which provide a comprehensive understanding of the model's behavior and its decision-making process [47]. For instance, in Jang *et al.* [48], a new unsupervised fault detection and diagnosis (FDD) method is proposed for industrial processes based on the SHAP values and adversarial auto-encoder (AAE). The proposed method assigns each variable a contribution value for a monitoring index calculated from the AAE model, allowing for more accurate diagnostic results. However, estimating the causal effects of process variables between fault status and physical behavior is challenging for the proposed method. In Akhlaghi *et al.* [49], SHAP is used to interpret the contribution of operating conditions on performance parameters of a guideless irregular dew point cooler (GIDPC). The SHAP values are visualized as forces to demonstrate the attribution of each feature on the prediction.

Likewise, local interpretable model-agnostic explanations (LIME) can effectively simplify complex models by focusing on individual predictions [34]. The mathematical expression of LIME is given as [50]

$$\xi(x) = \arg\min_{g \in M}\{L(f, g, \pi_x) + \Omega(g)\} \tag{16.9}$$

ξ signifies the optimal interpretation of the interpretable function *f*. Here, *M* refers to a collection of interpretable functions denoted as *g*. $_x$ represents a measure that defines the proximity of the interpretable representation, *x*. The function *L* delineates a loss function between functions *f* and *g* on *x*, providing a quantifiable gauge of the interpretability of *f*. Meanwhile, the term Ω is utilized to quantify the intricacy of the function *g*. Surrogate models have been successful in approximating complex models, facilitating better interpretability [40]. Let *f* be a neural network model. These approaches focus on creating an interpretable explanatory model g_s, which is designed to meet two main criteria: first, g_s should be clear and understandable; second, it should provide a good approximation of the original complex function *f*. A well-known class within these approaches is the additive feature attribution methods.

Wang *et al.* [51] introduced a novel genetic algorithm-based method, E-LIME, for evolving local interpretable explanations for deep convolutional neural networks (CNNs). The proposed approach successfully captures meaningful interpretable features, improving the confidence of users in the predictions. Potential limitations could include the computational cost and the requirement for manual fine-tuning, which were issues with the predecessor, LIME, that E-LIME aims to overcome. An interpretable hybrid experimental learning (HEL) model for trading behavior in the electricity market is proposed in Liu *et al.* [50]. HEL integrates economic experiments and ML, using generative adversarial networks and LIME to model and explain trading behavior. Unfortunately, the economic experiment component of HEL cannot guarantee the quality of human participants and their behaviors. Additionally, the outcomes are limited to the experimental environment and not adjusted by practical data.

3) Domain-specific interpretability methods: These methods serve as a crucial subset of XAI techniques, where domain knowledge or contextual specifics are integrated to render comprehensive interpretations. Represented as $d_{si} : M \times Ip \times DK \rightarrow Ex$, these methods operate on a model *m* from the set *M* of AI models, taking an input *ip* from the set *Ip*, and utilizing domain knowledge *DK*, which is a set of domain-specific information. A significant aspect of these methods is the involvement of detailed manual design and the necessity of expert input, further solidifying their efficacy. For instance, the knowledge-based approach [36] leverages explicit domain expertise to elucidate interpretations, providing a high degree of specificity and relevance. Similarly, the rule-based approach [41] demystifies complex AI models by generating human-understandable rules. Occlusion sensitivity [5] and gradient-weighted class activation mapping (GRAD-CAM) [5], typically used in image-based models, offer visual interpretations by highlighting influential regions in input data. To create an explanation map $h^c \in \mathbb{R}^{W \times H}$ for a particular concept *c*, with the width *W* and height *H* of the final convolutional layer in a CNN, one must first compute the gradient of the network's output with respect to concept *c* over the activations A^k of the

k-th feature map. This gradient is denoted as $\frac{\partial y^c}{\partial A^k_{i,j}}$. Following this, perform a global average pooling (GAP) across the gradient values, taking into account the height and width dimensions, to determine the importance weights of the neurons. The equation for calculating these weights is calculated as [52]

$$\omega_k^c = \frac{1}{W \times H} \sum_i \sum_j \frac{\partial y^c}{\partial A^k_{i,j}} \tag{16.10}$$

These weights ω_k^c indicate the significance of the k-th feature map in identifying the concept c. The explanation map h^c is subsequently obtained by computing a weighted sum of the feature map activations and applying the Rectified linear unit (ReLU) function to this sum, as follows [52]:

$$h^c = \text{ReLU}\left(\sum_k \omega_k^c A^k\right) \tag{16.11}$$

Deep learning important features (DeepLIFT) [43], conversely, quantifies the contribution of each input feature, thereby making model decisions more understandable. The attribution of DeepLIFT is expressed as [43]

$$C^{(t)}_{\Delta x_i \to \Delta y} = (x_i - x_i') \cdot \left.\frac{\partial y}{\partial x_i}\right|_{x=x'} \cdot g\left(\frac{x_i - x_i'}{x_i - \bar{x}}\right) \tag{16.12}$$

where x_i is the input feature, x_i' is the reference feature value, $\frac{\partial y}{\partial x_i}$ is the gradient of the output with respect to the input feature at the reference point, and g is a scaling function that adjusts the contribution score based on the input's deviation from the reference and the average input.

While DeepLIFT provides a granular view of individual feature contributions to model decisions, conditional entropy complements this perspective by revealing the interdependencies between features, thereby enriching our interpretability of model behavior as discussed by Sahoo *et al.* [38]. Thus, domain-specific interpretability methods make significant strides in rendering AI models more transparent and understandable.

In Chen and Zhang [53], a novel theory-guided deep-learning for electrical load forecasting (TgDLF) is presented. The proposed method automatically extracts features without complex engineering and its gradient-free characteristic, which allows for fewer constraints on activation and evaluation functions. However, the TgDLF faces difficulty in obtaining derivatives of elaborate and biomimetic activation functions. This hinders the building of a loss function based on non-differentiable evaluation criteria.

16.5 Successful XAI applications for smart grids

In this section, a plethora of XAI-based SG works are considered from the application perspective. These applications include PS stability and security assessment, energy management systems, and forecasting and predictive control, along with its critical role in FDD.

1) PS stability and security assessment: The ability of XAI to decipher complex ML algorithms has been instrumental in improving security assessments, enabling more accountable and interpretable electrical transient stability prediction [54–56]. XAI has also found significant use in predicting frequency stability indicators and deterministic frequency deviations, providing an effective tool to ensure stable power delivery. In terms of control mechanisms, XAI has enabled more understandable and reliable activation of secondary control power [57]. Furthermore, XAI applications in under voltage load shedding have made substantial strides, providing insights that lead to more efficient power management strategies [58].

Consider, for illustration, a class activation map (CAM)-based CNN outlined in Fang *et al.* [59], which assesses the contribution of different input features to the output results for fault classification and location on transmission lines. The activation of a given channel c in the GAP layer is denoted by a_c and is computed as the channel-wise average of the activations across all timesteps, given by $a_c = \frac{1}{T}\sum_{i=1}^{T} x_i^c$, where x_i^c represents the activation at timestep i for channel c. The class-specific input to the softmax layer for class k is then expressed as [59]

$$q_k = \sum_{c=1}^{C} w_c^k a_c = \sum_{c=1}^{C} w_c^k \cdot \frac{1}{T}\sum_{i=1}^{T} x_i^c = \frac{1}{T}\sum_{i=1}^{T}\sum_{c=1}^{C} w_c^k x_i^c \tag{16.13}$$

where w_c^k is the weight associated with the cth GAP unit for class k. The softmax function's output for class k is then $p_k = \exp(q_k)/\sum_l \exp(q_l)$. For constructing the CAM of a particular input for class k, the activation map at each timestep i is defined as $A_i^k = \sum_{c=1}^{C} w_c^k x_i^c$, which leads to $q_k = \frac{1}{T}\sum_{i=1}^{T} A_i^k$. This formulation demonstrates that A_i^k encapsulates the significance of each timestep i in ascertaining the input's classification as class k. However, the model's robustness could be compromised as fault-related transients in temporally contiguous sections within a signal may mask out clues for fault classification and location. In Zhang *et al.* [1], the authors proposed the deep-SHAP method-based deep reinforcement learning (DRL), which calculates the importance of input features based on a backpropagation strategy. Through the lens of deep-SHAP analysis, The proposed model demonstrates its ability to quantify the contribution of each input variable. However, the interpretability of DRL models remains a critical weakness, and the deep-SHAP method, while helpful, may not provide a complete solution [60]. Future research of this work could include combining symbolism and connectionism and building a heterogeneous information network for PSs to obtain better interpretability for end-users.

The traditional approaches, such as time domain simulations (TDS), present considerable challenges in deciphering the causal factors impacting stability thresholds. Consequently, in Hamilton and Papadopoulos [61], a SHAP methodology is utilized to elucidate the influence of power system variables on the estimations of stability limits within an ANN model. Nevertheless, addressing the considerable challenges that traditional TDS face in analyzing and understanding the causal factors impacting stability thresholds.

2) Energy management systems: Recent studies utilizing XAI methods have enhanced long-term cooling energy consumption predictions and building energy management performance [62]. These methods enable the development of more precise energy usage models and support the prediction of cooling system performance, including irregular dew point coolers [63]. Furthermore, XAI techniques have been employed for short-term forecasting of energy consumption [64,65], enabling more precise and adaptive energy management systems (EMS). Leveraging these approaches also aids in identifying energy usage patterns [66] and benchmarking energy performance levels, which are instrumental in improving the energy efficiency of human-in-the-loop systems through SG infrastructure. Other applications of XAI include clustering to inform incentive schemes for improving energy efficient response and identifying optimal design strategies for optimal energy solutions. These diverse applications demonstrate the transformative potential of XAI in advancing EMS [67].

Jang *et al.* [68] proposed a daily peak load forecasting method based on separate deep neural network (DNN) models for each day of the week. The model's advantage lies in its interpretability, facilitated by the SHAP method, and its accessibility for energy providers with limited computational resources. However, the model's dependence on the previous day's peak load and temperature-related variables for prediction may not always be accurate or available. An interpretable transformer model has been introduced for probabilistically forecasting real-time electricity prices [69]. The prediction interval normalized average width (PINAW) quantifies the conciseness of the prediction interval, which is indicative of its level of informativeness. The PINAW is computed as follows [69]:

$$E_{t_0+\tau}^{\text{PINAW}} = \frac{(U_{t_0+\tau} - L_{t_0+\tau})}{\lambda^{\text{RT}} - \lambda_{\text{RT}}} \qquad (16.14)$$

where λ^{RT} and λ_{RT} denote the maximum and minimum observed real-time price values within the test dataset, respectively. Normalizing $E_{t_0+\tau}^{\text{PINAW}}$ to a percentage metric allows for the evaluation of the prediction interval's sharpness. An optimal PINAW score close to zero signifies a highly informative prediction interval. Unfortunately, this post hoc method has limited potential when used on multivariate time series as they do not consider the temporal dependencies between input features.

3) Energy forecasting and predictive control: XAI methods have been instrumental in enhancing the predictability of photovoltaic (PV) power generation [70]. The authors have also been effectively used to strategize demand response for prosumers through improved PV power generation forecasting. Moreover, XAI methodologies have been utilized to predict whether the generated solar power will meet the load demand of a prosumer, and for developing probabilistic forecasts for PV power generation [71]. Additionally, XAI has played a pivotal role in solar irradiance [72] and solar radiation forecasting, reinforcing its crucial role in RES management. It has also made significant strides in load forecasting [73] and in predicting the electricity generation mix. More importantly, these AI-based methods are making inroads in managing grid control errors in renewable-rich grids and in controlling grid-connected voltage source inverters, thereby underscoring their critical importance in modern grid management and control [74].

Kim and Cho [75] proposed an explainable autoencoder for electrical consumption estimation. The proposed method provides a state transition mechanism that allows for more efficient control of predicted electric energy consumption values by adjusting conditions according to various situations, such as weather or economic information. The mean squared error of the model's predictions is 0.2113 for 15-min intervals, 0.2937 for 30-min intervals, 0.3450 for 45-min intervals, and 0.3840 for 60-min intervals. However, the model's limitations include its current application to only one individual household's energy consumption and its reliance on certain conditions for state transition. Recently, the DRL approach [76] has been employed for energy management in manufacturing systems. The model significantly improves training speed and stability, thereby reducing energy costs. It also offers an explainable mechanism, making it easier for human operators to understand the system-level tasks. However, the model is currently limited to manufacturing systems with fixed layouts, and handling continuous action spaces in other equipment types requires further investigation. An interpretable memristive LSTM network is proposed for residential load forecasting [77]. The interpretability of a memristive LSTM network leverages an attention mechanism, assessing the impact of input variables and temporal hidden states by embedding a mixture attention technique into the LSTM cells. The input relevance $Q(z_n|X_{0:n-1})$ for the n-th variable is expressed as

$$Q(z_n|X_{0:n-1}) = \sum_{i=1}^{n} \alpha_i^n P(z_n|h_i^m, h_{i-1}^n) \tag{16.15}$$

where α_i^n denotes the attention weight for the i-th timestep, h_i^m is the memristive hidden state affected by the i-th input, and $P(z_n|h_i^m, h_{i-1}^n)$ is the conditional probability of the input variable given the current and previous hidden states. Unfortunately, the model can characterize variable-wise features and temporal importance, but it may still be challenging to fully interpret the forecasting results.

Considering an interpretable an encoder–decoder for multi-step load forecasting. A global variable importance, denoted by I^n, is computed as the mean of the attention weights $\beta_t^{m,n}$ over all samples in the training set as [78]

$$I^n = \frac{1}{M(T+1)} \sum_{m=1}^{M} \sum_{t=t_0}^{T} \beta_t^{m,n}, \tag{16.16}$$

where M is the total count of samples in the training dataset and T represents the number of timesteps. Additionally, the variable-wise temporal importance $I_{T_k}^n$ for a particular feature at a given time is determined by summing the temporal attention weights across all prediction timesteps, reflecting the feature's impact on the model's output at each timestep. However, potential privacy leakage issues and the need for model compression hinder model deployment in IoT devices.

In the research detailed by Van Zyl *et al.* [79], a multi-headed CNN leverages SHAP and Grad-CAM for national demand forecasting. The study reveals that SHAP not only outperforms Grad-CAM in terms of forecasting accuracy but also excels at identifying non-contributing features, thus minimizing the risk of model overfitting. SHAP's computational intensity and occasional misprioritization of influential features, however, signal a need for cautious application, especially when it overlooks key predictors. Conversely, Grad-CAM's effectiveness is diminished due to its sensitivity to the scale of data and its propensity to ignore features with a negative influence on the forecast outcome. While SHAP provides post hoc interpretability by analyzing the contributions of input features to predictions, there is a need for an XAI that enhances both prediction and explanation. Thus, a deep autoencoder is proposed to explain the energy demand prediction results by manipulating the latent space [80]. This implies a focus on direct correlations, potentially overlooking the complex, multi-dimensional interactions within the data that could also influence energy demand predictions. The study proposes visual explanations at both sequence and point levels for load disaggregation and non-intrusive load monitoring.

4) Fault detection and diagnosis: XAI methods have seen considerable adoption in the domain of FDD in SGs. These methods enable easier identification of errors and faults in complex systems. Event identification is one such application where XAI plays a crucial role in identifying anomalies in system behavior [81]. Classifiers leveraging XAI are crucial for detecting disturbances, thereby preventing damage and ensuring the efficient operation of the grid. Furthermore, XAI methods also contribute significantly to the diagnosis of faults in power transformers, which is paramount for preventing any catastrophic system failures. Likewise, XAI methods can facilitate the detection and diagnosis of incipient faults in PV panels, contributing to the efficient operation of RES [34]. To amplify this effect, edge node-based explainable incipient fault detection systems for PV panels have also been proposed, enhancing the system's reliability and resilience [82]. Therefore, the utilization of XAI methods for FDD emerges as a promising area in the landscape of SG technology [83].

An LRP-based CNN model has been proposed for gearbox fault diagnosis to provide pixel-wise physical explanations for its decisions [84]. However, the proposed model relies on the quality of input images and the complexity of understanding the relevance scores. The method's effectiveness is also contingent on the accuracy of the DCNN's initial training. Li *et al.* [85] deploy a graph wavelet denoising network (GWDN) to effectively handle signal denoising and retain fault-related components of the signal, enhancing its interpretability. The GWDN method demonstrates its ability to resist adversarial attacks and smoothly adapt to parameter changes. However, adaptively selecting the most suitable hyperparameters for GWDN is a complex and costly task that requires further investigation. Li *et al.* [86] propose a multi-layer Grad-CAM technique for FDD. The proposed Grad-CAM uses information from multiple convolutional layers to produce a comprehensive activation map, offering higher resolution and better explainability of model results. Relative amplitude to maximum amplitude (RATM), relative amplitude to average amplitude (RATA), and comprehensive explainable indicator indicators are employed to quantitatively evaluate the explainability of DNNs. These indicators are computed as [86]

$$RATM = \frac{max(F(w))}{A_{max}}, RATA = \frac{max(F(w))}{A_{avg}}, \tag{16.17}$$

where $F(w)$ and A_{avg} denote the frequency spectrum of the comprehensive activation map and the average amplitude of the frequency spectrum for the comprehensive activation map, respectively. Unfortunately, the deepening of network layers dramatically decreases the feature resolution of the proposed Grad-CAM.

A model-driven deep unrolling method has been introduced to design ante hoc IDL models that resist noise attacks, thereby enhancing intelligent fault diagnosis within aero-engine systems [87]. The proposed method's general sparse coding accounts for noise distribution and complex prior terms simultaneously across different layers. However, the approach necessitates setting a multitude of predefined parameters, a task that can present as both challenging and potentially impractical. The study in Grezmak *et al.* [88] provided a pixel-level representation-based LRP of the contributions of individual values in the input signal to the diagnosis results. While LRP provides a pixel-level representation of the contributions of individual values in the input signal, LRP's effectiveness can vary depending on the specific architecture and design choices of the neural network. A feature visualization approach-based adversarial algorithm unrolling was introduced in An *et al.* [89] to verify the meaningfulness of the encoded features. Nonetheless, the adversarial autoencoder is opaque by nature which poses a challenge in understanding how the model is built and how it makes decisions.

5) Explainable defense strategies for cyberattacks in SG: SGs are inherently vulnerable to cyber threats due to their extensive connectivity and automation [90]. XAI serves as a pivotal element in fortifying these grids by not only detecting anomalies that signify potential cyber threats but also by elucidating the reasoning behind these detections [91]. This transparency enables operators to understand and trust AI-proposed defensive measures, which is crucial for the system's security [92]. With XAI, defense mechanisms become proactive rather than reactive, enabling real-time responses to unusual patterns and potential breaches with a clear understanding of their context and potential impact [93].

Moreover, XAI facilitates strategic decision-making by providing insights into the nature and probable source of the threats, allowing for targeted and efficient countermeasures. It enables a continuous learning loop where defensive strategies are refined based on explainable feedback, leading to more sophisticated and adaptive security protocols. Ultimately, the role of XAI in intelligent defense is not just to act as a shield but also to serve as a guide, offering clarity and direction in the intricate battleground of cyber warfare within SGs.

An innovative adversarial approach is presented to generate explanations for incorrect classifications made by intrusion detection systems (IDSs) [94]. The authors focused on understanding why certain cyber-security models make specific errors. By simulating minimal input modifications that would lead to correct classifications, the proposed approach highlights the most relevant features responsible for misclassifications. The use of visualizations and intuitive explanations ensures that the complex workings of IDSs are made transparent and understandable, even to those without extensive technical expertise. A novel black-box attack was introduced to test the consistency, correctness, and confidence of gradient-based XAI methods. Gradient-based methods and multilayer perceptrons are used to demonstrate how these explanations can be manipulated or attacked, underscoring the need for robust and secure XAI methods. Through the simulations, the vulnerability of XAI methods to such attacks highly requires a robust and secure XAI design in cybersecurity.

16.6 Challenges and future research directions

The research on XAI technologies in SGs is still in its early stages. There are many issues to be addressed to truly assist operators in understanding the decision-making process and offer fertile ground for future work. Regarding the managerial dimensions of explainability, seven challenges and future directions stemming from the incorporation of XAI-driven SG applications have been identified.

1) Crafting robust and interactive XAI for energy systems: Developing interactive explanations through human–machine teaming that can handle the complexity and non-linearity of PSs remains a significant challenge [10]. The manipulation of explanations by the wrong operators that have a limited awareness of the inner workings of the algorithms could lead to harmful decisions [95]. One way to mitigate this issue is through robustness testing and adversarial training, where the AI system is trained to resist manipulation by being exposed to manipulated inputs during training [96]. Additionally, transparency in AI systems can be promoted, where not only the output but the process of generating the output is made clear. Furthermore, there is a lack of consensus on the definition of XAI and explainability, which can lead to inconsistencies in how these tradeoffs are managed.

2) Sector-specific adaptations: Most XAI solutions are not ready-made or plug-and-play; they require customization for each unique SG application. Moreover, a deep understanding of the energy sector is essential to discern which AI model decisions necessitate additional explanations, determining which responses are substantive for the end-users, as opposed to those that are trivial or expected. Following article 22 of the General Data Protection Regulation (EU GDPR) [97], the explanations produced by XAI systems must be substantial, grounded, and sufficiently informative for stakeholders, including engineers, regulatory bodies, and energy specialists. As it stands, developing such XAI solutions for SGs necessitates a collaborative effort among experts in ML, energy informatics, data analysis, electrical engineers, and domain specialists. This collaboration is vital to ensure that XAI methods are fit for their intended purpose, contributing to the trustworthiness and dependability of the answers they provide.

3) Consensus XAI strategies in SGs: Due to the current scarcity of methodologies that encompass all the preferred characteristics of XAI for SG, a prominent role in the near to mid-term is anticipated to be filled by consensus (jury) strategies. These strategies aim to merge the strengths of individual XAI techniques to bolster model reliability within SG applications [98]. In the long term, jury XAI methods, leveraging diverse algorithms and data representations pertinent to energy systems, are expected to furnish a comprehensive perspective of the modeled energy distribution and consumption processes. Presently, many of the advanced ML models employed in SG optimization and management lack this full spectrum of XAI features.

4) Benchmarking human-like explanations in dynamic SGs: Evaluations of explanations often involve human participants to determine whether the explanations align with human reasoning or to assess whether they enhance human task performance [99]. Additionally, one can create a dataset to produce explanations based on human rationale as a benchmark. Nevertheless, SGs are massive, dynamic systems in which the operators require a deep understanding of the intricacies of electrical grids to gauge the AI explanations. Crafting a dataset to benchmark "human-like" explanations in such an ever-shifting electrical operation with changes in supply and demand, weather conditions, and even market prices is challenging.

5) Standardizing XAI in SGs: The lack of XAI standardization for the diverse range of users, including AI researchers, PS experts, energy policymakers, and consumers, presents a significant challenge. This diversity necessitates the development of standardized methods for explaining AI models that can cater to the specific needs of each user group [100]. Furthermore, there is a need for ground-truth explanations to be provided, ensuring that the explanations given by the AI models align with the actual reasons behind their decisions. Moreover, ensuring the scalability of these models to cope with the massive data influx from SGs is crucial. Ethical considerations surrounding explainability and data privacy are also key challenges that need comprehensive research for successful implementation and acceptance. It is of utmost importance to have a diverse set of checks and balances, including human oversight, to validate the explanations provided by AI systems [101]. Enabling machine-to-machine communication can significantly enhance efficiency and real-time response, thereby promoting the reliability and resilience of the PS. Understanding the human factors in XAI, such as user trust and decision-making based on AI explanations, will be pivotal in the journey toward truly human-centered AI for SGs.

6) Non-identifiability in DNN interpretation: Interpreting DNNs is often complicated by the issue of non-identifiability of the models. DNNs are characterized by having a number of parameters that typically exceed the number of available training samples. This situation is further exacerbated by the dependency of the model parameters on the random selection of training data and their initial values. Interpretations that are based on the model's structure are particularly vulnerable to this non-identifiability problem [102]. Therefore, careful consideration is required when interpreting the model, recognizing that certain features and properties may emerge by chance.

7) Cybersecurity risks in XAI-enabled SGs: Transparency in AI systems is a double-edged sword; it may inadvertently expose the system's inner workings to potential cyber threats. As XAI provides insights into the decision-making processes, it could reveal critical information that attackers could exploit [92]. For instance, XAI algorithms might allow malevolent entities to reverse-engineer the model, enabling them to identify vulnerabilities in power flow optimization or to understand how the power system responds to sudden disturbances, thereby facilitating targeted attacks [92]. This entails developing sophisticated methods of providing explanations that are informative yet do not compromise the system's integrity or the privacy of the data subjects. Effective strategies could involve layering the explanations, providing different levels of insight depending on the audience, and ensuring that all disclosures are coupled with robust security measures to safeguard against the exploitation of XAI transparency.

16.7 Conclusions

XAI has made significant progress in making complex models understandable to users. This allows for the use of strong nonlinear ML models in practical applications without having to compromise on predictivity for interpretability. This chapter introduces the technical concepts of XAI, presentations of various practical algorithms such as occlusion and integrated gradients, real-world examples of how to implement these explanation techniques, and a discussion of their successful applications. According to the reviewed literature, recent efforts have been made to use XAI for more than just validation and to develop more accurate models. The significance of XAI is crucial, especially in safety-critical fields such as large-scale PSs, where complete transparency is necessary to prevent harmful outcomes. Additionally, XAI has proven to be a valuable tool in various scientific fields, providing new insights and furthering our understanding. Despite the rapid growth in interest and progress in XAI, there are still many open problems and challenges that present opportunities for further research and development. Overall, it is believed that XAI will become an essential component in creating improved, transparent, safe, fair, and unbiased ML models in SG.

Acknowledgments

Acknowledging the contributions of researchers, engineers, and organizations at the forefront of AI-powered innovations in power conversion systems.

References

[1] Zhang K, Zhang J, Xu PD, *et al.* Explainable AI in deep reinforcement learning models for power system emergency control. *IEEE Transactions on Computational Social Systems*. 2021;9:419–427.

[2] Hutson M. *Has artificial intelligence become alchemy?* American Association for the Advancement of Science; 2018;360(6388):478.

[3] Xu C, Liao Z, Li C, *et al.* Review on interpretable machine learning in smart grid. *Energies*. 2022;15(12):4427.

[4] Gunning D and Aha D. DARPA's explainable artificial intelligence (XAI) program. *AI Magazine*. 2019;40(2):44–58.

[5] Machlev R, Perl M, Belikov J, *et al.* Measuring explainability and trustworthiness of power quality disturbances classifiers using XAI-explainable artificial intelligence. *IEEE Transactions on Industrial Informatics*. 2021;18:5127–5137.

[6] Tjoa E and Cuntai G. Quantifying explainability of saliency methods in deep neural networks with a synthetic dataset. *IEEE Transactions on Artificial Intelligence*. 2022.

[7] Machlev R, Heistrene L, Perl M, *et al.* Explainable Artificial Intelligence (XAI) techniques for energy and power systems: Review, challenges and opportunities. *Energy and AI*. 2022;9:100169.

[8] Chen Z, Xiao F, Guo F, and Yan J. Interpretable machine learning for building energy management: A state-of-the-art review. *Advances in Applied Energy*. 2023;9:100123.

[9] Alsaigh R, Mehmood R, and Katib I. AI explainability and governance in smart energy systems: A review. *Frontiers in Energy Research*. 2023;11:1071291.

[10] Ali S, Abuhmed T, El-Sappagh S, *et al.* Explainable artificial intelligence (XAI): What we know and what is left to attain trustworthy artificial intelligence. *Information Fusion*. 2023;99:101805.

[11] Shi W, Han X, Wang X, *et al.* Application of artificial intelligence and its interpretability analysis in power grid dispatch and control. In: *2023 8th Asia Conference on Power and Electrical Engineering (ACPEE)*. IEEE; 2023. p. 1704–1711.

[12] Rawal A, McCoy J, Rawat DB, *et al.* Recent advances in trustworthy explainable artificial intelligence: Status, challenges, and perspectives. *IEEE Transactions on Artificial Intelligence*. 2021;3(6):852–866.

[13] Salmon WC. *Scientific explanation and the causal structure of the world*. Princeton University Press; 1984.

[14] Ahmed I, Jeon G, and Piccialli F. From artificial intelligence to explainable artificial intelligence in industry 4.0: A survey on what, how, and where. *IEEE Transactions on Industrial Informatics*. 2022;18(8): 5031–5042.

[15] Yayla A, Haghnegahdar L, and Dincelli E. Explainable artificial intelligence for smart grid intrusion detection systems. *IT Professional*. 2022;24(5): 18–24.

[16] Kabir MH, Hasan KF, Hasan MK, *et al.* Explainable artificial intelligence for smart city application: A secure and trusted platform. In: *Explainable Artificial Intelligence for Cyber Security: Next Generation Artificial Intelligence*. Springer; 2022. p. 241–263.

[17] Phillips PJ, Hahn CA, Fontana PC, *et al.* *Four principles of explainable artificial intelligence*. Gaithersburg, Maryland. 2020;18.

[18] Arnold NI, Angelov P, and Atkinson PM. An improved explainable point cloud classifier (XPCC). *IEEE Transactions on Artificial Intelligence*. 2022;4(1):71–80.

[19] Hossain E, Khan I, Un-Noor F, *et al.* Application of big data and machine learning in smart grid, and associated security concerns: A review. *IEEE Access*. 2019;7:13960–13988.

[20] Crockett K, Colyer E, Gerber L, and Latham A Building trustworthy AI solutions: A case for practical solutions for small businesses. *IEEE Transactions on Artificial Intelligence*. 2021;4(4):778–791.

[21] Weber L, Lapuschkin S, Binder A, and Samek W Beyond explaining: Opportunities and challenges of XAI-based model improvement. *Information Fusion*. 2023;92:154–176.

[22] Yohanandhan RV, Elavarasan RM, Manoharan P, *et al.* Cyber-physical power system (CPPS): A review on modeling, simulation, and analysis with cyber security applications. *IEEE Access*. 2020;8:151019–151064.

[23] Miller T. Explanation in artificial intelligence: Insights from the social sciences. *Artificial Intelligence*. 2019;267:1–38.

[24] Montavon G, Binder A, Lapuschkin S, *et al.* Layer-wise relevance propagation: An overview. In: Samek W, Montavon G, Vedaldi A, Hansen L, and Müller KR. (eds), *Explainable AI: Interpreting, Explaining and Visualizing Deep Learning*. 2019, pp. 193–209.

[25] Yang M, Xu C, Bai Y, Ma M, and Su X. Investigating black-box model for wind power forecasting using local interpretable model-agnostic explanations algorithm: Why should a model be trusted? *CSEE Journal of Power and Energy Systems*. 2025;11(1):227–242.

[26] Chen M, Liu Q, Chen S, *et al.* XGBoost-based algorithm interpretation and application on post-fault transient stability status prediction of power system. *IEEE Access*. 2019;7:13149–13158.

[27] Wu S, Zheng L, Hu W, *et al.* Improved deep belief network and model interpretation method for power system transient stability assessment. *Journal of Modern Power Systems and Clean Energy*. 2020;8(1):27–37.

[28] Golizadeh Akhlaghi Y, Aslansefat K, Zhao X, *et al.* Hourly performance forecast of a dew point cooler using explainable artificial intelligence and evolutionary optimisations by 2050. *Applied Energy*. 2021;281:116062.

[29] Chakraborty D, Alam A, Chaudhuri S, *et al.* Scenario-based prediction of climate change impacts on building cooling energy consumption with explainable artificial intelligence. *Applied Energy*. 2021;291:116807.

[30] Alova G, Trotter P, and Money A. A machine-learning approach to predicting Africa's electricity mix based on planned power plants and their chances of success. *Nature Energy*. 2021;6(2):158–166.

[31] Zhang D, Li C, Shahidehpour M, *et al.* A bi-level machine learning method for fault diagnosis of oil-immersed transformers with feature explainability. *International Journal of Electrical Power & Energy Systems*. 2022;134:107356.

[32] Kruse J, Schäfer B, and Witthaut D. Revealing drivers and risks for power grid frequency stability with explainable AI. *Patterns*. 2021;2(11):1–26.

[33] Kruse J, Schafer B, and Witthaut D. Exploring deterministic frequency deviations with explainable AI. In: *2021 IEEE International Conference on Communications, Control, and Computing Technologies for Smart Grids (SmartGridComm)*. IEEE; 2021. p. 133–139.

[34] Sairam S, Srinivasan S, Marafioti G, *et al.* Explainable Incipient Fault Detection Systems for Photovoltaic Panels. arXiv preprint arXiv:201109843. 2020.

[35] Sarp S, Kuzlu M, Cali U, *et al.* An interpretable solar photovoltaic power generation forecasting approach using an explainable artificial intelligence tool. In: *2021 IEEE Power Energy Society Innovative Smart Grid Technologies Conference*; 2021. p. 1–5.

[36] Bahani K, Ali-Ou-Salah H, Moujabbir M, *et al.* A novel interpretable model for solar radiation prediction based on adaptive fuzzy clustering and linguistic hedges. In: *SITA'20: Proceedings of the 13th International Conference on Intelligent Systems: Theories and Applications*; 2020. p. 1–12.

[37] Toubeau JF, Bottieau J, Wang Y, and Vallée F. Interpretable probabilistic forecasting of imbalances in renewable-dominated electricity systems. *IEEE Transactions on Sustainable Energy*. 2022;13(2):1267–1277.

[38] Sahoo S, Wang H, and Blaabjerg F. On the explainability of black box data-driven controllers for power electronic converters. In: *2021 IEEE Energy Conversion Congress and Exposition*; 2021. p. 1366–1372.

[39] Chaibi M, Benghoulam E, Tarik L, *et al.* An interpretable machine learning model for daily global solar radiation prediction. *Energies*. 2021;14(21):7367.

[40] Zhang W, Liu F, Wen Y, *et al.* Toward explainable and interpretable building energy modeling: An explainable artificial intelligence approach. In: *Proceedings of the 8th ACM International Conference on Systems for Energy-Efficient Buildings, Cities, and Transportation*. Association for Computing Machinery; 2021. p. 255–258.

[41] Bhatia A, Garg V, Haves P, *et al.* Explainable clustering using hyperrectangles for building energy simulation data. In: *IOP Conference Series: Earth and Environmental Science*. vol. 238; 2019. p. 012068.

[42] Lee Y, Oh J, and Kim G. Interpretation of load forecasting using explainable artificial intelligence techniques. *Transactions of the Korean Institute of Electrical Engineers*. 2020;69(3):480–485.

[43] Lu Y, Murzakhanov I, and Chatzivasileiadis S. Neural network interpretability for forecasting of aggregated renewable generation. In: *IEEE International Conference on Communications, Control, and Computing Technologies for Smart Grids*; 2021. p. 282–288.

[44] Arjunan P, Poolla K, and Miller C. Energystar++: Towards more accurate and explanatory building energy benchmarking. *Applied Energy*. 2020;276:115413.

[45] Linardatos P, Papastefanopoulos V, and Kotsiantis S. Explainable AI: A review of machine learning interpretability methods. *Entropy*. 2020;23(1):18.

[46] Lundberg SM and Lee SI. A unified approach to interpreting model predictions. *Advances in Neural Information Processing Systems*. 2017;30.

[47] Javeed D, Gao T, Kumar P, and Jolfaei A. An explainable and resilient intrusion detection system for Industry 5.0. *IEEE Transactions on Consumer Electronics*. 2024;70(1):1342–1350.

[48] Jang K, Pilario KES, Lee N, Moon I, and Na J. Explainable artificial intelligence for fault diagnosis of industrial processes. *IEEE Transactions on Industrial Informatics*. 2025;21(1):4–11.

[49] Akhlaghi YG, Aslansefat K, Zhao X, *et al.* Hourly performance forecast of a dew point cooler using explainable artificial intelligence and evolutionary optimisations by 2050. *Applied Energy*. 2021;281:116062.

[50] Liu W, Zhao J, Qiu J, *et al.* Interpretable hybrid experimental learning for trading behavior modeling in electricity market. *IEEE Transactions on Power Systems*. 2022;38(2):1022–1032.

[51] Wang B, Pei W, Xue B, *et al.* Evolving local interpretable model-agnostic explanations for deep neural networks in image classification. In: *Proceedings of the Genetic and Evolutionary Computation Conference Companion*; 2021. p. 173–174.

[52] Selvaraju RR, Cogswell M, Das A, *et al.* Grad-cam: Visual explanations from deep networks via gradient-based localization. In: *Proceedings of the IEEE International Conference on Computer Vision*; 2017. p. 618–626.

[53] Chen Y and Zhang D. Theory-guided deep-learning for electrical load forecasting (TgDLF) via ensemble long short-term memory. *Advances in Applied Energy*. 2021;1:100004.

[54] Cremer J, Konstantelos I, and Strbac G. From optimization-based machine learning to interpretable security rules for operation. *IEEE Transactions on Power Systems*. 2019;34(5):3826–3836.

[55] Ren C, Xu Y, and Zhang R. An interpretable deep learning method for power system transient stability assessment via tree regularization. *IEEE Transactions on Power Systems*. 2021;37(5):3359–3369.

[56] Han T, Chen J, Wang L, *et al.* Interpretation of stability assessment machine learning models based on Shapley value. In: *2019 IEEE 3rd Conference on Energy Internet and Energy System Integration*. IEEE; 2019. p. 243–247.

[57] Machlev R, Perl M, Caciularu A, *et al.* Explaining the decisions of power quality disturbance classifiers using latent space features. *International Journal of Electrical Power & Energy Systems*. 2023;148:108949.

[58] Zhang K, Xu P, and Zhang J. Explainable AI in deep reinforcement learning models: A SHAP method applied in power system emergency control. In: *2020 IEEE 4th Conference on Energy Internet and Energy System Integration*. IEEE; 2020. p. 711–716.

[59] Fang J, Chen K, Liu C, and He J. An explainable and robust method for fault classification and location on transmission lines. *IEEE Transactions on Industrial Informatics*. 2023;19(10):10182–10191.

[60] Chen Z, Silvestri F, Tolomei G, Wang J, Zhu H, and Ahn H. Explain the explainer: Interpreting model-agnostic counterfactual explanations of a deep reinforcement learning agent. *IEEE Transactions on Artificial Intelligence*. 2024;5(4):1443–1457.

[61] Hamilton RI and Papadopoulos PN. Using SHAP values and machine learning to understand trends in the transient stability limit. *IEEE Transactions on Power Systems*. 2023.

[62] Wenninger S, Kaymakci C, and Wiethe C. Explainable long-term building energy consumption prediction using QLattice. *Applied Energy*. 2022;308:118300.

[63] Fan C, Xiao F, Yan C, *et al.* A novel methodology to explain and evaluate data-driven building energy performance models based on interpretable machine learning. *Applied Energy*. 2019;235:1551–1560.

[64] Gao Y and Ruan Y. Interpretable deep learning model for building energy consumption prediction based on attention mechanism. *Energy and Buildings*. 2021;252:111379.

[65] Li A, Xiao F, Zhang C, *et al.* Attention-based interpretable neural network for building cooling load prediction. Applied Energy. 2021;299:117238.

[66] Miller C. What's in the box?! Towards explainable machine learning applied to non-residential building smart meter classification. *Energy and Buildings*. 2019;199:523–536.

[67] Jiang R, Zeng S, Song Q, *et al.* Deep-chain echo state network with explainable temporal dependence for complex building energy prediction. *IEEE Transactions on Industrial Informatics*. 2022;19(1):426–435.

[68] Jang J, Jeong W, Kim S, *et al.* RAID: Robust and interpretable daily peak load forecasting via multiple deep neural networks and Shapley values. *Sustainability*. 2023;15(8):6951.

[69] Bottieau J, Wang Y, De Grève Z, Vallée F, and Toubeau JF Interpretable transformer model for capturing regime switching effects of real-time electricity prices. *IEEE Transactions on Power Systems*. 2023;38(3):2162–2176.

[70] Kuzlu M, Cali U, Sharma V, *et al.* Gaining insight into solar photovoltaic power generation forecasting utilizing explainable artificial intelligence tools. *IEEE Access*. 2020;8:187814–187823.

[71] Mitrentsis G and Lens H. An interpretable probabilistic model for short-term solar power forecasting using natural gradient boosting. *Applied Energy*. 2022;309:118473.

[72] Wang H, Cai R, Zhou B, *et al.* Solar irradiance forecasting based on direct explainable neural network. *Energy Conversion and Management*. 2020;226:113487.

[73] Gao J, Chen Y, Hu W, *et al.* An adaptive deep-learning load forecasting framework by integrating transformer and domain knowledge. *Advances in Applied Energy*. 2023;10:100142.

[74] Batic D, Stankovic V, Stankovic L. Towards transparent load disaggregation – A framework for quantitative evaluation of explainability using explainable AI. *IEEE Transactions on Consumer Electronics*. 2024;70(1):4345–4356.

[75] Kim JY and Cho SB. Electric energy consumption prediction by deep learning with state explainable autoencoder. *Energies*. 2019;12(4):739.

[76] Yun L, Wang D, and Li L. Explainable multi-agent deep reinforcement learning for real-time demand response towards sustainable manufacturing. *Applied Energy*. 2023;347:121324.

[77] Li C, Dong Z, Ding L, *et al.* Interpretable memristive LSTM network design for probabilistic residential load forecasting. *IEEE Transactions on Circuits and Systems I: Regular Papers*. 2022;69(6):2297–2310.

[78] Xu C, Li C, and Zhou X. Interpretable LSTM based on mixture attention mechanism for multi-step residential load forecasting. *Electronics*. 2022;11(14):2189.

[79] Van Zyl C, Ye X, and Naidoo R. Harnessing eXplainable artificial intelligence for feature selection in time series energy forecasting: A comparative analysis of Grad-CAM and SHAP. *Applied Energy*. 2024;353:122079.

[80] Kim JY and Cho SB. Explainable prediction of electric energy demand using a deep autoencoder with interpretable latent space. *Expert Systems with Applications*. 2021;186:115842.

[81] Santos OL, Dotta D, Wang M, *et al.* Performance analysis of a DNN classifier for power system events using an interpretability method. *International Journal of Electrical Power & Energy Systems*. 2022;136:107594.

[82] Sairam S, Seshadhri S, Marafioti G, *et al.* Edge-based explainable fault detection systems for photovoltaic panels on edge nodes. *Renewable Energy*. 2022;185:1425–1440.

[83] Li G, Chen L, Fan C, *et al.* Interpretation and explanation of convolutional neural network-based fault diagnosis model at the feature-level for building energy systems. *Energy and Buildings*. 2023;295:113326.

[84] Grezmak J, Wang P, Sun C, *et al.* Explainable convolutional neural network for gearbox fault diagnosis. *Procedia CIRP*. 2019;80:476–481.

[85] Li T, Sun C, Li S, Wang Z, Chen X, and Yan R. Explainable graph wavelet denoising network for intelligent fault diagnosis. *IEEE Transactions on Neural Networks and Learning Systems*. 2024;35(6):8535–8548.

[86] Li S, Li T, Sun C, *et al.* Multilayer Grad-CAM: An effective tool towards explainable deep neural networks for intelligent fault diagnosis. *Journal of Manufacturing Systems*. 2023;69:20–30.

[87] Zhao Z, Li T, An B, *et al.* Model-driven deep unrolling: Towards interpretable deep learning against noise attacks for intelligent fault diagnosis. *ISA Transactions*. 2022;129:644–662.

[88] Grezmak J, Zhang J, Wang P, *et al.* Interpretable convolutional neural network through layer-wise relevance propagation for machine fault diagnosis. *IEEE Sensors Journal*. 2019;20(6):3172–3181.

[89] An B, Wang S, Qin F, Zhao Z, Yan R, and Chen X. Adversarial algorithm unrolling network for interpretable mechanical anomaly detection. *IEEE Transactions on Neural Networks and Learning Systems*. 2024;35(5):6007–6020.

[90] Abou El Houda Z, Brik B, and Khoukhi L. "Why should I trust your IDs?": An explainable deep learning framework for intrusion detection systems in internet of things networks. *IEEE Open Journal of the Communications Society*. 2022;3:1164–1176.

[91] Chandre PR, Vanarote V, Patil R, *et al.* Explainable AI for intrusion prevention: A review of techniques and applications. In: *International Conference on Information and Communication Technology for Intelligent Systems*. Springer; 2023. p. 339–350.

[92] He K, Kim DD, and Asghar MR. Adversarial machine learning for network intrusion detection systems: A comprehensive survey. *IEEE Communications Surveys & Tutorials*. 2023;25(1):538–566.

[93] Tabassum S, Parvin N, Hossain N, *et al.* IoT network attack detection using XAI and Reliability Analysis. In: *2022 25th International Conference on Computer and Information Technology (ICCIT)*. IEEE; 2022. p. 176–181.

[94] Marino DL, Wickramasinghe CS, and Manic M. An adversarial approach for explainable AI in intrusion detection systems. In: *IECON 2018 – 44th Annual Conference of the IEEE Industrial Electronics Society*. IEEE; 2018. p. 3237–3243.

[95] Laurinec P and Lucká M. Interpretable multiple data streams clustering with clipped streams representation for the improvement of electricity consumption forecasting. *Data Mining and Knowledge Discovery*. 2019;33:413–445.

[96] Zhang W, Wen Y, Tseng KJ, *et al.* Demystifying thermal comfort in smart buildings: An interpretable machine learning approach. *IEEE Internet of Things Journal*. 2020;8(10):8021–8031.

[97] Bayamlıoğlu E. The right to contest automated decisions under the General Data Protection Regulation: Beyond the so-called "right to explanation". *Regulation & Governance*. 2022;16(4):1058–1078.

[98] Massaoudi M, Chihi I, Sidhom L, *et al.* Enhanced random forest model for robust short-term photovoltaic power forecasting using weather measurements. *Energies*. 2021;14(13):3992.

[99] Colin J, Fel T, Cadène R, *et al.* What I cannot predict, I do not understand: A human-centered evaluation framework for explainability methods. *Advances in Neural Information Processing Systems*. 2022;35:2832–2845.

[100] Miller C. What's in the box?! Towards explainable machine learning applied to non-residential building smart meter classification. *Energy and Buildings*. 2019;199:523–536.

[101] Hu M, Ge D, Telford R, *et al.* Classification and characterization of intraday load curves of PV and non-PV households using interpretable feature extraction and feature-based clustering. *Sustainable Cities and Society*. 2021;75:103380.

[102] Heil BJ, Hoffman MM, Markowetz F, *et al.* Reproducibility standards for machine learning in the life sciences. *Nature Methods*. 2021;18(10):1132–1135.

Chapter 17
Electric mobility integration: a deep dive into AI solutions

Ali Sharida[1], Anas Karaki[1], Sertac Bayhan[1] and Haitham Abu-Rub[2]

17.1 Introduction

Electric mobility, driven by the proliferation of electric vehicles (EVs), is rapidly emerging as a cornerstone of sustainable energy systems. The shift from traditional internal combustion engine vehicles to EVs represents a critical step toward reducing greenhouse gas emissions, minimizing dependence on fossil fuels, and achieving global climate targets. Beyond environmental benefits, the adoption of EVs also addresses pressing challenges in energy management and urban infrastructure, such as enhancing the efficiency of transportation networks [1].

However, integrating electric mobility into existing energy systems introduces significant challenges. EVs demand a robust and scalable charging infrastructure, impose additional load on the power grid, and require synchronization with renewable energy sources (RESs) to fully realize their environmental potential. As EV adoption grows, these challenges highlight the need for intelligent solutions to manage the intricate interplay between mobility and energy systems [2].

Artificial intelligence (AI) has emerged as a transformative enabler for addressing these complexities. AI-driven solutions provide unparalleled capabilities for real-time data analysis, predictive modeling, and autonomous decision-making. These capabilities make AI solutions critical to the efficient and scalable integration of EVs [3]. AI can revolutionize the electric mobility landscape through many applications including optimizing charging schedules [4], managing bi-directional energy flows [5], and improving the reliability of EV systems through predictive maintenance [6]. Its ability to learn from vast datasets and adapt to changing conditions makes AI an indispensable tool for a seamless transition toward electrified transportation. This chapter delves into the role of AI in electric mobility integration, focusing on key areas where AI can make a transformative impact. The discussion encompasses a range of topics, including the optimization of EV charging infrastructure,

[1]Qatar Environment and Energy Research Institute, Hamad Bin Khalifa University, Qatar
[2]College of Science and Engineering, Hamad Bin Khalifa University, Qatar

the management of bi-directional energy flows between vehicles and the grid, and the application of predictive analytics for maintaining EV systems.

The objectives of this chapter are threefold:

- To highlight the importance of AI in electric mobility.
- To explore key AI applications in EV ecosystems.
- To address challenges and future directions.

17.2 AI for EV charging infrastructure

The EV charger is composed of two critical power electronic converters. The first one is the AC–DC converter, which is also known as the active front-end rectifier (AFR). Its primary function is to regulate the DC link voltage and ensure that the operation maintains a unity power factor while converting the power from AC to DC. The second converter is DC–DC, which is responsible for regulating the battery voltage between 200 and 1000 V [1,7], power, and the charging current. Typically, multiple parallel converters (DC–DC and AC–DC) are employed as a solution for the higher power rating devices [1], as illustrated in Figure 17.1. A single converter is typically unable to meet the power requirements of medium or heavy-duty EVs. In addition to the power electronic converters, galvanic isolation is necessary to prevent the direct passage of electrical current between the power grid and the vehicle. This isolation could be either low frequency or high frequency (as illustrated in Figure 17.1(a) and (b)).

Numerous infrastructures and approaches were proposed for interfacing these units [1] as shown in Figure 17.1, which are (i) stand-alone station (Figure 17.1(a)), where each AFR is connected in series with a DC–DC converter having a single input and a single output [1]; (ii) charger link DC (CLDC) coupled station (Figure 17.1(b)), where DC–DC converters are connected in parallel with separate DC inputs and a common DC output [8]; (iii) intermediate link DC (ILDC) coupled station [9,10] (Figure 17.1(c)), where multiple AFRs are connected in parallel with a common AC input and common DC output; and (iv) ILDC and CLDC-coupled stations (Figure 17.1(d)), where the AFRs at the ILDC bus and the DC–DC converters at the CLDC bus are connected in parallel with common DC input and common DC output.

17.2.1 Control of EV charging infrastructures

As the EV charger is primarily composed of two cascaded converters, each one must accomplish a minimum of two objectives, namely voltage and current regulation, which translates to four control variables. To accomplish this, two cascaded control loops are required as illustrated in Figure 17.2. The primary objective of the outer control loop is to regulate the grid currents and the voltage of the middle DC link, while the objective of the inner loop is to regulate the voltage and currents at the charger's output terminal [11].

Although the control process looks simple consisting of only two-control loops, the design and operation of EV chargers are technically demanding due to the complexity of the control tasks involved. EV chargers operate in environments with

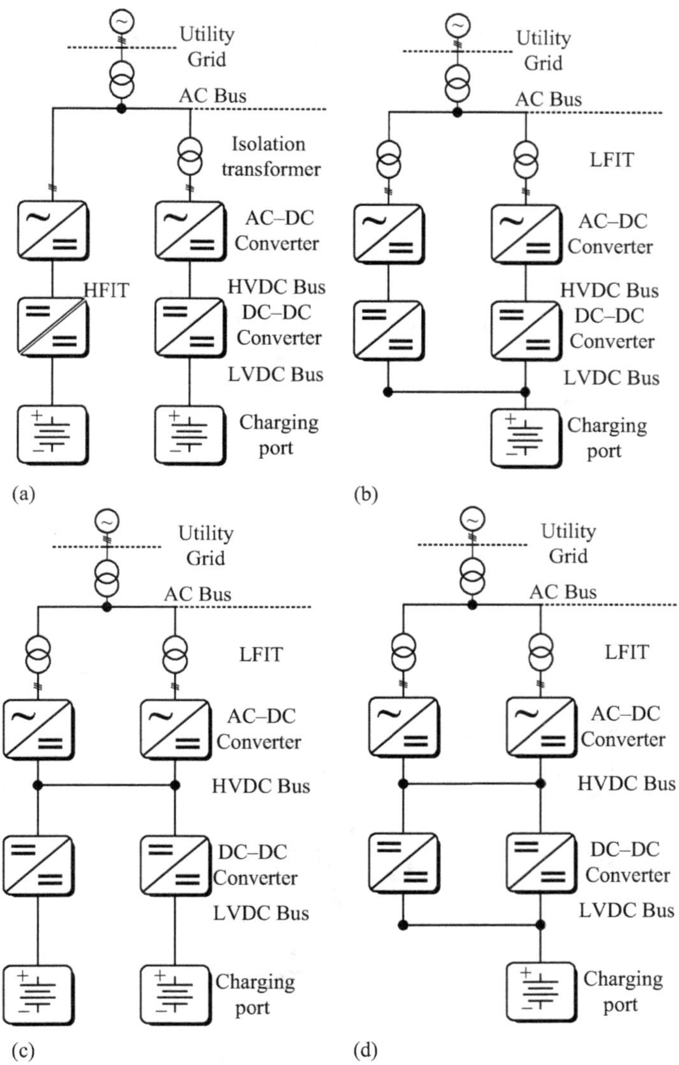

Figure 17.1 Charging station infrastructures: (a) stand-alone stations, (b) CLDC-coupled station, (c) ILDC-coupled station, and (d) ILDC- and CLDC-coupled stations.

varying grid conditions, such as fluctuations in voltage, frequency, and load demand. Managing these variations while maintaining precise control over multiple variables is challenging [12]. Moreover, the cascaded converters exhibit nonlinear behavior, and the control loops are tightly coupled. This makes the system highly sensitive to disturbances, such as grid instability or sudden changes in load. Conversely, with the increasing integration of EV chargers into the power grid, ensuring grid stability is critical. Chargers must comply with grid codes, minimize harmonic distortion, and

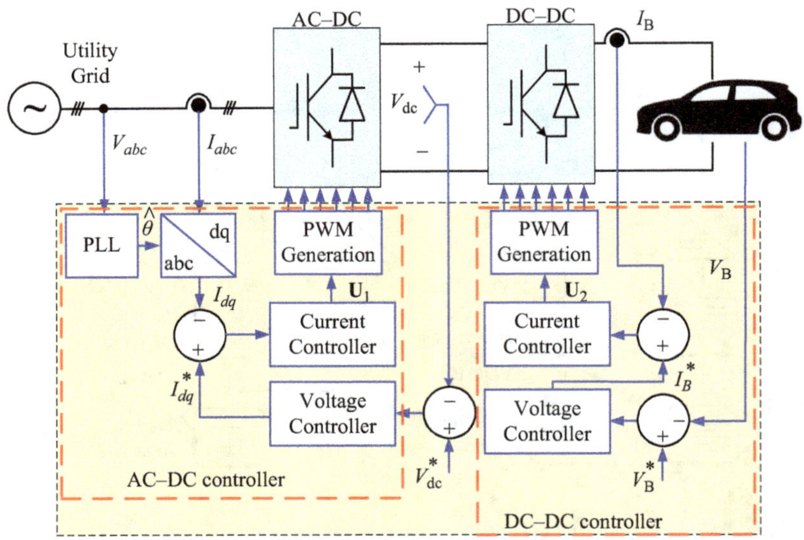

Figure 17.2 Block diagram of EV charger control system

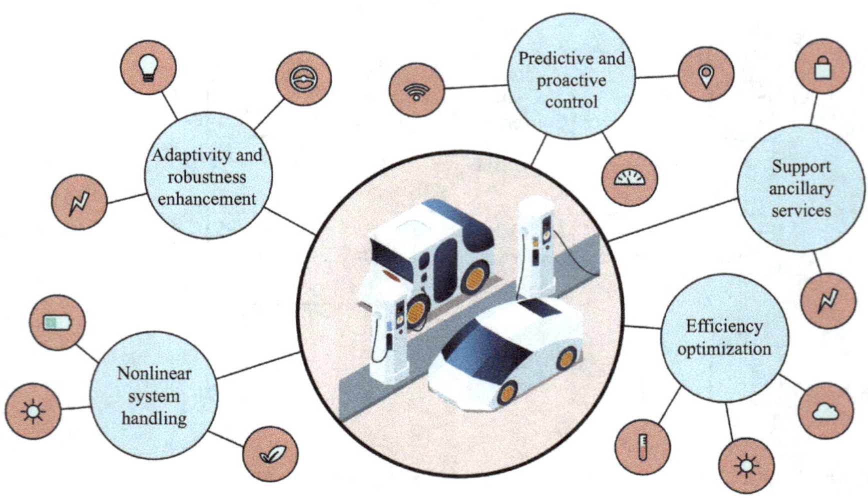

Figure 17.3 Applications of AI in EV charging control

participate in ancillary services, such as reactive power support or demand response (Figure 17.3)

AI offers transformative solutions to overcome the complexities associated with EV charger control. Unlike traditional control techniques, which rely on linear models and fixed parameters, AI-driven approaches are adaptive, data-driven, and capable of handling the nonlinear, multi-variable nature of EV chargers. AI algorithms, such as reinforcement learning and neural networks, can dynamically adapt to changing grid and load conditions to ensure optimal control performance even under challenging scenarios [13]. Moreover, AI enables predictive control by analyzing historical and real-time data [14]. For example, machine learning models can forecast grid disturbances or load changes, which allows the charger to proactively adjust its operations. In addition, AI excels in managing nonlinear and coupled dynamics. Advanced algorithms can model complex relationships between control variables to ensure stability and precision. Furthermore, AI-driven optimization algorithms can identify the most efficient operating points for the cascaded converters, which maximizes the energy conversion efficiency while minimizing losses.

In addition to the main charging objectives, AI can enable EV chargers to perform intelligent grid interactions, such as participating in demand response programs or optimizing reactive power injection to support grid stability [15].

To understand the role of AI in enhancing the reliability of the EV charging control system, consider a case study where the control system must achieve the primary charging requirements while supporting the inertia of the AC grid while minimizing the degradation of the battery.

The first phase in designing the control system is the mathematical modeling. The grid-side converter is modeled using a *dq*-frame representation for three-phase systems. The *d*- and *q*-axis voltage equations are given as [16]

$$V_d = R_g I_d + L_g \frac{dI_d}{dt} - \omega L_g I_q + u_d \tag{17.1}$$

$$V_q = R_g I_q + L_g \frac{dI_q}{dt} + \omega L_g I_d + u_q \tag{17.2}$$

where V_d, V_q are the *d*- and *q*-axis voltages, I_d, I_q are the *d*- and *q*-axis currents, R_g, L_g are the grid resistance and inductance, respectively, ω is the angular frequency, and u_{dq} is the control signal.

The DC link voltage can be modeled by the mean of the power balance formula as follows:

$$P_{\text{in}} = P_{\text{out}} + C_{dc} \frac{dV_{dc}}{dt} \tag{17.3}$$

where P_{in} is the power supplied from the grid-side converter, P_{out} is the power delivered to the load-side converter, C_{dc} is the capacitance of the DC link, and V_{dc} is the DC link voltage.

For the output side, the output voltage and current dynamics are modeled as

$$V_o = R_o I_o + L_o \frac{dI_o}{dt} \tag{17.4}$$

where V_o is the output voltage, I_o is the output current, and R_o, L_o are the load resistance and inductance, respectively.

The control objective here is to regulate the output voltage V_o and output current I_o to match the EV battery's charging requirements.

To mimic the behavior of conventional generators and provide frequency regulation, virtual inertia is introduced:

$$J_v \frac{d\Delta\omega}{dt} + D_v \Delta\omega = \Delta P \tag{17.5}$$

where J_v is the virtual moment of inertia, D_v is the virtual damping coefficient, $\Delta\omega$ is the frequency deviation, and ΔP is the power deviation injected into the grid.

For the EV's battery side, the battery degradation cost is modeled as

$$C_b = \int_0^T \left(\alpha I_o^2 + \beta \left(\frac{dV_o}{dt} \right)^2 \right) dt \tag{17.6}$$

where α and β are coefficients related to the battery's degradation and T is the charging duration.

To fulfill all of the aforementioned requirements, reinforcement learning (RL) will be adopted. In RL, an agent learns to make decisions by interacting with the environment to maximize a cumulative reward.

The key elements of RL technique in an EV charging system can be summarized as follows:

- Agent: The AI controller responsible for managing the EV charger's operations.
- Environment: The EV charger system, including its components (grid-side converter, load-side converter), grid conditions, and EV battery.
- State (s_t): A vector representing the system's current condition, including

$$s_t = [I_g, V_{dc}, V_o, I_o, \Delta\omega, \mathrm{SoC}] \tag{17.7}$$

  where I_g is the grid current, V_{dc} is the DC link voltage, V_o and I_o are the output voltage and current, respectively, $\Delta\omega$ is the grid frequency deviation, and SoC is the battery state of charge.
- Action (a_t): The control actions taken by the agent, such as

$$a_t = [\mathrm{PWM_{GSC}}, \mathrm{PWM_{LSC}}, Q_{\mathrm{grid}}] \tag{17.8}$$

  where $\mathrm{PWM_{GSC}}$ and $\mathrm{PWM_{LSC}}$ are the pulse-width modulation signals for the converters and Q_{grid} is the reactive power injected into the grid.
- Reward (r_t): The feedback signal that indicates how well the agent is performing. It is based on the deviation from the desired performance.

The reward function incorporates all the control objectives:

$$\begin{aligned} r_t = -\Big(& w_1 |I_g - I_g^*|^2 + w_2 |V_{dc} - V_{dc}^*|^2 + w_3 |V_o - V_o^*|^2 \\ & + w_4 |I_o - I_o^*|^2 + w_5 C_b + w_6 |\Delta\omega|^2 \Big) \end{aligned} \tag{17.9}$$

where $w_1, w_2, \ldots, w_6$ are weights that prioritize the importance of each objective, C_b is the battery degradation cost, and $\Delta\omega$ is the grid frequency deviation.

For continuous state and action spaces, a deep reinforcement learning algorithm such as deep deterministic policy gradient or proximal policy optimization (PPO) is suitable. These algorithms learn both the policy and value functions using deep neural networks.

The RL agent is trained in a simulated environment that acts as a virtual representation of the EV charging system, enabling the agent to interact with and learn from the system's dynamics. Initially, the agent explores the action space by performing random actions to gather diverse experiences (exploration), and over time, it shifts toward exploiting learned policies to maximize the cumulative reward. The neural network architecture used in the RL agent is designed with an input layer that receives the state vector s_t and an output layer that provides the action vector a_t. Between these layers, hidden layers with rectified linear unit activation functions process the data. This enables the agent to learn complex relationships within the system.

Then, the agent uses the Bellman equation for value function update:

$$Q(s_t, a_t) = r_t + \gamma \max_a Q(s_{t+1}, a) \tag{17.10}$$

where γ is the discount factor.

For policy updates, they can be performed using the policy gradient method as follows:

$$\nabla_\theta J(\pi_\theta) = \mathbb{E}\left[\nabla_\theta \log \pi_\theta(a_t|s_t) Q(s_t, a_t)\right] \tag{17.11}$$

Once the RL agent is trained, it can be deployed to the real system, where it continuously optimizes the control of the EV charger.

17.2.2 Optimizing charging profiles

Optimal charge profiling refers to designing a charging plan tailored to the specific needs of each EV while considering broader objectives. This involves determining the most efficient and effective charging schedule for EVs to minimize charging costs, minimize the impact of the charging process on grid stability, and minimize the degradation of the battery. As EV adoption rises and charging demand grows, traditional charging methods are becoming inadequate to address challenges such as grid overload, high electricity costs, and uneven charging patterns. AI-driven optimal charge profiling approaches provide advanced, data-driven solutions to these challenges by dynamically adapting charging schedules based on real-time data and predictive analytics.

Multiple AI-based techniques were proposed in the literature to achieve optimal charge profiling such as RL, optimization-based approaches, machine learning predictive models, multi-agent systems, and clustering algorithms [17,18].

RL enables EVs or charging stations to learn optimal charging strategies by interacting with the environment. The RL agent makes decisions about when and how much to charge to maximize a cumulative reward function. The reward function is carefully designed to balance a given set of objectives, and algorithms like deep Q-learning or PPO are commonly used for optimization.

Optimization methods such as linear programming, mixed-integer linear programming, and genetic algorithms are effective for solving complex charge scheduling problems. These methods use objective functions to minimize costs or emissions and incorporate constraints such as vehicle departure times, charger capacity, and grid limitations. As a result, they provide a globally optimal or near-optimal solution.

Multi-agent systems are usually employed in scenarios that involve multiple EVs and charging stations. They allow each EV or station to act as an independent agent.

These agents negotiate with one another to find optimal charging schedules that meet their individual needs while contributing to system-wide goals.

Finally, AI clustering techniques, like K-means or hierarchical clustering, group EVs with similar charging requirements (such as similar state of charge or departure times). This helps create tailored charge profiles for each group to improve charging efficiency and reduce charging time.

Despite the great advantages offered by AI-driven optimal charge profiling, AI-driven solutions face several challenges. First, they rely heavily on high-quality, real-time data from EVs, charging stations, and the grid, meaning that incomplete or inaccurate data can degrade the performance. Additionally, as the number of EVs grows, AI algorithms must scale to manage the increasing complexity of charge profiling. Integrating RESs adds further complexity, as synchronizing charging schedules with the intermittent nature of renewable energy requires accurate forecasting. Finally, ensuring fairness and equity in charging resource access, particularly during high-demand periods, remains a critical issue.

To understand the role of AI in optimizing the charging profiles, consider a case study where an EV charger uses a constant-current constant-voltage charging profile. It is desired to minimize charging time T_{charge}, maximizing energy efficiency η, and prolonging battery life by limiting stress caused by high currents, temperatures, or overcharging. These objectives can be formulated as follows:

$$Minimize\ T_{charge} = t_{end} - t_{start} \tag{17.12}$$

$$Maximize\ \eta = \frac{\int_{t_{start}}^{t_{end}} V(t)I(t)dt}{E_{battery}} \tag{17.13}$$

$$Minimze\ L_{battery} = \sum_{t} (\alpha I(t)^2 + \beta \Delta T(t) + \gamma (V(t) - V_{nom})^2) \tag{17.14}$$

where T_{charge} is the total charging time, t_{start} is the start time of the charging process, t_{end} is the time when the charge process completes, η is the charging efficiency, $E_{battery}$ is the stored energy in the battery, $I(t)$ is the charging current at time t, $\delta T(t)$ is the temperature rise above ambient at time t, $V(t)$ is the voltage at time t, V_{nom} is the nominal voltage of the battery, and α, β, and γ are the weights for current stress, thermal stress, and over-voltage stress, respectively.

After defining these objectives, they can be merged into one cost function as follows:

$$J = \omega_1 T_{charge} + \omega_2 L_{battery} + \omega_3 (1 - \eta) \tag{17.15}$$

where J is the equivalent cost function and $\omega_1 - \omega_3$ are the weighting factors.

After defining the cost function, RL can be used to minimize it. First, the components of the RL can be defined as follows:

- The AI agent learns optimal charging policies by balancing T_{charge}, $L_{battery}$, and η.
- State: Battery voltage, current, temperature, and SoC.
- Action: Adjust charging current or voltage dynamically.
- Reward: Negative of the cost function J.

Once the state, action, and reward are defined, the next steps involve designing, training, and deploying the RL model.

17.3 AI in bidirectional energy flow: V2G and G2V

As EVs continue to gain popularity, their integration with power grids has opened up opportunities to enhance grid stability, renewable energy utilization, and overall energy system efficiency. In this domain, there are two key concepts: vehicle-to-grid (V2G) and grid-to-vehicle (G2V). V2G refers to the bidirectional energy flow between an EV and the power grid. In this mode, EVs act as mobile energy storage units, capable of returning stored energy to the grid when required. The key objective of V2G is to support the grid through ancillary services. Conversely, G2V represents the traditional unidirectional energy flow where the grid supplies energy to charge EVs. However, advanced G2V systems optimize charging to minimize costs, align with renewable energy availability, and reduce grid stress.

These paradigms redefine the role of EVs in the energy ecosystem, leveraging AI to optimize operations and unlock new potentials [19]. On the first hand, AI leverages historical and real-time data to forecast energy demand, renewable generation, and user behavior. This enables precise, generation, and demand predictions. This allows the control system to forecast when the grid might require EV energy support (V2G) and optimal charging times during periods of high renewable energy production (G2V). By aligning EV charging and discharging with these forecasts, AI helps mitigate the challenges posed by the intermittent nature of RES.

On the other hand, AI algorithms enable real-time optimization by dynamically scheduling when and how much energy to charge (G2V) or discharge (V2G) based on grid conditions, electricity prices, and user preferences. They also optimize resource allocation by prioritizing which EVs will participate in ancillary services, considering factors such as SoC, location, and user preferences.

Moreover, AI plays a crucial role in battery health management by monitoring and predicting battery aging, which ensures that frequent charging and discharging cycles associated with V2G operations do not compromise EV longevity [20]. For example, AI can adjust discharge rates or limit charging cycles to minimize battery degradation while maximizing the vehicle's participation in grid services.

17.4 AI-based predictive maintenance for EVs

Predictive maintenance has emerged as a cornerstone of modern EV management systems. This functionality leverages advanced technologies to improve reliability, extend component life, and minimize downtime. Moreover, predictive maintenance ensures that the EV operates efficiently and sustainably while reducing costs associated with unexpected failures. This section explores four key dimensions of predictive maintenance (i) condition monitoring of EV components, (ii) extending battery life through predictive analytics, (iii) proactive solutions enabled by IoT, and (iv) real-time fault diagnosis.

Condition monitoring is critical for assessing the health and performance of essential EV components, including batteries, motors, inverters, and braking systems. AI enhances this process through analyzing real-time data from onboard sensors. Then, AI identifies anomalies and predicts potential failures before they occur. Machine learning models, such as support vector machines (SVM) and neural networks, detect abnormal patterns, behavior, or deviations in operating parameters. Predictive maintenance mainly targets the mechanical and electrical systems of the EVs. For mechanical systems, techniques like vibration and acoustic analysis are usually utilized to identify issues in motors and bearings by analyzing changes in sound or vibration frequencies [21]. These approaches are particularly useful for detecting overheating in motors, friction in bearings, and faults in regenerative braking systems. By enabling early detection, AI-driven condition monitoring enhances safety, reduces maintenance costs, and ensures uninterrupted vehicle operation.

Similar to mechanical predictive maintenance, predictive maintenance for power electronic systems can be achieved by continuously monitoring the controlled states such as currents, voltages, and the temperature of the operating components and observing abnormal patterns in their behavior. These patterns can indicate potential failures in power electronic components on the component or module levels including open-circuit, short-circuit, and passive component faults. This allows for timely intervention and minimizes the risk of catastrophic failures in the EV's electrical systems [22].

Moreover, as the battery is the most critical and costly component, the longevity of the EV's battery is a central focus of predictive maintenance. AI-driven predictive analytics monitors key battery parameters, such as capacity fade and temperature, to estimate its state of health and remaining useful life. Then, techniques like deep learning and reinforcement learning are usually utilized to optimize the charging profiles to reduce wear while maximizing the charging performance. In addition, digital twin technology is usually used to simulate the performance of the battery under various scenarios to predict degradation and guide the decisions of predictive maintenance.

AI-driven predictive maintenance utilizes IoT to collect data from all EV components. This enhances the predictive maintenance functionality by connecting vehicles to a network of sensors, cloud platforms, and diagnostic tools. Embedded sensors within EVs continuously monitor parameters such as temperature, vibration, current, and voltage. Then, transmit real-time data to cloud-based systems for further analysis. AI algorithms process this data to identify patterns, detect anomalies, and predict potential issues before they lead to failures. Edge computing plays a crucial role in this field, as it enables instant responses to critical conditions directly at the vehicle level. This minimizes response time and ensures prompt actions. Moreover, IoT-enabled systems provide proactive maintenance solutions. These solutions include remote diagnostics, automated alerts for emerging issues, and predictive scheduling of maintenance tasks based on the performance of key components.

Finally, fault diagnosis is another critical component of predictive maintenance. Fault diagnosis focuses on detecting, isolating, and tolerating issues in EV systems before they lead to failures. AI-powered techniques enhance fault diagnosis by analyzing sensor data to identify abnormal patterns or deviations in system behavior that might indicate underlying issues [21]. Machine learning models, such as decision trees, SVM, and neural networks, can be used to classify faults based on historical data and real-time sensor readings. Advanced signal processing methods are also employed to analyze electrical signals and waveforms, which identify issues like partial discharges in insulation or misalignment in motors. Fault diagnosis techniques are particularly valuable in detecting problems in critical EV components, such as the battery, motor, inverters, and power electronics. For example, it can detect early signs of battery imbalances, overheating, or internal short circuits, motor faults like rotor damage or bearing wear, and inverter failures like open and short circuits.

17.5 Challenges in AI integration

As AI continues to transform the EV industry, the integration of AI into EV systems introduces a range of challenges that need to be addressed for successful deployment and operation. While AI has the potential to significantly enhance EV performance, reliability, and user experience, these advancements also come with critical hurdles. In general, there are three primary challenges in AI integration for EV solutions (i) data security and privacy concerns, (ii) the need for interoperability and standardization, and (iii) the limitations of computational and infrastructural resources.

17.5.1 Data security and privacy

One of the most significant challenges in AI integration for EVs is managing data security and privacy. AI systems rely heavily on vast amounts of data collected from EVs, including real-time driving patterns, battery performance, environmental conditions, user behavior, and vehicle diagnostics. This data is crucial for AI models to learn, adapt, and optimize vehicle performance, but it also raises concerns about the security and privacy of sensitive information.

The sheer volume of personal and operational data collected by EVs creates vulnerabilities that can be exploited by malicious actors. For example, GPS tracking data can reveal personal travel habits, and battery usage patterns may expose sensitive information about users' daily routines. Furthermore, data communication between vehicles, cloud platforms, and external systems increases the potential attack surfaces for cyber threats, which can compromise the integrity and confidentiality of the information.

To address these concerns, robust encryption techniques, secure data storage solutions, privacy-preserving mechanisms, and advanced authentication mechanisms are necessary to protect sensitive data from unauthorized access or breaches. Additionally, AI models should be designed to anonymize data and ensure that personally identifiable information is either encrypted or not used for analysis. The implementation of privacy-by-design principles, where privacy considerations are integrated into the development of AI-driven systems from the outset, is crucial for mitigating privacy risks. Governments and regulatory bodies must also establish clear guidelines and standards for data collection, usage, and sharing.

17.5.2 Interoperability and standards

The successful deployment of AI-powered EV solutions requires seamless integration with various systems, including power grids, charging infrastructure, V2G technologies, and RESs. However, the lack of standardized communication protocols and interoperability between different platforms remains a significant barrier.

EV manufacturers, charging station operators, utility companies, and renewable energy providers all utilize different technologies, protocols, and data formats. This creates a fragmented ecosystem where systems may not easily communicate with each other or share data in real-time, which limits the effectiveness of AI-driven optimization strategies. For instance, V2G scenario, where EVs can supply power back to the grid, AI algorithms need to gather data from both the EV and the grid to make real-time decisions. However, if the EV and grid management systems use incompatible data formats or protocols, the AI system will struggle to efficiently manage power flows, optimize charging schedules, or participate in ancillary services.

To address these challenges, the development of standardized communication protocols, data formats, and application programming interfaces is essential. Industry-wide collaboration is required to create interoperability frameworks that allow various EV and energy systems to work together. International organizations, such as the International Electrotechnical Commission and the Open Charge Alliance, have already initiated standardization efforts for electric vehicle charging infrastructure and communication protocols. Further advancement of these standards will be necessary to ensure that AI systems can be effectively integrated across diverse platforms and environments.

17.5.3 Resource constraints: computational and infrastructural limitations

Another challenge in the integration of AI into EV systems lies in the resource constraints in terms of computational power and infrastructure. AI algorithms, especially those used for real-time decision-making, predictive maintenance, and energy optimization, require significant computational resources. Training complex AI models, such as deep learning networks, necessitates high-performance computing (HPC) systems, while real-time applications demand low-latency processing capabilities.

In many cases, the computational requirements of these AI models may exceed the capabilities of the onboard systems in EVs. EVs typically have limited processing power, as the primary focus is on ensuring efficient power management, safety, and vehicle dynamics. As a result, AI models may need to be offloaded to cloud-based servers or edge computing systems, which adds complexity and potential delays due to data transmission and communication between the vehicle and external systems. Furthermore, the availability of high-bandwidth, low-latency communication networks, such as 5G, is essential for enabling real-time AI processing and decision-making.

Additionally, the deployment of AI-driven EV solutions requires significant infrastructure investments, including smart charging stations, V2G communication systems, and cloud-based platforms for data storage and processing. These infrastructural limitations can delay the wide-scale adoption of AI technologies and increase costs for both manufacturers and consumers.

References

[1] Rivera S, Goetz SM, Kouro S, *et al.* Charging infrastructure and grid integration for electromobility. *Proceedings of the IEEE*. 2022;111(4):371–396.

[2] Srdic S and Lukic S. Toward extreme fast charging: Challenges and opportunities in directly connecting to medium-voltage line. *IEEE Electrification Magazine*. 2019;7(1):22–31.

[3] Sanguesa JA, Torres-Sanz V, Garrido P, *et al.* A review on electric vehicles: Technologies and challenges. *Smart Cities*. 2021;4(1):372–404.

[4] Titus F, Thanikanti SB, Deb S, *et al.* Charge scheduling optimization of plug-in electric vehicle in a PV powered grid-connected charging station based on day-ahead solar energy forecasting in Australia. *Sustainability*. 2022;14(6):3498.

[5] Shamami MS, Alam MS, Ahmad F, *et al.* Artificial intelligence-based performance optimization of electric vehicle-to-home (V2H) energy management system. *SAE International Journal of Sustainable Transportation, Energy, Environment, & Policy*. 2020;1(13-01-02-0007):115–125.

[6] Ricciardi Celsi L and Valli A. Applied control and artificial intelligence for energy management: An overview of trends in EV charging, cyber-physical security and predictive maintenance. *Energies*. 2023;16(12):4678.

[7] Blaabjerg F, Wang H, Vernica I, *et al.* Reliability of power electronic systems for EV/HEV applications. *Proceedings of the IEEE*. 2020;109(6):1060–1076.

[8] Götz S and Reber V. Modular power electronics system for charging an electrically operated vehicle. Google Patents; 2019. US Patent 10,391,870.

[9] Tu H, Feng H, Srdic S, *et al.* Extreme fast charging of electric vehicles: A technology overview. *IEEE Transactions on Transportation Electrification*. 2019;5(4):861–878.

[10] Deb N, Singh R, Brooks RR, *et al.* A review of extremely fast charging stations for electric vehicles. *Energies*. 2021;14(22):7566.

[11] Sharida A, Bayhan S, and Abu-Rub H. A novel multi-objectives sliding mode control for bidirectional EV charging stations. In: *2024 4th International Conference on Smart Grid and Renewable Energy (SGRE)*. IEEE; 2024. p. 1–6.

[12] Sharida A, Bayindir AB, Bayhan S, *et al.* Enhanced inverse model predictive control for EV chargers: Solution for rectifier-side. *IEEE Open Journal of the Industrial Electronics Society*. 2024;5:795–806.

[13] Zhang F, Yang Q, and An D. CDDPG: A deep-reinforcement-learning-based approach for electric vehicle charging control. *IEEE Internet of Things Journal*. 2020;8(5):3075–3087.

[14] D'Amore G, Cabrera-Tobar A, Petrone G, *et al.* Integrating model predictive control and deep learning for the management of an EV charging station. *Mathematics and Computers in Simulation*. 2024;224:33–48.

[15] Chen Q and Folly KA. Application of artificial intelligence for EV charging and discharging scheduling and dynamic pricing: A review. *Energies*. 2022;16(1):146.

[16] Sharida A, Bayhan S, and Abu-Rub H. Fault-tolerant self-tuning control for three-phase three-level T-type rectifier. *IEEE Transactions on Power Electronics*. 2023;38(6):7049–7058.

[17] Shahriar S, Al-Ali AR, Osman AH, *et al.* Machine learning approaches for EV charging behavior: A review. *IEEE Access*. 2020;8:168980–168993.

[18] Yaghoubi E, Yaghoubi E, Khamees A, *et al.* A systematic review and meta-analysis of machine learning, deep learning, and ensemble learning approaches in predicting EV charging behavior. *Engineering Applications of Artificial Intelligence*. 2024;135:108789.

[19] Escoto M, Guerrero A, Ghorbani E, *et al.* Optimization challenges in vehicle-to-grid (V2G) systems and artificial intelligence solving methods. *Applied Sciences*. 2024;14(12):5211.

[20] Ghalkhani M and Habibi S. Review of the Li-ion battery, thermal management, and AI-based battery management system for EV application. *Energies*. 2022;16(1):185.

[21] Lang W, Hu Y, Gong C, *et al.* Artificial intelligence-based technique for fault detection and diagnosis of EV motors: A review. *IEEE Transactions on Transportation Electrification*. 2021;8(1):384–406.

[22] Zhang X, Hu Y, Gong C, *et al.* Artificial intelligence technique-based EV powertrain condition monitoring and fault diagnosis: A review. *IEEE Sensors Journal*. 2023;23(15):16481–16500.

Chapter 18

Optimization problems related to electric vehicle adoption

Raka Jovanovic[1] *and Sertac Bayhan*[1]

This chapter provides a comprehensive exploration of the multifaceted optimization challenges that arise with the rapid integration of electric vehicles (EVs) into urban transportation systems. With the accelerating adoption of EVs comes an urgent need for effective infrastructure planning, scheduling, and fleet management to support these vehicles efficiently, sustainably, and economically. This chapter systematically addresses these challenges across three primary areas of optimization.

The first section delves into infrastructure optimization, specifically the design and strategic placement of charging stations to enhance accessibility, reduce waiting times, and ensure efficient use of resources. This includes considerations of urban density, projected EV adoption rates, and geographical accessibility, all of which are critical for developing a robust and effective charging network.

In the second section, the focus shifts to the optimization of EV charging schedules to address grid-related challenges, particularly the mitigation of the "duck curve" effect. This phenomenon arises from the steep increase in energy demand during peak evening hours, as many EVs charge simultaneously. By exploring advanced scheduling algorithms, this section highlights strategies to balance charging demand, reduce peak load stress, and leverage grid flexibility through demand-side management (DSM).

Finally, the chapter examines the operational management of EV fleets, with a particular emphasis on electric buses (EBs). Managing EV fleets poses unique logistical and scheduling challenges, especially as EBs often operate under constrained charging infrastructure while needing to maintain rigorous schedules. This section explores methods for optimizing fleet schedules, coordinating charging times, and maximizing fleet availability within infrastructure and operational limits.

Through these three areas, this chapter aims to provide a holistic view of the optimization solutions necessary for the seamless integration of EVs into urban systems, emphasizing strategies that promote reliability, cost-effectiveness, and minimal environmental impact.

[1]Qatar Environment and Energy Research Institute, Hamad Bin Khalifa University, Qatar

18.1 Optimizing infrastructure

In recent years, the global push for reducing greenhouse gas emissions and curbing pollution has driven a rapid expansion in the adoption of EVs. EVs are recognized as a crucial element in the ongoing efforts to decarbonize the transportation sector, contributing significantly to cleaner air in urban environments and aligning with international commitments to sustainability goals [1]. This has been particularly noticeable in major cities worldwide, where local and national governments have proactively incentivized EV adoption through a variety of policy measures. These include direct purchase subsidies, free parking privileges, access to restricted lanes, and – most importantly – the development of a robust public charging infrastructure [2]. The availability of reliable charging networks is seen as a primary driver for EV adoption, as it alleviates concerns over range limitations and charging convenience, thereby reducing "range anxiety" among EV users [3,4].

The emergence of public charging stations as a pivotal enabler of EV usage presents unique challenges for infrastructure planning and optimization. Unlike traditional internal combustion engine vehicles (ICEVs), which require refueling exclusively at gas stations, EVs have the advantage of multiple charging options that vary by location and charging speed [5]. Charging options generally fall into three categories: Level 1 (slow charging) for residential use, Level 2 (medium-speed charging) at locations such as office parking lots, and Level 3 (fast charging) at public charging stations. Level 1 charging, while widely accessible, is time-consuming and typically suitable only for overnight use, as it takes multiple hours to replenish an EV's battery. Level 2 chargers, found in public and semi-public spaces such as workplaces and shopping centers, offer moderately faster charging and can be integrated into demand management systems to help balance grid loads during peak periods [6–8]. Fast-charging Level 3 stations, on the contrary, provide a significantly faster solution, capable of replenishing an EV battery in about 30 min, though these come with considerably higher infrastructure costs – sometimes up to ten times the cost of a Level 2 station [9–11].

With the high costs associated with Level 3 charging stations, optimizing their deployment and capacity is essential to ensure economic feasibility and to maximize utilization. Research on EV charging infrastructure optimization often models the problem as a location-allocation problem, aiming to strategically place charging stations where they will serve the greatest number of EV users [12–14].

Two main types of EV charging infrastructure networks are typically considered: inter-city networks for long-distance travel and intra-city networks for urban commuting. For long-distance travel, the placement of charging stations is geared toward minimizing the need for EVs to make significant detours from main routes, thus ensuring that stations are conveniently located along high-traffic corridors [3,4]. Conversely, for urban networks, station density is a critical factor, as EV drivers generally travel shorter distances and can more easily access a denser network of chargers. Urban network planning thus bears a resemblance to classical covering problems, where the goal is to ensure that each area within the city has convenient access to a charging station without requiring users to deviate far from their usual routes [15,16].

The efficient and strategic placement of charging stations is essential not only for meeting demand but also for maximizing system profitability and minimizing operational costs. Many studies also highlight the impact of battery characteristics on charging infrastructure needs. The limited range of many EVs, often less than 250 km, combined with the current distribution of charging stations, has led to challenges in ensuring continuous service for all users.

Optimization of charging networks is usually done through different mathematical models. A simplified mixed integer programming (MIP) model for the general problem of optimizing locations of EV charging stations is provided in the following text. The model considers a set of candidate locations for charging stations, I, each with an associated installation cost, C_i, and a set of demand nodes, J, where each demand node $j \in J$ has a demand, D_j. The distance or cost of serving demand from node j to station i is given by d_{ij}. The decision variables are x_i, a binary variable indicating whether a station is built at location i, and y_{ij}, a continuous variable representing the fraction of demand from node j served by station i. This simplified model aims to minimize the total cost of station installation and demand allocation, which can be formalized using the following objective function

$$\text{Minimize} \quad \sum_{i \in I} C_i x_i + \sum_{i \in I} \sum_{j \in J} d_{ij} D_j y_{ij} \tag{18.1}$$

The objective function minimizes the total cost, which includes the installation cost of charging stations, $\sum_{i \in I} C_i x_i$, and the transportation or allocation cost of satisfying demand, $\sum_{i \in I} \sum_{j \in J} d_{ij} D_j y_{ij}$. Note that the transportation cost does not necessarily represent actual costs, but rather serves as a measure of user satisfaction.

To fully specify the model, the constraints are also provided

$$\sum_{i \in I} y_{ij} = 1 \qquad \forall j \in J, \tag{18.2}$$

$$y_{ij} \leq x_i \qquad \forall i \in I, \forall j \in J, \tag{18.3}$$

$$x_i \in \{0, 1\} \qquad \forall i \in I, \tag{18.4}$$

$$y_{ij} \geq 0 \qquad \forall i \in I, \forall j \in J \tag{18.5}$$

The first constraint given in (18.2) ensures that the demand at each node j is fully satisfied by allocating it to one or more charging stations. Equation (18.3) links the allocation variable y_{ij} to the binary decision variable x_i, ensuring that demand can only be assigned to locations where charging stations are built. The constraint given in (18.4) and (18.5) define specify the decision variables. This formulation guarantees feasibility and optimality for the simplified charging station location problem. Note that in practical applications, while the models typically incorporate additional details to better reflect real-world complexities, they still retain a similar core structure to the simplified model presented here.

These models often result in NP-hard optimization problems, which necessitates the use of advanced heuristics and metaheuristic techniques to find near-optimal solutions efficiently [17,18]. Metaheuristic algorithms such as genetic algorithms (GA), differential evolution (DE), and particle swarm optimization (PSO) have been extensively applied to optimize station placement. For instance, GA minimized social costs, including construction and environmental expenses, as demonstrated in a study in Ireland [19]. Similarly, DE and Harris Hawks optimization reduced power losses and maintained voltage stability under demand uncertainties [20]. In urban scenarios, PSO improved energy efficiency, achieving up to 43% savings compared to traditional methods like greedy algorithms [21].

The integration of renewable energy sources and storage systems into charging stations is another key focus. Studies have explored combining photovoltaic (PV) systems with EV stations, using methods like simulated annealing (SA), GA, and cuckoo search algorithms to enhance voltage stability and network reliability [22]. A robust optimization model incorporating wind and solar energy addressed uncertainties in demand and supply, ensuring balanced load fluctuations in transportation and power networks [23]. Multi-objective optimization models have tackled goals such as maximizing traffic network coverage, minimizing energy losses, and improving voltage profiles. A hybrid approach combining PSO with direct search optimized station locations and incorporated reactive power compensation through shunt capacitors, ensuring improved grid reliability and spatial accessibility for EV users [24].

Another significant aspect of optimizing EV infrastructure is the consideration of charging station capacity, which involves determining the number of charging outlets at each station. The determination of optimal capacities is influenced by factors such as expected demand, population density, and proximity to key destinations such as shopping centers, workplaces, and highways. While in some cases, the locations of charging stations are predetermined by existing infrastructure, many studies assume flexible placement to enable a higher degree of optimization. Some models account for dynamic demand patterns, proposing capacities based on anticipated peak times and daily travel trends to reduce queuing times and improve the overall quality of service [17].

Optimizing the location and capacity of EV charging stations is crucial for supporting widespread EV adoption and maintaining high service quality. Various studies have proposed advanced methods to address this challenge effectively. Jovanovic *et al.* developed a greedy algorithm with local search to minimize service drop rates by utilizing an M/M/c/c loss queue model. Their approach considered real-world factors such as population density and existing infrastructure, resulting in significant improvements in charging network efficiency [25]. Similarly, Brown *et al.* proposed an MIP model that integrates population density and grid data for optimizing infrastructure expansion. This model was successfully applied to cities like Doha and San Francisco, demonstrating its practical utility [26].

Other researchers have explored hierarchical and multi-layered frameworks. Kong *et al.* introduced a hierarchical optimization model that combines flow capturing, queuing theory, and battery dynamics to improve both system performance and user satisfaction. The model was applied to diverse networks, showcasing its flexibility and effectiveness [27]. In parallel, hybrid heuristic methods have been employed to tackle the complexity of EV infrastructure planning. For example, a hybrid approach combining GA-PSO optimized the siting and sizing of charging stations, addressing issues such as voltage stability and service quality in dynamic urban environments [28,29].

Advanced optimization techniques also integrate fuzzy logic and multi-objective frameworks to enhance decision-making. Studies incorporating fuzzy quality of service models considered uncertainties in EV arrival and service times, providing robust plans to improve customer satisfaction [30]. Additionally, optimization methods that account for charging satisfaction and renewable energy integration used clustering and multi-objective GA-PSO to determine optimal locations and capacities while supporting sustainable energy goals [31]. Research has also focused on design and profitability, with metaheuristics like PSO, salp swarm algorithm, and arithmetic optimization algorithm used to optimize fast-charging station design. These models maximized profitability by determining the optimal number of chargers and renewable energy integration. Among the methods, PSO demonstrated the highest net present value, highlighting its effectiveness [32]. These diverse methodologies highlight the importance of combining mathematical rigor with heuristic approaches and real-world data to design efficient and user-friendly EV charging infrastructure.

In summary, the transition to EVs presents an unprecedented opportunity to reshape the transportation landscape toward a more sustainable and environmentally friendly future. However, this transition is accompanied by significant challenges, particularly in the planning and optimization of EV charging infrastructure. From strategic location planning and capacity allocation to the integration of renewable energy sources, researchers have developed a diverse array of mathematical and heuristic models to address these complexities. These efforts underscore the critical role of advanced optimization techniques in ensuring that charging networks are not only economically viable but also capable of supporting widespread EV adoption with high service quality and user satisfaction. By leveraging these innovative approaches, the foundation is set for a cleaner, more efficient transportation ecosystem that aligns with global sustainability goals.

18.2 Optimizing EV charge scheduling

The rapid expansion of solar power generation has introduced unique challenges to electricity grids, primarily due to the timing mismatch between peak solar energy production and peak electricity demand – a phenomenon commonly referred to as the "duck curve." This effect is marked by a sharp increase in electricity demand during the late afternoon and early evening, following a period of reduced demand and high solar output earlier in the day. Such fluctuations require grid operators to rapidly increase conventional, often fossil-fuel-based power generation to meet demand spikes, causing operational inefficiencies and higher costs for utilities [33,34]. With ongoing growth in solar capacity, the duck curve effect is expected to intensify, underscoring the need for effective strategies to smooth energy transitions and enhance grid stability.

Concurrently, many governments are implementing policies to accelerate the adoption of EVs as part of their commitments to achieving carbon neutrality and reducing emissions from the transportation sector. These policies have led to a surge in EVs on the road, a trend that is anticipated to continue as EVs become increasingly dominant in ground transportation. Unlike conventional vehicles, EVs offer flexibility in their charging schedules, presenting an opportunity to align EV charging patterns with periods of peak solar production. By scheduling EV charging during midday, when solar energy generation is at its highest, grid operators can help mitigate the duck curve by reducing the need to ramp up non-renewable generation sources. This approach forms part of a broader suite of DSM strategies, which also include demand response programs for residential and industrial users and the deployment of energy storage solutions. Together, these measures can reduce grid operational costs and facilitate the integration of renewable energy [35].

Among the available charging options, Level 2 chargers, which typically operate between 5 and 7 kW, are particularly well-suited for managing the duck curve. These chargers are frequently deployed in the workplace and commercial parking areas, where vehicles remain parked for extended durations, allowing for flexible and optimizable charging schedules throughout the day. In contrast, Level 1 chargers, often used in residential settings, have lower power output and are typically used for overnight charging, which does not align with peak solar availability. Meanwhile, Level 3 fast chargers, although powerful, are generally used for short charging sessions due to their high-speed nature, making them less effective for DSM purposes [10,11]. As a result, research efforts have increasingly focused on optimizing the use of Level 2 chargers in workplace and park-and-ride (PR) locations, where charging schedules can be adjusted to coincide with solar production patterns, thereby helping to flatten the duck curve [6–8].

This general framework is best understood base on an MIP model for which an outline is provided in the following text, for which details can be found in Jovanovic and Bayram [36]. The model is built around the following assumptions and notations. Scheduling occurs over a time window $\mathscr{T}$, divided into discrete intervals $\{1, \ldots, T\}$. The charging is managed for a set of vehicles $\mathscr{C} = \{1, \ldots, M\}$, which utilize a potential parking with chargers. Each EV's charging can be activated or paused during its stay, according to a flexible schedule. The model parameters, assumed to be constant over $\mathscr{T}$, are defined as follows:

- Parameter q_t represents the net base load at each time period $t \in \mathscr{T}$, calculated as the base demand minus available solar generation.
- For each vehicle $i \in \mathscr{C}$, arrival and departure times, denoted by a_i and d_i, respectively, are predetermined.
- Each EV i has a battery capacity f_i, with an initial charge level b_i^a at arrival, where $0 \leq b_i^a \leq f_i$.
- The charging rate per period is given by parameter s, reflecting the consistent charging power of the facility's chargers.

Key constraints are integrated into the model to ensure practical operation. Each EV's battery charge is limited to its full capacity, and a minimum required charge r must be delivered to each vehicle before departure. Although a maximum system-wide charging limit per time period could be imposed, this model assumes a sufficiently high limit to accommodate the entire station's charging demand without restriction. The objective of this scheduling model is to mitigate the "duck curve" by minimizing fluctuations in grid load. This is achieved by optimizing the combined load of q_t and the additional energy required for EV charging, aiming to keep successive time periods as balanced as possible in terms of total energy demand.

This MIP model has been adapted for large-scale EV charging at PR facilities in Jovanovic and Bayram [36]. Using real-world data, the results in this work demonstrate that optimized charging schedules can reduce power system ramping needs even up to 25% in the case of high levels of EV and PV adoption, leading to potential maintenance and operational cost savings. Jovanovic *et al.* [37] analyzed this system through an online model (OLM) for scheduling EV charging at PR facilities. Their model leverages a neural network in combination with a heuristic function, trained on optimally solved instances of a full information model (FIM), to make dynamic charging decisions based on real-time information. Computational experiments across varying levels of solar generation and EV adoption demonstrate that the OLM can generate charging schedules comparable in quality to those produced by FIM, with additional benefits in adaptability.

Extensive research has explored optimal scheduling techniques for EV charging within grid-to-vehicle (G2V) systems to enhance solar utilization and minimize reliance on conventional power sources. In recent years, the scope of this research has expanded to include vehicle-to-grid (V2G) systems, where EVs can discharge electricity back into the grid during times of high demand. Recent advances in V2G technology highlight its potential for supporting grid stability, particularly by using EVs as bidirectional energy storage. A smart V2G scheduling model optimized for EVs paired with PV sources has been shown to reduce grid stress and energy costs through a controlled charging algorithm that considers real-time solar generation and transformer loads [38]. This approach, which leverages data from urban neighborhoods, underscores that while PV support alone is insufficient, combining it with V2G effectively mitigates transformer stress during peak hours, particularly as EV adoption increases.

Other models have explored game-theoretic strategies, framing interactions between distribution operators and EV clusters as a Stackelberg game to minimize ramping and grid load fluctuations, achieving up to a 39% reduction in system ramp requirements [39]. Additionally, a smart charge scheduling framework for EVs that integrates G2V and V2G modes offers further flexibility in managing demand and optimizing energy costs [40]. By incorporating stochastic connection patterns and real-time tariffs, V2G-based charging infrastructures in urban settings can both enhance load stability and accommodate demand fluctuations, smoothing peak loads while enabling efficient utilization of renewable sources like PV [41,42].

However, V2G systems must carefully account for battery degradation, as frequent charge–discharge cycles can lead to accelerated wear on EV batteries. Battery degradation is a nonlinear process that complicates optimization efforts; accordingly, many studies employ advanced techniques such as gradient-based algorithms, heuristics, and metaheuristic approaches to solve these complex optimization problems [43–46]. Alternatively, linear and quadratic MIP formulations have been developed to incorporate battery degradation costs using piecewise-linear approximations, as demonstrated in Farzin *et al.* [47].

An effective approach based on MIP is proposed in Jovanovic *et al.* [48] effectively evaluates battery degradation in the context of smart charging for V2G systems at PR facilities for reducing the "duck curve." Results show that V2G, which allows EVs to discharge energy back into the grid, significantly reduces the need for power ramp-down in the mornings, outperforming unidirectional G2V systems. However, the benefits decrease in the evening as fewer EVs remain on-site, limiting discharge potential. The study also reveals that while V2G systems can achieve substantial demand reduction, they incur added battery wear, with up to 9% additional degradation for maximum impact. Importantly, around half of the ramp-down reduction can be achieved with just 3% extra degradation, suggesting an opportunity to optimize for grid support while minimizing battery costs.

Multi-objective optimization frameworks are increasingly being used to balance multiple goals within EV charging management, such as minimizing grid ramp-up requirements and maximizing the satisfaction of EV charging demands. For example, in facilities with limited maximum charging capacity (MCC), trade-offs may need to be made between maintaining grid stability and fulfilling all user charging requests. These trade-offs can be addressed through multi-objective optimization models that enable grid operators to achieve a balanced solution that improves both grid reliability and user satisfaction [49,50].

An interesting multi-objective study is presented in Jovanovic *et al.* [51], which optimizes EV charging schedules at away-from-home facilities to help flatten the duck curve, using a quadratic programming model. Key findings show that near-optimal duck curve flattening requires an MCC about 2–3 times the minimum necessary to meet total demand. Lower MCC can reduce ramp-up requirements but may increase peak loads unless some charging requests are dropped. Reducing the satisfaction of charge requests can also help manage ramp-up needs, indicating that selective reductions may be needed in practice.

The recent availability of detailed real-world data has significantly improved the accuracy and practical relevance of EV charging optimization models. Datasets from smart charging providers and open-access sources like ACN-Data from the California Institute of Technology and Pecan Street Dataport allow researchers to validate their models with actual charging behavior, providing insights into how optimized charging can alleviate peak demand pressures, enhance grid stability, and increase renewable energy utilization [52,53]. These datasets enable more realistic case studies that demonstrate the potential benefits of optimized charging, such as decreased operational costs, reduced need for conventional power generation, and more effective integration of renewable sources.

In summary, the alignment of EV charging with renewable energy production, particularly through the strategic use of Level 2 charging stations, presents a promising solution for addressing the duck curve. Mathematical models and optimization techniques, ranging from MIP to heuristic methods, enable effective scheduling of EV charging while accounting for factors such as battery degradation and grid load balancing. As more comprehensive datasets become available, these models continue to evolve, offering grid operators valuable tools to support high levels of renewable energy penetration while ensuring reliable grid operations.

18.3 Challenges in the use of EVs in public transport

The shift toward EVs is revolutionizing the transportation sector, with EBs and EVs in taxi and ride-hailing services playing pivotal roles in advancing sustainable urban mobility. These modes of transport offer significant environmental benefits, including reductions in emissions and noise pollution, while presenting unique operational challenges that require innovative optimization approaches.

In public transport, EBs bring complexities such as range limitations, battery degradation, and the need for efficient charging infrastructure, all of which complicate traditional vehicle scheduling methods. Likewise, EVs in taxi and ride-hailing fleets face barriers like limited charging accessibility and prolonged downtime, impacting fleet utilization. Addressing these challenges calls for a combination of advanced optimization models, simulation techniques, and machine learning, enabling dynamic solutions that enhance efficiency, sustainability, and service quality.

The following subsections delve into specific optimization strategies for EBs in public transport and EVs in taxi and ride-hailing fleets, highlighting current research, methodologies, and practical applications to overcome these challenges.

18.3.1 Optimizing the use of EBs in public transport

Among the various types of EVs, EBs have seen considerable adoption in public transport systems, particularly in cities across China, where over 400,000 EBs are in operation, representing more than 14% of the nation's public transport fleet [54]. The shift toward EBs is motivated by their environmental benefits, including reductions in air and noise pollution, and their potential to contribute to net-zero emission goals [55–57]. Additionally, the operational costs of EBs are significantly lower than those of diesel buses (DBs), as the high daily mileage of EBs helps offset the initial investment in EVs through savings on fuel costs [57].

As EBs gain popularity, various approaches have emerged to tackle practical issues, such as charging infrastructure [58,59] and operational concerns [60,61], including battery degradation and scheduling. These issues are not only operational but also require optimization to ensure efficiency in fleet management. In particular, battery degradation is a critical factor that can affect the long-term performance and cost-effectiveness of the fleet. Optimizing scheduling to minimize battery wear while maintaining service quality becomes an essential part of the problem. These infrastructural and operational concerns are often modeled together [62,63], creating highly complex optimization models.

The integration of EBs into urban transport presents new operational challenges, primarily due to their limited range and long charging times compared to DBs. The vehicle scheduling problem (VSP) [64,65] is a common model for DB fleet optimization, aiming to assign trips to vehicles while minimizing costs such as the number of buses used. With EBs, the problem becomes more complex due to additional constraints like range and charging schedules, leading to the electric vehicle scheduling problem (E-VSP) [66]. The E-VSP is widely used to represent the scheduling of EBs in public transport systems. The goal of the E-VSP is to assign trips to EBs in a way that minimizes costs – such as the number of vehicles used or total energy consumption – while satisfying constraints like route feasibility, adherence to schedules, and efficient charging management. These added complexities make the E-VSP more challenging and necessitate sophisticated optimization techniques. Variations of the E-VSP consider different operational setups, such as single and multiple depots [67–69], and mixed fleet configurations involving both EBs and DBs [70,71]. Each of these setups introduces unique constraints, such as the allocation of resources (e.g., buses and charging stations) and the management of energy consumption.

Notably, robustness against traffic conditions has been explored in the literature [67], as transportation systems are inherently affected by dynamic environmental factors such as traffic congestion, road conditions, and weather. This leads to stochastic models, where uncertainty is incorporated into the optimization process to make the scheduling resilient to unpredictable variations. Additionally, charging constraints – such as the limited availability of charging slots or restrictions on charging power during peak hours – have been modeled in some studies [67,69]. These constraints further complicate the optimization task, as they introduce a need to balance charging station usage with fleet availability, all while minimizing operational costs.

Most of the E-VSP models are solved using MIPs [67–69], a powerful optimization technique that can handle the binary and continuous decision variables typical of VSPs. MIP is well-suited for problems involving discrete decisions, such as the assignment of buses to specific trips, and continuous variables, like fuel or energy consumption. However, solving large-scale instances of MIP problems is computationally expensive, especially when additional real-world constraints are incorporated, leading to a demand for more efficient algorithms.

To address the computational challenges, some studies have turned to heuristic methods. These include hybrid approaches combining greedy algorithms with SA [69], which provides a flexible search procedure for large, complex solution spaces, and GAs [70], which mimic natural selection to iteratively evolve better solutions. Large-scale public transports have also been solved using the greedy randomized adaptive search procedure (GRASP) metaheursitic [72]. It is interesting to mention that a metaheuristic method, combination of heuristics and MIP, has been applied to this type of problem using the metaheuristic fixed set search [73]. While these heuristic methods may not guarantee an optimal solution, they can provide good approximations within a reasonable amount of time, making them particularly useful in real-time applications.

Further advancements in optimization techniques are focused on integrating machine learning and optimization. By learning from historical data, predictive models can be developed to anticipate demand, battery use and optimize fleet scheduling dynamically [74]. Additionally, machine learning can be combined with optimization algorithms to fine-tune parameters and constraints, improving the accuracy and efficiency of solutions [75]. This integration of data-driven methods with traditional optimization techniques holds great potential for solving increasingly complex transportation problems as the adoption of EBs continues to grow.

In summary, the integration of EBs into urban transportation systems has introduced a range of challenges that extend beyond traditional VSPs. Addressing these challenges requires a combination of advanced optimization techniques, such as MIP and heuristics, as well as innovative approaches like metaheuristics and machine learning. The growing adoption of EBs underscores the importance of developing robust models that account for operational constraints, charging infrastructure, battery degradation, and dynamic environmental factors. By leveraging predictive capabilities and real-time optimization, these methodologies offer promising solutions to enhance the efficiency and sustainability of public transport systems. As the field continues to evolve, the synergy between data-driven methods and optimization is expected to play a pivotal role in addressing the complexities of EB fleet management and supporting the transition toward greener urban mobility.

18.3.2 Optimizing the use of EVs in taxi and ride hailing services

EVs in taxi and ride-hailing fleets face unique operational challenges that are crucial to address for widespread adoption. Pilot studies, such as Uber's project in London, reveal significant barriers, including limited access to home charging and insufficient public charging infrastructure, both of which hinder the operational efficiency of EV fleets compared to traditional ICEVs [76,77]. Optimizing charging infrastructure is essential to overcome these limitations. Research has focused on the strategic placement of charging stations [78] and the development of adaptive charging schedules to reduce downtime and improve fleet utilization [79,80].

In Shenzhen, China, where the entire taxi fleet has transitioned to EVs, drivers report substantial revenue losses due to extended queuing times and delays at charging stations, highlighting the critical need for optimizing charging management [81]. Such optimization is particularly essential in high-demand urban areas, where fleet efficiency is closely tied to charging accessibility and turnaround time.

To address these operational complexities, simulation methods, particularly agent-based models (ABM) and discrete event simulation (DES), are extensively used for optimization in EV dispatching, charging management, and fleet allocation. ABMs are adept at simulating complex interactions within autonomous EV fleets, providing valuable insights into dispatching efficiency and advanced control systems. These models are especially effective when combined with reinforcement learning techniques to dynamically improve fleet responses to varying demand patterns and environmental conditions [82–84].

DES is also widely applied in urban fleet management due to its ability to capture detailed usage patterns and operational dependencies, making it well-suited for evaluating various strategies in ride-sharing and public transportation systems [85,86]. Furthermore, DES is often integrated with optimization algorithms, such as GAs, to solve fleet allocation and routing problems in EV sharing systems [87]. A notable approach involves combining DES with heuristic methods to evaluate the effectiveness of taxi relief stands as potential charging points, thereby optimizing infrastructure use and reducing downtime in EV taxi and ride-hailing fleets [78].

The integration of these simulation methods with optimization techniques provides powerful tools for tackling EV fleet management challenges, enabling both the assessment and improvement of charging and dispatching strategies under real-world constraints. Such approaches not only enhance operational efficiency but also support sustainable urban transportation solutions by maximizing the utility of existing infrastructure and reducing the overall environmental impact of fleet operations.

18.4 Conclusions and further research

In conclusion, optimizing EV adoption and infrastructure is crucial for the successful integration of EVs into transportation systems worldwide. The optimization of EV infrastructure, such as charging station placement and capacity, plays a critical role in reducing range anxiety and ensuring the widespread adoption of EVs. Research in this field has made significant strides, particularly with the use of advanced optimization models, such as MIP and metaheuristic techniques like GA, PSO, and DE. These methods have been employed to effectively design and implement EV charging infrastructure, balancing the trade-offs between installation costs, energy efficiency, and accessibility.

The integration of renewable energy sources such as solar and wind into the charging network further adds complexity to these optimization problems but also enhances their sustainability and economic viability. Models that combine renewable energy sources with energy storage systems not only contribute to grid stability but also provide a pathway to reducing reliance on fossil fuel-based energy. Additionally, research has demonstrated the importance of considering dynamic demand patterns and flexible charging station capacities to ensure that charging stations can meet varying demands and optimize service quality throughout the day. As the adoption of EVs grows, so too does the need for ongoing optimization of infrastructure and service models, particularly to address the nuances of different types of networks – urban, inter-city, and long-distance corridors.

The development of sophisticated EV charge scheduling models, including G2V and V2G systems, further highlights the promise of EVs in supporting renewable energy utilization and contributing to energy system flexibility. These models focus on optimizing charging schedules in real-time, allowing for greater integration of solar and other renewable resources, which are vital for reducing grid stress during peak hours. However, these systems also introduce new challenges, particularly concerning battery degradation, and necessitate advanced optimization techniques to strike a balance between energy demand management and the long-term health of EV batteries.

Despite the significant progress made, several challenges remain. The complexities involved in the interaction between EV adoption, infrastructure design, and grid integration continue to evolve as new technologies and data sources emerge. A key challenge is the need for optimization models that can efficiently handle the growing variability in energy demand and supply as the number of EVs on the road increases. Moreover, the adoption of EVs in public transport, such as EBs, introduces additional layers of complexity, including fleet management, range limitations, and the need for optimized charging schedules to ensure service reliability.

Optimizing the use of EVs in taxi and ride-hailing services is critical for addressing the unique operational challenges these fleets face, particularly in urban environments. The limitations of charging infrastructure, such as limited home charging access and insufficient public charging stations, remain significant barriers to the widespread adoption of EVs in these sectors. However, research has made notable progress in addressing these challenges through the strategic placement of charging stations and the development of adaptive charging schedules that reduce fleet downtime. Simulation methods, particularly ABM and DES, have proven effective in optimizing EV fleet management, including dispatching, charging, and fleet allocation. By integrating these methods with optimization algorithms, such as GAs and reinforcement learning, fleet efficiency can be significantly improved, even in high-demand urban settings.

Future research should focus on several key areas to further improve the optimization of EV adoption and infrastructure. One promising direction involves the development of hybrid models that integrate traditional optimization techniques with machine learning and artificial intelligence methods. These approaches can be used to better predict demand patterns, improve real-time decision-making, and optimize charging schedules dynamically based on evolving conditions. Machine learning models can also enhance forecasting capabilities, allowing for more accurate predictions of traffic patterns, energy consumption, and charging station utilization.

Additionally, research into advanced multi-objective optimization models that balance multiple factors – such as cost, energy efficiency, accessibility, and user satisfaction – will be critical as EV adoption continues to expand. These models can incorporate uncertainties in demand and supply, as well as the integration of new energy sources, to develop more robust and flexible charging networks. Another important avenue for future research is the exploration of new charging technologies, such as wireless charging and ultra-fast charging, which could dramatically reduce charging times and further improve the convenience of EVs for users.

As the shift toward EVs in public transport accelerates, researchers must also explore the unique operational challenges faced by fleets of EBs and other commercial EVs. Optimizing charging infrastructure and fleet scheduling for EBs, particularly in urban environments, will require the development of specialized models that account for not only charging station location and capacity but also fleet management, operational constraints, and real-time traffic conditions. Furthermore, research into battery life cycle management and the development of low-maintenance, long-lasting batteries will be essential to reducing the long-term costs associated with fleet electrification.

Incorporating real-time data from various sources – such as smart grids, EV charging stations, and user behavior analytics – will also be crucial for the next generation of optimization models. Big data analytics, combined with edge computing and Internet of Things (IoT) technologies, can enable more responsive and adaptive systems that optimize energy use and service delivery based on real-time conditions.

Lastly, future research should focus on the social, economic, and policy aspects of EV adoption. Understanding the impacts of EV infrastructure deployment on various communities, as well as the economic incentives needed to encourage adoption, will help policymakers design more effective strategies for supporting the transition to electric mobility. Research into equitable access to charging stations and the potential for creating a fair distribution of infrastructure across urban and rural areas will be critical for ensuring that the benefits of EV adoption are shared broadly.

Overall, while significant progress has been made in optimizing EV adoption and infrastructure, there remains much to explore and improve. By addressing the challenges and leveraging emerging technologies, future research can continue to advance the transition to a more sustainable and efficient electric transportation ecosystem.

References

[1] Langbroek JHM, Franklin JP, and Susilo YO. The effect of policy incentives on electric vehicle adoption. *Energy Policy*. 2016;94:94–103.

[2] Mehrjerdi H and Hemmati R. Electric vehicle charging station with multilevel charging infrastructure and hybrid solar-battery-diesel generation incorporating comfort of drivers. *Journal of Energy Storage*. 2019;26:100924.

[3] Jochem P, Szimba E, and Reuter-Oppermann M. How many fast-charging stations do we need along European highways? *Transportation Research Part D: Transport and Environment*. 2019;73:120–129.

[4] Napoli G, Polimeni A, Micari S, *et al.* Optimal allocation of electric vehicle charging stations in a highway network: Part 1. Methodology and test application. *Journal of Energy Storage*. 2020;27:101102.

[5] Daina N, Sivakumar A, and Polak JW. Electric vehicle charging choices: Modelling and implications for smart charging services. *Transportation Research Part C: Emerging Technologies*. 2017;81:36–56.

[6] Ferguson B, Nagaraj V, Kara EC, *et al.* Optimal planning of workplace electric vehicle charging infrastructure with smart charging opportunities. In: *2018 21st International Conference on Intelligent Transportation Systems (ITSC)*. IEEE; 2018. p. 1149–1154.

[7] Levinson RS and West TH. Impact of convenient away-from-home charging infrastructure. *Transportation Research Part D: Transport and Environment*. 2018;65:288–299.

[8] Wu X. Role of workplace charging opportunities on adoption of plug-in electric vehicles – Analysis based on GPS-based longitudinal travel data. *Energy Policy*. 2018;114:367–379.

[9] Jovanovic R and Bayram IS. Scheduling Electric Vehicle Charging at Park-and-Ride Facilities to Flatten Duck Curves. In: *2019 IEEE Vehicle Power and Propulsion Conference (VPPC)*; 2019. p. 1–5.

[10] Meyer D and Wang J. Integrating ultra-fast charging stations within the power grids of smart cities: A review. *IET Smart Grid*. 2018;1(1):3–10.

[11] Rahman I, Vasant PM, Singh BSM, *et al.* Review of recent trends in optimization techniques for plug-in hybrid, and electric vehicle charging infrastructures. *Renewable and Sustainable Energy Reviews*. 2016;58:1039–1047.

[12] Motoaki Y. Location-allocation of electric vehicle fast chargers—Research and practice. *World Electric Vehicle Journal*. 2019;10(1):12. https://doi.org/10.3390/wevj10010012.

[13] Chraibi K, Chraibi A, Chaker I, *et al.* Hybrid metaheuristic to solve location problem for electric vehicles charging stations. In: *2018 IEEE International Conference on Technology Management, Operations and Decisions (ICTMOD)*. IEEE; 2018. p. 12–17.

[14] Ren X, Zhang H, Hu R, *et al.* Location of electric vehicle charging stations: A perspective using the grey decision-making model. *Energy*. 2019;173:548–553.

[15] Efthymiou D, Chrysostomou K, Morfoulaki M, *et al.* Electric vehicles charging infrastructure location: A genetic algorithm approach. *European Transport Research Review*. 2017;9(2):27.

[16] Răboacă MS, Băncescu I, Preda V, *et al.* An optimization model for the temporary locations of mobile charging stations. *Mathematics*. 2020;8(3):453.

[17] Choi BG, Oh BC, Choi S, *et al.* Selecting locations of electric vehicle charging stations based on the traffic load eliminating method. *Energies*. 2020;13(7):1650.

[18] Guo C, Yang J, and Yang L. Planning of electric vehicle charging infrastructure for urban areas with tight land supply. *Energies*. 2018;11(9):2314.

[19] Zhou G, Zhu Z, and Luo S. Location optimization of electric vehicle charging stations: Based on cost model and genetic algorithm. *Energy*. 2022;247:123437.

[20] Pal A, Bhattacharya A, and Chakraborty AK. Allocation of electric vehicle charging station considering uncertainties. *Sustainable Energy, Grids and Networks*. 2021;25:100422.

[21] Aljaidi M, Samara G, Singla MK, *et al.* A particle swarm optimizer-based optimization approach for locating electric vehicles charging stations in smart cities. *International Journal of Hydrogen Energy*. 2024;87:1047–1055.

[22] Yenchamchalit K, Kongjeen Y, Prabpal P, *et al.* Optimal placement of distributed photovoltaic systems and electric vehicle charging stations using metaheuristic optimization techniques. *Symmetry*. 2021;13(12):2378.

[23] Li C, Zhang L, Ou Z, *et al.* Robust model of electric vehicle charging station location considering renewable energy and storage equipment. *Energy*. 2022;238:121713.

[24] Muthukannan S and Karthikaikannan D. Multiobjective planning strategy for the placement of electric-vehicle charging stations using hybrid optimization algorithm. *IEEE Access*. 2022;10:48088–48101.

[25] Jovanovic R, Bayhan S, Bayram IS. Capacity optimization of EV charging networks: A greedy algorithmic approach. In: *2022 3rd International Conference on Smart Grid and Renewable Energy (SGRE)*; 2022. p. 1–6.

[26] Brown P, Contestabile M, and Jovanovic R. A mixed integer program for optimizing the expansion of electrical vehicle charging infrastructure. In: Simian D, Stoica LF, editors. *Modelling and Development of Intelligent Systems*. Cham: Springer International Publishing; 2021. p. 302–314.

[27] Kong C, Jovanovic R, Bayram IS, and Devetsikiotis M. A hierarchical optimization model for a network of electric vehicle charging stations. *Energies*. 2017;10(5):675. https://doi.org/10.3390/en10050675.

[28] Awasthi A, Venkitusamy K, Padmanaban S, *et al.* Optimal planning of electric vehicle charging station at the distribution system using hybrid optimization algorithm. *Energy*. 2017;133:70–78.

[29] Choi M, Van Fan Y, Lee D, *et al.* Location and capacity optimization of EV charging stations using genetic algorithms and fuzzy analytic hierarchy process. *Clean Technologies and Environmental Policy*. 2024;p. 1–14.

[30] Zhao Z, Xu M, and Lee CK. Capacity planning for an electric vehicle charging station considering fuzzy quality of service and multiple charging options. *IEEE Transactions on Vehicular Technology*. 2021;70(12):12529–12541.

[31] Liu Jp, Zhang Tx, Zhu J, *et al.* Allocation optimization of electric vehicle charging station (EVCS) considering with charging satisfaction and distributed renewables integration. *Energy*. 2018;164:560–574.

[32] Antarasee P, Premrudeepreechacharn S, Siritaratiwat A, *et al.* Optimal design of electric vehicle fast-charging station's structure using metaheuristic algorithms. *Sustainability*. 2022;15(1):771.

[33] Obi M and Bass R. Trends and challenges of grid-connected photovoltaic systems – A review. *Renewable and Sustainable Energy Reviews*. 2016;58:1082–1094.

[34] Denholm P, O'Connell M, Brinkman G, *et al.* Overgeneration from solar energy in California. A field guide to the duck chart. National Renewable Energy Lab. (NREL), Golden, CO (United States); 2015.

[35] Zehir MA, Batman A, and Bagriyanik M. Review and comparison of demand response options for more effective use of renewable energy at consumer level. *Renewable and Sustainable Energy Reviews*. 2016;56:631–642.

[36] Jovanovic R and Bayram IS. Scheduling electric vehicle charging at park-and-ride facilities to flatten duck curves. In: *2019 IEEE Vehicle Power and Propulsion Conference (VPPC)*; 2019. p. 1–5.

[37] Jovanovic R, Bayhan S, and Bayram IS. An online model for scheduling electric vehicle charging at park-and-ride facilities for flattening solar duck curves. In: *2020 International Joint Conference on Neural Networks (IJCNN)*; 2020. p. 1–8.

[38] Yayla S and Ozdemir E. Smart vehicle to grid energy management algorithm for electric vehicle supported by photovoltaic energy sources. *Electric Power Components and Systems*. 2024;0(0):1–17.

[39] Zhang J, Che L, Wang L, *et al.* Game-theory based V2G coordination strategy for providing ramping flexibility in power systems. *Energies*. 2020;13(19):5008.

[40] Das S, Acharjee P, and Bhattacharya A. Charging scheduling of electric vehicle incorporating grid-to-vehicle and vehicle-to-grid technology considering in smart grid. *IEEE Transactions on Industry Applications*. 2020;57(2):1688–1702.

[41] Jian L, Zheng Y, Xiao X, *et al.* Optimal scheduling for vehicle-to-grid operation with stochastic connection of plug-in electric vehicles to smart grid. *Applied Energy*. 2015;146:150–161.

[42] Niu Z, An K, and Ma W. Vehicle-to-grid enabled charging infrastructure planning and operations considering demand uncertainties. *Transportation Research Part D: Transport and Environment*. 2024;127:103918.

[43] Shi Y, Xu B, Tan Y, *et al.* A convex cycle-based degradation model for battery energy storage planning and operation. In: *2018 Annual American Control Conference (ACC)*; 2018. p. 4590–4596.

[44] Shi Y, Xu B, Tan Y, *et al.* Optimal battery control under cycle aging mechanisms in pay for performance settings. *IEEE Transactions on Automatic Control*. 2019;64(6):2324–2339.

[45] Wei Z, Li Y, and Cai L. Electric vehicle charging scheme for a park-and-charge system considering battery degradation costs. *IEEE Transactions on Intelligent Vehicles*. 2018;3(3):361–373.

[46] Yang Q, Li J, Cao W, *et al.* An improved vehicle to the grid method with battery longevity management in a microgrid application. *Energy*. 2020;198:117374.

[47] Farzin H, Fotuhi-Firuzabad M, and Moeini-Aghtaie M. A practical scheme to involve degradation cost of lithium-ion batteries in vehicle-to-grid applications. *IEEE Transactions on Sustainable Energy*. 2016;7(4):1730–1738.

[48] Jovanovic R, Bayhan S, and Bayram IS. Optimization of electric vehicle charge scheduling with consideration of battery degradation. In: *2022 24th European Conference on Power Electronics and Applications (EPE'22 ECCE Europe)*; 2022. p. 1–11.

[49] Develder C, Sadeghianpourhamami N, Strobbe M, *et al.* Quantifying flexibility in EV charging as DR potential: Analysis of two real-world data sets. In: *2016 IEEE International Conference on Smart Grid Communications (SmartGridComm)*. IEEE; 2016. p. 600–605.

[50] Flammini MG, Prettico G, Julea A, *et al.* Statistical characterisation of the real transaction data gathered from electric vehicle charging stations. *Electric Power Systems Research*. 2019;166:136–150.

[51] Jovanovic R, Bayhan S, and Bayram IS. A multiobjective analysis of the potential of scheduling electrical vehicle charging for flattening the duck curve. *Journal of Computational Science*. 2021;48:101262. Available from: https://www.sciencedirect.com/science/article/pii/S1877750320305603.

[52] Lee ZJ, Li T, and Low SH. ACN-data: Analysis and applications of an open EV charging dataset. In: *Proceedings of the Tenth ACM International Conference on Future Energy Systems*; 2019. p. 139–149.

[53] Pecan Street Inc. Pecan Street Dataport; 2019. Available from: https://dataport.cloud.

[54] Song Z, Liu Y, Gao H, *et al.* The underlying reasons behind the development of public electric buses in China: The Beijing case. *Sustainability*. 2020;12(2):688.

[55] Topal O and Nakir İ. Total cost of ownership based economic analysis of diesel, CNG and electric bus concepts for the public transport in Istanbul City. *Energies*. 2018;11(9):2369.

[56] Quarles N, Kockelman KM, and Mohamed M. Costs and benefits of electrifying and automating bus transit fleets. *Sustainability*. 2020;12(10):3977.

[57] Feng W and Figliozzi M. Vehicle technologies and bus fleet replacement optimization: Problem properties and sensitivity analysis utilizing real-world data. *Public Transport* 2014;6(1–2):137–157.

[58] Liu Z and Song Z. Robust planning of dynamic wireless charging infrastructure for battery electric buses. *Transportation Research Part C Emerging Technologies*. 2017;83:77–103.

[59] Xylia M, Leduc S, Patrizio P, *et al.* Locating charging infrastructure for electric buses in Stockholm. *Transportation Research Part C Emerging Technologies*. 2017;78:183–200.

[60] Wang J, Kang L, and Liu Y. Optimal scheduling for electric bus fleets based on dynamic programming approach by considering battery capacity fade. *Renewable and Sustainable Energy Reviews*. 2020;130:109978.

[61] Ji J, Bie Y, and Shen B. Vehicle scheduling model for an electric bus line. In: *Smart Transportation Systems 2020*. Springer; 2020. p. 29–39.

[62] Liu T and Ceder AA. Battery-electric transit vehicle scheduling with optimal number of stationary chargers. *Transportation Research Part C Emerging Technologies*. 2020;114:118–139.

[63] Alwesabi Y, Wang Y, Avalos R, *et al.* Electric bus scheduling under single depot dynamic wireless charging infrastructure planning. *Energy*. 2020;213:118855.

[64] Foster BA and Ryan DM. An integer programming approach to the vehicle scheduling problem. *Journal of the Operational Research Society*. 1976;27(2):367–384.

[65] Desaulniers G and Hickman MD. Public transit. *Handbooks in Operations Research and Management Science*. 2007;14:69–127.

[66] Wen M, Linde E, Ropke S, *et al.* An adaptive large neighborhood search heuristic for the electric vehicle scheduling problem. *Computers & Operations Research* 2016;76:73–83.

[67] Tang X, Lin X, and He F. Robust scheduling strategies of electric buses under stochastic traffic conditions. *Transportation Research Part C Emerging Technologies*. 2019;105:163–182.

[68] van Kooten Niekerk ME, van den Akker J, and Hoogeveen J. Scheduling electric vehicles. *Public Transport* 2017;9(1–2):155–176.

[69] Zhou G, Xie D, Zhao X, *et al.* Collaborative optimization of vehicle and charging scheduling for a bus fleet mixed with electric and traditional buses. *IEEE Access*. 2020;8:8056–8072.

[70] Yao E, Liu T, Lu T, *et al.* Optimization of electric vehicle scheduling with multiple vehicle types in public transport. *Sustainable Cities and Society.* 2020;52:101862.
[71] Li L, Lo HK, and Xiao F. Mixed bus fleet scheduling under range and refueling constraints. *Transportation Research Part C Emerging Technologies.* 2019;104:443–462.
[72] Jovanovic R, Bayram IS, Bayhan S, and Voβ S. A GRASP approach for solving large-scale electric bus scheduling problems. *Energies*. 2021;14(20):6610. https://doi.org/10.3390/en14206610.
[73] Jovanovic R, Bayhan S, and Voß S. Matheuristic fixed set search applied to electric bus fleet scheduling. In: Sellmann M, Tierney K, editors. *Learning and Intelligent Optimization*. Cham: Springer; 2023. p. 393–407.
[74] Abdelaty H, Al-Obaidi A, Mohamed M, *et al.* Machine learning prediction models for battery-electric bus energy consumption in transit. *Transportation Research Part D: Transport and Environment*. 2021;96:102868.
[75] Gerbaux J, Desaulniers G, and Cappart Q. A machine-learning-based column generation heuristic for electric bus scheduling. *Computers & Operations Research*. 2025;173:106848.
[76] George SR and Zafar M. *Electrifying the ride-sourcing sector in California.* California Public Utilities Commission. 2018.
[77] Hagman J and Langbroek JH. Conditions for electric vehicle taxi: A case study in the Greater Stockholm region. *International Journal of Sustainable Transportation*. 2019;13(6):450–459.
[78] Jovanovic R, Safak Bayram I, Bayhan S, *et al.* Exploring the potential of placing charging stations at relief stands for EV fleets in ride-hailing and taxi services. *IEEE Access*. 2024;12:122249–122261.
[79] Lu JL, Yeh MY, Hsu YC, *et al.* Operating electric taxi fleets: A new dispatching strategy with charging plans. In: *2012 IEEE International Electric Vehicle Conference*. IEEE; 2012. p. 1–8.
[80] Tian Z, Jung T, Wang Y, *et al.* Real-time charging station recommendation system for electric-vehicle taxis. *IEEE Transactions on Intelligent Transportation Systems*. 2016;17(11):3098–3109.
[81] Bauer GS, Zheng C, Shaheen S, *et al.* Leveraging big data and coordinated charging for effective taxi fleet electrification: The 100% EV conversion of Shenzhen, China. *IEEE Transactions on Intelligent Transportation Systems.* 2022;23(8):10343–10353.
[82] Yi Z, Smart J. A framework for integrated dispatching and charging management of an autonomous electric vehicle ride-hailing fleet. *Transportation Research Part D: Transport and Environment*. 2021;95:102822.
[83] Shi J, Gao Y, Wang W, *et al.* Operating electric vehicle fleet for ride-hailing services with reinforcement learning. *IEEE Transactions on Intelligent Transportation Systems*. 2019;21(11):4822–4834.
[84] Farhan J and Chen TD. Impact of ridesharing on operational efficiency of shared autonomous electric vehicle fleet. *Transportation Research Part C: Emerging Technologies*. 2018;93:310–321.

[85] Jefferies D and Göhlich D. A comprehensive TCO evaluation method for electric bus systems based on discrete-event simulation including bus scheduling and charging infrastructure optimisation. *World Electric Vehicle Journal*. 2020;11(3):56.

[86] Yi Z and Smart J. A framework for integrated dispatching and charging management of an autonomous electric vehicle ride-hailing fleet. *Transportation Research Part D: Transport and Environment*. 2021;95:102822.

[87] Li HC, Lu CC, Eccarius T, *et al.* Genetic algorithm with an event-based simulator for solving the fleet allocation problem in an electric vehicle sharing system. *Asian Transport Studies*. 2022;8:100060.

Chapter 19
Predictive photovoltaic maintenance strategies

Antonio Sanfilippo[1], Dhanup Pillai[1], Juan Lopez Garcia[1] and Veronica Bermudez Benito[2]

The objective of predictive photovoltaic (PV) maintenance is the estimation of future asset anomalies, asset degradation, asset faults, and remaining asset life of a PV plant with the objective of prescribing maintenance that reduces asset failure to increase productivity and extend lifespan. The goal of this chapter is to review operation and management activities relevant to predictive PV maintenance and review commercial tools available in the market and novel solutions emerging in the scientific literature.

19.1 Introduction

Planning appropriate operation and maintenance (O&M) strategies is essential in optimizing PV performance and minimizing downtime due to faults [1]. For example, Livera *et al.* [2] demonstrate that O&M contributes a net economic gain of approximately 4.17 e/kW/year in PV system simulations that include component malfunctions and failures. This result is corroborated by a 2021 industry benchmark study [3] showing that the use of optimized O&M strategies in PV plants enables an average 5.27% recovery of energy. Altogether, the global non-residential PV O&M market is expected to reach 15 bn USD by 2030 from 4 bn USD in 2022, as shown in Figure 19.1.

The operation and management of PV plants involves several tasks to ensure performance and reliability of service through monitoring, optimization, control, documentation and reporting, security, and maintenance activities [5].

Monitoring activities provide situation awareness through the measurement and analysis of observed parameters of plant performance and reliability, and their expected change in the future with reference to power generation, faults occurrence, and grid code compliance. Power generation is concerned with the recording and forecasting of PV productivity as a function of solar radiation and other meteorological and environmental factors (e.g., wind speed and direction, temperature, aerosol, soil, and snow deposition). Power plant fault monitoring focuses on the detection and analysis of data that indicate reduced performance relative to the degradation of

[1]Qatar Environment and Energy Research Institute, Hamad Bin Khalifa University, Qatar
[2]BERBETIN, Antibes, France

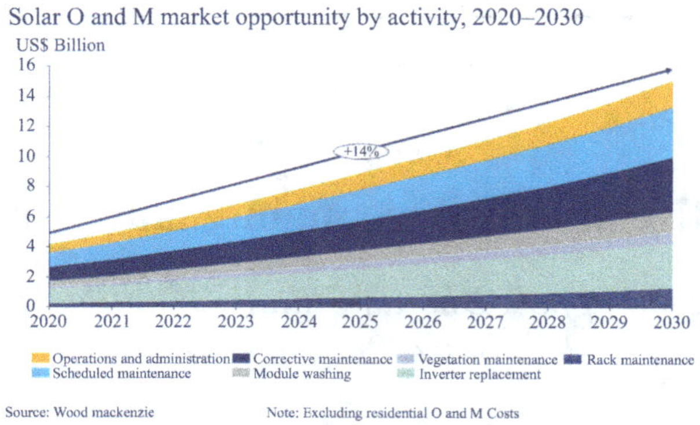

Figure 19.1 PV O&M market 2022–2030 [4]

PV modules due to hotspots, cracks, delamination, bubbles and yellowing/browning, shading and soiling, string faults, potential induced degradation (PID), inverter clipping, ground faults, line–line faults, DC arc faults, and AC overvoltage [6,7], as detailed in the list below.

- PID is caused by harmful current leakages due to a strong voltage difference between the solar cell and the module frame.
- Inverter clipping occurs when the PV system's power production exceeds the inverter's maximum output rating.
- Ground faults occur when current leaks from the PV system to the ground.
- Line–line faults result in short circuits between different modules.
- DC arc faults are caused by sudden high-power discharge between two conductors.
- AC overvoltages are voltage increases beyond the inverter's threshold.

Monitoring activities related to grid code compliance concentrate on the recording and analysis of data about active power, reactive power, current, frequency, voltage, power factor, and ramp.

Optimization focuses on the reduction of production losses and costs with reference to the response time to faults and disturbances and the analysis of costs incurred for interventions.

Control is concerned with the implementation of actions aimed to ensure compliance with grid codes with reference to power quality and voltage regulation through the management of parameters such as active power, reactive power, current, frequency, voltage, power factor, and ramp.

Documentation and reporting provide detailed records of all PV system characteristics and changes during the life of the PV plant, including layouts, electrical diagrams, modifications, alarms, and O&M activities.

Power plant security provides protection against intrusion through surveillance and alerts.

PV maintenance activities focus on automated digital solutions aimed to improve PV performance. O&M contracts for PV systems typically feature three types of maintenance: preventive, corrective, and predictive maintenance. Preventive maintenance comprises regular visual and physical inspections to verify that all the key PV plant components are in good working order [8] and activate proactive loss and degradation mitigating measures such as anti-soiling coating [7,9]. Corrective maintenance focuses on temporary and permanent fault repairs to address malfunctions detected through remote monitoring or onsite inspections [1] including the scheduling of cleaning events [10] based on economic models [1,11,12]. Predictive maintenance, also known as condition monitoring, is carried out before faults occur to anticipate system failure based on forecasts derived from the analysis and evaluation of parameters relative to power generation, faults occurrence, and grid code compliance (see monitoring activities above). Traditionally, PV O&M has been based on descriptive analytics, but recently greater attention has been devoted to methods focused on predictive analytics and big data [13], as shown in Figure 19.2.

Predictive maintenance is accomplished through the continuous analysis of historical and real-time data relating to asset performance (e.g., technical and environmental conditions, usage, maintenance history, similar equipment elsewhere) using modeling methods such as regression analysis, anomaly detection, and forecasting to schedule and optimize maintenance activities. These include scheduling cleaning events and the replacement of equipment at risk of failure before failure occurs causing plant damage. For example, advance knowledge about likely power losses from soiling or snow deposition from available meteorological and environmental data can be used to optimize the O&M schedule proactively to increase productivity and extend plant lifespan by reducing wear-out [14,15].

According to a survey of 268 respondents from various industrial sectors, the use of predictive maintenance has achieved 12% cost reduction, 9% uptime improvement, 14% in the reduction of safety, health, environment, and quality risks, and 20% in the lifetime extension of aging asset [16]. For the PV sector, these potential advantages are equally if not more compelling. For example, a 1.6% loss of production for a 100 MWac PV plant with a 35 €/MWh power purchase agreement rate located at a site with 2,500 hours of solar radiation per year could result in economic losses of up to 140,000 €/year [17].

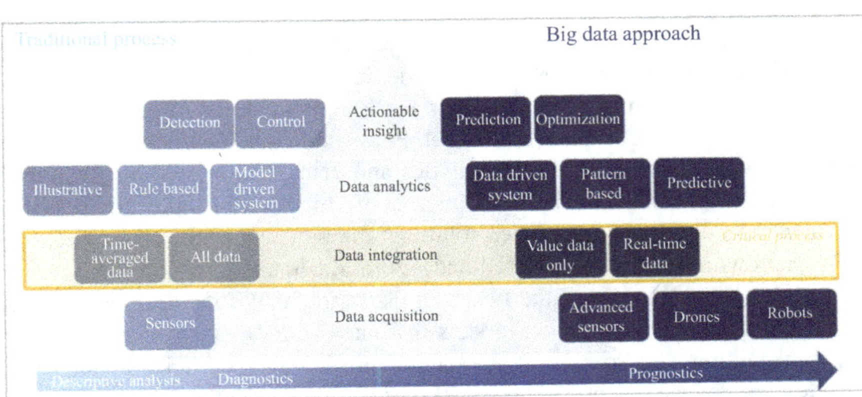

Figure 19.2 Evolution of data-driven PV O&M practices [13]

19.2 Commercial solutions in predictive maintenance

Despite the great potential impact in anticipating PV plant system faults for the reduction of the levelized cost of energy, the use of predictive maintenance is still uncommon in the PV sector. However, a significant amount of research has been devoted to the use of artificial intelligence (AI) for predictive PV maintenance (see Section 19.3) and industries around the globe are now starting to use the methods and techniques emerging from this endeavor to improve PV plant O&M.

In 2019, Huawei launched a new smart PV management system (AI Boost Smart *I–V* Curve Diagnosis 3.0) [18] that scans PV strings using Huawei's smart PV inverters and performs data analysis on the *I–V* curve of PV modules to identify faulty strings and output a diagnosis report. The *I–V* curve represents measured current (*I*) and power as a function of voltage (*V*) and is used to test PV modules for reliable power generation during manufacturing and installation and to assess operation during maintenance. The comparison of the measured and expected *I–V* curves helps identify faults such as a damaged solar cell, shaded areas, increased temperature strains, or defective bypass diodes. This PV management solution helps improving O&M efficiency through the life cycle of a PV plant in both utility and residential/commercial applications. In its current implementation, Huawei's AI Boost Smart *I–V* Curve Diagnosis seems to primarily offer preventive and corrective maintenance services, although the data collected could in principle be used to enable predictive maintenance. In 2020, Huawei Technologies announced the use of drone inspections, robot-based automatic O&M, digital technologies to enable preventative maintenance [19].

In partnership with Modelon Impact, ENGIE has developed and deployed a digital twin for predictive maintenance of solar power plants [20]. ENGIE Energía Chile has implemented a pilot project for the El Aguila PV Plant in Chile (7600 solar panels and numerous inverter components) where the Modelon Impact digital twin has helped determine the precise location of several broken fuses at the moment of failure, and schedule maintenance accordingly. Pairing real-world data with a virtual representation of the PV system provides a bird's eye view of the entire power system's health, down to the component level, that can be operated remotely. While in its pilot implementation, the digital twin has focused on preventive and corrective maintenance, equipped with the appropriate forecasting models it could also provide the ability to anticipate component failures and needed maintenance.

Microsoft offers customized solutions for PV maintenance [21] that use core Microsoft Office components (Word, Excel, PowerPoint, and Outlook), SharePoint, and Teams with Power BI for data analytics and visualization and the Azure IoT Hub to enable communication between Internet of Things (IoT) applications and the associated devices. These customized solutions combine legacy SCADA systems at the PV plant with Microsoft Power BI and Azure Machine Learning to deliver analytic insights. The Azure Machine Learning platform enables the development and deployment of anomaly detection models that analyze PV plant data and forecast equipment failures to help prevent unexpected breakdowns, optimize maintenance schedules, and reduce downtime. Power BI provides data visualization and analytics that enable operators to create interactive dashboards and reports that offer insights into equipment performance, maintenance trends, and key performance indicators. These visualizations enable operators to make informed decisions, allocate resources effectively, and continuously improve the maintenance strategies of PV plants.

Several startup companies applying AI to the solar industry are developing predictive maintenance capabilities. SmartHelio uses advanced learning models that integrate with legacy SCADA systems at the power plant to calculate daily losses due to diverse factors (e.g., soiling, temperature, degradation, cabling length, shading) and establish the best time to perform maintenance activities (Figure 19.3). Suncast (https://www.suncast.cl) applies AI and machine learning to forecast solar energy to maximize performance and estimate PV soiling to optimize PV cleaning scheduling. Their alert system assesses losses, schedules cleaning, and generates weather alarms. ENCORED provides a customized plant management system based on machine learning generation forecasting, alarming, and optimization algorithms [23]. Using as input the result of generation forecasting and observed PV generation, alarming algorithms detect inverter anomalies which result in alert actions. Optimization finds the best battery charge/discharge scheduling by using the error between predicted

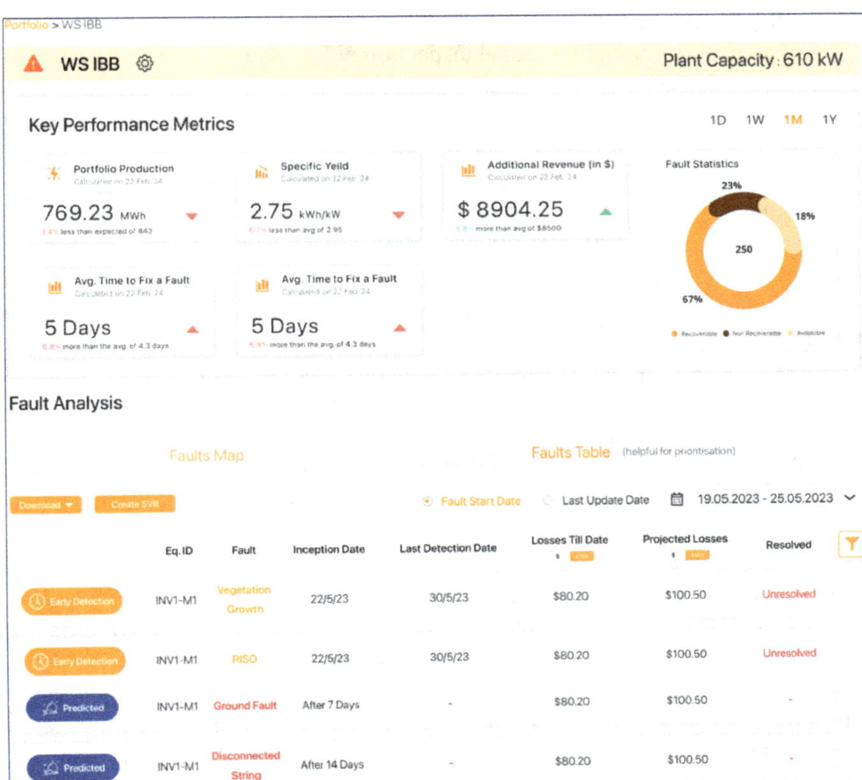

Figure 19.3 SmartHelio Autopilot platform [22]

and actual loads to minimize the sum of battery cost, electricity price subject to peak demand learning from [24].

Overall, the current commercial landscape of current predictive PV maintenance solutions is somewhat sparse and restricted in scope. Only a limited number of solutions are available, and their reach is either limited to specific functions, primarily aimed at preventive and corrective maintenance, still in a pilot mode, or focused on the development of customized solutions. A comprehensive platform for preventive PV plant management that offers a comprehensive array of functionalities is still not available in the market.

19.3 Emerging solutions in predictive maintenance

The need for PV predictive maintenance has been discussed in detail in the scientific literature with reference to several factors including PV panel faults and malfunction, PV performance and profitability, and asset aging (see [25] for a survey). Several methodologies have been developed based on manual diagnostics, failure mode and effects analysis (FMEA), machine learning and forecasting, and real-time sensors [25]. These methodologies have been used in the development of Predictive Energy Management and Control Systems for PV grid-tied systems with battery storage (PEMCSs) and decision support systems (DSSs) [26].

Manual diagnostics include visual inspection of PV systems and components for shade analysis [27] and soiling [28], aerial and ground-based infrared thermographic inspection of PV panels [29,30], *I–V* curve measurements of PV generators [31], and insulation resistance analysis of PV modules [32].

FMEA focuses on the identification of PV system components at risk of failure as a preventive PV maintenance measure by correlating the failure modes of each component with their causes (e.g., climate condition data) and resultant consequences [33–39].

Machine learning and forecasting approaches have primarily focused on PV productivity through the estimation of PV power using solar irradiance as training and evaluation data [40–44]. In predictive maintenance models, solar irradiance data have often been combined with other measurements such as cell temperature, ambient temperature, and power output [45–48].

Real-time sensors have been used to orient panels towards the sun (tracking) [49] and measure a variety of parameters including solar irradiance, temperature, short-circuit current, open-circuit voltage, maximum power, voltage and current at maximum power, fill factor and efficiency of PV arrays using a variety of wired and wireless networking technologies [50–57].

Several PEMCS solutions are based on a model predictive control aimed to manage energy from a grid-tied PV plant that integrates energy storage to optimize revenues [58–61]. For example, the model predictive control proposed in [61] anticipates states of full charge of the energy storage system (ESS) with reference to varying meteorological conditions to manage PV plant response so as to minimize economic penalties due to deviations from the plant commitments to the electricity market. As shown in Figure 19.4, the model predictive control computes the

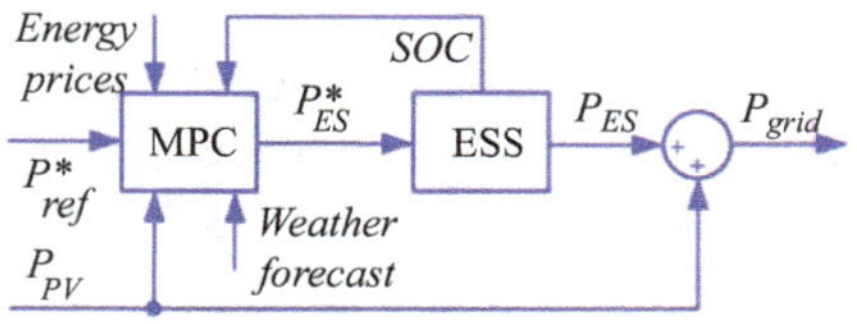

Figure 19.4 Model predictive control scheme [61]

power exchange of the ESS (P_{ES}) with the grid (P_{grid}) as a function of the plant's power reference (P_{ref}), the actual PV panels production (P_{PV}), state of charge (SOC) of the energy storage component, the predicted plant's power production (P^*_{ref}), the predicted power exchange of the ESS (P^*_{ES}), weather forecasts, and future energy prices.

The objective of model predictive control is to optimize the economic performance of a PV power plant that integrates energy storage. This optimization is carried out by maximizing the objective function in 19.1, where

$\lambda(k)$ is a weighting sequence
N is the given horizon
$c(t+k)$ is the imbalance cost at time $(t+k)$
T is the sampling period
$P_{grid(t+k)}$ is the power fed to the grid at $(t+k)$
$P_{ref(t+k)}$ is the hourly power committed by the PV plant at $(t+k)$
$c_{soc(t+N+1)}$ is the value of the energy stored in the ESS at time $(t+N+1)$
$E_{ES(t+N+1)}$ is the energy stored in the ESS at instant $(t+N+1)$.

$$J_N = \sum_{k=0}^{N} \lambda(k)c(t+k)T(P_{\text{grid}}(t+k) - P_{\text{ref}}(t+k)) + c_{\text{soc}}(t+N+1)E_{\text{ES}}(t+N+1) \tag{19.1}$$

The optimization problem is expressed as $J_N^{\text{opt}} = \max J_N$ subject to the constraints below for $k = 0, \ldots, N$.

$$P_{\text{grid}}(t+k) = P_{\text{PV}}(t+k) + P_{\text{ES}}(t+k)$$

$$E_{\text{ES}}(t+k+1) = E_{\text{ES}}(t+k) - TP_{\text{ES}}(t+k)$$

$$P_{\text{ES,min}} \leq P_{\text{ES}}(t+k) \leq P_{\text{ES,max}}$$

$$E_{\text{ES,min}} \leq E_{\text{ES}}(t+k) \leq E_{\text{ES,max}}$$

The PEMCS discussed in the literature [62] integrates load shedding as an additional consideration with the aim of increasing PV productivity and enhancing energy supply by minimizing PV curtailment, loss of load in load-shedding periods, and unnecessary battery charging from the grid. The PEMCS operates in three modes: grid-tied where the PV system, battery storage, and grid all interact with one another (Mode-1); standalone where there is no access to the grid (Mode-2), and a non-PEMCS mode where energy is supplied from PEG and PV (if available) with no access to battery storage (Mode-3). The PEMCS is presented as a two part solution: predictive optimization and predictive algorithm. Predictive optimization is executed at Mode-1 to avoid energy shortage due to surplus during future load-shedding periods. This is achieved through the objective function in 19.2 that minimizes the forecasted curtailed (or shortage) power $PCF(k)$, which is the difference between forecasted PV and the load, at a given interval Δt subject to efficiency charging/discharging constraints relative to predicted power from/to BESS

$$J(U,k) = \sum_{k=LSi}^{LSf} PCF(k)\Delta t \tag{19.2}$$

The predictive algorithm runs in Mode-1 to minimize shortage/curtailment during Mode-2 by injecting the forecasted surplus in Mode-2 into the grid. Figure 19.5 provides a description of the PEMCS workflow.

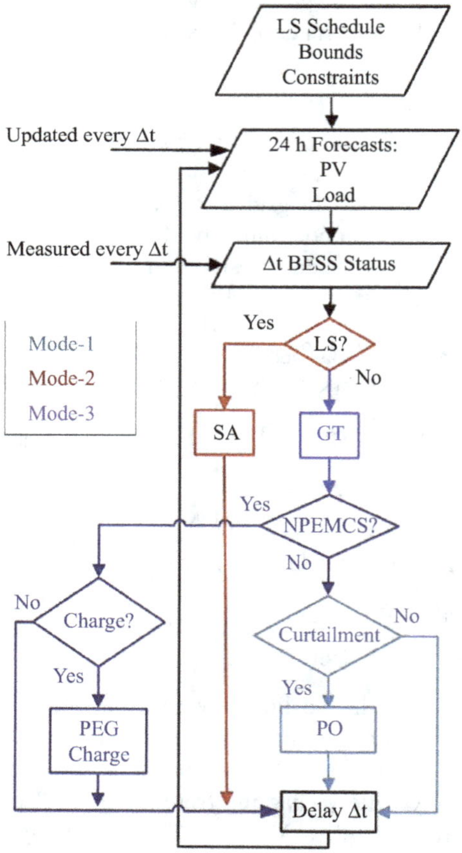

Figure 19.5 PEMCS workflow. LS, load shedding; BESS, battery energy storage stations; PEG, power electric grid; SA, standalone; GT, grid-tied; PO, predictive optimization [62]

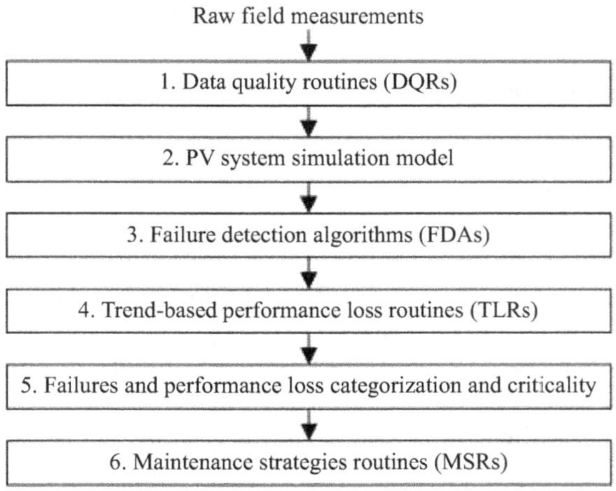

Figure 19.6 Flowchart of DSS for PV plant O&M [2]

DSSs for PV plant preventive management are a more recent development and can be seen as the evolution and convergence of earlier approaches. A DSS that integrates PV plant preventive, corrective, and predictive maintenance is described in Livera *et al.* [2].

As shown in Figure 19.6, the approach proposed in Livera *et al.* [2] comprises six processes. First, raw meteorological and PV operational data collected in the field are filtered out to eliminate invalid entries (step 1), as detailed in Livera *et al.* [63]. These measurements include: in-plane irradiance; ambient temperature; wind speed and direction; back-surface module temperature; array DC current and voltage used to calculate the DC power and AC output power; array energy yield, final PV system yield, reference yield, and performance ratio [64], and solar position parameters of solar azimuth and elevation angles derived through solar position algorithms [65]. Filtered measurements are used to train a power forecasting model that simulates PV performance in the absence of faults (step 2) using the eXtreme Gradient Boosting (XGBoost) algorithm [66] as described in Livera *et al.* [67]. The power forecasting model provides the reference to detect failures by comparing measured against predicted DC power production (step 23). Faults and losses are detected when the difference between the predicted and measured power exceeds a given threshold. Identified faults that cause gradual or seasonal power losses are then used to identify "trend-based" performance losses (step 4) due to soiling and degradation using the RdTools open-source python library [68]. Failures are classified into reversible (e.g., soiling) and irreversible (e.g., degradation) through anomaly detection and change-point models and criticality evaluation is then performed for the detected failures by using analytical methods such as FMECA [36] and multi-criteria decision analysis [69] (step 5). Finally, suggestions for maintenance actions are generated using net present value as a cost–benefit analysis to evaluate the economic impact of O&M measures (step 6).

19.4 Conclusions

As solar PV takes the lead as the fastest growing renewable energy sector and a key enabler of a net-zero emissions future [70], the need for O&M strategies that optimize PV performance, minimize downtime due to faults, and extend asset lifespan becomes more pressing. Preventive PV maintenance is emerging as the next-generation O&M framework for PV systems. AI plays a major role in enabling this development through the use of IoT sensors and autonomous vehicles to automate data collection in the field, and machine learning to identify faults and losses before they occur. By providing the means to address failures proactively, predictive PV maintenance can be highly instrumental in reducing the escalation of faults and losses into irreversible changes (i.e., degradation).

While a significant amount of research has been devoted to the use of AI for predictive PV maintenance, the current commercial landscape is somewhat sparse and restricted in scope. A comprehensive platform for preventive PV plant management that offers a comprehensive array of functionalities is still not available in the market. The development of such a solution is now needed to increase PV productivity through proactive loss avoidance and lower PV plant O&M costs by mitigating degradation.

An increasing number of studies investigating aspects of predictive PV maintenance have appeared in the scientific literature during the last two decades. The vast majority of these studies tends to provide component solutions with reference to manual diagnostics, FMEA, PV productivity forecasting, and real-time sensors. Some overarching proposals that build on component solutions for data gathering, predictive modeling, and optimization have appeared in the form of PEMCS for grid-tied PV plants that integrate energy storage. A more recent trend is now emerging integrating PEMCS with decision support. The uptake of these latest developments in the commercial landscape is likely to provide a much needed new generation of PV O&M tools for utility, residential, and commercial PV applications.

References

[1] Andrews, R., Ara, A., Assmus, J.H., *et al.* 2019. In *SolarPower Europe: Operation & Maintenance Best Practice Guidelines*, Version 4.0.

[2] Livera, A., Theristis, M., Micheli, L., Fernandez, E.F., Stein, J.S. and Georghiou, G.E., 2022. "Operation and maintenance decision support system for photovoltaic systems", *IEEE Access* 10, pp. 42481–42496.

[3] *Solar Asset Optimization, Industry Benchmark Study*, Raycatch, Tel Aviv, Israel, 2021, https://raycatch.com/white-paper/ (accessed June 1, 2025).

[4] da Fonseca, L.G., 2022. The solar aftermarket is growing fast – but new entrants could be left in the cold. Available from: https://www.woodmac.com/news/opinion/the-solar-aftermarket-is-growing-fast–but-new-entrants-could-be-left-in-the-cold/ (accessed June 1, 2025).

[5] Europe, S., 2021. "Operation & Maintenance Best Practices Guidelines, Version 5.0", SolarPower Europe, [Online]. Available from: https://www.

solarpowereurope.org/insights/thematic-reports/o-and-m-best-practice-guide lines-version-5-0 (accessed June 1, 2025).

[6] Shimshon, R., Green, M., Ulbrich, C., Graniero, P., Louwen, A., and Jahn, U., 2021. "The Use of Advanced Algorithms in PV Failure Monitoring", Report IEA-PVPS T13-19:2021. [Online]. Available from: https://iea-pvps.org/wp-content/uploads/2021/10/Final-Report-IEA-PVPS-T13-19_2021_PV-Failure-Monitoring.pdf (accessed June 1, 2025).

[7] Gonzalo, A.P., Marugán, A.P. and Márquez, F.P.G., 2020. Survey of maintenance management for photovoltaic power systems. *Renewable and Sustainable Energy Reviews*, 134, p. 110347.

[8] Deli, K. and Noel, D., 2018. *On-Field Operation and Maintenance of Photovoltaic Systems in Cameroon*, London, UK: IntechOpen.

[9] Ilse, K., Micheli, L., Figgis, B. W., *et al.*, 2019. "Techno-economic assessment of soiling losses and mitigation strategies for solar power generation", *Joule*, 3(10), pp. 2303–2321.

[10] Cristaldi, L., Faifer, M., Rossi, M., *et al.*, 2012. "Economical evaluation of PV system losses due to the dust and pollution", *Proceedings under IEEE International Instrumentation and Measurement Technology Conference*, pp. 614–618.

[11] Micheli, L., Theristis, M., Livera, A., *et al.*, 2021. "Improved PV soiling extraction through the detection of cleanings and change points", *IEEE Journal of Photovoltaics*, 11(2), pp. 519–526.

[12] Micheli, L., Fernández, E.F., Aguilera, J.T. and Almonacid, F., 2021. "Economics of seasonal photovoltaic soiling and cleaning optimization scenarios", *Energy*, 215.

[13] Mackenzie, W., 2019. The Increasing Role of Digital in the Solar PV O&M Space. Available: https://www.woodmac.com/reports/power-markets-the-increasing-role-of-digital-in-the-solar-pv-oandm-space-348152/ (accessed June 1, 2025).

[14] Micheli, L., Fernández, E.F., Müller, M. and Almonacid, F., 2020. "Extracting and generating PV soiling profiles for analysis forecasting and cleaning optimization", *IEEE Journal of Photovoltaics*, 10(1), pp. 197–205.

[15] Micheli, L., Fernández, E.F. and Almonacid, F., 2021. "Photovoltaic cleaning optimization through the analysis of historical time series of environmental parameters", *Solar Energy*, 227, pp. 645–654.

[16] Mulders, M. and Haarman, M., 2018. "Predictive maintenance 4.0. Beyond the hype: PdM 4.0 delivers results", PwC, Maininnovation.

[17] DNV GL and GreenPowerMonitor, 2021. "Predictive Maintenance of Solar PV Plants: The Time is Now", DNV GL and GreenPowerMonitor Position Paper [Online]. Available from: https://www.dnv.com/Publications/predictive-maintenance-of-solar-pv-plants-196282 (accessed June 1, 2025).

[18] What Is the Smart I-V Curve Diagnosis 3.0? Available from: https://solar.huawei.com/au/news-room/au/2019/news-20190710#:~:text=Huawei%20Smart%20I%2DV%20Curve%20Diagnosis,covering%2014%20types%20of%20faults (accessed June 1, 2025).

[19] Shaw, V., Gifford, J., Hutchins, M., Bellini, E. and Willuhn, M., 2020. "PV is entering the AI era. Special edition developed in partnership with Huawei", PV Magazine Special, 78538. Available from: https://solar. https://solar.huawei.com/~/media/Solar/minisite/2020HWFusionSolar/Assets/resource/Huawei-SpecialEdition2020.pdf (accessed June 1, 2025).

[20] ENGIE Collaborates with Modelon to Build Solar PV Plant Digital Twin for Predictive Maintenance. Available from: https://wizsp.com/streamlining-pv-plant-maintenance-with-microsoft-365-custom-solutions/ (accessed June 1, 2025).

[21] Efficient PV Plant Maintenance with Microsoft 365. Available from: https://wizsp.com/streamlining-pv-plant-maintenance-with-microsoft-365-custom-solutions (accessed June 1, 2025).

[22] New SmartHelio tool includes Climate Risk Assessment for solar projects. Available from: https://smarthelio.com/new-smarthelio-tool-includes-climate-risk-assessment-for-solar-projects/ (accessed June 1, 2025).

[23] ENCORED Renewable Asset Management. Available from: https://s3.ap-northeast-2.amazonaws.com/www.enertalk.co.kr/brochure/Encored_Renewable+Asset+Management_20200224.pdf (accessed June 1, 2025).

[24] Kim, M., Kim, K., Choi, H., Lee, S. and Kim, H., 2019. "Practical operation strategies for energy storage system under uncertainty", *Energies*, 12(6), p. 1098.

[25] Bosman, L.B., Leon-Salas, W.D., Hutzel, W. and Soto, E.A., 2020. "PV system predictive maintenance: Challenges, current approaches, and opportunities", *Energies*, 13(6), p. 1398.

[26] Osmani, K., Haddad, A., Lemenand, T., Castanier, B. and Ramadan, M., 2020. "A review on maintenance strategies for PV systems", *Science of the Total Environment*, 746, p. 141753.

[27] Bulanyi, P. and Zhang, R., 2014, May. Shading analysis & improvement for distributed residential grid-connected photovoltaics systems. In *The 52nd Annual Conference of the Australian Solar Council.*

[28] Figgis, B., Ennaoui, A., Ahzi, S. and Rémond, Y., 2016, November. Review of PV soiling measurement methods. In *2016 International Renewable and Sustainable Energy Conference (IRSEC)* (pp. 176–180). IEEE.

[29] Grimaccia, F.A.M.M.M.E.A., Aghaei, M., Mussetta, M., Leva, S. and Quater, P.B., 2015. "Planning for PV plant performance monitoring by means of unmanned aerial systems (UAS)", *International Journal of Energy and Environmental Engineering*, 6, pp. 47–54.

[30] Koch, S., Weber, T., Sobottka, C., Fladung, A., Clemens, P. and Berghold, J., 2016. Outdoor electroluminescence imaging of crystalline photovoltaic modules: Comparative study between manual ground-level inspections and drone-based aerial surveys. In *32nd European Photovoltaic Solar Energy Conference and Exhibition* (pp. 1736–1740).

[31] Aranda, E.D., Galan, J.A.G., De Cardona, M.S. and Marquez, J.M.A., 2009. "Measuring the IV curve of PV generators. *IEEE Industrial Electronics Magazine*", 3(3), pp. 4–14.

[32] Dechthummarong, C., Wiengmoon, B., Chenvidhya, D., Jivacate, C. and Kirtikara, K., 2010. "Physical deterioration of encapsulation and electrical insulation properties of PV modules after long-term operation in Thailand", *Solar Energy Materials and Solar Cells*, 94(9), pp. 1437–1440.

[33] Catelani, M., Ciani, L., Cristaldi, L., Faifer, M. and Lazzaroni, M., "Electrical performances optimization of photovoltaic modules with FMECA approach", *Measurement*, 46(10), pp. 3898–3909, 2013.

[34] Kuitche, J.M., Pan, R. and TamizhMani, G., 2014. "Investigation of dominant failure mode(s) for field-aged crystalline silicon PV modules under desert climatic conditions", *IEEE Journal of Photovoltaics*, 4(3), pp. 814–826.

[35] Shrestha, S.M., Mallineni, J.K., Yedidi, K.R., Knisely, B., Tatapudi, S., Kuitche, J. and TamizhMani, G., 2014. "Determination of dominant failure modes using FMECA on the field deployed c-Si modules under hot-dry desert climate", *IEEE Journal of Photovoltaics*, 5(1), pp. 174–182.

[36] Colli, A., 2015. "Failure mode and effect analysis for photovoltaic systems", *Renewable and Sustainable Energy Reviews*, 50, pp. 804–809.

[37] Villarini, M., Cesarotti, V., Alfonsi, L. and Introna, V., 2017. "Optimization of photovoltaic maintenance plan by means of a FMEA approach based on real data", *Energy Conversion and Management*, 152, pp. 1–12.

[38] Gallardo-Saavedra, S., Hernández-Callejo, L. and Duque-Pérez, O., 2019. "Quantitative failure rates and modes analysis in photovoltaic plants", *Energy*, 183, pp. 825–836.

[39] Livera, A., Theristis, M., Makrides, G. and Georghiou, G.E., 2019. "Recent advances in failure diagnosis techniques based on performance data analysis for grid-connected photovoltaic systems", *Renewable Energy*, 133, pp. 126–143.

[40] Diagne, M., David, M., Lauret, P., Boland, J., Schmutz, N., 2013. "Review of solar irradiance forecasting methods and a proposition for smallscale insular grids", *Renewable and Sustainable Energy Reviews* 27, pp. 65–76.

[41] Inman, R.H., Pedro, H.T. and Coimbra, C.F., 2013. "Solar forecasting methods for renewable energy integration", *Progress in Energy and Combustion Science* 39 (6), pp. 535–576.

[42] Voyant, C., Notton, G., Kalogirou, S., Nivet, M.L., Paoli, C., Motte, F. and Fouilloy, A., 2017. "Machine learning methods for solar radiation forecasting: A review", *Renewable Energy*, 105, pp. 569–582.

[43] Nespoli, A., Ogliari, E., Leva, S., Massi Pavan, A., Mellit, A., Lughi, V. and Dolara, A., 2019. "Day-ahead photovoltaic forecasting: A comparison of the most effective techniques", *Energies*, 12(9), p. 1621.

[44] Mellit, A., Massi Pavan, A., Ogliari, E., Leva, S. and Lughi, V., 2020. "Advanced methods for photovoltaic output power forecasting: A review", *Applied Sciences*, 10(2), p. 487.

[45] Mellit, A., Arab, A.H., Khorissi, N. and Salhi, H., 2007, June. An ANFIS-based forecasting for solar radiation data from sunshine duration and ambient temperature. In *2007 IEEE Power Engineering Society General Meeting* (pp. 1–6). IEEE.

[46] Polo, F.A.O., Bermejo, J.F., Fernández, J.F.G. and Márquez, A.C., 2015. "Failure mode prediction and energy forecasting of PV plants to assist dynamic maintenance tasks by ANN based models", *Renewable Energy*, 81, pp. 227–238.

[47] De Benedetti, M., Leonardi, F., Messina, F., Santoro, C. and Vasilakos, A., 2018. "Anomaly detection and predictive maintenance for photovoltaic systems", *Neurocomputing*, 310, pp. 59–68.

[48] Ferrero Bermejo, J., Gómez Fernández, J.F., Olivencia Polo, F. and Crespo Márquez, A., 2019. "A review of the use of artificial neural network models for energy and reliability prediction. A study of the solar PV, hydraulic and wind energy sources", *Applied Sciences*, 9(9), p. 1844.

[49] Devabhaktuni, V., Alam, M., Depuru, S.S.S.R., Green II, R.C., Nims, D. and Near, C., 2013. "Solar energy: Trends and enabling technologies", *Renewable and Sustainable Energy Reviews*, 19, pp. 555–564.

[50] Xiaoli, X. and Daoe, Q., 2011, April. Remote monitoring and control of photovoltaic system using wireless sensor network. In *2011 International Conference on Electric Information and Control Engineering* (pp. 633–638). IEEE.

[51] Rivai, A. and Rahim, N.A., 2013, November. A low-cost photovoltaic (PV) array monitoring system. In *2013 IEEE Conference on Clean Energy and Technology (CEAT)* (pp. 169–174). IEEE.

[52] Papageorgas, P., Piromalis, D., Antonakoglou, K., Vokas, G., Tseles, D. and Arvanitis, K.G., 2013. "Smart solar panels: In-situ monitoring of photovoltaic panels based on wired and wireless sensor networks", *Energy Procedia*, 36, pp. 535–545.

[53] Shariff, F., Abd Rahim, N. and Ping, H.W., 2013. Photovoltaic remote monitoring system based on GSM. In *2013 IEEE Conference on Clean Energy and Technology (CEAT)* (pp. 379–383). IEEE.

[54] Prieto, M.J., Pernía, A.M., Nuño, F., Díaz, J. and Villegas, P.J., 2014. "Development of a wireless sensor network for individual monitoring of panels in a photovoltaic plant", *Sensors*, 14(2), pp. 2379–2396.

[55] Guerriero, P., Di Napoli, F., Vallone, G., d'Alessandro, V. and Daliento, S., 2015. "Monitoring and diagnostics of PV plants by a wireless self-powered sensor for individual panels", *IEEE Journal of Photovoltaics*, 6(1), pp. 286–294.

[56] Adhya, S., Saha, D., Das, A., Jana, J. and Saha, H., 2016, January. An IoT based smart solar photovoltaic remote monitoring and control unit. In *2016 2nd International Conference on Control, Instrumentation, Energy & Communication (CIEC)* (pp. 432–436). IEEE.

[57] Moreno-Garcia, I.M., Palacios-Garcia, E.J., Pallares-Lopez, V., *et al.* 2016. "Real-time monitoring system for a utility-scale photovoltaic power plant", *Sensors*, 16(6), p. 770.

[58] Costa, L.M., Bourry, F., Juban, J. and Kariniotakis, G., 2008. "Management of energy storage coordinatedwith wind power under electricity market

conditions", *Proceedings of the 10th International Conference on Probabilistic Methods Applied to Power Systems*, pp. 1–8.
[59] Khalid, M. and Savkin, A., 2010. "A model predictive control approach to the problem of wind power smoothing with controlled battery storage", *Renewable Energy*, 35, pp. 1520–1526.
[60] Dicorato, M., Forte, G., Pisani, M. and Trovato, M., 2012. "Planning and operating combined wind-storagesystem in electricity market", *IEEE Transactions on Sustainable Energy*, 3(2), pp. 209–217.
[61] Perez, E., Beltran, H., Aparicio, N. and Rodriguez, P., 2012. "Predictive power control for PV plants with energy storage", *IEEE Transactions on Sustainable Energy*, 4(2), pp. 482–490.
[62] Syed, I.M. and Raahemifar, K., 2016. "Predictive energy management and control system for PV system connected to power electric grid with periodic load shedding", *Solar Energy*, 136, pp. 278–287.
[63] A. Livera, M. Theristis, E. Koumpli, *et al.*, 2021. "Data processing and quality verification for improved photovoltaic performance and reliability analytics", *Progress in Photovoltaics: Research and Applications*, 29(2), pp. 143–158.
[64] Theristis, M., Venizelou, V., Makrides, G. and Georghiou, G.E., 2018. Energy yield in photovoltaic systems. In *McEvoy's Handbook of Photovoltaics* (pp. 671–713). New York: Academic Press.
[65] Reda, I. and Andreas, A., 2004. "Solar position algorithm for solar radiation applications", *Solar Energy*, 76(5), pp. 577–589.
[66] Chen, T. and Guestrin, C., 2016, August. XGboost: A scalable tree boosting system. In *Proceedings of the 22nd ACM SIGKDD International Conference on Knowledge Discovery and Data Mining* (pp. 785–794).
[67] Livera, A., Paphitis, G., Theristis, M., Lopez-Lorente, J., Makrides, G. and Georghiou, G.E., 2022. Photovoltaic system health-state architecture for data-driven failure detection. In *Solar* (Vol. 2, No. 1, pp. 81–98). MDPI.
[68] Deceglie, M.G., Jordan, D., Shinn, A. and Deline, C., 2018. "RdTools: An open source Python library for PV degradation analysis degradation rate", pp. 1–15.
[69] Herz, M., Friesen, G., Jahn, U., Köntgens, M., Lindig, S. and Moser, D., 2021. Quantification of technical risks in PV power systems 2021. International Energy Agency, Paris, France, Tech. Rep. IEA-PVPS T13-23, 2021.
[70] REN21. 2024. Renewables 2024 Global Status Report Collection. https://www.ren21.net/gsr-2024/ (accessed June 1, 2025).

Chapter 20
AI and robotic techniques for PV inspection

Eduardo Feo Flushing[1] *and Gianni A. Di Caro*[1]

20.1 Introduction

The growing interest in renewable energy sources, particularly solar energy, has highlighted the importance of maintaining the efficiency and reliability of photovoltaic (PV) systems. PV systems are subjected to various environmental and operational stresses that can cause defects over time. If not detected and addressed promptly, these defects can lead to significant energy production losses and, in some cases, safety hazards. Therefore, regular inspections and maintenance are essential to ensure the long-term efficiency and reliability of PV systems [1].

Faults in PV modules can arise from various sources, including manufacturing defects, environmental factors, and operational wear and tear. For instance, defects such as microcracks, potential-induced degradation, and hot spots can severely impact the performance of PV modules [2]. These faults are not always visible to the naked eye and require advanced diagnostic techniques for accurate detection. External factors like shading, soiling, and mechanical damage can further complicate the inspection process. Regular and detailed inspections are crucial for identifying and mitigating these issues before they escalate into more significant problems, ensuring PV systems operate efficiently and safely over their intended lifespan.

Traditional inspection methods, such as manual visual inspections and electrical testing, have proven inadequate for the comprehensive and efficient maintenance of large-scale PV installations [3]. They are often time-consuming, labor-intensive, and prone to human error, making them unsuitable for the dynamic and extensive monitoring required for optimal PV system performance. The increasing scale of PV installations underscores the need for innovative and automated inspection solutions. Large-scale PV plants, which can span extensive regions and easily comprise thousands of PV modules, present a significant challenge for traditional inspection methods (e.g., one of the solar farms in Rajasthan, India, comprises 10 million solar panels and extends over about 57 km^2 [4]). The logistical difficulties of manually accessing and inspecting each module are substantial, leading to inefficiencies and potential oversight. Moreover, the dynamic nature of environmental conditions

[1]Department of Computer Science, Carnegie Mellon University, Qatar

means that defects can develop and change rapidly, which requires frequent and comprehensive inspections to ensure system performance.

Inspection tasks involve executing a well-coordinated sequence of operations: planning how to survey the plant's elements, positioning sensors close to the panels, employing advanced sensing technologies to gather accurate data, and processing these data to identify potential issues. Data processing can either be integrated into a closed-loop system, enabling real-time adaptation of the sensing process, or performed offline, where insights are derived post-inspection for later adjustments. Integrating AI and robotic technologies offers a promising solution to implement this chain of operations while addressing existing challenges effectively. The combined use of advanced imaging technologies, machine learning algorithms, and autonomous robotic systems can significantly improve the detection and diagnosis of PV module faults in vast installations, ultimately improving the reliability and performance of solar energy systems.

This chapter provides an overview of the application of artificial intelligence (AI) and robotics in inspecting PV systems, showcasing their potential to address these challenges. Unlike manual inspections, the combination of AI and robotics offers scalable, accurate, and efficient solutions for detecting and diagnosing faults, enabling timely intervention and optimizing energy production.

While several reviews in the literature focus on specific aspects of AI and robotics for PV inspection, this chapter adopts a more integrated perspective. It does not aim to serve as a comprehensive review of all existing works – a daunting task given the rapid advancements in the field – but rather as a concise overview of the predominant tasks and most effective techniques. The chapter also highlights key research gaps and identifies potential underexplored applications, aiming to inspire future developments in this transformative domain.

Our approach involves diving deeper into the systematic processes that underpin AI- and robotics-driven PV inspections, from sensor-level decisions and data collection strategies to advanced data processing techniques. By framing these discussions within the context of the challenges faced by modern PV installations, this chapter underscores the importance of embracing innovative, technology-driven approaches to ensure the long-term sustainability and efficiency of solar energy systems.

20.1.1 Recent survey papers on AI and robotics for PV inspection

Research on AI and robotics for PV system inspections has grown significantly in recent years, with numerous review papers categorizing advancements and identifying emerging trends. These reviews collectively underscore the transformative potential of AI and robotics in enabling efficient fault detection, predictive maintenance, and optimization of PV systems, particularly for large-scale installations. Recent efforts emphasize integrating imaging technologies, machine learning algorithms, and unmanned aerial vehicles (UAVs) to effectively implement the required chain of operations of inspection tasks and overcome the limitations of traditional inspection methods.

A recurring theme in existing reviews is the importance of fault detection for ensuring the reliability and efficiency of PV systems. Several works focus on the application of advanced imaging modalities, such as electroluminescence (EL), infrared thermography (IRT), and photoluminescence (PL), to identify defects that are not visible by conventional means or regular visual inspection. For instance, Jalal *et al.* [5] explore deep learning techniques for visual fault diagnosis, emphasizing the role of imaging technologies in improving defect detection accuracy. Similarly, Tang *et al.* [6] highlight the advantages of using aerial drones equipped with imaging sensors – both UAVs and remotely piloted drones – over Internet of Things (IoT)-based approaches, where sensors are typically installed directly on the panels. Integrating imaging techniques via UAVs with IoT data collection (e.g., electrical and environmental data) and data streaming methods is emphasized as beneficial, if not essential. Additionally, the study stresses the importance of tailoring approaches to the unique characteristics of different imaging modalities. Overall, these imaging-based methods underscore the critical role of advanced diagnostic tools in addressing the complexities and challenges of modern PV systems.

AI and machine learning techniques feature prominently across these reviews as enablers of more efficient and accurate fault detection. El-Banby *et al.* [7] highlight the role of AI in thermography-based fault detection, categorizing faults and emphasizing the need for timely detection to sustain system performance. Sohani *et al.* [8] broaden this perspective by reviewing machine learning applications, including fault detection, performance prediction, and energy management, demonstrating the versatility of AI in PV system optimization.

Adopting UAVs for PV inspections is another key focus area, reflecting a shift toward automation and scalability. Mustafa Abro *et al.* [9,10] examine UAV-based inspection strategies, discussing their potential to provide high-resolution imaging and efficient coverage of large-scale installations. These works also address practical challenges, such as path planning, battery life, and integrating UAVs with other inspection technologies. Milidonis *et al.* [11] extend this discussion to concentrating solar thermal systems, highlighting UAV applications in fault inspection, mirror soiling detection, and flight path optimization. These studies illustrate how UAVs can complement AI and imaging technologies to streamline inspection workflows.

While some reviews adopt a focused approach, such as El Bany *et al.* [5,7] on specific fault detection methods, others, like Aghaei *et al.* [12] and Abdulla *et al.* [3], take a broader view, encompassing operation and maintenance strategies. These broader reviews emphasize the need for systematic approaches integrating AI, IoT, and robotics to address the multifaceted challenges of PV system inspection and maintenance. The role of autonomous systems in combining visual inspection with real-time analytics emerges as a promising direction for future research.

A common conclusion in these reviews is that several gaps remain in the literature despite technological advancements and challenges. For example, while imaging technologies and UAV-based inspections have been extensively studied, integrating AI for real-time analytics and decision-making requires further exploration. Reviews such as those by Yahya *et al.* [13] and Mateo Romero *et al.* [14] highlight the need for advanced algorithms and hybrid approaches to enhance the accuracy and efficiency of PV inspections. Table 20.1 summarizes recent reviews on applications of AI and robotics to PV inspection.

Table 20.1 Summary of key survey papers on AI and robotics for PV inspection

Ref.	Year	No. works	Scope and applications	Research gaps identified
[7]	2023	54	Specific techniques for fault detection and maintenance, focusing on thermography and AI integration	Lack of standard benchmarks for thermographic data
[5]	2023	229	Imaging-based fault diagnosis using EL, IRT, and PL with deep learning for improved detection accuracy	Integration of multimodal imaging for diagnosis
[6]	2024	153	Broad overview of imaging, IoT-enabled UAV inspections, and tailored approaches for scalable PV maintenance	Scalability of UAV inspections in large PV farms
[3]	2024	249	Comprehensive strategies for predictive maintenance, combining AI with system reliability practices	Lack of integration between AI and maintenance tools
[9]	2023	108	UAV-specific methodologies focusing on predictive maintenance and fault detection in large-scale PV systems	Battery life, flight path optimization
[13]	2023	128	Image acquisition, pre-processing, and defect detection using various imaging techniques	Comprehensive datasets, advanced imaging techniques and machine learning models

20.1.2 A systematic perspective on PV panel inspection: Sensing, data collection, data processing

Compared to existing reviews, this chapter takes a more systematic approach, offering readers a concise and focused overview of how AI and robotics are applied to PV inspections. Building on the observation that inspection tasks are typically organized as a coordinated sequence of operations, as outlined in the Introduction, the chapter frames the discussion of the complex landscape of AI and robotics applications around three interrelated pillars: sensor selection, data collection strategies, and data processing.

This framework is designed to guide readers through the critical elements of the inspection process, ensuring a coherent discussion while providing practical insights to support decision-making for real-world implementations. The chapter systematically explores these pillars and emphasizes how these technologies streamline workflows, overcome key challenges, and optimize inspection outcomes. Furthermore, this approach helps identify emerging opportunities and research gaps, equipping readers with actionable insights into how AI and robotics revolutionize PV inspection practices.

The process of a practical PV inspection task begins with selecting the appropriate sensors to capture relevant data. Visual sensors, such as high-resolution cameras, enable the detection of surface-level defects like cracks, discoloration, and soiling. Thermal cameras are indispensable for identifying temperature anomalies that signify electrical faults or shading issues. Electrical sensors complement these by monitoring voltage, current, and temperature to identify performance issues. In Section 20.2, we show that AI is critical in optimizing sensor configurations and adapting them to specific environmental conditions, defect types, and inspection goals.

Once the sensors are selected, an effective data collection strategy is vital for comprehensive inspection. Static inspection methods involve fixed cameras regularly capturing images of specific areas and providing continuous monitoring. Drone-based inspections offer a comprehensive overview of the PV plant, capturing high-resolution photos and videos efficiently. Dynamic inspection employs mobile data collection devices with sensors to autonomously navigate the PV plant, allowing for close-up inspections of individual panels. In Section 20.3, we examine the main strategies that involve AI and robotics in collecting data for PV inspection.

In Section 20.4, we focus on the last stage of the inspection process: data processing. In this stage, image processing and computer vision techniques, including image segmentation, feature extraction, and machine learning algorithms, can help to detect and classify potential defects.

AI and robotics can be utilized independently or synergistically within these phases of operations. In sensor selection, AI can optimize the choice of sensors based on environmental conditions, panel types, and desired inspection granularity. Robotics automates data collection, ensuring consistent and comprehensive coverage of the PV plant, while AI optimizes robotic path planning and task scheduling. In data processing, AI algorithms efficiently process the acquired data, identifying defects and providing actionable insights. The combination of AI and robotics facilitates early detection of potential issues, maximizing energy output and minimizing downtime.

20.2 Sensor-level decisions in PV inspection

The success of PV panel inspection depends mainly on the ability to capture accurate and meaningful data regarding potential defects. This process begins with crucial decisions at the sensor level, including the selection, design, composition, and calibration of the sensors. These decisions establish a foundation for high-quality raw data acquisition, directly impacting the subsequent analysis needed for defect detection and system maintenance.

Sensor-level decisions also encompass the preprocessing strategies of the raw data gathered by the sensors. This includes using intelligent sensing techniques that optimize the data quality by improving sensor performance, filtering out noise, and prioritizing relevant features. For instance, when gathering images of PV panels, preprocessing may involve image segmentation (e.g., aerial imagery) to identify the regions of interest, such as the PV panels or their individual cells. It may also include combining multiple images when capturing all areas of interest in a single image is neither possible nor practical.

Sensor-level tasks are critical steps to combine raw sensing and data processing by isolating meaningful areas and highlighting the most relevant data for further analysis. This initial stage ensures the resulting dataset is cohesive and ready for practical analysis, bridging the gap between data acquisition and actionable insights.

There is growing evidence that AI is being integrated to optimize sensor performance and data acquisition for many tasks at the sensor level. To systematically survey the advancements in sensor-level decisions for PV inspection, we have identified three major tasks highlighting the integration of AI and robotics in this domain. The first task focuses on automated parameter tuning and sensor calibration, optimizing sensor configurations and performance to meet specific inspection requirements. The second task involves PV panel segmentation, primarily using deep learning techniques, to enable precise identification of areas of interest within complex environments. The third task addresses the creation of orthomosaics for PV inspection. Orthomosaics – high-resolution images generated by stitching multiple individual photos together – provide a seamless and geometrically accurate representation of large-scale installations. This technique is essential for inspecting PV systems, enabling detailed analysis across extensive areas.

The remainder of this section will discuss these tasks in detail, highlighting the most representative works and their contributions to improving the accuracy, efficiency, and scalability of PV inspection processes.

20.2.1 Automated sensor calibration and parameter tuning

The application of AI and robotics for sensor-level decision-making and automated data preprocessing in PV panel inspections remains a relatively underexplored area. Yet, it holds immense potential to revolutionize the field. By automating traditionally manual tasks such as parameter tuning and sensor calibration, these technologies can significantly enhance the accuracy and efficiency of inspections.

For instance, Cardoso *et al.* [15] noted that a significant limitation of using RGB, thermal, and hyperspectral cameras to inspect PV installations is the requirement for manual parameter adjustments, including thresholds in edge detection and kernel sizes in morphological operations. Carlson *et al.* [16] found that automatic gain control algorithms of commercial infrared cameras hinder the accurate identification of PV panels. Switching to fixed manual gain and offset settings improved quantitative analysis. This reliance on manual tuning introduces subjectivity and requires user intervention.

The work in Vergura [17] explores critical sensor-level decisions for UAV-based thermal inspections of PV modules, focusing on factors like infrared wavelength selection, emissivity settings, drone distance, orientation, and flight speed. It highlights the impact of these parameters on radiometric accuracy and emphasizes the need for careful calibration to minimize errors.

20.2.2 PV panel segmentation

PV panel segmentation is crucial in ensuring that subsequent analyses focus exclusively on the regions of interest, such as PV panels, while effectively filtering out irrelevant data (e.g., terrain data). Recent advancements in AI and computer vision have revolutionized this process, enabling highly accurate and efficient segmentation, even in complex scenarios. These innovations have been particularly impactful in applications involving thermal or visible light imagery captured by UAVs, where variability in environmental conditions and image resolutions poses significant challenges.

The literature on PV panel segmentation is extensive, particularly with the growing adoption of deep learning methods. This section highlights and discusses key studies that represent significant advancements in the field. By leveraging deep learning techniques, these modern approaches have greatly improved the precision and scalability of PV panel segmentation in remote sensing applications.

Bommes *et al.* [18] developed a semi-automated computer vision tool to extract PV modules from thermographic UAV videos. This tool addresses the challenge of identifying PV panels in thermal imagery, where temperature differences between panels and their surroundings are exploited. The tool segments PV modules from the background and facilitates the creation of large, high-quality datasets for further anomaly detection. One of the key contributions of this work is its ability to generalize across different PV plants, demonstrating robustness in diverse environmental conditions. This ensures the accurate identification of panels, setting a strong foundation for subsequent thermal anomaly detection and defect classification.

Similarly, Di Tommaso *et al.* [19] introduced a multi-stage model based on the YOLOv3 (You Only Look Once) deep learning framework, designed explicitly for PV panel extraction from aerial imagery. The model integrates two detection phases: first, to locate the PV panels within the image, and second, to identify and classify potential defects. This pipeline supports visible light and infrared imagery, ensuring versatility across different sensing modalities. Technical highlights include the model's real-time processing capabilities, average inference time of less than one second per image, and ability to handle large-scale PV farms efficiently. This approach significantly enhances operational efficiency, enabling rapid identification of defects and prioritization of maintenance activities.

Vlaminck *et al.* [20] use a region-based convolutional neural network (R-CNN) framework to extend the capabilities of aerial image analysis by focusing on extracting and classifying anomalies in PV power plants. Unlike traditional methods that process entire images, R-CNN employs region proposals to isolate specific areas of interest, such as individual PV modules, before applying deep learning models for defect detection. This approach ensures high accuracy in identifying anomalies, such as hotspots, cracks, and soiling, while reducing false positives. A notable technical contribution of this work is its integration of aerial imagery with advanced object detection algorithms, which allows for more precise localization of defects, even in complex or cluttered environments.

Guo *et al.* [21] introduced a robust framework to tackle the challenges of PV panel detection, particularly in scenarios with significant data imbalance and resolution disparities. The proposed segmentation approach employs a deep learning architecture to address variations in PV panel sizes, shapes, and spatial distributions across diverse environments. One of the key problems highlighted in the study is the imbalance in spatial resolution within training datasets, where high-resolution imagery offers more detailed information crucial for deep learning model training. Still, such data is often costly and limited in scope. Conversely, datasets from lower-resolution sensors, while more readily available, may lack the granularity required for precise segmentation. To address these challenges, the framework leverages advanced data augmentation techniques and customized loss functions to mitigate the effects of resolution and class imbalances. A notable innovation is the integration of multi-scale feature extraction, which allows the model to capture fine-grained details from high-resolution imagery and broader contextual information from lower-resolution data. This capability is particularly beneficial for detecting small-scale PV panels within complex or cluttered imagery, ensuring robust performance across various scenarios. The model achieves superior accuracy and generalizability by effectively handling resolution disparities and dataset imbalances, marking a significant advancement in PV panel segmentation for real-world remote sensing applications.

Gasparyan *et al.* [22] propose a hyperspectral solar segmentation network method to address challenges such as low image quality, varying resolutions, and computational inefficiencies. Their approach uses a Chebyshev Transformation to enhance image quality and hyperspectral synthetic decomposition to optimize band selection. Both techniques are integrated into a deep neural network architecture. The reported results outperformed state-of-the-art methods, including CNN and transformer-based networks. Key novelties include demonstrating the application of hyperspectral decomposition to refine spectral details and using Chebyshev transformers to improve scalability and computational efficiency.

Other applications of deep learning methods to refine the segmentation of PV panels include leveraging contextual understanding, particularly prior knowledge of the color of PV panels, to enhance the accuracy and shape regularization of segmented regions [23]. In addition, using vision-transformer-based methods for PV segmentation tasks has recently shown promising results [24].

While most studies focus on visible light imagery, some have explored segmenting PV panels in the infrared domain. One of the earliest deep learning-based approaches for PV segmentation in the infrared spectrum is Deep Res-UNet [25], a semantic segmentation neural network to extract PV panels in aerial infrared images. Building on such foundational work, more advanced methods have been developed, including the approach by Shen *et al.* [26], which presents a modified U-Net architecture specifically tailored for segmenting PV panels in complex infrared scenes. This adaptation improves performance in scenarios where PV panels are surrounded by challenging environmental features, such as objects with varying temperatures or emissivity, that complicate segmentation tasks. The authors benchmark their method against [25], demonstrating superior performance in these challenging environments.

20.2.3 Orthomosaics and image stitching

Orthomosaics from multiple aerial photographs are invaluable in extensive PV plant inspections. They provide expansive, high-resolution views of PV installations, facilitating precise PV panel detection, segmentation, and comprehensive analysis [27]. Most approaches for creating orthomosaics in PV plant inspections rely on computer vision techniques for image stitching [28,29]. These methods highlight the reliance on image stitching as a critical step for further processing and analysis.

Deep learning algorithms have the potential to automate and enhance image stitching, ensuring seamless integration of images captured from different angles and under varying lighting conditions. For instance, using AI-driven feature matching techniques could allow for accurate overlapping image alignment, resulting in high-quality orthomosaics that accurately represent the spatial characteristics of PV panels [30]. Deep learning frameworks have demonstrated the potential to surpass traditional methods in estimating homography transformations between overlapping infrared images of PV panels captured at close range – a crucial task for stitching multiple images from the panels without relying on precise GPS localization [31].

In PV plant inspections, UAVs with visible and IRT sensors are increasingly utilized. These UAVs can capture comprehensive datasets with thermal information critical for identifying anomalies such as hotspots and calculating the temperature and efficiency of PV panels [32]. Deep learning can also facilitate the fusion of visible and IR images, enhancing the quality and informational content of the resulting orthomosaics. For example, integrating thermal data into orthomosaics enables the visualization of temperature distributions across the PV array, aiding in the early detection of potential issues [33].

Integrating AI and robotics into creating orthomosaics for PV plant inspections presents significant advantages, yet it remains underexplored. Traditionally, computer vision algorithms have been the dominant approach for creating orthomosaics and stitching images, relying on feature matching and geometric corrections. While these methods have been effective, they have inherent limitations, such as sensitivity to lighting conditions, occlusions, and computational inefficiencies. Despite growing evidence that AI-based methods can outperform traditional techniques – offering greater accuracy, robustness, and adaptability – many PV inspection workflows still rely on conventional approaches.

One promising direction for integrating AI and robotics is optimizing UAV trajectories to enhance the real-time construction of orthomosaics. Unlike traditional methods, where orthomosaics are generated offline after all images have been collected, AI-driven systems can dynamically adjust flight paths based on real-time image analysis. This ensures better coverage, reduces redundant data collection, and improves the overall quality of the final orthomosaic. Additionally, AI can enhance multi-spectral data fusion, allowing for more precise defect detection by intelligently aligning and processing images from different sensors. These advancements can make PV inspections more efficient, accurate, and scalable, ultimately optimizing solar energy production and improving system maintenance.

20.3 Data collection strategies for PV inspection

Inspecting PV plants requires effective data collection strategies to ensure accurate, efficient, and scalable monitoring of large-scale installations. With the increasing deployment of solar farms in remote and expansive areas, manual inspection methods are becoming less practical due to their time-consuming nature, high costs, and safety concerns. Autonomous robotic systems, particularly UAVs and remotely operated aerial vehicles, have emerged as a transformative solution, efficiently collecting high-quality data across vast infrastructures. With advanced sensors, UAVs can capture visible light, thermal, and hyperspectral images, enabling comprehensive assessments of PV panel performance and condition [9].

Several studies have explored using UAVs, either individually [34] or in fleets [35], to inspect PV plants. These systems leverage their maneuverability to capture detailed imagery from different, highly informative perspectives, overcoming limitations posed by stationary cameras or ground-based robots. Furthermore, UAVs guided by advanced path-planning algorithms can ensure efficient PV installation coverage while minimizing travel time and energy consumption [36]. Integrating AI and robotics into these systems streamlines data collection and paves the way for automated processing and analysis, ultimately enhancing the accuracy and reliability of PV inspections.

Robust data collection strategies are critical to exploiting aerial imaging capabilities, especially in extensive and intricate PV plants and power facilities. These strategies must ensure thorough data acquisition and optimize efficiency, dependability, and scalability. Robotic systems, whether operating individually or in multi-robot configurations, rely on meticulously crafted data collection workflows to tackle issues related to coverage, sensor accuracy, and environmental changes.

For single-robot systems, data collection strategies involve precise path optimization and adaptive data acquisition techniques [37–39]. In PV inspections, UAVs must follow efficient flight paths that balance coverage and energy consumption while maintaining safe distances from obstacles [40,41]. Additionally, the integration of machine learning algorithms for real-time image processing and boundary detection enhances the accuracy of inspections, allowing UAVs to work in closed-loop with data acquisition, adapting their paths dynamically based on the identified locations of dirt or defects [42].

Specific mission goals of data collection strategies include capturing high-resolution imagery at appropriate angles and distances to improve defect detection and thermal analysis [43]. Additionally, strategies can emphasize minimizing errors caused by environmental factors like wind turbulence or electromagnetic interference by dynamically adjusting flight parameters, such as altitude and speed [44]. Incorporating redundancy and error-checking mechanisms further ensures data reliability, reducing the need for repeated flights and enhancing overall inspection efficiency [35].

In this section, we first discuss works on path planning, a prominent task in the context of data collection strategies. We then consider research works using multi-robot systems, where additional challenges arise in coordinating the data collection efforts across a fleet of autonomous and possibly diverse agents. Finally, we highlight general operational issues of autonomous aerial platforms.

20.3.1 Path planning

When using one or more UAVs, thorough path planning is crucial for optimizing the efficiency and effectiveness of the inspection process. Various criteria are employed to ensure that the UAVs cover the entire area of the PV plant while minimizing energy consumption and time. Standard methods include coverage path planning (CPP) techniques [41], which systematically calculate flight paths to ensure that all modules are inspected without leaving gaps. For instance, Pérez-González *et al.* [45] propose three CPP methods – boustrophedon exact cellular decomposition, grid-based spanning tree coverage, and wavefront coverage – demonstrating their effectiveness in automating UAV flight paths for PV inspections.

Other path optimization approaches for UAV-based inspections include modeling the problem as a Traveling Salesman Problem (TSP), which seeks to determine the shortest route that visits a set of locations and returns to the origin point. This method is particularly relevant in PV plant inspections to minimize travel distance while ensuring comprehensive coverage of all modules. Salahat *et al.* effectively utilized the TSP framework to optimize waypoint planning for autonomous aerial inspections, demonstrating its applicability in enhancing operational efficiency in large-scale solar farms [39]. Heuristic approaches such as particle swarm optimization (PSO) have also been employed to address UAV path planning challenges, taking into account factors such as flight attitude and gimbal limitations to improve the reliability of the inspection system [38].

Another notable example of path optimization is presented in [46], which introduces a multi-module system that manages boundary detection, path planning, and dynamic flight control. This coordinated approach allows UAVs to operate efficiently over extensive solar arrays, eliminating the need for manually programmed routes. The software autonomously adapts flight paths in real time based on the plant's layout, ensuring comprehensive coverage and reducing inspection time. The system enables seamless transitions between data collection tasks and flight maneuvers by communicating via the MAVLink protocol with multi-rotor flight controllers like Pixhawk.

Further emphasizing robust path planning, Morais at al. [47] demonstrate how vertical takeoff and landing capabilities can be combined with fixed-wing flight. This dual-mode operation tackles the unique requirements of floating PV installations, where environmental constraints such as wind and moving water surfaces demand adaptable flight strategies. Once the UAV transitions to a fixed-wing mode, it can traverse more considerable distances more efficiently, thus optimizing the overall duration and energy consumption of the inspection mission. These works collectively underline the importance of flexible navigation algorithms that can handle diverse operating environments on land-based fields or offshore floating solar plants.

20.3.2 Multi-robot systems

Due to the vast spatial distribution of PV panels, inspecting large-scale PV installations presents significant challenges. To address this, researchers have explored using multi-robot systems (MRS), which can leverage the scalability and parallelism offered by coordinated teams of aerial and ground vehicles.

In MRS, typical tasks include path planning, navigation, and allocation algorithms to allow a fleet of unmanned vehicles to cover large PV sites efficiently [35].

Existing works explore optimization techniques to plan inspection paths and task assignments for multiple agents, such as linear temporal logic (LTL) [36], and market-based approaches [35]. Proposed algorithms consider factors like battery life, charging stations, temporal constraints, and obstacle avoidance to ensure the feasibility and efficiency of robotic inspection missions.

The use of heterogeneous MRS, particularly of teams combining UAVs and unmanned ground vehicles (UGVs), has gained attention due to their ability to combine the strengths of different platforms for enhanced operational efficiency. UAVs excel in aerial inspections, covering large areas quickly, while UGVs provide logistical support, such as energy supply and landing platforms [48]. This collaboration addresses the limitations of each platform, optimizing inspection processes in large-scale PV installations where traditional methods are impractical.

In this context of heterogeneous MRS, De Benedetti *et al.* [35] proposed a market-based task assignment algorithm that optimally distributes inspection tasks among a fleet of UAVs and UGVs. The goal was to minimize the overall energy consumption while accounting for battery degradation and vehicle deterioration. The authors used a decentralized approach where each robot computes the cost of executing individual tasks, and a central auctioneer assigns tasks to the robot with the lowest bid.

Similarly, Liao and Liu [48] developed a path-planning method that prioritizes meeting points between UAVs and UGVs. In their scenarios, the UAVs perform aerial inspections, while the UGVs provide landing platforms and energy supplies for the UAVs. They use an improved genetic algorithm to plan the inspection path for each UAV. The resulting paths are then clustered to determine the optimal number and location of meeting points. An adaptive PSO algorithm was employed to find the best locations for the meeting points within the road network.

Focusing on long-term missions, Huang *et al.* [36] introduced LTL-based path planning algorithms that consider the need for charging, multiple visits to PV equipment, a single visit to communication equipment, and the avoidance of restricted regions when using a fleet of UAVs. The authors also proposed a heuristic algorithm to optimally deploy a charging station for the long-term monitoring of PV solar farms.

A common challenge highlighted in these works is the need to address the complex and dynamic environments of large-scale PV plants. The works emphasize the importance of accurate positioning and navigation, as well as the ability to handle obstacles, no-fly zones, and the changing orientation of the PV modules due to sun tracking [35,36,39]. Addressing these challenges is crucial to ensure the reliability and effectiveness of MRS for PV inspection tasks.

20.3.3 Operational issues of robotic platforms for PV inspection

Beyond navigation and path planning, the success of using robotic platforms for PV inspections relies heavily on integrating reliable hardware platforms with software mechanisms for real-time decision-making. In [9], the authors highlight several practical challenges for autonomous flights, including regulatory frameworks, operational safety, and real-time communications. UAV platforms must be equipped with sensors capable of monitoring flight conditions and adapting accordingly, such as adjusting altitude to avoid obstacles or modifying flight speed to ensure high-resolution data capture. These considerations become even more critical for large solar farms or remote locations, where GPS signals may be unreliable [39], and weather conditions can change unexpectedly [42].

System modularity is another key aspect. The diverse range of sensors used for PV inspection supports a compartmentalized design, enabling quick upgrades or replacements of sensors, cameras, or onboard processors without requiring a complete UAV overhaul [12,49]. This modularity is particularly valuable in dynamic environments where equipment requirements evolve over the lifespan of a PV plant.

Different UAV platforms offer unique advantages: multirotor drones provide stability and maneuverability for close-up inspections while fixed-wing aircraft are better suited for conducting broad surveys over vast landscapes [50]. In either case, tailoring the UAV hardware to the specific inspection mission can significantly enhance the efficiency of data collection.

These studies underscore that well-chosen platforms and robust operational protocols are critical for advancing robotics in PV inspections, regardless of the deployment environment.

20.4 AI techniques for data processing in PV inspection

At the core of AI-powered PV inspection lies the ability to analyze diverse data modalities – such as EL, PL, and IRT to identify a wide range of defects. For example, EL imaging reveals micro-cracks [51] and cell degradation [52], while IRT detects hot spots [53] and thermal anomalies indicative of electrical issues or shading.

The adoption of AI in PV inspection supports the scalability of solar energy solutions and advances their sustainability by improving reliability and minimizing energy losses. This section explores the various AI techniques employed for data processing in PV inspection, including supervised and unsupervised learning methods, state-of-the-art imaging modalities, and emerging approaches such as generative adversarial networks (GANs). By understanding these technologies, we can better address the challenges and opportunities in maintaining the efficiency and longevity of solar energy systems.

This section explores key AI techniques used in PV inspection, structured to first introduce the imaging modalities that provide the foundational data for inspection methods, including EL, PL, and IRT. With these imaging techniques established, the discussion moves to fault detection and classification, focusing on identifying and categorizing defects for effective maintenance. Next, it examines hot spot detection, a crucial process for preventing power losses and safety risks using deep learning-based object detection techniques. Anomaly detection follows, addressing deviations from normal PV module behavior to enable proactive fault management. Finally, the section concludes with synthetic data augmentation using generative adversarial networks, which enhances AI model performance by generating diverse and realistic training samples, overcoming data scarcity challenges in PV inspection.

By organizing the discussion this way, we create a clear progression – from data acquisition to imaging modalities, AI-driven fault detection and classification, proactive anomaly identification, and methods for enhancing model robustness.

20.4.1 Imaging techniques for AI-driven PV inspection

Different imaging methods can significantly contribute to identifying a wide range of defects in PV modules, each offering unique advantages and insights. Visible light imaging allows for detecting surface-level defects such as discoloration, physical damage, and soiling, providing a straightforward and cost-effective method for routine inspections. IRT is instrumental in identifying thermal anomalies like hot spots, which can indicate electrical issues or shading effects, thereby preventing potential fire hazards and efficiency losses. EL imaging excels in detecting micro-cracks, cell degradation, and other subtle defects not visible through conventional means, making it essential for in-depth diagnostic assessments. PL imaging further enhances defect detection by identifying issues related to series resistance and minority carrier lifetime, offering detailed insights into the internal quality and performance of PV cells. Together, these imaging techniques provide a comprehensive diagnostic toolkit, enabling accurate and efficient identification of various defects, ultimately ensuring the optimal performance and longevity of PV systems.

The data from IR images often require different preprocessing techniques, such as normalization, colorization, and temperature calibration, to ensure that the models can accurately interpret the thermal data [54,55]. Furthermore, the architecture may need to be adapted to handle the inherent noise and variability of thermal images [35,49].

EL imaging provides high-resolution images that can reveal defects not visible through conventional means. Deep learning models leveraged for analyzing EL images often utilize architectures capable of handling high-resolution data and distinguishing subtle variations in brightness that indicate defects [56]. The literature highlights the need for specialized deep learning architectures that can effectively leverage the high-resolution and nuanced information provided by EL imaging for comprehensive fault detection and classification in PV modules. Several deep learning architectures are employed for defect detection in PV modules through EL image analysis. CNNs are widely used because they can effectively extract local features and detect subtle brightness variations indicative of defects [51,57]. Region-based CNNs (R-CNNs) extend this approach by incorporating region proposal networks to identify and classify defects in EL images automatically [58].

The deep learning models applied to PL images may require different preprocessing and augmentation techniques to account for the lower signal strength compared to other imaging types. This can influence the choice of architecture, as models may need to be more robust to noise and variations in illumination [59]. The comparative analysis of PL and EL imaging techniques in the literature suggests that while both can visualize similar faults, the choice of architecture may depend on the specific characteristics of the images being analyzed [59].

In summary, the type of imaging used in PV module inspection directly impacts the design and effectiveness of deep learning architectures. Works like [51] discuss the implications of image type on model performance, emphasizing the need for specialized architectures that can effectively process the unique features of each imaging modality.

20.4.2 Fault detection and classification

Ensuring the efficiency and reliability of solar energy systems heavily relies on fault detection and classification in PV modules. These two tasks, though related, serve distinct purposes. Fault detection identifies the presence of anomalies or defects in PV systems, while fault classification determines the specific type of fault, often at the cell level.

Fault detection involves locating defects such as hot spots, shading, and material degradation that impact the performance of PV modules. Several studies have leveraged both standard [60] and advanced deep learning architectures for this purpose. For instance, Wu and Hao [61] utilized a faster R-CNN framework combined with thermal imaging to localize defects in PV modules effectively. Advanced architectures have refined detection capabilities, significantly enhancing fault detection by employing attention mechanisms [62].

Fault classification, distinct from detection, focuses on identifying the specific type of fault, typically at the cell level. This task is crucial for targeted maintenance and system optimization. Various studies have proposed methodologies to classify faults in PV modules using advanced machine learning techniques, particularly deep learning models. For example, [63] addresses the classification of 11 different fault types, achieving an impressive average accuracy of 93.51%. This work highlights the importance of comprehensive datasets encompassing diverse fault types to improve classification performance.

Fonseca Alves *et al.* [64] classify anomalies into eight distinct classes, achieving a testing classification accuracy of 78.85%. The study underscores the challenges posed by high within- and between-class variations, which can complicate classification. The authors also emphasize the role of data augmentation techniques in enhancing their CNN model's performance.

20.4.3 Hot spot detection

Hot spot detection is crucial for maintaining the efficiency and reliability of PV systems, as these defects, often caused by shading, material degradation, or electrical mismatches, can result in significant energy losses and safety hazards. Recent advancements in machine learning and deep learning techniques have greatly enhanced hot spot detection accuracy and efficiency.

Object detection models are pivotal in automating the identification and localization of hot spot defects in PV modules. Among these, the YOLO framework excels with its balance of speed and accuracy, making it ideal for real-time fault detection in large-scale PV installations. YOLO's single-stage detection approach processes high-resolution images efficiently, ensuring timely identification of hot spots that could cause power losses or safety hazards. Studies such as Lei *et al.* [65] demonstrate its effectiveness when combined with semantic segmentation, enhancing the precision of hot spot localization in thermal images. Similarly, Shaban [66] highlights YOLOv5's ability to detect and classify multiple defect types within PV modules, showcasing its scalability and versatility in addressing complex diagnostic tasks.

YOLO's adaptability to various imaging modalities, including IRT and EL imaging, further enhances its applicability. For instance, Ozturk *et al.* [56] demonstrate YOLO's robustness in analyzing EL images to detect and categorize defects that affect power output, while Lei *et al.* [65] show its effectiveness in handling thermal data under challenging PV inspection conditions. These attributes, combined with YOLO's real-time processing capabilities, make it an invaluable tool for optimizing PV maintenance. Future research, such as integrating attention mechanisms and feature fusion strategies outlined in Su *et al.* [67], can further improve YOLO's performance by addressing challenges like data imbalance and enhancing its robustness in operational conditions.

Challenges in PV hot spot detection systems have driven the exploration of novel neural network architectures and strategies to enhance detection accuracy, efficiency, and adaptability. These cutting-edge designs address critical challenges in PV monitoring through innovative mechanisms and frameworks. One notable contribution is the bi-branch collaborative training algorithm introduced in Hao *et al.* [68]. This framework employs a teacher–student paradigm using a knowledge distillation mechanism, allowing lightweight yet highly accurate detection systems. By incorporating a YOLOX-based student network optimized for speed and a HorNet-enhanced teacher network for robust feature representation, their model achieves high accuracy, demonstrating its applicability in real-world UAV-based PV inspections.

APM^{2}Det [69] addresses challenges in angle distortion and small-scale characteristics in infrared images by introducing an angle perception module and a model migration mechanism. The angle perception module mitigates distortions from varying camera angles, ensuring consistent feature capture, while model migration leverages pre-trained weights to enhance performance on small-scale datasets. These innovations enable APM^{2}Det to perform robustly under diverse imaging conditions, making it suitable for real-world applications.

The selective kernel mechanism is another area of innovation, as highlighted in SK-FRCNN [61]. This model improves traditional faster R-CNN networks by integrating a selective kernel attention mechanism, enabling adaptive feature extraction across scales. Adding a region-of-interest align layer further enhances localization precision, addressing the challenges of small-scale fault detection in complex environmental conditions. Experimental results reveal an accuracy improvement over baseline models, making SK-FRCNN a practical choice for field applications.

These advancements above highlight the diverse strategies employed in modern PV hot spot detection. Each deep learning architecture addresses unique challenges in detection precision, environmental adaptability, and computational efficiency, paving the way for more reliable and scalable PV monitoring solutions.

20.4.4 Anomaly detection

In the context of PV systems, anomaly detection is a critical process aimed at identifying defects, inefficiencies, or faults that could jeopardize the performance and reliability of solar energy systems. This process seeks to uncover deviations from expected operational patterns that may signal potential system malfunctions or indicate the need for maintenance interventions.

The practical implementation of anomaly detection enhances the operational efficiency of PV systems and contributes to the longevity and sustainability of solar energy production. Advances in machine learning have significantly improved anomaly detection in PV systems, and many works [70–73] have reported successful applications of AI in this area. Despite their shared focus, studies vary in methodologies and data types, reflecting the diversity of challenges in this domain.

Many approaches utilize unsupervised [74] or semi-supervised learning to address the common issue of limited labeled data. For example, weakly supervised methods like feature map conversion and hypersphere transformation [75] and positive and unlabeled (PU) learning with generative adversarial networks [76] demonstrate robust anomaly detection using minimal labeled samples. Similarly, deep learning models such as SeMaCNN [71] accurately identify defects in EL images. Pipelines combining weakly supervised and unsupervised approaches have also been used to create annotated datasets for anomaly detection [74].

Time-series data offers another avenue for unsupervised learning. Methods like LSTM autoencoders [77] reconstruct standard operational patterns to identify anomalies. At the same time, spectral clustering [70] adapts to environmental variations in PV power plants without requiring additional weather data.

20.4.5 Synthetic data augmentation with GANs

While CNNs have significantly advanced the automation of defect detection, these models require large, annotated datasets to achieve high accuracy. The scarcity of annotated defect images presents a substantial bottleneck. GANs offer a compelling solution by generating high-quality synthetic images that enrich training datasets, thereby enhancing the performance of defect detection models [57,78–80].

The motivation for using GANs in PV inspection stems primarily from the need to overcome the limitations of small and imbalanced datasets. As highlighted by Chen *et al.* [78], the paucity of defective PV images constrains the training of deep learning models. They propose using a Wasserstein GAN (WGAN) to synthesize realistic defect images, demonstrating through extensive experiments that including these synthetic images significantly improves the classification accuracy of CNN-based models. Their work lays the foundation for employing GANs as an effective data augmentation strategy in PV inspection. By leveraging the Wasserstein distance metric, WGANs ensure stable training and better convergence, which is critical for generating high-quality synthetic images.

Building on this idea, Zhao *et al.* [57] integrate GAN-generated images with traditional data augmentation techniques to enhance datasets of EL images. They employ a high-resolution network (HRNet) for defect identification, achieving notable improvements in defect classification. This approach demonstrates the potential of combining multiple augmentation methods to tackle the scarcity of annotated datasets and highlights the adaptability of GANs in complementing other techniques.

HRNet further ensures the preservation of fine-grained details in EL images, which is crucial for accurate defect detection.

Jiang *et al.* [79] present further advancements, introducing an innovative feature moving method combined with a YOLOv5S model. This approach aligns the features of GAN-generated images with those of real samples, effectively reducing the distribution gap between synthetic and real data. The improved alignment results in higher defect recognition accuracy, underscoring the importance of ensuring the representational fidelity of synthetic data in real-world applications. Their study also demonstrates how integrating GANs with state-of-the-art object detection models can enhance overall system performance, making it a promising direction for future research.

Luo *et al.* [80] propose the AC-PG GAN model, which progressively generates high-resolution EL images for PV inspection. Their experiments demonstrate that this approach enhances the quality of generated images and significantly improves classification accuracy, with gains of up to 14% in specific defect categories. This study highlights the advantages of progressive generation methods in refining the resolution and diversity of synthetic images, further solidifying the role of GANs in advancing PV inspection technologies. The AC-PG GAN model also employs attention mechanisms to focus on critical regions of the PV modules, improving the relevance and utility of the generated images.

Despite these promising advancements in using GANs to generate synthetic data and help advance the field of AI-driven PV inspection, several challenges remain. The training of GANs is often unstable and computationally expensive, posing barriers to their scalability and practical deployment. Additionally, while GANs can generate realistic images, ensuring the diversity and accuracy of these images to represent the full range of defect types remains an open research problem. Future work should explore hybrid approaches that integrate GANs with other generative models, such as variational autoencoders, to enhance the robustness of data augmentation. Moreover, addressing computational efficiency through techniques like model pruning or quantization could make GANs more viable for real-time applications. The validation of these methods on larger and more diverse datasets, coupled with their testing in real-world operational settings, is essential for their widespread adoption in the energy industry.

20.5 Challenges and future directions

Despite significant advancements in the application of AI and robotics in PV inspection tasks, including fault detection and maintenance of PV modules, numerous challenges still hinder the efficient and reliable operation of PV systems. Addressing these issues is essential for improving the accuracy, scalability, and integration of AI- and robotics-based inspection technologies. This section explores key challenges and outlines possible future directions.

One of the most significant challenges in AI-driven PV fault detection is the scarcity of labeled data, particularly for rare fault types. Deep learning models, which have demonstrated high accuracy in fault detection, rely heavily on large, well-balanced datasets to perform effectively. Techniques such as data augmentation and GANs have shown promise in artificially expanding datasets, enhancing model generalization, and addressing data imbalance [78,80]. Additionally, weakly supervised learning approaches have leveraged both labeled and unlabeled data to improve anomaly detection capabilities while mitigating the dependency on extensive labeled datasets [71,74]. Yet it remains an open question whether these approaches generalize well to various imaging modalities and environmental contexts.

Deploying fault detection systems in large-scale PV installations necessitates real-time processing and scalability. Traditional deep learning models often incur high computational costs, making them unsuitable for resource-constrained environments. Recent studies have proposed lightweight CNNs to reduce computational requirements while maintaining detection accuracy [81,82]. Edge computing frameworks have enhanced real-time processing by enabling localized data analysis on UAVs or other edge devices, reducing latency and the need for centralized data transmission [73]. Additional work in this area is needed to show that these systems can be utilized in cost-effective, resource-limited devices to support large-scale PV installations.

Effective integration of fault detection systems with predictive maintenance frameworks and IoT technologies can potentially improve the reliability and efficiency of PV operations. Future research should focus on predictive models that identify potential failures based on historical and real-time data, enabling proactive maintenance strategies. Moreover, advancing the autonomy of UAVs for inspection tasks could enhance the efficiency of maintenance operations [73,83].

While deep learning models have achieved impressive accuracy in fault detection, their interpretability and robustness remain areas for improvement. Techniques such as attention mechanisms and feature fusion have been explored to enhance model interpretability and detection accuracy, particularly for subtle defects in complex environments [58,79]. Additionally, noise reduction techniques are critical for improving model reliability under diverse operating conditions.

Using robotics and autonomous systems for PV inspection also presents several research gaps and challenges that must be addressed to enhance their effectiveness and reliability. One of the significant challenges in robotic systems is optimizing energy consumption and battery management. The current literature highlights the impact of battery degradation on operational efficiency and the need for strategies to maximize recharge cycles during inspections [35,36]. Future work should focus on dynamic energy-sharing frameworks between ground and aerial robots and adaptive energy-saving algorithms. Furthermore, deploying robotic systems across large-scale PV installations remains cost-prohibitive [35]. Future research should also focus on modular, low-cost robotic platforms and scalable inspection algorithms to enhance affordability without compromising performance.

Path planning for collaborative ground-air robotic teams in dynamic environments is another persistent challenge. Effective coordination is challenging when considering obstacles, no-fly zones, and the need for efficient meeting points for energy resupply [48]. Novel approaches could include multi-agent reinforcement learning for real-time adaptive path planning and developing robust coordination algorithms that handle heterogeneous robot capabilities.

Fusing data from multiple sensors (e.g., thermal, visual, and LiDAR) is critical for comprehensive inspections. However, data alignment, synchronization, and processing challenges persist [39]. Developing standardized protocols for sensor integration and employing AI techniques for multi-modal data fusion could improve inspection accuracy.

Effective collaboration between human operators and robotic systems is essential for the success of autonomous PV inspections. Current gaps include the design of intuitive user interfaces and interaction protocols [84]. Future research should explore augmented reality interfaces and adaptive collaboration frameworks that enhance operator understanding and control. Emerging challenges include integrating real-time data processing with autonomous decision-making and creating standardized benchmarks for evaluating multi-robot inspection systems. Future directions should also emphasize interdisciplinary research combining robotics, artificial intelligence, and domain-specific knowledge to develop robust, scalable, and adaptable solutions for PV inspection.

20.6 Conclusion

This chapter presented a systematic view of the application of AI and robotics in PV inspection, focusing on the key tasks involved – sensor selection, data collection strategies, and data processing. By structuring the discussion around these fundamental components, we provided a coherent framework highlighting both the technological advancements and the practical challenges faced in real-world implementations. Integrating AI and robotics has proven to be a powerful approach for addressing the limitations of traditional inspection methods, enabling more accurate, scalable, and efficient monitoring of PV systems.

The application of AI and robotics for PV inspection is becoming an increasingly active area of research and development. Numerous survey papers are published each year, each with a distinct focus, reflecting the rapid advancements and growing interest in this field. However, while these surveys provide valuable insights into state-of-the-art techniques, there remains a gap between theoretical advancements and practical deployment. This chapter aimed to bridge that gap by focusing on the practitioner's perspective, emphasizing the technical challenges that arise in real-world applications, and discussing how AI and robotics can be effectively implemented in PV inspections.

Despite the progress, significant challenges remain. Ensuring the reliability and robustness of AI-driven inspection systems, addressing data limitations, improving real-time processing capabilities, and optimizing robotic autonomy for large-scale PV farms are critical issues that need further research. The chapter underscored these challenges, highlighting the need for continued innovation in multi-modal data fusion, adaptive AI models, and collaborative robotic inspection strategies.

By presenting a structured and application-oriented overview, this chapter serves as a resource for researchers and practitioners looking to implement AI and robotics in

PV inspection. As this field continues to evolve, bridging the gap between theoretical advancements and practical deployment will be crucial in realizing the full potential of these technologies in ensuring the efficiency, reliability, and sustainability of solar energy systems.

References

[1] Özer T and Türkmen Ö. An approach based on deep learning methods to detect the condition of solar panels in solar power plants. *Computers and Electrical Engineering*. 2024;116:109143.

[2] Tang W, Yang Q, Xiong K, *et al.* Deep learning based automatic defect identification of photovoltaic module using electroluminescence images. *Solar Energy*. 2020;201:453–460.

[3] Abdulla H, Sleptchenko A, and Nayfeh A. Photovoltaic systems operation and maintenance: A review and future directions. *Renewable and Sustainable Energy Reviews*. 2024;195:114342.

[4] Energy W. Bhadla Solar Park: The World's Largest Solar Power Plant; 2025. Available from: https://www.world-energy.org/article/41570.html (Accessed: 13-01-2025).

[5] Jalal M, Khalil IU, and Haq AU. Deep learning approaches for visual faults diagnosis of photovoltaic systems: State-of-the-Art review. *Results in Engineering*. 2024;23:102622.

[6] Tang W, Yang Q, Dai Z, *et al.* Module defect detection and diagnosis for intelligent maintenance of solar photovoltaic plants: Techniques, systems and perspectives. *Energy*. 2024;297:131222.

[7] El-Banby GM, Moawad NM, Abouzalm BA, *et al.* Photovoltaic system fault detection techniques: A review. *Neural Computing and Applications*. 2023;35(35):24829–24842.

[8] Sohani A, Sayyaadi H, Cornaro C, *et al.* Using machine learning in photovoltaics to create smarter and cleaner energy generation systems: A comprehensive review. *Journal of Cleaner Production*. 2022;364:132701.

[9] Mustafa Abro GE, Ali A, Ali Memon S, *et al.* Strategies and challenges for unmanned aerial vehicle-based continuous inspection and predictive maintenance of solar modules. *IEEE Access*. 2024;12:176615–176629.

[10] Michail A, Livera A, Tziolis G, *et al.* A comprehensive review of unmanned aerial vehicle-based approaches to support photovoltaic plant diagnosis. *Heliyon*. 2024;10(1):e23983.

[11] Milidonis K, Eliades A, Grigoriev V, *et al.* Unmanned aerial vehicles (UAVs) in the planning, operation and maintenance of concentrating solar thermal systems: A review. *Solar Energy*. 2023;254:182–194.

[12] Aghaei M, Kolahi M, Nedaei A, *et al.* Autonomous intelligent monitoring of photovoltaic systems: An in-depth multidisciplinary review. *Progress in Photovoltaics: Research and Applications*. 2024;33:381–409.

[13] Yahya Z, Imane S, Hicham H, *et al.* Applied imagery pattern recognition for photovoltaic modules' inspection: A review on methods, challenges

and future development. *Sustainable Energy Technologies and Assessments*. 2022;52:102071.

[14] Mateo Romero HF, González Rebollo MÁ, Cardeñoso-Payo V, *et al.* Applications of artificial intelligence to photovoltaic systems: A review. *Applied Sciences*. 2022;12(19):10056.

[15] Cardoso A, Jurado-Rodríguez D, López A, *et al.* Automated detection and tracking of photovoltaic modules from 3D remote sensing data. *Applied Energy*. 2024;367:123242.

[16] Carlson J, Menicucci D, Vorobieff P, *et al.* Infrared imaging method for flyby assessment of solar thermal panel operation in field settings. *Applied Thermal Engineering*. 2014;70(1):163–171.

[17] Vergura S. Correct settings of a joint unmanned aerial vehicle and infrared camera system for the detection of faulty photovoltaic modules. *IEEE Journal of Photovoltaics*. 2021;11(1):124–130.

[18] Bommes L, Pickel T, Buerhop-Lutz C, *et al.* Computer vision tool for detection, mapping, and fault classification of photovoltaics modules in aerial IR videos. *Progress in Photovoltaics: Research and Applications*. 2021;29(12):1236–1251.

[19] Di Tommaso A, Betti A, Fontanelli G, *et al.* A multi-stage model based on YOLOv3 for defect detection in PV panels based on IR and visible imaging by unmanned aerial vehicle. *Renewable Energy*. 2022;193:941–962.

[20] Vlaminck M, Heidbuchel R, Philips W, *et al.* Region-based CNN for anomaly detection in PV power plants using aerial imagery. *Sensors*. 2022;22(3):1244.

[21] Guo Z, Zhuang Z, Tan H, *et al.* Accurate and generalizable photovoltaic panel segmentation using deep learning for imbalanced datasets. *Renewable Energy*. 2023;219:119471.

[22] Gasparyan HA, Davtyan TA, and Agaian SS. A novel framework for solar panel segmentation from remote sensing images: Utilizing Chebyshev transformer and hyperspectral decomposition. *IEEE Transactions on Geoscience and Remote Sensing*. 2024;62:1–11.

[23] Tan H, Guo Z, Zhang H, *et al.* Enhancing PV panel segmentation in remote sensing images with constraint refinement modules. *Applied Energy*. 2023;350:121757.

[24] Guo Z, Lu J, Chen Q, *et al.* TransPV: Refining photovoltaic panel detection accuracy through a vision transformer-based deep learning model. *Applied Energy*. 2024;355:122282.

[25] Zhang H, Hong X, Zhou S, *et al.* Infrared image segmentation for photovoltaic panels based on Res-UNet. In: *Chinese Conference on Pattern Recognition and Computer Vision* (PRCV); 2019. p. 611–622.

[26] Shen Y, Fan T, Lai G, *et al.* Modified U-Net based photovoltaic array extraction from complex scene in aerial infrared thermal imagery. *Solar Energy*. 2022;240:90–103.

[27] Oliveira AKVD, Bracht MK, Aghaei M, *et al.* Automatic fault detection of utility-scale photovoltaic solar generators applying aerial infrared thermography and orthomosaicking. *Solar Energy*. 2023;252:272–283.

[28] Aghaei M, Leva S, and Grimaccia F. PV power plant inspection by image mosaicing techniques for IR real-time images. In: *IEEE 43rd Photovoltaic Specialists Conference (PVSC)*; 2016. p. 3100–3105.

[29] Ismail H, Rahmani A, Aljasmi N, *et al.* Stitching approach for PV panel detection. In: *2020 Advances in Science and Engineering Technology International Conferences (ASET)*; 2020. p. 1–4.

[30] Xiao J, Zhang N, Tortei D, *et al.* STHN: Deep homography estimation for UAV thermal geo-localization with satellite imagery. arXiv preprint arXiv:2405.20470; 2024.

[31] Dahmani H, Yaqoob M, Ansari MY, *et al.* Thermal homography in photovoltaic panels: Evaluating deep learning and feature-based methods. In: *IEEE Texas Power and Energy Conference (TPEC)*; 2025. p. 1–6.

[32] Akay SS, Özcan O, Öcan O, *et al.* Efficiency analysis of solar farms by UAV-based thermal monitoring. *Engineering Science and Technology: An International Journal*. 2024;53:101688.

[33] Debaque B, Perreault H, Mercier JP, *et al.* Thermal and visible image registration using deep homography. In: *25th International Conference on Information Fusion (FUSION)*; 2022. p. 1–8.

[34] Tanda G and Migliazzi M. Infrared thermography monitoring of solar photovoltaic systems: A comparison between UAV and aircraft remote sensing platforms. *Thermal Science and Engineering Progress*. 2024;48:102379.

[35] De Benedetti MM, Bascetta L, Falsone A, *et al.* Automated photovoltaic power plant inspection via unmanned vehicles. *IEEE Transactions on Control Systems Technology*. 2024;32(2):399–412.

[36] Huang Y, Chen Z, Chu J, *et al.* Long-term path planning with optimal deployment of a charging station for monitoring photovoltaic solar farms. *Scientific Reports*. 2024;14(1):17279.

[37] Kong X, Xi Z, Wei S, *et al.* Infrared vision based automatic navigation and inspection strategy for photovoltaic power plant using UAVs. In: *Chinese Control and Decision Conference (CCDC)*; 2019. p. 347–352.

[38] Luo X, Li X, Yang Q, *et al.* Optimal path planning for UAV based inspection system of large-scale photovoltaic farm. In: *Chinese Automation Congress (CAC)*; 2017. p. 4495–4500.

[39] Salahat E, Asselineau CA, Coventry J, *et al.* Waypoint planning for autonomous aerial inspection of large-scale solar farms. In: *45th Annual Conference of the IEEE Industrial Electronics Society (IECON)*; 2019. p. 763–769.

[40] Savva A, Zacharia A, Makrigiorgis R, *et al.* ICARUS: Automatic autonomous power infrastructure inspection with UAVs. In: *International Conference on Unmanned Aircraft Systems (ICUAS)*; 2021. p. 918–926.

[41] Ding Y, Cao R, Liang S, *et al.* Density-based optimal UAV path planning for photovoltaic farm inspection in complex topography. In: *Chinese Control And Decision Conference (CCDC)*; 2020. p. 3931–3936.

[42] Mahdavipour Z. Defect inspection of photovoltaic solar modules using aerial electroluminescence (EL): A review. *Solar Energy Materials and Solar Cells*. 2024;278:113210.

[43] Kirsten Vidal De Oliveira A, Aghaei M, and Rüther R. Aerial infrared thermography for low-cost and fast fault detection in utility-scale PV power plants. *Solar Energy*. 2020;211:712–724.

[44] Romanov AM, Gyrichidi N, Volkova MA, *et al.* Automated mission planning for aerial large-scale power plant thermal inspection. *Journal of Field Robotics*. 2024;41(5):1313–1348.

[45] Pérez-González A, Benítez-Montoya N, Jaramillo-Duque Á, *et al.* Coverage path planning with semantic segmentation for UAV in PV plants. *Applied Sciences*. 2021;11(24):12093.

[46] Moradi Sizkouhi AM, Esmailifar SM, Aghaei M, *et al.* RoboPV: An integrated software package for autonomous aerial monitoring of large scale PV plants. *Energy Conversion and Management*. 2022;254:115217.

[47] Morais R, Martins JJ, Lima P, *et al.* Novel approach for offshore photovoltaic panels inspection with VTOL UAV. In: *OCEANS* 2024. IEEE; 2024. p. 1–6.

[48] Liao C and Liu Z. Cooperative path planning of ground-air robots for distributed photovoltaic inspection. *Journal of Physics: Conference Series*. 2023;2658(1):012015.

[49] Addabbo P, Angrisano A, Bernardi ML, *et al.* UAV system for photovoltaic plant inspection. *IEEE Aerospace and Electronic Systems Magazine*. 2018;33(8):58–67.

[50] Cabreira T, Brisolara L, and Ferreira Jr PR. Survey on coverage path planning with unmanned aerial vehicles. *Drones*. 2019;3(1):4.

[51] Munawer Al-Otum H. Classification of anomalies in electroluminescence images of solar PV modules using CNN-based deep learning. *Solar Energy*. 2024;278:112803.

[52] Karimi AM, Fada JS, Parrilla NA, *et al.* Generalized and mechanistic PV module performance prediction from computer vision and machine learning on electroluminescence images. *IEEE Journal of Photovoltaics*. 2020;10(3):878–887.

[53] Zhao S, Chen H, Wang C, *et al.* SNCF-Net: Scale-aware neighborhood correlation feature network for hotspot defect detection of photovoltaic farms. *Measurement*. 2023;206:112342.

[54] Ansar E, Sewil S, Maksood FZ, *et al.* Performance evaluation of feature detectors and descriptors with close-range solar panel images. In: *IEEE International Conference on Visual Communications and Image Processing (VCIP)*; 2024. p. 1–5.

[55] Pathak SP and Patil SA. Evaluation of effect of pre-processing techniques in solar panel fault detection. *IEEE Access*. 2023;11:72848–72860.

[56] Ozturk E, Ogliari E, Sakwa M, *et al.* Photovoltaic modules fault detection, power output, and parameter estimation: A deep learning approach based on electroluminescence images. *Energy Conversion and Management*. 2024;319:118866.

[57] Zhao X, Song C, Zhang H, *et al.* HRNet-based automatic identification of photovoltaic module defects using electroluminescence images. *Energy*. 2023;267:126605.
[58] Zhang X, Hao Y, Shangguan H, *et al.* Detection of surface defects on solar cells by fusing multi-channel convolution neural networks. *Infrared Physics & Technology*. 2020;108:103334.
[59] Høiaas I, Grujic K, Imenes AG, *et al.* Inspection and condition monitoring of large-scale photovoltaic power plants: A review of imaging technologies. *Renewable and Sustainable Energy Reviews*. 2022;161:112353.
[60] Ali MU, Khan HF, Masud M, *et al.* A machine learning framework to identify the hotspot in photovoltaic module using infrared thermography. *Solar Energy*. 2020;208:643–651.
[61] Wu X and Hao X. SK-FRCNN: A fault detection method for hot spots on photovoltaic panels. *IEEE Access*. 2023;11:121379–121386.
[62] Su B, Chen H, Liu K, *et al.* RCAG-Net: Residual channelwise attention gate network for hot spot defect detection of photovoltaic farms. *IEEE Transactions on Instrumentation and Measurement*. 2021;70:1–14.
[63] Korkmaz D and Acikgoz H. An efficient fault classification method in solar photovoltaic modules using transfer learning and multi-scale convolutional neural network. *Engineering Applications of Artificial Intelligence*. 2022;113:104959.
[64] Fonseca Alves RH, Deus Júnior GAD, Marra EG, *et al.* Automatic fault classification in photovoltaic modules using Convolutional Neural Networks. *Renewable Energy*. 2021;179:502–516.
[65] Lei Y, Wang X, An A, *et al.* Deeplab-YOLO: a method for detecting hot-spot defects in infrared image PV panels by combining segmentation and detection. *Journal of Real-Time Image Processing*. 2024;21(2):52.
[66] Shaban WM. Detection and classification of photovoltaic module defects based on artificial intelligence. *Neural Computing and Applications*. 2024;36(27):16769–16796.
[67] Su B, Chen H, and Zhou Z. BAF-detector: An efficient CNN-based detector for photovoltaic cell defect detection. *IEEE Transactions on Industrial Electronics*. 2022;69(3):3161–3171.
[68] Hao S, He T, Ma X, *et al.* KDBiDet: A bi-branch collaborative training algorithm based on knowledge distillation for photovoltaic hot-spot detection systems. *IEEE Transactions on Instrumentation and Measurement*. 2024;73:1–15.
[69] He T, Hao S, Zhang X, *et al.* APM^2 Det: A photovoltaic hot-spot fault detection network based on angle perception and model migration. *IEEE Transactions on Dielectrics and Electrical Insulation*. 2024.
[70] Ma W, Ma M, Zhang Z, *et al.* Anomaly detection of mountain photovoltaic power plant based on spectral clustering. *IEEE Journal of Photovoltaics*. 2023;13(4):621–631.
[71] Korovin A, Vasilev A, Egorov F, *et al.* Anomaly detection in electroluminescence images of heterojunction solar cells. *Solar Energy*. 2023;259:130–136.

[72] Miraftabzadeh SM, Longo M, Foiadelli F, *et al.* Anomaly detection in photovoltaic systems via deep learning autoencoder. In: *International Conference on Smart Energy Systems and Technologies (SEST)*; 2023. p. 1–6.

[73] Chang Z, Zhang AJ, Wang H, *et al.* Photovoltaic cell anomaly detection enabled by scale distribution alignment learning and multiscale linear attention framework. *IEEE Internet of Things Journal*. 2024;11(16):27816–27827.

[74] Otamendi U, Martinez I, Olaizola IG, *et al.* A scalable framework for annotating photovoltaic cell defects in electroluminescence images. *IEEE Transactions on Industrial Informatics*. 2023;19(9):9361–9369. Available from: https://ieeexplore.ieee.org/document/9982488/.

[75] Chang Z, Jia K, Han T, *et al.* Towards more reliable photovoltaic energy conversion systems: A weakly-supervised learning perspective on anomaly detection. *Energy Conversion and Management*. 2024;316:118845.

[76] Sevetlidis V, Pavlidis G, Mouroutsos SG, *et al.* Positive and unlabelled learning with data approximation for photovoltaic defect detection. In: *IEEE International Conference on Imaging Systems and Techniques (IST)*; 2024. p. 1–6.

[77] Syamsuddin A, Adhi AC, Kusumawardhani A, *et al.* Predictive maintenance based on anomaly detection in photovoltaic system using SCADA data and machine learning. *Results in Engineering*. 2024;24:103589.

[78] Chen L, Yang Q, and Yan W. Generative adversarial network based data augmentation for PV module defect pattern analysis. In: *Chinese Control Conference (CCC)*. IEEE; 2019. p. 8422–8427.

[79] Jiang F, Huang Q, Zhang S, *et al.* An enhancement generative adversarial networks based on feature moving for solar panel defect identification. *IEEE Sensors Journal*. 2023;23(19):23744–23752.

[80] Luo Z, Cheng SY, and Zheng QY. GAN-based augmentation for improving CNN performance of classification of defective photovoltaic module cells in electroluminescence images. *IOP Conference Series: Earth and Environmental Science*. 2019;354(1):012106.

[81] Akram MW, Li G, Jin Y, *et al.* CNN based automatic detection of photovoltaic cell defects in electroluminescence images. *Energy*. 2019;189:116319.

[82] Saeed F, Aldera S, Al-Shamma'a AA, *et al.* Rapid adaptation in photovoltaic defect detection: Integrating CLIP with YOLOv8n for efficient learning. *Energy Reports*. 2024;12:5383–5395.

[83] Hong F, Song J, Meng H, *et al.* A novel framework on intelligent detection for module defects of PV plant combining the visible and infrared images. *Solar Energy*. 2022;236:406–416.

[84] Cannella F and Di Francia G. PV large utility visual inspection via unmanned vehicles. In: *20th IEEE/ASME International Conference on Mechatronic and Embedded Systems and Applications (MESA)*; 2024. p. 1–8.

Chapter 21

Toward a blockchain-based smart microgrid: a peer-to-peer renewable energy trading framework

Ameni Boumaiza[1], Antonio Sanfilippo[1] and Dragan Boscovic[2]

Peer-to-peer (P2P) energy trading mechanisms incorporate significant potential toward the revolution of the energy sector. This entails allowing the clients to become manufacturers and sellers of the surplus energy. The process is perfect in boosting the incorporation of renewable energy and helping people have control over how the utility of their energy and the levels of expenditure. For example, through the inclusion of P2P energy transactions, there is enhancement of safety, transparency, as well as efficiency in the parameters of transactions alongside energy distributions. P2P energy transaction also reduces carbon emissions by inventing renewable sources of energy and reducing dependency on conventional central electricity plants. The process further enhances energy resilience by establishing a decentralized network that counters disruptions and blackouts. Generally, P2P energy trading is suitable for transforming the energy landscape and establishing an equitable, reliable, and efficient energy system. This chapter explores the future of managing renewable energy that enables clients to create and distribute energy and thus fostering a sustainable and efficient energy ecosystem. The main focus of this chapter is to examine the integration of energy consumption and generation forecasting, social simulation, and blockchain technology in order to create a testbed for smart cities.

21.1 Introduction

Almost half of the global energy marketplace is accounted for by renewable energy that enhances the establishment of eco-friendly aspects. According to Zaidi and Hong [1], the energy transaction portrays a strategic milestone toward enhancing energy reliance and environmental sustenance. The microgrids are coupled with conventional large-scale grids to ascertain an efficient, less carbon-emitting, and

[1]Qatar Environment and Energy Research Institute, Hamad Bin Khalifa University, Qatar
[2]W. P. Carey Department of Information Systems, Arizona State University, USA

reliable energy production process. Microgrid is used for a distribution system that disintegrates from the centralized production approach and enhances localized energy manufacturing within societies that rely on renewable resources like wind, solar, and hydropower.

Blockchain technology is ideal for the transforming environment as it revolutionizes power transaction. The concept was launched as a platform for fostering digital currency transactions, but the technology has developed advanced implications for the energy industry. Zaidi and Hong [1] indicate that the technology facet has been transparent, reliable, and decentralized enabling a seamless regulation of the energy trade in environmentally fluid circumstances. Blockchain facilitates a transparent and reliable energy transfer that establishes an autonomous trading environment.

A fundamental outcome is that energy clients transform into co-creators within the market. Andoni *et al.* [2] state that clients of conventional energy can now establish, exchange, and consume energy while organizations can generate independent power by installing small-scale wind turbines and rooftop solar panels for selling to other organizations cheaply. Additionally, consumption trends are redesigned more collectively by democratizing the energy production. This research seeks to extensively explore blockchain technology's transformation of energy transactions within the microgrids. The study further analyzes the blockchain technology and determines how it facilitates P2P energy transactions, integration of distributed energy resources, and smart contracts.

21.2 Background

21.2.1 Blockchain technology in energy transaction

Blockchain technology is a significant contributor to digital transformation, initially centered around cryptocurrency and has further advanced into the digital space. According to Andoni *et al.* [2], blockchain is articulated as a shared ledger technology that enhances information storage through computer networks that makes it challenging to duplicate and imitate. For instance, blockchain facilitates a centralized energy trading platform where transaction documents are maintained on several computers within the decentralized computers. Decentralization fosters transparency and enables participants to perceive the transactions and their utilization process. The decentralized systems are immutable upon entering the system, which makes it challenging to reverse (Figure 21.1). Blockchain is applicable on most cases within the transactional processes, which lead to reduced operational expenses, elimination of middlemen, and an increased transactional process (see Figure 21.2). Also, blockchain is integral in enhancing a direct exchange between manufacturers and clients, which allows energy to be transacted easily and transparently [2]. The process is crucial in the microgrid system where energy production is localized. The underlying complexities of energy transaction in the microgrids are handled by blockchain technology per the variations and uncertainty of renewable energy sources that ascertain transparent and safe transactions.

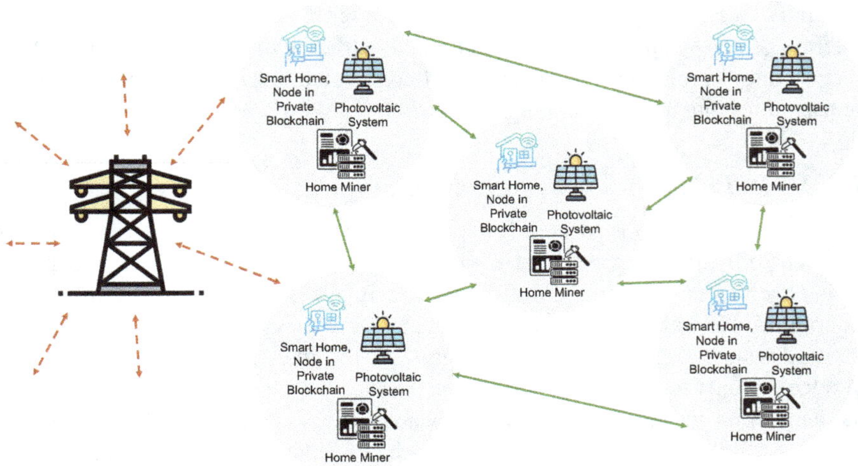

Figure 21.1 *A computer network that facilitates data sharing through decentralized ledger*

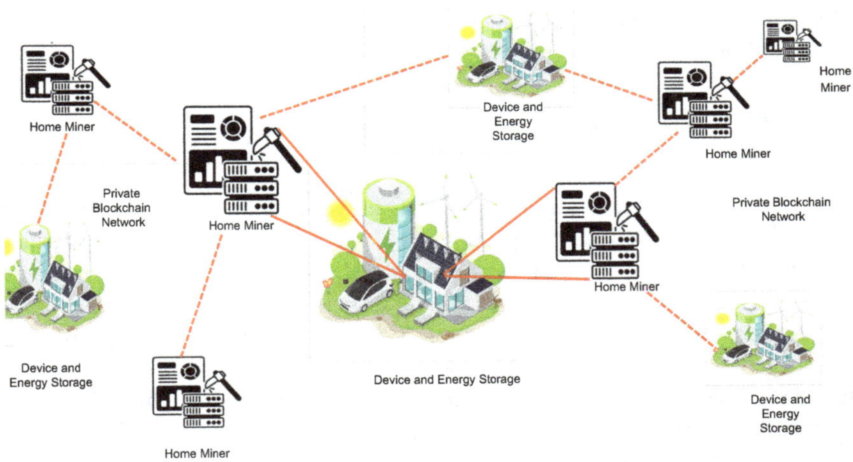

Figure 21.2 *A computer network that facilitates data sharing through decentralized ledger*

21.2.1.1 Smart contracts in energy transactions

The smart contracts consist of the self-executing contracts launched using computer codes defining the engagement parameters. Long *et al.* [3] indicate that the contracts are ideal for transactions based on the energy transmissions and safety measures. Also, the smart contracts are integrated into the blockchain once the initial conditions are met without third-party intervention to ensure compliance with the agreement terms. The contracts avail automatic performance of energy acquisition and exchange attributed to satisfactory preconditions. The process escalates efficiencies through the reduction of costs and time in the traditional contract administration process. Also, the procedures strengthen the security and confidence level of the electricity trade that ascertains integrity within the microgrids.

21.2.2 P2P energy transaction in microgrids

P2P energy transaction is located in the energy sphere within the microgrid's edge where an individual can sell and buyers exist without a third party. The model has been efficient in the microgrids and along the localized grids that operate autonomously and also within the principal power system powered by biomass, solar cells, and wind energy. According to Long *et al.* [3], the P2P microgrid participants can sell electricity to other participants with the excess solar power generated from the solar panels and surplus wind power. Renewable energy is the primary factor enhancing the use of the system as it enhances energy independence resilience that empowers societies. Long *et al.* [3] further state that the P2P trading comprises blockchain and smart contracts that ascertains the safety and transparency features while automating the required procedures as the energy demand and supply inclines to the dynamic changes within the microgrids. Generally, blockchain-centric microgrids transform energy interchange, and the platform provides a safe and straightforward mechanism that promotes new green energy systems within the peer networks.

21.2.3 Blockchain-centric smart renewable energy transaction paradigm

Blockchain technology facilitates new options for renewable energy transactions in the smart contracts and microgrids. According to Long *et al.* [3], the paradigm extensively depends on the blockchain's shared qualities that support the P2P transaction in RE among the suppliers and purchasers. The underlying model is influential in exchanging the monetary energy transactions and transfers the real-time energy units, which are assessed and documented on the distributed ledger. This ideology of transacting in clean energy is to advance an efficient smart-grid market, which suits clean energy sources like wind and solar power. Liu *et al.* [6] assert that the blockchain technology further ensures transparency of the transactions without disruptions, which automates issues attributed to renewable transactions like energy billing, tracking, and settlement. Blockchain can handle the volatility and intermittence of renewables, which makes the power grids durable and ideal for the modern world.

21.2.4 Responsibility of home miners, Internet of Things (IoT), and energy storage systems (ESS) in smart homes

Home miners are crucial in transacting energy in the blockchain technology as they have a fundamental control over the smart homes by determining how energy is exchanged in the blockchain systematics. According to Liu *et al.* [6], the miners affirm and document energy transactions on the blockchain to enhance the transaction's reliability. They are a substantive interface between the actual energy flow and the digital counterpart for an immediate energy transaction. Introducing IoT devices enhances blockchain-centric energy management within the smart homes. The actual data on energy consumption, storage, and production are crucial for efficient energy management provided by the IoT devices. The approaches enhance automatic control of the home-based energy sharing and ensuring an absolute efficiency and energy savings mechanism. Liu *et al.* [6] further state that energy storage systems (ESS) are ideal for smoothing energy consumption and production in the smart homes. The ESS accumulate power excess energy from an oversupply generation and discharges once demands are high and inputs are low. The process escalates the DERs uptime and facilitates smooth relationships with local microgrids.

21.2.5 Managing energy in blockchain-centric smart homes

Including blockchain technology in the smart homes portrays fundamental progress in energy conservation. The process implies that energy management surpasses consumption within the homes. Buth *et al.* [8] state that it generates an active participation in energy provision and exchange across the microgrid environment. The homeowner has an absolute mandate to monitor the energy utility through the blockchain technology and provide a substantive role in managing the energy metrics. Smart contracts enable homeowners to sell the excess energy to the neighbors while making excess purchases.

Additionally, the process yields economic benefits which improves the sustainability of the overall energy system. According to Buth *et al.* [8], blockchain's safety measures ascertain complete documentation of the described transactions, which further improves the confidence levels of individuals involved in the microgrid. Incorporating blockchain technology for transaction and management of renewable energy is a crucial pattern in the energy market for the smart houses. The underlying systems are pivotal in engaging the home miners, IoT, and ESS into a blockchain set that improves efficiency, user autonomy, and sustainability in energy systematization.

21.2.6 Trading procedure

P2P energy transaction involves a paradigm shift away from the centralized conventional energy sharing toward the decentralized facets, which leads to energy markets' democratization. Tushar *et al.* [7] indicate that prosumers have the ability to launch electricity using wind turbines and solar panels. The prosumers are then able to transact the surplus renewables with additional consumers within the network. The process skips the traditional centralized power supply schemes that involve the immediate energy transaction between a society or microgrid. The prosumers' renewable energy generation begins the P2P trading process and once the energy is converted into the microgrid, it is purchased by other microgrid users. Tushar *et al.* [7] further state that clients purchase the fuel per the demand levels for lower prices than most of the electricity suppliers. The sales take place between a prosumer-to-multiple clients and a single consumer-to-several prosumers per the costs.

21.2.7 Smart contract execution within the private blockchain networks

Smart contract enactment within the private chains is crucial in the security and automation of P2P energy transactions. Smart contracts enable instant energy delivery upon the receipt and accompanies clients who have significantly agreed to buy a pre-determined energy per unit price. Transactions are transparent following the incorporation of smart contracts within the private blockchain network. According to Boumaiza [10], the networks portray meaningful permission that allows a credible and integral assurance among the participants. Energy transaction is ideal between the prosumers and clients certified to ensure a genuine mechanism from the reliable sources. The validation process is crucial for upholding the P2P energy market authenticity and enhancing stakeholder confidence. Blockchain technology and intelligent contracts have the potential for contemporary energy sharing through the P2P energy exchange. The process enriches the individuals towards active engagement in the energy market usage of renewable energy sources and further encourages the enhancement of a modern energy system.

21.3 Literature review

21.3.1 P2P dispatch technologies and market procedures

In the context of power systems, the shift towards P2P energy trading has become crucial for researchers and practitioners. Cali and Cakir [4] emphasize the necessity of a robust framework for designing P2P markets. Their analytical approach highlights the fundamental concept of social inclusivity, which is essential for engaging consumers in P2P energy exchanges. While their work underscores the benefits of incorporating societal aspects into P2P trading, it also points to a reliance on existing auction frameworks that may limit the adaptability of these systems to the dynamic nature of energy trading.

The P2P model proposed by Cali and Cakir [4] incorporates a continuous double auction (CDA) mechanism, facilitating microgrid interactions through an hourly trading framework. However, Khorasany *et al.* [5] critique the CDA model for its inability to integrate energy products tied to virtual commitments, resulting in complexities that arise from market bids influenced by varying timeframes. This limitation is particularly pronounced in the case of non-linear consumer behaviors associated with devices like washing machines and dryers, which could skew the results of energy trading.

Moreover, the literature distinguishes non-collaborative P2P models that impact user behavior across diverse smart home configurations [5]. Here, the method of decentralized P2P bilateral dispatch aims to optimize costs for consumers, yet it is constrained by static energy storage solutions. As Liu *et al.* [6] point out, there is an urgent need for optimization models that address the complexities of product interdependencies in transient systems, calling for a refined framework to support dynamic energy transactions.

21.3.2 Blockchain energy markets

The evolution of blockchain technology has catalyzed innovative projects in localized energy transactions, drawing significant scholarly attention. Tushar *et al.* [7] argue that various global startups are actively enhancing decentralized P2P energy trading systems. These platforms, akin to power retailers, leverage artificial intelligence (AI) to optimize cost savings for users.

However, despite the promising nature of platforms like Pylon and Volt markets, as discussed by Buth *et al.* [8], challenges remain. The blockchain-enabled energy transaction system is often constrained by the activity dependency multi-layer model complexity [9], which complicates market clearance [10]. Integrating AI into the dispatch phase compounds these challenges by creating barriers that hinder inter-virtual dependencies and disrupt the natural flow of energy trading.

The limitations of current P2P and blockchain market models reveal a critical gap – the need for a more fluid and interconnected framework that accommodates the diverse and dynamic nature of energy markets [11]. Recent advancements in technology must be harnessed to create a more resilient and efficient trading environment.

21.3.3 Novel approach and contributions

In response to these identified limitations, our study introduces a novel approach to P2P energy trading that seeks to mitigate the drawbacks highlighted in the current literature. By leveraging advanced AI algorithms integrated within a decentralized architecture, our approach aims to enhance real-time decision-making, optimize market clearance, and improve the scalability of transactions while addressing inter-virtual product complexities.

This study not only expands upon the current frameworks by incorporating adaptive mechanisms for energy dispatch but also proposes a holistic model for decentralized interactions that promote collaborative trading among consumers. Through this innovation, we aim to transform the landscape of P2P energy markets, making them more effective, inclusive, and responsive to the changing needs of energy consumers.

21.4 Decentralized energy transactional platform

21.4.1 Datasets

This study incorporates a dataset involving 3 years of energy usage data that emanates from the Education City Community in Doha, Qatar (Figure 21.3). The data spans from 2017 and 2020 and encompasses the timeframe affected by the COVID-19 pandemic. The dataset further covers the energy usage for different loads. Qatar Foundation housing compound, consisting of lot#1 and lot#2, is a large residential complex with 623 homes of various types, including villas and flats. These homes are powered by a main substation, which receives electricity at 33 kV and distributes it to 15 secondary substations. These secondary substations then bring down the voltage to 415 V, ready for distribution to the homes. It is worth noting that each substation

Figure 21.3 Study's test-bed: education city community housing lot#1&lot#2

serves a specific area within the compound, with 6 substations providing power to 232 homes in lot#1 and 9 substations serving 391 homes in lot#2. This means that all homes connected to the same substation share the same power lines, creating a complex network of electricity distribution within the compound. In order to better understand and manage this network, extensive data was collected and utilized in this study. This data includes information on power transmission, buildings, generated rooftop photovoltaic (RPV) energy, and hourly electricity demand profiles. The power transmission layer includes the locations of the main and secondary substations, as well as the power line routes connecting them. This information is crucial in understanding the physical layout of the electricity distribution system within the compound. The building layer includes detailed information on the type and size of each home, as well as its precise location within the compound. This data is important in determining the electricity demand for each home, as different types and sizes of homes may have varying energy consumption patterns. The generated RPV energy layer adds another layer of complexity to the data, as it includes the capacity and location of each RPV system installed on the rooftop of a home. This renewable energy source can significantly impact the electricity demand and distribution within the compound.

The final layer of data used in this study is the hourly electricity demand profiles for each type of home in the compound. These profiles take into account various factors such as the number of occupants, appliances used, and weather conditions, providing a comprehensive understanding of the electricity consumption patterns within the compound. By combining all of these layers of data, the agent-based model used in this study can accurately simulate and predict the electricity usage and distribution within the Qatar Foundation housing compound. This allows for better management and optimization of the power supply, leading to increased efficiency and sustainability.

21.4.2 Methods

Scaling techniques like RobustScaler, MinMaxScaler, and StandardScaler were used to enhance the data training process's scalability. The process comprised a comparative assessment of the readings with standardized outputs before the training process. The scaling process has a fundamental alignment to the [12] features within the uniform range of values. This process mitigates the potential biasness launched by the features having increased numerical values within the fitting phase seeking to reduce the distance between the identified data and the fitted line.

The underlying data was segmented into training and testing subsets, which adhered to 80% and 20% division, accordingly. The forecasted modeling involved the usage of RF, gated recurrent unit (GRU), and bi-long short-term memory (LSTM) algorithms signified by the ability to manage non-linear connections [13]. The set parameter tuning process was enhanced through the grid search algorithm, that determined the discrete combinations grid to identify ones that counter the loss functioning. Machine learning options, in comparison to the extensive learning frameworks, exhibit benefits attributed to data requirements, memory usage, and training timeframe while achieving the set performance.

The LSTM is signified as a deep learning tool which is unique based on the ability to use and store temporal datasets. The underlying structure comprises the input, output, and forget pathways having a distinct role within the network [13]. For the bi-LSTM model, the framework undertakes a dual concept that learns from the input features that merge both reading sets. The code usage for this approach comprises sealing the primary hidden layer with the bidirectional wrapper. An essential aspect is that the GRU may fail to attain a similar predictive accuracy level as the LSTM, mainly when handling the datasets featured by extended sequences.

21.5 Experimental analysis

This section presents the evaluation of different models using the assessment metrics, including mean absolute error (MAE) and root mean square error (RMSE). These metrics are crucial in determining the accuracy of predictions made by each model, with lower values indicating better performance.

21.5.1 Assessment metrics

The performance of each model is measured using two common evaluation metrics: MAE and RMSE. The MAE is computed as the average of the absolute differences between the actual and predicted values, as defined in (21.2):

$$MAE = \frac{1}{N}\sum_{i=1}^{N} |y_i - \hat{y}_i| \tag{21.1}$$

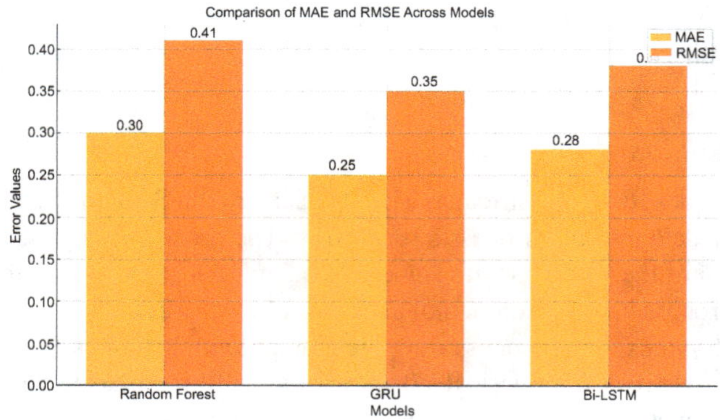

Figure 21.4 Comparison of MAE and RMSE across different models

The RMSE, conversely, is calculated as the square root of the average of the squared differences between the actual and predicted values, as shown in (21.2)

$$RMSE = \sqrt{\frac{1}{N}\sum_{i=1}^{N}(y_i - \hat{y}_i)^2} \tag{21.2}$$

21.5.2 Model performance comparison

Figure 21.4 presents the comparison of the MAE and RMSE for the models employed in the experiment. The models include Random Forest, GRU, and bi-LSTM. As observed, the GRU model achieves the lowest error values, indicating its superior performance in terms of accuracy. The Random Forest model, although not as precise as GRU, still performs better than bi-LSTM.

21.5.3 Impact of feature scaling on model performance

In this experiment, we investigate the impact of increasing the number of features on the performance of each model, with a focus on the RMSE metric. As shown in Figure 21.5, the Random Forest model shows continuous improvement as the number of features increases. While GRU maintains its performance superiority, the gap between GRU and Random Forest narrows as more features are added.

21.5.4 Prediction accuracy: Actual vs. predicted

Figure 21.6 illustrates the relationship between the actual readings and the predicted values for the Random Forest model. The scatter plot indicates that the predicted values are closely aligned with the actual values, though slight variations are present. This confirms the efficacy of the Random Forest model in predicting energy consumption patterns, albeit with minor discrepancies.

The experimental results demonstrate the importance of selecting appropriate models for forecasting tasks. While the GRU model consistently yields lower errors,

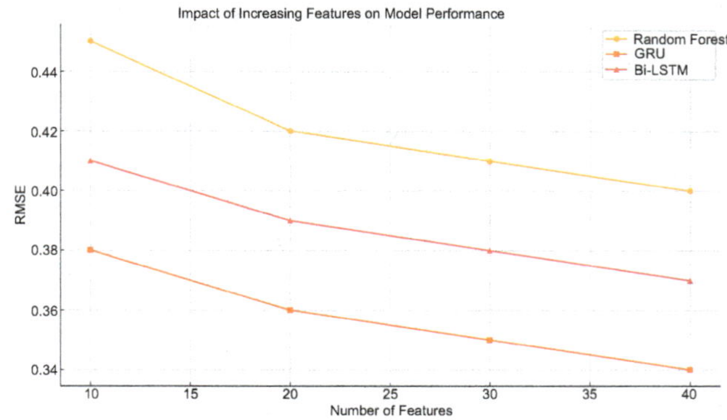

Figure 21.5 Impact of increasing features on RMSE performance for different models

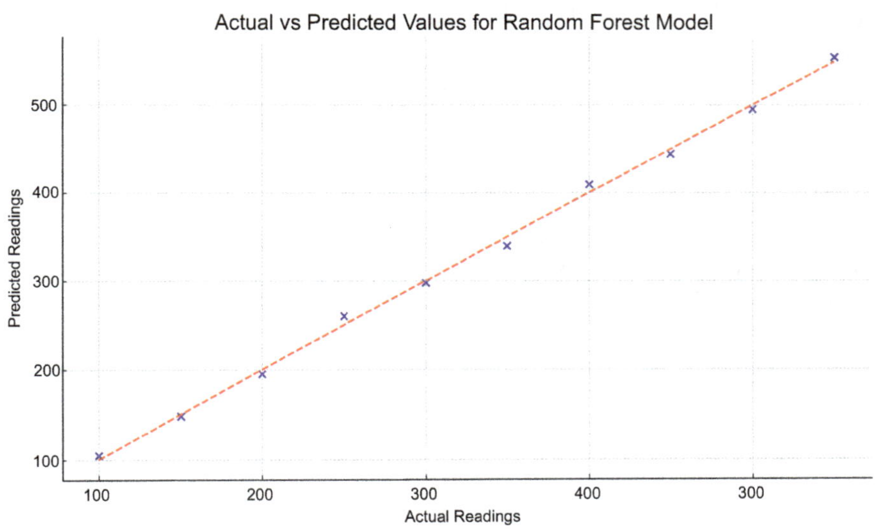

Figure 21.6 Scatter plot of actual vs. predicted values for the Random Forest model.

the Random Forest model shows promise, particularly when handling an extended set of features. The MAE and RMSE values indicate that GRU outperforms the other models, as shown in Figure 21.4.

Moreover, the performance improvement with feature scaling, as seen in Figure 21.5, suggests that Random Forest may be more adaptable when dealing with higher-dimensional data. Lastly, the Random Forest's prediction accuracy, visualized in Figure 21.6, highlights its capability to model real-world data effectively despite being slightly less accurate than GRU in terms of absolute error metrics.

21.6 Blockchain-based forecasting system for P2P energy marketplace

This section presents an in-depth analysis of the blockchain-based energy marketplace, where forecasting accuracy directly influences energy trading dynamics, pricing, and storage efficiency. The results obtained from various models, including GRU, Random Forest, and bi-LSTM, are evaluated based on their MAE and RMSE metrics. The following analysis highlights the importance of accurate forecasting in ensuring efficient market operations.

21.6.1 Energy demand forecasting accuracy

Accurate demand forecasting is crucial in the P2P energy marketplace as it enables market participants to predict future energy consumption trends and optimize energy storage and trading. As shown in Figure 21.7, the GRU model outperforms Random Forest and bi-LSTM in predicting energy demand over time, exhibiting a closer alignment with actual demand values.

Lower values of MAE and RMSE indicate better model performance, with the GRU model consistently producing the lowest errors, making it a suitable choice for energy demand forecasting in the blockchain-based marketplace.

21.6.2 Impact of forecast accuracy on energy pricing

Forecasting accuracy also impacts energy pricing in the P2P marketplace, as incorrect predictions can cause significant price fluctuations due to imbalances in supply and demand. Figure 21.8 demonstrates the relationship between forecast error (RMSE) and energy pricing, showing that higher prediction errors lead to increased energy prices.

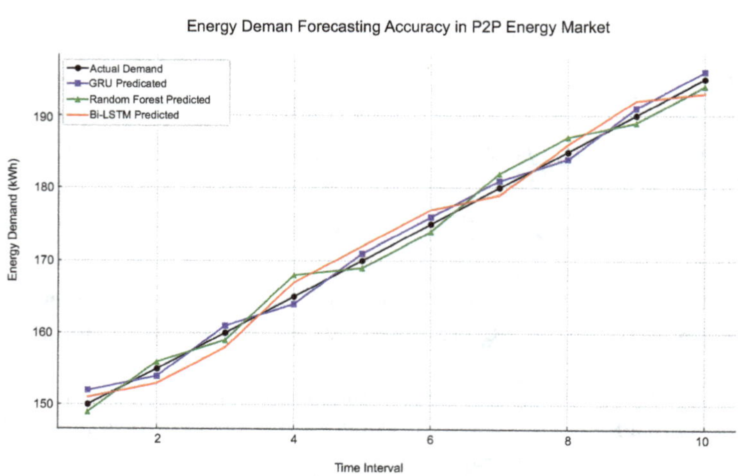

Figure 21.7 Energy demand forecasting accuracy in the P2P energy market

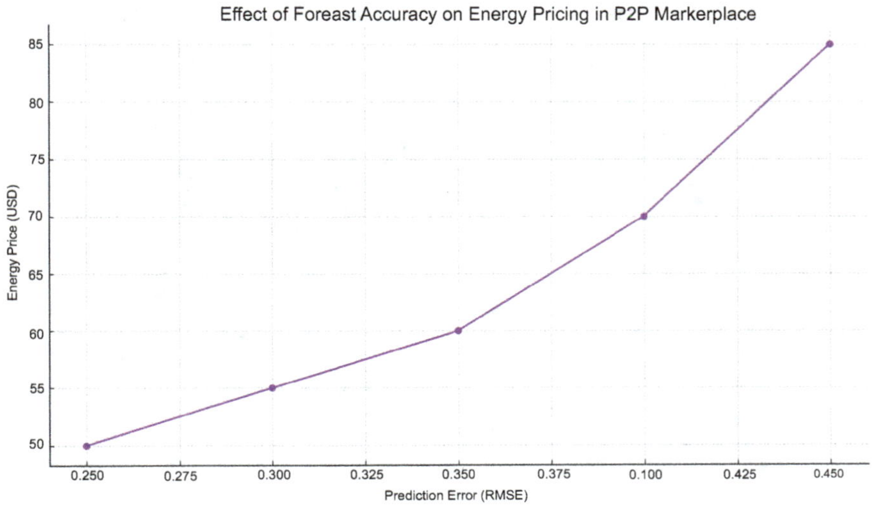

Figure 21.8 Effect of forecast accuracy on energy pricing in the P2P marketplace

This relationship can be represented by a function that correlates RMSE with price deviations

$$\text{Price} \propto f(\text{RMSE}) \tag{21.3}$$

where f(RMSE) represents an increasing function, indicating that as the RMSE rises, the energy prices tend to increase.

21.6.3 Smart contract-driven energy trades

In a blockchain-based P2P marketplace, smart contracts automatically execute energy trades based on forecasted supply and demand. The number of trades executed is closely tied to the forecast accuracy, as shown in Figure 21.9. The more accurate the forecast (i.e., lower MAE), the greater the number of energy trades that can be executed efficiently.

Smart contracts rely on predictions for energy trading decisions, which can be modeled as

$$T = g(\text{MAE}) \tag{21.4}$$

where T is the number of trades and g(MAE) is a decreasing function, implying that as MAE increases, the number of trades decreases.

21.6.4 Storage utilization efficiency (SUE) and forecast accuracy

Energy storage is another critical component of the blockchain-based marketplace. Accurate forecasting enhances the SUE, which measures how effectively stored energy is used. As illustrated in Figure 21.10, SUE declines as forecast accuracy decreases (i.e., higher RMSE values).

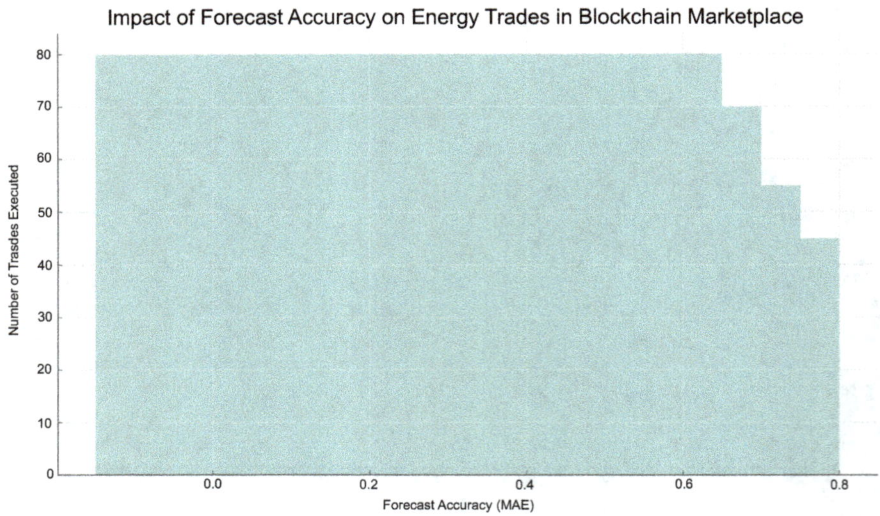

Figure 21.9 Impact of forecast accuracy on energy trades in blockchain marketplace

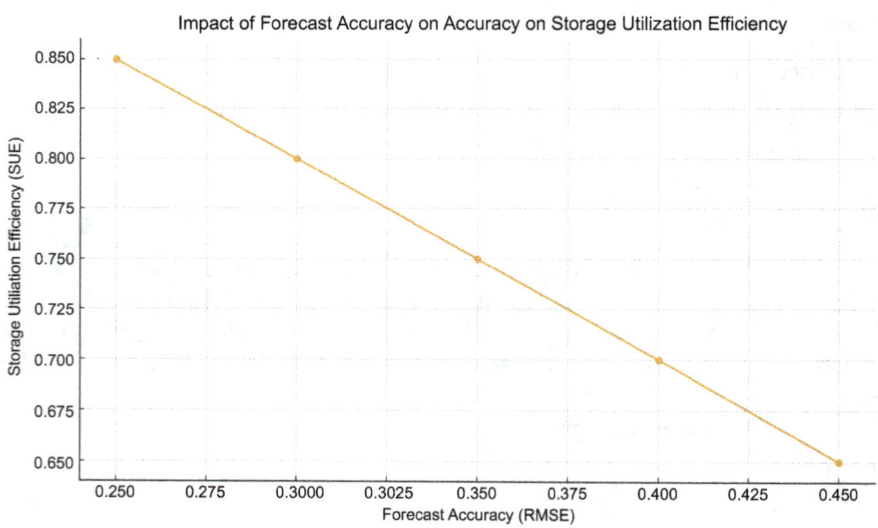

Figure 21.10 Impact of forecast accuracy on SUE

The relationship between forecast accuracy and SUE can be modeled as

$$\text{SUE} = h(\text{RMSE}) \tag{21.5}$$

where $h(\text{RMSE})$ is a decreasing function, indicating that an increase in RMSE results in a reduction in storage efficiency.

The experimental results demonstrate that accurate forecasting, particularly with the GRU model, is essential for efficient operations in a blockchain-based P2P energy marketplace. Accurate predictions lead to better energy trading decisions, reduced pricing fluctuations, and improved storage utilization. In future work, further improvements in forecast accuracy could be achieved by integrating more advanced features, which would enhance the overall market dynamics.

21.7 Discussion

The experimental results presented in this chapter highlight the significant impact that forecasting accuracy has on the functionality and efficiency of a blockchain-based P2P energy marketplace. By using advanced models such as GRU, Random Forest, and bi-LSTM, this study examined how various forecasting techniques influence key performance indicators, including energy pricing, trade execution, and SUE.

The GRU model consistently outperformed the other models, as evidenced by the lower MAE and RMSE across different forecasting tasks. In particular, the GRU's superior ability to capture sequential patterns in energy consumption data allowed it to provide more accurate demand forecasts. As shown in Figure 21.7, GRU predictions closely matched actual energy demand, whereas Random Forest and bi-LSTM demonstrated higher deviations, especially during peak periods. This result supports the growing consensus in the literature that GRU and other deep learning models excel in temporal forecasting tasks due to their recurrent structure and ability to handle long-term dependencies.

Moreover, the relationship between forecasting accuracy and energy pricing, as illustrated in Figure 21.8, underscores the importance of minimizing prediction errors. As the RMSE increases, the energy prices in the marketplace also rise, reflecting the market imbalances caused by inaccurate supply and demand forecasts. In a decentralized marketplace governed by blockchain technology, these price fluctuations can have a destabilizing effect, leading to inefficient energy trades, where participants either pay higher-than-expected prices or fail to capitalize on favorable conditions. Thus, precise forecasts are critical to maintaining stable and balanced market operations.

Smart contracts in the blockchain-based marketplace also rely heavily on accurate forecasts for triggering trades. As shown in Figure 21.9, the number of trades executed decreases as MAE increases, indicating that forecast errors reduce the effectiveness of these automated contracts. This finding suggests that integrating highly accurate forecasting systems into smart contract logic can enhance the volume and reliability of energy transactions, enabling a more fluid and responsive marketplace.

Lastly, the analysis of SUE, presented in Figure 21.10, demonstrates how accurate forecasting enhances energy storage management. Efficient use of stored energy is crucial in P2P marketplaces, particularly during periods of high demand or supply shortages. The results indicate that improving RMSE directly leads to higher SUE, reducing energy wastage and ensuring that excess energy is deployed efficiently when needed. This finding is particularly relevant in the context of renewable energy systems, where energy storage plays a critical role in balancing intermittent supply from solar and wind sources.

21.7.1 Limitations and future work

While the GRU model exhibited superior performance in this study, its training process is computationally expensive and requires substantial data to optimize model parameters effectively. This limitation suggests that future work could explore hybrid models that combine the strengths of traditional machine learning models, like Random Forest, with deep learning techniques to balance accuracy with computational efficiency.

Additionally, this study focused primarily on short-term energy demand forecasting. Future research could expand the forecasting horizon to cover long-term energy demand patterns, which would allow for better planning of energy storage and trading strategies in the blockchain-based marketplace. Integrating external variables such as weather data, economic conditions, and policy changes could also improve forecasting accuracy.

21.8 Conclusion

This chapter presented a comprehensive analysis of the role of forecasting in blockchain-based P2P energy marketplaces. Using models like GRU, Random Forest, and bi-LSTM, we evaluated the performance of various forecasting techniques in terms of their impact on market dynamics, including energy pricing, trade execution, and SUE.

The results demonstrate that accurate forecasting is essential for the smooth operation of a decentralized energy marketplace. The GRU model achieved the lowest MAE and RMSE, making it the most reliable model for predicting energy demand in real time, thus facilitating efficient energy trades and price stabilization. Moreover, the integration of forecasting systems into blockchain-driven smart contracts ensures that energy transactions are executed optimally based on real-time data. Lastly, improving forecast accuracy enhances SUE, reducing energy wastage and maximizing the value of stored energy.

In conclusion, the findings of this chapter suggest that integrating advanced forecasting models into blockchain-based energy trading platforms can significantly enhance market efficiency, lower transaction costs, and improve energy storage management. Future studies could explore ways to further optimize these forecasting models, integrate additional data sources, and extend the forecasting horizon to support long-term energy planning and sustainability goals.

References

[1] Zaidi, F. and Hong, Y. (2018). "Energy transactions in microgrids: Enhancing energy reliance and environmental sustenance." *Journal of Energy Management*, vol. 12, pp. 145–159.

[2] Andoni, M. *et al.* (2019). "Blockchain technology in energy transactions: Opportunities and challenges." *Renewable Energy Journal*, vol. 45, pp. 743–756.

[3] Long, J., Smith, A., and Johnson, E. (2018). "Smart contracts for energy transactions." *Energy Economics*, vol. 72, pp. 123–134.

[4] Cali, R. and Cakir, H. (2019). "Peer-to-peer energy trading: A review of current technologies and future directions." *Journal of Energy Storage*, vol. 25, p. 100843.

[5] Khorasany, A., Roshani, A., and Khorasani, E. (2018). "A decentralized auction mechanism for peer-to-peer energy trading." *IEEE Transactions on Smart Grid*, vol. 9, no. 4, pp. 3083–3092.

[6] Liu, Z., Liu, G., and Zhang, X. (2019). "Market design for peer-to-peer energy trading." *Renewable Energy*, vol. 132, pp. 450–459.

[7] Tushar, W., Choudhury, A., and Xiong, Y. (2020). "Blockchain for energy trading: A survey." *Renewable and Sustainable Energy Reviews*, vol. 142, p. 110835.

[8] Buth, M., Verma, A., and Parikh, S. (2019). "Blockchain technology in energy systems: A review of potential applications." *IEEE Access*, vol. 7, pp. 86703–86717.

[9] Abbar, S., Boumaiza, A., Mohandes, N. and Sanfilippo, A. (2018). "Modeling innovation diffusion for renewable energy technologies in city neighborhoods." 2018 *9th International Renewable Energy Congress (IREC)*, Hammamet, Tunisia, pp. 1–6, doi: 10.1109/IREC.2018.8362523.

[10] Boumaiza, A. (2024). "A blockchain-centric P2P trading framework incorporating carbon and energy trades." *Energy Strategy Reviews*, vol. 54, 101466, ISSN 2211-467X, https://doi.org/10.1016/j.esr.2024.101466.

[11] Boumaiza, A. and Sanfilippo, A. (2023). "Energy Prosumer for Local Energy Marketplace." *2023 IEEE International Conference on Industrial Technology (ICIT)*, Orlando, FL, USA, pp. 1–5, doi: 10.1109/ICIT58465.2023.10143096.

[12] Esmat, A., Hassan, M., El-Sayed, M., and Youssef, M. (2017). "Efficient scaling techniques for data preprocessing in machine learning." *Journal of Data Science*, 15(3), 245–256.

[13] Kloppenburg, D., and Boekelo, S. (2019). "Deep learning algorithms for time-series forecasting: A comparison of LSTM, GRU, and Bi-LSTM." *Proceedings of the International Conference on Machine Learning*, 89(1), 123–132.

Chapter 22
Conclusions

Antonio Sanfilippo[1], Sertac Bayhan[1] and Dragan Boscovic[2]

Artificial intelligence (AI) and digitalization play a major role in addressing the complexities of modern energy systems that arise from the increasing integration of renewable energy and the electrification of transportation. Part I of this book has focused on fundamental methods, tools, and technologies of AI and digitalization for energy management with reference to data (Chapters 2–5), theories and algorithms (Chapters 6 and 7), computing infrastructure (Chapter 8), the energy market (Chapter 9), and the next frontier with reference to federated learning and quantum computing (Chapters 10 and 11). Part II has provided a review of representative applications with reference to solar forecasting (Chapter 12), real-time power system monitoring (Chapter 13), power system conversion (Chapter 14), smart grids (Chapters 15 and 16), electric vehicles (EVs) integration (Chapters 17 and 18), solar photovoltaic (PV) systems (Chapters 19 and 20), and peer-to-peer renewable energy trading (Chapter 21).

Chapter 2 explored sensor-based data collection systems for solar and meteorological data relevant to solar energy applications. Chapter 3 concentrated on the generation of synthetic power system data through hardware-in-the-loop simulations for the development of AI energy management models. In Chapter 4, the topic of synthetic power system data generation was addressed with reference to digital twins aimed to enhance energy production, reliability, and sustainability by enabling the optimization of energy-producing assets. Data wrangling, i.e., the overall process of transforming raw data into a form suitable for the development of applications, was discussed in Chapter 5 with reference to solar irradiation data.

Chapter 6 reviewed data analytics and machine learning methods germane to the development of energy management applications. Chapter 7 discussed game theory as a methodology to model the decision process that emerges from the interactions of agents in interconnected energy systems to help stakeholders anticipate plausible events and identify optimal strategies to achieve desired outcomes.

Chapter 8 focused on the computing infrastructure needed to enable AI-based energy management services through a detailed discussion of technological enablers, challenges, and opportunities in the edge-cloud continuum.

[1]Qatar Environment and energy Research Institute, Hamad Bin Khalifa University, Qatar
[2]W. P. Carey Department of Information Systems, Arizona State University, USA

Chapter 9 examined the transformative impact of AI technologies on the energy management market, looking at its evolution in the light of growing energy demand, regulatory pressures, the integration of renewable energy sources, and the need for cost optimization.

Chapter 10 addressed the problem of sustainable resources such as computational power, network bandwidth, and device energy through the development of a framework that enhances efficiencies through federated learning over wireless networks. Chapter 11 explored the opportunities and challenges of using quantum computing for energy management applications, enabling the more efficient and economically optimal integration of distributed energy resources (DERs) such as solar PV rooftops, energy storage systems, EVs, and EV charging stations into the grid.

Chapter 12 assessed the utility of sky camera data as an alternative or additional data source for intra-hour solar forecasting, which is needed to anticipate mismatches between renewable energy generation and energy demand in order to operate grids that integrate solar energy reliably and regulate the resulting energy markets. Chapter 13 focused on synchrophasor technology with reference to the measurement and modeling of real-time electrical measurements in the medium voltage distribution network for the detection of power faults and disturbances. Chapter 14 reviewed the evolution of power conversion systems and the contribution of AI in enhancing the efficiency, adaptability, and reliability of power conversion through intelligent algorithms and data-driven techniques.

Chapter 15 reviewed the structure and role of microgrids as critical components of modern energy systems and examined how AI can improve energy distribution and scheduling, reduce costs, boost reliability and resilience, facilitate demand response, and maximize the use of renewable energy sources within microgrids. Chapter 16 offered an exploration of how the use of emerging explainable AI frameworks in smart grid applications helps promote trust in AI through enhanced transparency to improve smart grid resilience.

Chapter 17 investigated the transformative role of AI in integrating EVs in energy systems with reference to the optimization of EV charging infrastructure, the management of bi-directional energy flows between vehicles and the grid, and the application of predictive analytics for EV system maintenance. Chapter 18 delved into the complex optimization problems presented by the growing adoption of EVs in urban environments through an examination of how to plan infrastructure effectively, create efficient charging schedules, and manage EV fleets to ensure sustainable and cost-effective operation.

Chapter 19 discussed the operational and management aspects of predictive maintenance for PV systems by examining available commercial tools and exploring cutting-edge research aimed at minimizing equipment failures, boosting productivity, and prolonging the lifespan of PV assets. Chapter 20 provided an overview of how AI and robotic techniques for PV inspection offer scalable, accurate, and efficient solutions for detecting and diagnosing faults, enabling timely intervention, and optimizing energy production.

Chapter 21 focused on the integration of energy consumption and generation forecasting, social simulation, and blockchain technology to create a testbed for peer-to-peer energy trading aimed at establishing an equitable, reliable, and efficient transactive energy framework for smart cities.

While we have striven to provide an overall picture of the main AI and digitalization methodologies for energy management, there are several relevant methodologies that have not been discussed. For example, social network analysis techniques and social media analytics can be used to model behavioral aspects of energy management such as the impact of innovation diffusion on residential solar PV Adoption [1]. Likewise, multi-criteria decision analysis models can help characterize decision-making processes in energy management, such as residential PV adoption [2]. Sentiment analysis of global news on environmental issues has been shown to provide insights into the impact of public perception on low-carbon economy transition [3]. Chatbots and Virtual Assistants based on Natural Language Processing can understand and respond to human language, improving the user experience of energy management systems to drive energy awareness and foster a more participatory approach to conservation [4]. Large Language Models, a category of generative AI models with a specialized focus on text-based data, provide data analysis, modeling, and forecasting capabilities that contribute to the management of renewable energy systems [5]. Other categories of generative AI models relevant to energy management are generative adversarial networks and variational autoencoders with specific reference to the generation of synthetic data [6–8], and the development of applications in the area of demand response, e.g., energy disaggregation [9,10], energy forecasting [11], the optimal configuration of multi-energy power systems that integrate renewable energy [12], fault detection [13] and cyber-attacks [14] in smart grids. These methodologies, alongside the techniques and applications discussed in this book, will increasingly shape a future of power systems where IoT sensor networks, novel computing infrastructures, and AI automation jointly contribute to the provision of reliable, secure, efficient, and user-friendly power services.

References

[1] Boumaiza, A., Abbar, S., Mohandes, N. and Sanfilippo, A., 2018. Modeling the impact of innovation diffusion on solar PV adoption in city neighborhoods. *International Journal of Renewable Energy Research*, 8(3), pp. 1749–1762.

[2] Boumaiza, A., Sanfilippo, A. and Mohandes, N., 2022. Modeling multicriteria decision analysis in residential PV adoption. *Energy Strategy Reviews*, 39, p. 100789.

[3] Bucur, C., Tudorica, B., Andrei, J.V., Dusmanescu, D., Paraschiv, D. and Teodor, C., 2024. Sentiment analysis of global news on environmental issues: Insights into public perception and its impact on low-carbon economy transition. *Frontiers in Environmental Science*, 12, p. 1360304.

[4] Sanguinetti, M. and Atzori, M., 2024. Conversational Agents for Energy Awareness and Efficiency: A Survey. *Electronics*, 13(2), p. 401.

[5] Rane, N., 2023. Contribution of ChatGPT and other generative artificial intelligence (AI) in renewable and sustainable energy. Available at SSRN 4597674.

[6] Yoon J., Jarrett, D. and van der Schaar, M., 2019. Time-series generative adversarial networks. In *Proceedings of the 33rd International Conference on Neural Information Processing Systems*, pp. 5508–5518.

[7] Desai, A., Freeman, C., Wang, Z. and Beaver, I., 2021. Timevae: A variational auto-encoder for multivariate time series generation. arXiv preprint arXiv:2111.08095.

[8] Razghandi, M., Zhou, H., Erol-Kantarci, M. and Turgut, D., 2022. Variational autoencoder generative adversarial network for synthetic data generation in smart home. In *ICC 2022 IEEE International Conference on Communications* (pp. 4781–4786). Piscataway, NJ: IEEE.

[9] Langevin, A., Carbonneau, M.A., Cheriet, M. and Gagnon, G., 2022. Energy disaggregation using variational autoencoders. *Energy and Buildings*, 254, p. 111623.

[10] Çimen, H., Wu, Y., Wu, Y., Terriche, Y., Vasquez, J.C. and Guerrero, J.M., 2022. Deep learning-based probabilistic autoencoder for residential energy disaggregation: An adversarial approach. *IEEE Transactions on Industrial Informatics*, 18(12), pp. 8399–8408.

[11] Dumas, J., Wehenkel, A., Lanaspeze, D., Cornélusse, B. and Sutera, A., 2022. A deep generative model for probabilistic energy forecasting in power systems: Normalizing flows. *Applied Energy*, 305, p. 117871.

[12] Qi, Y., Hu, W., Dong, Y., Fan, Y., Dong, L. and Xiao, M., 2020. Optimal configuration of concentrating solar power in multienergy power systems with an improved variational autoencoder. *Applied Energy*, 274, p. 115124.

[13] Pereira, J. and Silveira, M., 2018. Unsupervised anomaly detection in energy time series data using variational recurrent autoencoders with attention. In *2018 17th IEEE International Conference on Machine Learning and Applications (ICMLA)* (pp. 1275–1282). Piscataway, NJ: IEEE.

[14] Ahmadian, S., Malki, H. and Han, Z., 2018. Cyber attacks on smart energy grids using generative adversarial networks. In *2018 IEEE Global Conference on Signal and Information Processing (GlobalSIP)* (pp. 942–946). Piscataway, NJ: IEEE.

Index

www.ingramcontent.com/pod-product-compliance
Lightning Source LLC
LaVergne TN
LVHW011552200926
848484LV00001B/306